Structural Impact

STRUCTURAL IMPACT

NORMAN JONES

A. A. Griffith Professor of Mechanical Engineering
The University of Liverpool

CAMBRIDGE
UNIVERSITY PRESS

PUBLISHED BY THE PRESS SYNDICATE OF THE UNIVERSITY OF CAMBRIDGE
The Pitt Building, Trumpington Street, Cambridge CB2 1RP, United Kingdom

CAMBRIDGE UNIVERSITY PRESS
The Edinburgh Building, Cambridge CB2 2RU, United Kingdom
40 West 20th Street, New York, NY 10011-4211, USA
10 Stamford Road, Oakleigh, Melbourne 3166, Australia

First published 1989
First paperback edition (with corrections) 1997

Printed in the United Kingdom at the University Press, Cambridge

Typeset in Times 10/14 pt.

A catalogue record for this book is available from the British Library

Library of Congress Cataloguing in Publication data
Jones, Norman,
Structural impact/Norman Jones.
p. cm.
Bibliography: p.
Includes index.
ISBN 0 521 30180 7
1. Structural dynamics. 2. Impact. 3. Girders. 4. Plates
(Engineering) 5. Shells (Engineering) I. Title
TA654.J66   1989
624.1′71–dc 19   88-31794   CIP

ISBN 0 521 30180 7 hardback
ISBN 0 521 62890 3 paperback

To Jenny, Alison and Catherine

Contents

Preface

Impact events occur in a wide variety of circumstances, from the everyday occurrence of striking a nail with a hammer to the protection of spacecraft against meteoroid impact. All too frequently, we see the results of impact on our roads. Newspapers and television report spectacular accidents which often involve impact loadings, such as the collisions of aircraft, buses, trains, and ships, together with the results of impact or blast loadings on pressure vessels and buildings due to accidental explosions and other accidents. The general public is becoming increasingly concerned about safety, including, for example, the integrity of nuclear transportation casks in various accident scenarios involving impact loads.

Clearly, impact is a large field, which embraces both simple structures (e.g., nails) and complex systems, such as the protection of nuclear power plants. The materials which are impacted include bricks, concrete, ductile and brittle metals, and polymer composites. Moreover, on the one hand, the impact velocities may be low and give rise to a quasi-static response, and, on the other hand, they may be sufficiently large to cause the properties of the target material to change significantly.

In this book, I have concentrated on the impact behaviour of ductile structures and, in particular, beams, plates, and shells. Most complex engineering systems are constructed largely of these simple structural members, so that an understanding of their response is an essential prerequisite for revealing the dynamic behaviour of a more complex system. The topic still remains a large one, and so I have specialised it further by focusing on large impact loads producing plastic strains which dominate the elastic effects.

A dynamic load causes elastic and plastic stress waves to propagate through the thickness of this class of structures, as well as producing an overall structural response. The propagation of stress waves through the structural thickness can cause failure by spalling when the shock or impact loads are sufficiently severe. This phenomenon occurs in the same order of time that it takes a stress wave to propagate through the thickness of the structure. Thus, this type of failure usually occurs within microseconds of initial impact, and is sometimes referred to as the 'early time response' to distinguish it from the gross structural behaviour which occurs at later times. The early time response of structural members is not considered further.

This book focuses on the long-term behaviour of structures (typically of the order of milliseconds for small structures), for which the external dynamic load is assumed to impart momentum instantaneously to the middle surface of a structure (i.e., transverse wave propagation is disregarded). It is customary practice to uncouple the early time wave propagation behaviour from the long-time or gross structural response because the time durations of these two phenomena usually differ by a few orders of magnitude. Obviously, a gross structural analysis cannot be used to predict the detailed behaviour through the structural thickness. It is necessary, therefore, to establish separately whether or not failure due to spalling can occur in a given case.

Although the static plastic behaviour of structures was first studied last century (e.g., J. A. Ewing, *The Strength of Materials*, Cambridge University Press, 1899), systematic investigations on the dynamic plastic behaviour is much more recent. Serious studies appear to have commenced during the Second World War, when, for example, J. F. Baker designed the Morrison air raid shelters to protect people from falls of masonry in their own homes, G. I. Taylor studied the dynamic response of thin plates, and Pippard and Chitty examined the dynamic behaviour of cylindrical shells for research into submarine hulls. Considerable research activity and progress has been reported over the past forty years, some of which is discussed in this book, or cited in the references. Generally speaking, this body of work seeks the response of a structural member when subjected to a known impact load. However, the theoretical solutions may also be used for diagnostic or forensic purposes. Lord Penney and his colleagues, for example, estimated the nuclear explosive yields at Hiroshima and Nagasaki by calculating the impact loads required to cause the permanent damage which was observed for bent or snapped poles, squashed empty drums or cans, tops of office cabinets pushed in by the blast, etc. Coupling between the external impact loading and the structural response is a difficult topic, which is not well understood, and is disregarded in this book.

Despite restricting our attention to the impact behaviour of beams, plates and shells subjected to large impact loads, the field is still large, active, and growing rapidly. The results of studies in this area are being used to guide the development of rational design procedures which avoid the destructive action of earthquakes on buildings, and to improve the collision protection of passengers in automobiles, trains, buses, and aircraft. The collision protection of automobiles and buses is achieved by improving the interior and exterior energy-absorbing capabilities of the vehicles, and incorporating the principles of structural crashworthiness into the design of highway safety systems and roadside furniture.

Theoretical methods have been used to design impact absorbers of various types, as well as to assess the safety of reactor tubes subjected to violent transient pressure pulses, which can arise in certain circumstances in sodium-cooled fast-breeder reactors. The slamming damage which is sustained by the bottom plating of ships and hydrovehicles, has been estimated using these methods which have also been employed to design buildings in order to withstand internal gaseous explosions. The response of re-entry vehicles, structural crashworthiness of offshore platforms, safety calculations for industrial plant, various military applications, interpretation of constitutive equations from dynamic ring tests, and even the denting of aircraft surfaces due to hail, have been studied using the various methods to be discussed in this book. Many other practical applications have been made, and, no doubt, more will be found as engineers strive to design new and efficient structures which must be as light and safe as possible, yet withstand the large dynamic loads arising in many practical situations without catastrophic failure, or absorb the external dynamic energy in a controlled and predictable fashion.

It is the aim of this book to equip the reader with a clear understanding of the impact behaviour of some simple structures. The dynamic response of the particular cases studied may be adequate to predict the response of various practical problems, particularly when recognising the lack of information on the impact loading characteristics, and the shortage of data on the properties of materials under dynamic loads. If simple methods of analysis are inadequate for a particular problem, then the understanding gained from this book provides a foundation for a reader to make further progress with other solution strategies. In particular, this understanding and insight is indispensable for the efficient use and interpretation of numerical codes which play an increasingly important role in engineering design.

It is assumed that a reader has not studied previously the static plastic behaviour of structures, though a knowledge of elementary strength of materials is a prerequisite. Thus, Chapter 1 introduces some basic concepts of plasticity theory, from an engineering viewpoint, including the limit theorems of plasticity, and examines the static plastic collapse behaviour of several beams. Yield conditions for biaxial stress states are introduced in Chapter 2, together with the important normality condition which requires the generalised strain rate vector to remain normal to the associated portion of the yield curve during plastic flow. Theoretical solutions are presented for the static collapse behaviour of circular plates, rectangular plates, and cylindrical shells. These two chapters contain all the basic elements from the theory of static plasticity which are required for the remainder of the book.

The rigid-plastic approximations, which have been developed for the static plastic behaviour of structures in Chapters 1 and 2, are also used to obtain the dynamic plastic response of structures. The static plastic collapse load is the largest possible external load which may act on a perfectly plastic structure according to the limit theorems of plasticity. Thus, a structure is not in static equilibrium for larger external loads, so that inertia forces are generated, and motion commences. This motion continues until all the external energy has been consumed by internal plastic work. It is evident that the permanent displacements and response duration are of particular interest for dynamic loads.

Chapters 3 to 5 examine, respectively, the dynamic plastic behaviour of beams, plates, and shells, many of which were studied in Chapters 1 and 2 for static loads. The energy consumed in the plastic deformation dominates the elastic energy, so that a rigid-plastic method of analysis, which neglects elastic effects, is suitable.

The yield conditions in Chapters 1 to 5 have ignored the influence of transverse shear forces which were retained in the equilibrium equations. However, transverse shear forces are more important for dynamic loads than for similar problems subjected to static loads. In fact, failures may occur due to excessive transverse shear forces such as at hard points in structures loaded dynamically. Thus, Chapter 6 retains the transverse shear force in the yield conditions for dynamically loaded beams, circular plates, and cylindrical shells, and contains some comments on the influence of rotatory inertia.

The theoretical solutions in Chapters 1 to 6 have been developed for infinitesimal displacements, since the equilibrium equations were derived in the initial undeformed configuration. This appears to be an unreasonable simplification when elastic effects are neglected and the external dynamic loads produce plastic strains and permanent deformations. However, there are some structural problems for which the displacements may be taken as infinitesimal without loss of accuracy. Nevertheless, there are other problems for which the influence of finite-displacements plays a significant role in the dynamic response.

The first part of Chapter 7 examines the influence of finite-displacements, or geometry changes, on the static plastic behaviour of beams, circular plates, and rectangular plates. An approximate kinematic method of analysis is also introduced in Chapter 7, and is used to examine the dynamic plastic response of beams, circular plates, rectangular plates, and circular membranes. Comparisons are made with the corresponding experimental results, and a simple procedure is introduced to estimate the threshold impact energy which is required to cause structural failure due to material failure.

The properties of many materials under dynamic loading conditions are different from the corresponding static values. In particular, the stress–strain relations are sensitive to the speed of a test, a phenomenon which is known as strain rate sensitivity, or viscoplasticity. The strain-rate-sensitive properties of various materials, under several dynamic loadings, are discussed in Chapter 8. The well-known Cowper–Symonds constitutive equation is introduced, and theoretical solutions are obtained for several structural problems, using various simplifications and approximations.

The theoretical solutions in Chapters 1 to 8 have been developed for structures undergoing a stable response. However, an unstable response may develop in many practical problems. Thus, we study, in Chapter 9, the behaviour of a circular tube subjected to a dynamic axial load. This produces many axisymmetric convolutions, or wrinkles, in the tube, and gives rise to a fluctuating resistance about a mean crushing force. This phenomenon is known as dynamic progressive buckling, because the deformations form progressively with time from one end of a tube. The inertia forces in a tube are not significant, and are neglected, but the phenomenon of material strain rate sensitivity must be retained for a strain-rate-sensitive material. The mode of deformation is taken to be the same as for static loads. It transpires that considerable energy may be absorbed in an axially crushed tube before bottoming-out. Thus, some comments on structural crashworthiness are given in Chapter 9.

The phenomenon of dynamic progressive buckling in Chapter 9 develops, typically, at tens of metres per second. At higher impact velocities, the inertia forces in a tube become important, and the mode of deformation may change into a more highly wrinkled form. This phenomenon is known as dynamic plastic buckling, and is studied in Chapter 10 for axially loaded columns, rings, and cylindrical shells.

Finally, Chapter 11 examines similitude, or geometrically similar scaling, which is important for relating the results of impact experiments on small-scale models to the response of geometrically similar full-scale prototypes.

The various phenomena associated with structural impact are introduced in this book in a simple, yet rigorous manner, as far as it is possible. The book should be useful, therefore, as a textbook for undergraduate and postgraduate students in universities and polytechnics who are pursuing upper-level courses, projects and research investigations into structural impact, dynamic plasticity, or advanced strength of materials. However, the book has been written with the designer in mind, so that it should be of value for a wide range of industries which have an interest in, and responsibility for, safety assessment and the evaluation of the response of structures subjected to dynamic loadings.

Many postgraduate students and visitors, who have worked with me over the past twenty years at the Massachusetts Institute of Technology and the University of Liverpool, have contributed in various ways to this book. In particular, I wish to acknowledge the valuable assistance of Professor W. Abramowicz, Dr R. S. Birch, Mrs S. E. Birch, Dr J. C. Gibbings, Dr W. S. Jouri, Dr Jianhui Liu, Dr R. A. W. Mines, Mr Wen He Ming, Dr J. G. de Oliveira, Mr Jiang Ping, Professor W. Q. Shen, Dr W. J. Stronge, Professor H. Vaughan, Professor T. Wierzbicki, Professor Jilin Yu and Professor T. X. Yu. I take full responsibility for any errors in this book and I should be grateful if readers would inform me of any they find.

Thanks are also due to Dr George Abrahamson, Mr E. Booth, Professor W. Johnson, Professor S. B. Menkes and again Dr R. S. Birch for their assistance with the photographs, and to the American Society of Mechanical Engineers, Pergamon Press and Her Majesty's Stationery Office, for permission to reproduce some photographs. I wish also to record my appreciation to Mr F. Cummins, Mr H. Parker and Mrs A. Green for their assistance with the figures, to Mr R. Coates for some assistance with computing, and last, but not least, my secretary, Mrs M. White, who typed the drafts and final copy over the extended gestation period of this book.

May 1988   Norman Jones

Preface to paperback edition

The author is grateful to all those readers who have pointed out misprints in the first edition and who have suggested improvements to the text. In particular, I am indebted to Mr Jiang Ping, Professor Jilin Yu and Professor Wang LiLi, who read it most thoroughly when preparing the Chinese edition, to my Ph.D. students, Dr M. Alves, Mr Q. M. Li and Mr C. C. Yang, and to Mr M. Moussouros.

February 1997   Norman Jones

1

Static plastic behaviour of beams

1.1 Introduction

Many ductile materials which are used in engineering practice have a considerable reserve capacity beyond the initial yield condition. The uniaxial yield strain of mild steel, for example, is 0.001 approximately, whereas this material ruptures, in a standard static uniaxial tensile test, at an engineering strain of 0.3, approximately. This reserve strength may be utilised in a structural design to provide a more realistic estimate of the safety factor against failure for various extreme loads. Thus, the static plastic behaviour of structures has been studied extensively and is introduced in many textbooks.[1.1-1.11] An interested reader is referred to these textbooks for a deeper presentation of the subject than is possible in this book, which is concerned primarily with the influence of dynamic loadings. However, the methods of dynamic structural plasticity presented in this book owe a substantial debt to the theoretical foundation of static structural plasticity, which is, therefore, reviewed briefly in this chapter and the following one.

A considerable body of literature is available on the static behaviour of structures made from ductile materials which may be idealised as perfectly plastic. This simplification allows the principal characteristics and overall features of the structural response to be obtained fairly simply for many important practical cases. Moreover, the static collapse loads predicted by these simplified methods often provide good estimates of the corresponding experimental values. Indeed, the design codes in several industries now permit the use of plasticity theory for the design of various structures and components. The theoretical background of these methods, which were developed primarily to examine the static loading of structures made from perfectly plastic materials, are valuable for studies into the response of structures subjected to dynamic loads. Thus, this chapter and the next focus on the static behaviour of structures which are made from perfectly plastic materials.

The basic equations which govern the static behaviour of beams are introduced in the next section. The plastic collapse, or limit moment, is also derived in § 1.2 for a beam with a solid rectangular cross-section which is subjected to a pure bending moment. However, considerable effort is required sometimes to obtain the exact collapse load of a beam which is subjected to a more general form of loading. Thus, the lower and upper bound theorems of plastic collapse are proved in § 1.3. These theorems provide a simple yet rigorous procedure for bounding the exact plastic collapse load of a beam which is subjected to any form of external loading, as illustrated in § 1.4 to § 1.7 for several cases.

A heuristic approach is introduced in § 1.8 and used to obtain the exact static plastic collapse load of a partially loaded beam. Some experimental results are reported in § 1.9 and a few final remarks are given in § 1.10.

1.2 Basic equations for beams

Beams are defined as structural members having a length which is large compared with the corresponding width and depth. It is observed in this circumstance that the lateral, or transverse, shear stresses are small compared with the axial, or longitudinal, stresses. Moreover, it is reasonable to replace the actual force distribution across the depth of a beam by a lateral, or transverse, shear force Q ($Q = \int_A \sigma_{xz}\, dA$) and a bending moment M ($M = \int_A \sigma_x z\, dA$), as shown in Figure 1.1. The actual strain field is then described in terms of the curvature change of the longitudinal axis.† These assumptions lead to considerable simplifications in analyses, and are the usual ones which are incorporated in the engineering theory of elastic beams. It has been shown by Hodge[1.2] that these approximations are also acceptable for the behaviour of perfectly plastic beams.

The moment and lateral force equilibrium equations for the beam in Figure 1.1 are

$$dM/dx = Q \tag{1.1}$$

and

$$dQ/dx = -p, \tag{1.2}$$

respectively, when the response is time-independent and where p is the external load per unit length. The corresponding change in curvature of the longitudinal axis is

$$\kappa = -d^2w/dx^2, \tag{1.3}‡$$

provided $dw/dx \ll 1$.

---

† The bending moment M produces a curvature change and is known as a generalised stress,[1.3] whereas the transverse shear force Q is a reaction since it does not produce any deformation of a beam.

‡ The curvature change is known as a generalised strain.[1.3] The product of generalised stress and the corresponding generalised strain rate gives a positive (or zero) energy dissipation rate (i.e., $M\dot{\kappa} \geqslant 0$, where (˙) is a time derivative).

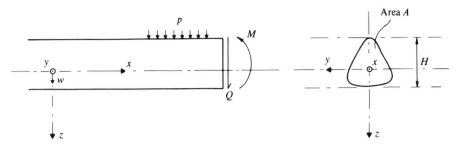

Figure 1.1 Notation for a beam.

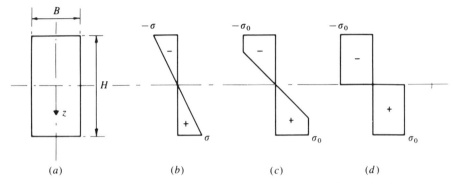

(a)　　　　(b)　　　　(c)　　　　(d)

Figure 1.2 Development of the plastic zones in an elastic, perfectly plastic beam with a rectangular shaped cross-section and subjected to a pure bending moment: (a) rectangular cross-section; (b) elastic stress distribution; (c) elastic, perfectly plastic stress distribution; (d) fully plastic stress distribution. + and − denote tensile and compressive stresses, respectively, for a pure bending moment acting, as shown in Figure 1.1.

Now consider a beam with a solid rectangular cross-section of breadth B and depth H, as shown in Figure 1.2(a). This beam is made from the elastic, perfectly plastic material in Figure 1.3 and is subjected to a pure bending moment M. Initially, the stress distribution across the depth of this beam is linear (see Figure 1.2(b)), so that the corresponding M–κ relation is also linear with a slope EI, as shown in Figure 1.4. (E is Young's modulus and I is the second moment of area for the cross-section.) If the applied bending moment is increased beyond the magnitude of the yield moment

$$M_y = 2I\sigma_0/H = \sigma_0 BH^2/6, \tag{1.4}†$$

then yielding occurs in outer zones, as indicated in Figure 1.2(c), while the associated M–κ relation becomes non-linear. The applied bending moment can be increased

† The elastic behaviour of beams is considered in many textbooks, e.g., Venkatraman and Patel.[1.12]

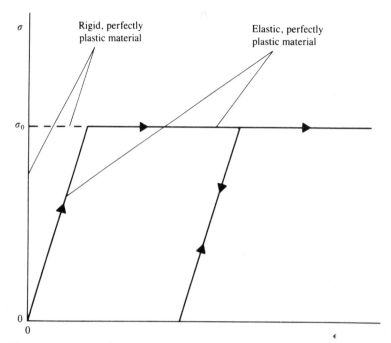

Figure 1.3 Elastic, perfectly plastic and rigid, perfectly plastic uniaxial stress–strain idealisations.

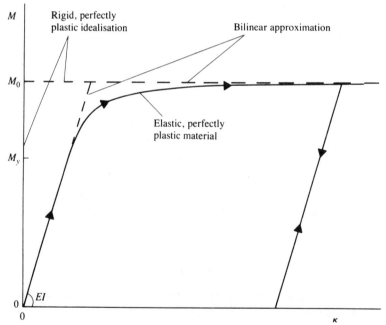

Figure 1.4 Moment–curvature characteristics for a beam with a rectangular cross-section.

further until the entire cross-section has yielded plastically and the strength of the beam has been exhausted, as shown in Figure 1.2(d). This maximum bending moment is known as the limit or collapse moment for the cross-section and can be written as

$$M_0 = (\sigma_0 BH/2)H/2 = \sigma_0 BH^2/4,\qquad\qquad (1.5)\dagger$$

which follows from Figure 1.2(d). In order to simplify theoretical calculations on the plastic behaviour of beams with solid rectangular cross-sections, the moment–curvature relation is often replaced by the rigid, perfectly plastic or bilinear approximations illustrated in Figure 1.4.

It is observed from equations (1.4) and (1.5) for a beam with a solid rectangular cross-section that

$$M_0 = 1.5 M_y.\qquad\qquad (1.6)$$

The factor 1.5 is known as a shape factor and depends on the cross-sectional shape of a beam, as shown in Table 1.1.

The preceding discussion was developed for a beam which was subjected to a pure bending moment. In general, the loading on an actual beam would produce a much more complicated distribution of bending moments, which would be accompanied by lateral shear forces, as indicated by equations (1.1) and (1.2). However, it has been observed by Hodge[1.2] that the influence of these lateral shear forces on the magnitude of the plastic collapse moment of a cross-section may be disregarded for many structures which may be meaningfully called beams.‡ Thus, the limit moment at any location on an actual beam with a solid rectangular cross-section would be given by equation (1.5). In passing, it should be remarked that a kinematically admissible§ collapse mechanism must form in order to develop the maximum strength of an actual beam. Consequently, the collapse load may well be larger than the shape factor times the load necessary to produce initial yielding of a beam (i.e., M_y).*

---

† The stress field $\sigma_x = \pm\sigma_0$, $\sigma_y = \sigma_z = \sigma_{xy} = \sigma_{xz} = \sigma_{yz} = 0$, which accompanies the limit moment M_0, satisfies the equilibrium equations for a three-dimensional continuum, even though there is a discontinuity in σ_x at $z = 0$. It is evident from Figure 1.4 that the change in curvature κ, which is defined by equation (1.3), is positive when the limit moment M_0 is reached. This gives rise to a strain field ($\epsilon_x = z\kappa$, $\epsilon_y = \epsilon_z = -\nu z\kappa$, $\epsilon_{xy} = \epsilon_{xz} = \epsilon_{yz} = 0$) which satisfies the compatibility equations for a three-dimensional continuum.

‡ Transverse shear effects are sometimes important for the static loading of beams with open cross-sections, and design methods are available to cater for the combined influence of a transverse shear force and a bending moment on the plastic yielding of a beam cross-section.[1.7] However, transverse shear effects are potentially more important for dynamic loadings, as discussed in Chapter 6.

§ A kinematically admissible collapse mechanism is a displacement field which satisfies the displacement boundary conditions, gives strains which satisfy the plastic incompressibility condition (constant volume) and allows the external loads to do positive work.

* See footnote of equation (1.32) for a specific example.

Table 1.1 *Shape factors*

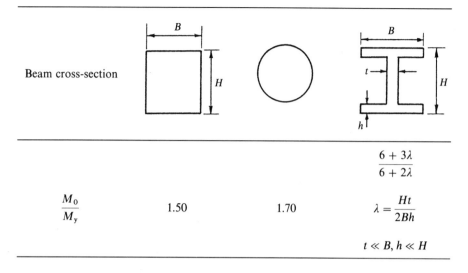

Beam cross-section			
$\dfrac{M_0}{M_y}$	1.50	1.70	$\dfrac{6+3\lambda}{6+2\lambda}$ $\lambda = \dfrac{Ht}{2Bh}$ $t \ll B, h \ll H$

1.3 Plastic collapse theorems for beams

1.3.1 Introduction

It was shown in the previous section that M_0 given by equation (1.6) is the plastic collapse or limit moment for a perfectly plastic beam with a solid rectangular cross-section when subjected to a pure bending moment. Clearly, the static load carrying capacity of the cross-section is exhausted and collapses, as illustrated in Figure 1.4. However, what is the carrying capacity of a beam with an external load which produces a bending moment distribution which varies along the axis?

The limit theorems of plasticity have been developed in order to provide simple estimates for the static collapse loads of perfectly plastic beams subjected to any form of external static loading. The lower and upper bound theorems of plasticity, which uncouple the static (equilibrium) and kinematic (deformation) requirements of a theoretical solution, are introduced in the next two sections.

1.3.2 Lower bound theorem

1.3.2(a) Statement of theorem

If any system of bending moments, which is in equilibrium with the applied loads and which nowhere violates the yield condition, can be found, then the beam will not collapse, or is at the point of collapse (incipient collapse).

1.3.2(b) Proof of theorem

It is assumed that a set of external concentrated and distributed loads denoted by $F(x)$

just causes collapse (incipient collapse) of a beam. The associated collapse mechanism for the beam is characterised by a velocity profile $\dot{w}(x)$ and rotation rates $\dot{\theta}$ with $\dot{\theta}_i$ at i discrete locations (hinges). The bending moment distribution at collapse is $M(x)$ and M_i at the plastic hinges.

Now, the principle of virtual velocities† gives

$$\sum M_i \dot{\theta}_i = \int F\dot{w}\,dx, \tag{1.7}$$

since M and F form an equilibrium set, while $\dot{\theta}$ and $\dot{w}$ are a kinematic set. All plastic hinge locations within the span of a beam and at the supports are included in the summation $\sum$, while the integral on the right-hand side of equation (1.7) extends over the entire beam.

The lower bound theorem of plasticity seeks to determine the multiplier λ^l so that the external load $\lambda^l F(x)$ does not cause collapse and is safely supported by a beam.‡ The associated bending moment distribution $M^s(x)$ is statically admissible when it satisfies the equilibrium equations (1.1) and (1.2) and nowhere exceeds the yield moment M_0 for the beam cross-section.

It is evident that M^s and $\lambda^l F$ are in equilibrium, and therefore the principle of virtual velocities predicts

$$\sum M_i^s \dot{\theta}_i = \int \lambda^l F\dot{w}\,dx, \tag{1.8}$$

which, when subtracted from equation (1.7), gives

$$(1 - \lambda^l)\int F\dot{w}\,dx = \sum (M_i - M_i^s)\dot{\theta}_i. \tag{1.9}$$

The generalised stress (M) and generalised strain rate $(\dot{\theta})$ are defined to give a non-negative energy dissipation rate $(M\dot{\theta} \geqslant 0)$, as observed in the footnote to equation (1.3).§ Moreover, $|M^s| \leqslant |M|$ throughout a beam according to the definition of a statically admissible bending moment field. Thus,

$$(M_i - M_i^s)\dot{\theta}_i \geqslant 0, \tag{1.10}$$

and, therefore, equation (1.9) predicts that

$$(1 - \lambda^l)\int F\dot{w}\,dx \geqslant 0,$$

or

$$\lambda^l \leqslant 1, \tag{1.11}$$

since the external work rate $\int F\dot{w}\,dx \geqslant 0$.

---

† The principle of virtual velocities is discussed in Appendix 3.

‡ This is known as proportional loading because only a proportional combination of loads are considered.

§ If $M_i = M_0$, then $\dot{\theta}_i \geqslant 0$, while $\dot{\theta}_i \leqslant 0$ when $M_i = -M_0$. In both cases $M_i\dot{\theta}_i \geqslant 0$.

Equation (1.11) constitutes the proof of the lower bound theorem for beams which is stated in § 1.3.2(a).

1.3.3 Upper bound theorem

1.3.3(a) Statement of theorem

If the work rate of a system of applied loads during any kinematically admissible collapse of a beam is equated to the corresponding internal energy dissipation rate, then that system of loads will cause collapse, or incipient collapse, of the beam.

1.3.3(b) Proof of theorem

It is assumed that a beam collapses under a load $\lambda^u F(x)$ with a bending moment field $M^k(x)$ and an associated kinematically admissible velocity field $\dot{w}^k(x)$, which has rotation rates $\dot{\theta}_j^k$ at j discrete locations (plastic hinges). Thus, equating the external work rate to the internal energy dissipation during a kinematically admissible collapse gives

$$\sum M_j^k \dot{\theta}_j^k = \int \lambda^u F \dot{w}^k \, dx, \qquad (1.12)\dagger$$

where M_j^k is the bending moment at the plastic hinges in the kinematically admissible collapse mechanism. Moreover,

$$\sum M_j \dot{\theta}_j^k = \int F \dot{w}^k \, dx, \qquad (1.13)$$

according to the principle of virtual velocities when using the equilibrium set (M, F) for the exact solution discussed in § 1.3.2(b). Subtracting equation (1.13) from equation (1.12) leads to

$$(\lambda^u - 1) \int F \dot{w}^k \, dx = \sum (M_j^k - M_j) \dot{\theta}_j^k. \qquad (1.14)$$

It is evident that $|M_j| \leqslant |M_j^k|$, where $M_j^k = \pm M_0$, and therefore $(M_j^k - M_j)\dot{\theta}_j^k \geqslant 0$, so that equation (1.14) requires

$$\lambda^u \geqslant 1 \qquad (1.15)$$

because $\int F \dot{w}^k \, dx \geqslant 0$. Inequality (1.15) constitutes the proof of the upper bound theorem for perfectly plastic beams which is stated in § 1.3.3(a).

1.3.4 Exact static collapse load

The inequalities (1.11) and (1.15) may be written

$$\lambda^l \leqslant 1 \leqslant \lambda^u. \qquad (1.16)$$

If

$$\lambda^l = \lambda^u = 1 \qquad (1.17)$$

---

† Equation (1.12) is a definition for λ^u.

then a theoretical solution is simultaneously statically admissible (i.e., satisfies the requirements of the lower bound theorem) and kinematically admissible (i.e., satisfies the requirements of the upper bound theorem) and is, therefore, exact.

1.4 Static plastic collapse of a cantilever

The limit theorems of plasticity, which were introduced in the previous section, are now used to obtain the static collapse load of the cantilever beam in Figure 1.5(a). The cantilever beam is made from a perfectly plastic material, is statically determinate, and has a linear bending distribution with a maximum value.

$$M = -PL \tag{1.18}$$

at $x = 0$. Thus, the elastic stress distribution according to the elementary beam bending theory is $\sigma_x = zM/I$. This expression may be used to predict that a load

$$P_E = 2\sigma_0 I/HL \tag{1.19}$$

could be supported in a wholly elastic manner when the beam cross-section is symmetric about the y-axis, where I is the second moment of area for the cross-section and σ_0 is the uniaxial plastic flow stress.

The bending moment distribution $P_E(L - x)$, which is associated with the load P_E, satisfies the requirements of the lower bound theorem of plasticity in § 1.3.2. However, we observe from equation (1.18) that a higher lower bound is

$$P^1 = M_0/L, \tag{1.20}$$

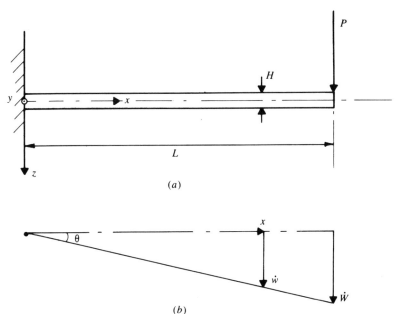

(a)

(b)

Figure 1.5 (a) Cantilever beam with an end load. (b) Transverse velocity profile for a cantilever beam with a plastic hinge at the support.

since it gives $M = -M_0$ at $x = 0$ and produces a bending moment distribution which nowhere violates the yield condition for a beam with the rigid, perfectly plastic, or bilinear approximation characteristics in Figure 1.4. The extent of the plastic flow region, which is associated with P^l, is sketched in Figure 1.6 for a beam with a rectangular cross-section. In this case, the stress distribution across the beam cross-section at the support ($x = 0$) would be similar to that in Figure 1.2(d) but with the signs of the stresses reversed. By way of contrast, the plastic flow would be confined to the extreme elements ($z = \pm H/2$) located at $x = 0$ (i.e., locations A and B in Figure 1.6) when the beam is subjected to the load P_E.

The transverse velocity field, which is shown in Figure 1.5(b), may be employed to calculate an upper bound to the exact collapse load according to the method outlined in § 1.3.3. Therefore,

$$M_0 \dot{\theta} = P^u L \dot{\theta}$$

or

$$P^u = M_0/L. \tag{1.21}$$

Thus, the exact static collapse or limit load of the perfectly plastic cantilever beam, which is illustrated in Figure 1.5(a), is $P_c = M_0/L$, since both the lower and the upper bound calculations predict the same result. It is observed that the cantilever beam may support a load which is 50 per cent larger than the maximum elastic value P_E given by equation (1.19) when the cross-section is rectangular ($I = BH^3/12$, $M_0 = BH^2\sigma_0/4$). In the case of a beam with a circular cross-section, then, according to Table 1.1, the plastic collapse load would be 1.70 times the initial yield value, which is predicted by an elementary linear elastic analysis.

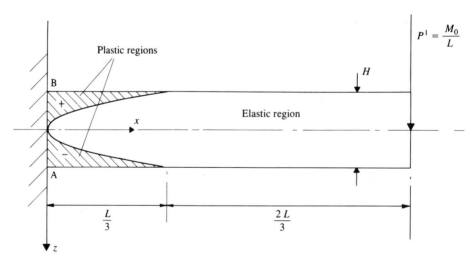

Figure 1.6 Elastic and plastic regions at collapse of the cantilever beam in Figure 1.5 with a rectangular cross-section.

The limit theorems of plasticity are valid for beams made from elastic, perfectly plastic or rigid, perfectly plastic materials. In other words, the exact static collapse load is identical for beams which are made from either material. In fact, it is evident from equation (1.5) in § 1.2 that the fully plastic bending moment M_0 is independent of the modulus of elasticity for the material. It is clear, therefore, that the limit theorems of plasticity bound the exact static plastic collapse load of a beam without any consideration of the complex elastic–plastic behaviour illustrated, for example, in Figure 1.6 for a cantilever beam with a rectangular cross-section.

1.5 Static plastic collapse of a simply supported beam

The limit theorems of plasticity in § 1.3 are now used to obtain the limit load of the simply supported beam in Figure 1.7(a) which is made from a rigid, perfectly plastic material.

If a plastic hinge forms at the beam centre owing to the action of a uniformly

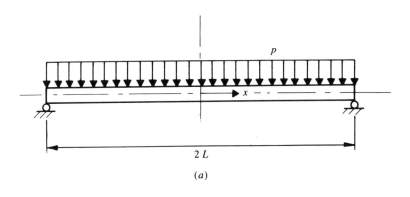

(a)

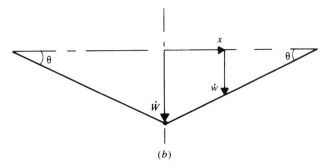

(b)

Figure 1.7 (a) Simply supported beam subjected to a uniformly distributed loading. (b) Transverse velocity profile for a simply supported beam with a plastic hinge at the mid-span.

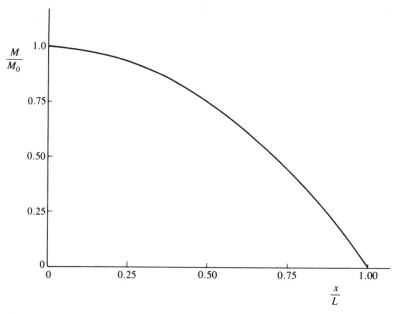

Figure 1.8 Bending moment distribution in one-half ($0 \leqslant x \leqslant L$) of the simply supported beam in Figure 1.7(a) according to equation (1.26).

distributed pressure p^u as shown in Figure 1.7(b), then an upper bound calculation (i.e., external work rate equals internal work rate) gives

$$2(p^u L)(L\dot{\theta}/2) = M_0 2\dot{\theta}$$

or

$$p^u = 2M_0/L^2. \tag{1.22}$$

The bending moment distribution in the region $0 \leqslant x \leqslant L$ of the beam in Figure 1.7(a) is

$$M = p(L^2 - x^2)/2, \tag{1.23}†$$

which has the largest value

$$M = pL^2/2 \tag{1.24}$$

at the beam centre. Thus, the bending moment distribution (equation (1.23)) is statically admissible (i.e., $-M_0 \leqslant M \leqslant M_0$) for a pressure

$$p^l = 2M_0/L^2, \tag{1.25}$$

which when substituted into equation (1.23) gives

$$M/M_0 = 1 - (x/L)^2, \tag{1.26}$$

as shown in Figure 1.8.

† This expression may be obtained from a free body diagram for the portion of the beam of length $L - x$ or from the solution of the equilibrium equations (1.1) and (1.2) with $M = 0$ at $x = L$ and $Q = 0$ at $x = 0$.

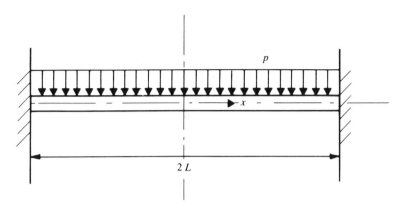

Figure 1.9 Fully clamped beam which is subjected to a uniformly distributed pressure.

It is evident from equations (1.22) and (1.25) that

$$p_c = 2M_0/L^2 \qquad (1.27)$$

is the exact collapse pressure, since it satisfies simultaneously the requirements of the upper and lower bound theorems of plasticity.

It should be noted that equation (1.27) may be used for uniform beams with any cross-sectional shape which is symmetrical with respect to the plane of bending.

1.6 Static plastic collapse of a fully clamped beam

The fully clamped beam shown in Figure 1.9 is statically indeterminate, unlike the simply supported case in § 1.5. However, the transverse velocity field sketched in Figure 1.7(b) could be used to predict an upper bound to the collapse pressure of the fully clamped case, provided the internal energy dissipation in the support hinges ($M_0 2\dot{\theta}$) were considered. Thus,

$$2(p^u L)(L\dot{\theta}/2) = M_0 2\dot{\theta} + M_0 2\dot{\theta},$$

or

$$p^u = 4M_0/L^2. \qquad (1.28)$$

If plastic hinges form at the supports, then the beam becomes statically determinate with a bending moment distribution

$$M = -M_0 + p(L^2 - x^2)/2. \qquad (1.29)\dagger$$

Thus, it may be shown that a lower bound to the exact collapse pressure

$$p^l = 4M_0/L^2 \qquad (1.30)$$

is obtained when the maximum value of M in equation (1.29) (at the beam mid-

---

† This expression may be obtained from a free body diagram of the beam or from the solution of equations (1.1) and (1.2) with $M = -M_0$ at $x = L$ and $Q = 0$ at $x = 0$.

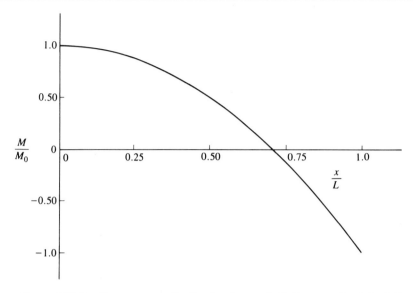

Figure 1.10 Bending moment distribution in one-half ($0 \leqslant x \leqslant L$) of the fully clamped beam in Figure 1.9 according to equation (1.31).

span) equals M_0. Equations (1.29) and (1.30) predict the associated bending moment distribution

$$M/M_0 = 1 - 2x^2/L^2, \tag{1.31}$$

which is statically admissible as shown in Figure 1.10.

The exact collapse pressure of a uniformly loaded, fully clamped beam according to equations (1.28) and (1.30) is, therefore,

$$\bar{p}_c = 4M_0/L^2. \tag{1.32}†$$

1.7 Static plastic collapse of a beam subjected to a concentrated load

Consider a simply supported beam which is subjected to a concentrated load at the mid-span as shown in Figure 1.11. It is assumed that the beam collapses with the kinematically admissible transverse velocity field in Figure 1.7(b). Thus, the upper bound theorem in § 1.3.3 gives

$$P^u L\dot{\theta} = M_0 2\dot{\theta}$$

or

$$P^u = 2M_0/L. \tag{1.33}$$

Now, the bending moment in a simply supported beam subjected to a concentrated load P^1 is

$$M = P^1(L - x)/2, \qquad 0 \leqslant x \leqslant L, \tag{1.34}$$

† This is double the pressure which causes the yield stress to be reached initially in the outermost elements at the supports when the beam has a rectangular cross-section.

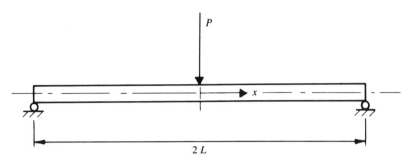

Figure 1.11 Simply supported beam which is subjected to a concentrated load at the mid-span.

according to equations (1.1) and (1.2), and is statically admissible provided the maximum value $P^1 L/2$ at the mid-span ($x = 0$) does not exceed the plastic collapse moment M_0 for the cross-section. The largest lower bound to the exact collapse load is, therefore,

$$P^1 = 2M_0/L. \tag{1.35}$$

Equations (1.33) and (1.35) are identical and, therefore, the exact static collapse load is

$$P_c = 2M_0/L. \tag{1.36}$$

It is left as an exercise for a reader to show that the exact collapse load is

$$\bar{P}_c = 4M_0/L \tag{1.37}$$

when both supports in Figure 1.11 are fully clamped.

1.8 Static plastic collapse of a partially loaded beam

The bound method which is outlined in § 1.3 and used in §§ 1.4 to 1.7 to obtain the static plastic collapse loads for several specific beam problems suggests an alternative approach for seeking an exact theoretical solution.

An exact solution is simultaneously kinematically and statically admissible, as observed in § 1.3.4. Thus, the first step in the alternative approach is to postulate a kinematically admissible collapse mechanism which allows the generalised strain rates to be obtained and the associated plastic regimes to be identified when employing the normality requirement. The equilibrium equations (1.1) and (1.2) are then solved and the boundary, continuity and symmetry conditions used to evaluate the unknown quantities. If the associated bending moment distribution nowhere violates the yield condition, then the theoretical solution is statically admissible and, therefore, exact. On the other hand, any yield violation would suggest the form of another kinematically admissible velocity profile, which might avoid the difficulty. This procedure is then repeated until an exact theoretical solution is found.

The above heuristic approach is now illustrated for the partially loaded beam with simple supports which is illustrated in Figure 1.12(a).

It is assumed that the beam collapses with the kinematically admissible transverse velocity field in Figure 1.12(b) which may be written in the form.

$$\dot{w} = \dot{W}(1 - x/L), \qquad 0 \leqslant x \leqslant L, \tag{1.38}$$

and, clearly, demands $M = M_0$ at $x = 0$. In view of the symmetry of the loading and the response about the mid-span ($x = 0$) of the beam in Figure 1.12(a), only the right-hand part of the beam need be examined in the subsequent analysis.

Equations (1.1) and (1.2) may be combined to give the governing equation

$$d^2M/dx^2 = -p, \tag{1.39}$$

which, for the beam in Figure 1.12(a), becomes

$$d^2M/dx^2 = -p_0, \qquad 0 \leqslant x \leqslant L_1. \tag{1.40}$$

Now, integrating equation (1.40) predicts the bending moment distribution

$$M = -p_0x^2/2 + M_0, \qquad 0 \leqslant x \leqslant L_1, \tag{1.41}$$

when satisfying the symmetry condition $Q = dM/dx = 0$ at $x = 0$ and the requirement that $M = M_0$ for a plastic hinge to develop at $x = 0$.

Equation (1.39) for the outer region becomes

$$d^2M/dx^2 = 0, \qquad L_1 \leqslant x \leqslant L, \tag{1.42}$$

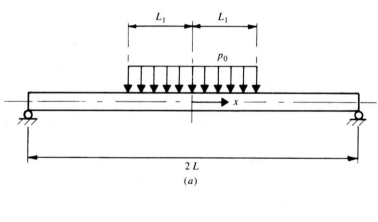

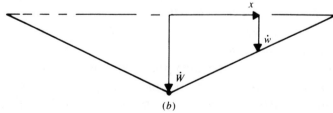

Figure 1.12 (a) Simply supported beam subjected to a pressure distributed uniformly within a central region of length $2L_1$. (b) Transverse velocity profile.

which gives

$$M = M_0 + p_0 L_1^2/2 - p_0 L_1 x, \qquad L_1 \leqslant x \leqslant L, \qquad (1.43)$$

when ensuring that the transverse shear force and bending moment distributions in the loaded and unloaded parts of the beam are continuous at $x = L_1$.

Finally, the simply supported boundary condition demands $M = 0$ at $x = L$, which is satisfied by equation (1.43) when

$$p_0 = 2M_0/\{L_1(2L - L_1)\}. \qquad (1.44)$$

Substituting equation (1.44) into equations (1.41) and (1.43) gives the bending moment distributions

$$M/M_0 = 1 - x^2/\{L_1(2L - L_1)\}, \qquad 0 \leqslant x \leqslant L_1, \qquad (1.45a)$$

and

$$M/M_0 = 1 + (L_1 - 2x)/(2L - L_1), \qquad L_1 \leqslant x \leqslant L, \qquad (1.45b)$$

which do not violate the yield condition, as shown in Figure 1.13, and are, therefore, statically admissible for $0 \leqslant x \leqslant L$ and $0 \leqslant L_1 \leqslant L$.

The theoretical solution is both kinematically and statically admissible so that equation (1.44) is the exact static collapse pressure.

It is evident that equation (1.44), with $L_1 = L$, reduces to equation (1.27) for a simply supported beam which is loaded over the entire span. Moreover, if equation (1.44) is rewritten to give the total load $P = p_0 2L_1 = 4M_0/(2L - L_1)$, then, when

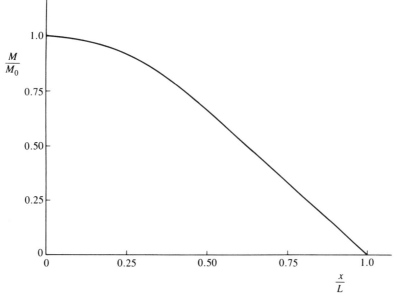

Figure 1.13 Bending moment distribution in one-half ($0 \leqslant x \leqslant L$) of the simply supported beam in Figure 1.12(a) according to equations (1.45a, b) with $L_1/L = 0.5$.

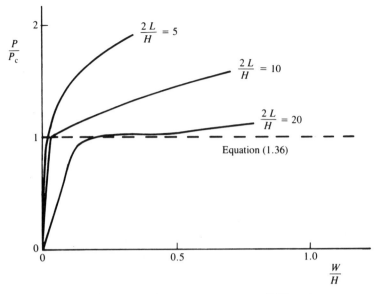

Figure 1.14 Experimental results of Haythornthwaite[1.13] on freely supported beams subjected to a centrally located concentrated load.

$L_1 \to 0$, $P = P_c = 2M_0/L$, which agrees with equation (1.36) for a concentrated load.

It is left as an exercise for a reader to use the plastic limit theorems in § 1.3 to obtain the static collapse pressure given by equation (1.44).

1.9 Experiments on beams

Haythornthwaite[1.13] reported the results of some experimental investigations which were conducted into the behaviour of laterally loaded steel beams with solid rectangular cross-sections. It was observed that the maximum centrally located concentrated load which could be carried by the freely supported thin beams in Figure 1.11 agreed quite closely with the corresponding limit load predicted by equation (1.36). However, a noticeable difference was observed between the theoretical and experimental results for the thick solid beams, although equation (1.36) still predicted the onset of large lateral displacements, as shown in Figure 1.14. This discrepancy is due probably to the influence of material strain hardening, since a decrease in L/H for constant values of L and W/H gives a larger deflection W which leads to larger curvature changes and strains.† The transverse shear force might become important for further decreases in L/H, although Hodge[1.2] and

† The strains are increased further because the beam thickness H increases as L/H decreases for a constant span L.

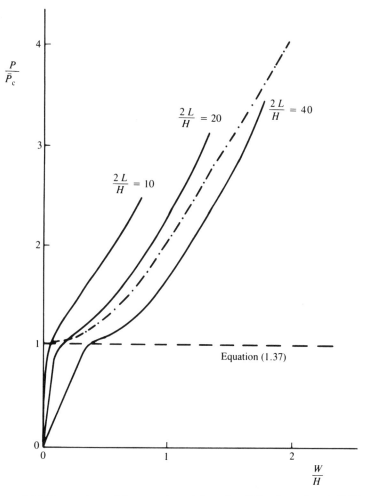

Figure 1.15 Comparison of the experimental results of Haythornthwaite[1.13] on axially restrained and fully clamped beams with the theoretical predictions of rigid, perfectly plastic analyses. ——: experimental results;[1.13] – – –: limit load, equation (1.37); —·—·—: theoretical predictions which include finite-deflection effects[1.13] (see equations (7.39)).

others have indicated that transverse shear forces are only significant for very thick solid beams with dimensions which would normally be considered as lying outside the range of simple beam theory.†

Haythornthwaite[1.13] conducted further tests on similar beams which were fully clamped and restrained axially at the supports and observed the lateral load-deflection characteristics shown in Figure 1.15. It is clear that the actual centrally

† Generally speaking, transverse shear effects are more important for beams with open cross-sections.

located, concentrated load-carrying capacity is now a function of the maximum lateral deflection W and, moreover, may increase to become many times larger than the static limit load $\bar{P}_c$ which is given by equation (1.37). This increase of strength is due principally to the favourable influence of membrane forces which arise during deformation but are neglected in the limit theorems. However, Haythornthwaite[1.13] has examined this phenomenon and predicted the enhanced load-carrying capacity, as shown in Figure 1.15 and discussed in § 7.2.5.

The influence of finite-deflections, or geometry changes, is also important in the dynamic plastic response of structures as discussed in Chapter 7.

1.10 Final remarks

A brief introduction to the vast literature extant on the static plastic behaviour of beams has been presented in this chapter. This branch of plasticity has been particularly well developed because of the importance of beams and frames in various civil engineering structures. However, the general procedures which were presented here can be used for structures in other branches of engineering. The textbooks listed in the references for this chapter contain numerous examples of and references to the experimental and theoretical behaviour of a large selection of beams and frames.

PROBLEMS

1.1 Derive the shape factor in Table 1.1 for a beam with an I-shaped cross-section.

1.2 Derive the shape factor for a beam with a hollow square box section, with outer dimensions H and a uniform wall thickness h, where $h \ll H$. Obtain the same result from the answer to Problem 1.1.

1.3 Derive the equation for the moment–curvature curve in Figure 1.4 for an elastic, perfectly plastic beam with a solid rectangular cross-section. What is the percentage difference between the curve and the fully plastic limit moment, M_0, when the curvature is 5 times larger than the value associated with the yield moment, M_y? Plot a curve showing the percentage difference between the exact curve in Figure 1.4 and the bilinear approximation with dimensionless curvature (i.e., curvature divided by the curvature at yield).

1.4 Obtain the uniformly distributed pressure which causes initial plastic yielding of the fully clamped beam shown in Figure 1.9, and show that it is one-half of the exact collapse pressure (equation (1.32)) when the beam has a solid rectangular cross-section.

1.5 Use the limit theorems of plasticity to bound the collapse load for the simply supported beam in Figure 1.11, with the concentrated load located at a distance a from the left-hand support where $a \leqslant L$.

1.6 Repeat Problem 1.5 when the beam is fully clamped at both supports and show that equation (1.37) is obtained when $a = L$.

1.7 Bound the static collapse load of the beam shown in Figure 1.12, using the lower and upper bound theorems of plasticity.

1.8 Repeat Problem 1.7, but for a beam with fully clamped supports.

1.9 Use the limit theorems of plasticity to bound the plastic collapse load for the beam in Figure 1.12 when the left-hand edge of the distributed load over a length $2L_1$ is at a distance a from the left-hand support (i.e., $2L - (a + 2L_1) \geqslant a$). Show that you can recover your answer to Problem 1.5.

1.10 Show that a uniform pressure of magnitude $(1.5 + \sqrt{2})M_0/L^2$ is the exact collapse pressure for a uniform, rigid, perfectly plastic beam of length $2L$ fully clamped at one end and simply supported at the other. M_0 is the plastic collapse moment for a typical cross-section of the beam.

1.11 Repeat Problem 1.10 using the upper and lower bound theorems of plasticity. Compare the prediction obtained by using the limit theorems with the exact solution. How could your bounds be improved?

1.12 Obtain the distribution of the transverse shear force, Q, for the simply supported beam illustrated in Figure 1.11. Show that the influence of the transverse shear force is likely to be unimportant for beams with solid rectangular cross-sections unless they are very short. Assume that the yield shear stress is $\sigma_0/2$, where σ_0 is the uniaxial tensile yield stress.

2

Static plastic behaviour of plates and shells

2.1 Introduction

An introduction to the static plastic behaviour of beams was presented in the preceding chapter, together with some theoretical solutions for several problems. It was observed that the idealisation of a perfectly plastic material, which is shown in Figures 1.3 and 1.4, is particularly attractive and simplifies considerably any theoretical calculations for the static plastic collapse load of a beam. Moreover, Figure 1.14 shows that the theoretical predictions for the static plastic collapse load give reasonable agreement with the corresponding experimental results on freely supported steel beams subjected to a central concentrated load. Good agreement between experimental results and the corresponding theoretical predictions for the static plastic collapse loads of many beams and frames may be found in the articles cited in the previous chapter.

The general concepts introduced in the previous chapter for beams are now used to study the static plastic collapse behaviour of plates and shells, which are important practical structures found throughout engineering. However, the theoretical analyses are more complex than for beams because the plastic flow in plates and shells is controlled by multi-dimensional yield criteria.

The generalised stresses and strains for plates and shells are introduced in the next section, while some basic concepts, which are associated mainly with multi-dimensional yield criteria, are discussed in § 2.3. The plastic collapse theorems are stated in § 2.4, together with some useful practical corollaries. Sections 2.5 to 2.11 contain the basic equations and some theoretical solutions for the static plastic collapse loads of circular plates, rectangular plates and axisymmetric cylindrical shells. A comparison of some experimental results and theoretical predictions for circular and rectangular plates and axisymmetric cylindrical shells is made in § 2.12. The chapter concludes with some final remarks in § 2.13.

2.2 Generalised stresses and strains

The simplifications, which were introduced for beams in § 1.2, are now developed further in order to obtain the static plastic collapse of plates and shells. Thus, the lateral, or transverse, dimensions of plate and shell structures are assumed to be small compared with the extent of the associated middle surface. In this circumstance, the stress σ_z in Figure 2.1, which is taken as normal to the middle surface (i.e., through the thickness of a plate or shell), may be neglected with little sacrifice in accuracy. Moreover, in common with the elastic theory of plates and shells, the remaining stresses in Figure 2.1 are replaced by the ten stress resultants, or generalised stresses,

$$N_x = \int_{-H/2}^{H/2} \sigma_x \, dz, \qquad N_y = \int_{-H/2}^{H/2} \sigma_y \, dz, \qquad N_{xy} = N_{yx} = \int_{-H/2}^{H/2} \sigma_{xy} \, dz,$$

$$M_x = \int_{-H/2}^{H/2} \sigma_x z \, dz, \qquad M_y = \int_{-H/2}^{H/2} \sigma_y z \, dz, \qquad M_{xy} = M_{yx} = \int_{-H/2}^{H/2} \sigma_{xy} z \, dz,$$

$$Q_x = \int_{-H/2}^{H/2} \sigma_{xz} \, dz \quad \text{and} \quad Q_y = \int_{-H/2}^{H/2} \sigma_{yz} \, dz, \qquad (2.1a\text{--}j)\dagger$$

where H is the uniform thickness of a plate or shell, as shown in Figure 2.2. Thus, an actual plate, or shell, is idealized as a (two-dimensional) surface with the ten generalised stresses in Figure 2.2, rather than the three-dimensional body in Figure 2.1 with the nine stresses $\sigma_x, \sigma_y, \sigma_{xy} = \sigma_{yx}, \sigma_{xz} = \sigma_{zx}, \sigma_{yz} = \sigma_{zy}$ and $\sigma_z = 0$. It should be noted that the thickness H of a plate, or shell, enters into a specific problem only through the definitions of the generalised stresses.

In order to remain consistent with the above model, the deformation of a plate or shell is expressed in terms of the displacements of the middle surface. Thus, the corresponding strains and curvature changes of this surface,

$$\epsilon_x, \epsilon_y, \epsilon_{xy} = \epsilon_{yx}, \kappa_x, \kappa_y, \kappa_{xy} = \kappa_{yx}, \gamma_x \text{ and } \gamma_y, \qquad (2.2a\text{--}j)$$

are associated with equations (2.1a–j).

The generalised stresses and strains, which are given by equations (2.1) and (2.2), respectively, are defined so that the product of each generalised stress and the corresponding generalised strain rate is a positive (or zero) energy dissipation rate.[(2.1)]‡ In other words, the total energy dissipation rate (per unit surface area),

$$\dot{D} = N_x \dot{\epsilon}_x + N_y \dot{\epsilon}_y + N_{xy} \dot{\epsilon}_{xy} + N_{yx} \dot{\epsilon}_{yx} + M_x \dot{\kappa}_x + M_y \dot{\kappa}_y + M_{xy} \dot{\kappa}_{xy}$$
$$+ M_{yx} \dot{\kappa}_{yx} + Q_x \dot{\gamma}_x + Q_y \dot{\gamma}_y, \qquad (2.3)\S$$

is positive (or zero).

† It should be noted that this general case reduces to a uniform beam of unit width having the generalised stresses M_x and Q_x which may be identified with M and Q in § 1.2.

‡ It was noted in the footnote to equation (1.3) that the generalised stress M and the associated curvature rate $\dot{\kappa}$ for beams were defined to give $M\dot{\kappa} \geqslant 0$.

§ Normally, $N_{xy}\dot{\epsilon}_{xy} + N_{yx}\dot{\epsilon}_{yx}$ in equation (2.3) is replaced by the term $N_{xy}\dot{\gamma}_{xy}$ when $N_{yx} = N_{xy}$ and $\dot{\gamma}_{xy} = 2\dot{\epsilon}_{xy}$, where $\dot{\gamma}_{xy}$ and $\dot{\epsilon}_{xy}$ are the engineering and tensorial shear strains, respectively.

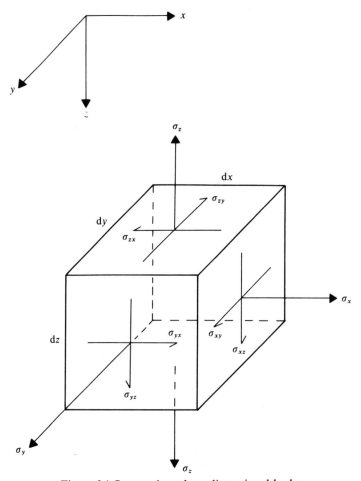

Figure 2.1 Stresses in a three-dimensional body.

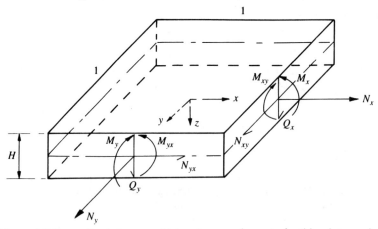

Figure 2.2 Generalised stresses which act on an element of a thin plate or shell of thickness H.

2.3 Basic concepts
2.3.1 Yield condition

It is evident from § 2.2 that, in the most general case, the plastic yielding of a structure is controlled by a combination of the ten generalised stresses which are defined by equations (2.1) and shown in Figure 2.2. In other words, the limit moment in Figure 1.4, which is required to develop a fully plastic hinge in a beam, is not a sufficient criterion for the plastic behaviour of plates and shells. It is necessary to develop yield surfaces in a space having the generalised stresses as coordinates, and the moment–curvature behaviour in Figure 1.4, therefore, gives the behaviour along one of these axes. Thus, no plastic yielding is possible when the generalised stresses lie within the yield surface, while a combination of generalised stresses which lie on the surface leads to plastic flow. The generalised stresses cannot lie outside the yield surface when the material is perfectly plastic.

Many studies have been published on the development of yield surfaces for various combinations of the generalised stresses. However, yield criteria are simply introduced throughout this book as they are required for particular problems. Some examples are the two-dimensional yield criteria in Figures 2.4 and 2.6 and the three-dimensional yield condition in Figure 6.15. A reader who is interested in this topic is referred either to one of the textbooks cited in Chapter 1, or to Hodge,[2.2] who has made a particular study of the yield conditions for plates and shells.

2.3.2 Drucker's stability postulate

Drucker[2.3] postulated that the work done by a set of external forces which act on a body must be positive during application and positive, or zero, over the complete cycle of application and removal of the external forces.† This is an appealingly simple requirement, which would appear to be valid for most practical materials. Indeed, the concept has already been used for beams, according to the footnote ‡ on p. 23, and was employed in the proof of the limit theorems presented in §§ 1.3.2 and 1.3.3.

2.3.3 Convexity of a yield condition

It may be shown when using Drucker's stability postulate that a multi-dimensional yield surface must be convex. A convex two-dimensional yield condition is sketched in Figure 2.3(a). However, the yield condition in Figure 2.3(b) is not convex and is, therefore, not admissible, according to Drucker's stability postulate. If the initial yielding of the material has been obtained experimentally at points C and D in Figure 2.3(b), then convexity demands that the portion of the yield condition lying between points C and D must lie to the right-hand side of, or on, a straight line which joins points C and D.

† A more precise definition of the postulate is given in reference 2.3.

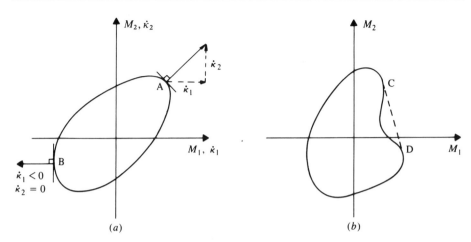

Figure 2.3 (*a*) Convex yield surface with normality rule. (*b*) Non-convex yield surface.

The convexity requirement provides considerable assistance for constructing a yield surface when incomplete experimental information is available for a particular material.

2.3.4 Normality of the generalised strain rate

Now, when a combination of generalised stresses lie on the surface of a yield condition, then plastic flow occurs with a non-negative energy dissipation rate ($\dot{D} \geqslant 0$). It is customary for the associated generalised strain rates to be plotted along the corresponding axes of the yield surface, as indicated in Figure 2.3(*a*).

Another consequence of Drucker's stability postulate in § 2.3.2 is the requirement that the generalised strain rate vector, which is associated with plastic flow, must be normal to the corresponding generalised stress point on the yield surface. In other words, if plastic flow occurs at the point A on the two-dimensional yield curve in Figure 2.3(*a*), then the generalised strain rate vector must be normal to the yield surface at point A, as shown. This vector has two generalised strain rate components along the two axes $\dot{\kappa}_1$ and $\dot{\kappa}_2$. However, the normality rule gives only the direction of the generalised strain rates and not the magnitude. This limitation is consistent with the calculations for beams in Chapter 1, since the generalised stress M is known at collapse (e.g., equations (1.26) and (1.31)), while the magnitude of the generalised strain rate $\dot{\kappa}$ and hence the transverse velocity $\dot{w}$ is not.

The normality rule of plasticity may render invaluable assistance for the solution of a particular problem. For example, once a kinematically admissible velocity field has been selected, then the evaluation of the generalised strain rates may suggest immediately which part of the yield surface controls plastic flow in that region of a structure. Thus, if $\dot{\kappa}_1 < 0$ and $\dot{\kappa}_2 = 0$ for a particular problem, which is controlled by the yield condition in Figure 2.3(a), then the normality rule requires that the associated generalised stresses lie at the point B.

2.3.5 Influence of transverse shear forces

The plate and shell structures examined in this chapter are assumed to be sufficiently thin so that the influence of the lateral or transverse shear strains (γ_x and γ_y) may be neglected. However, the lateral shear forces Q_x and Q_y are retained in the governing equations as reactions which are necessary for equilibrium. Nevertheless, they do not enter into the yield conditions according to the normality requirements of plasticity, because the associated generalised strain rates ($\dot{\gamma}_x$ and $\dot{\gamma}_y$) do not exist.†

Transverse shear forces may play a more important role when a structure is loaded dynamically, as discussed in Chapter 6.

2.4 Plastic collapse theorems

2.4.1 Introduction

The plastic collapse theorems were introduced in § 1.3. It was shown that the lower and upper bound theorems provide rigorous bounds on the exact collapse, or limit, load of a perfectly plastic beam.

The lower bound theorem proves that the external load associated with any statically admissible set of generalised stresses which do not violate the yield condition will be supported safely by a beam. The upper bound theorem, on the other hand, shows that the external loads associated with any kinematically admissible collapse pattern are greater than or equal to the exact collapse load.

The plastic collapse theorems have been proved for a three-dimensional continuum, which, therefore, means that they are valid for any structural member. The theorems are stated for structures in the next two sections without proof since they are proved in § 1.3 for beams, while many of the books cited in the previous chapter contain proofs for continua and structures.

2.4.2 Lower bound theorem

If any system of generalised stresses can be found throughout a structure which is in equilibrium with the applied loads and which nowhere violates the yield condition, then the structure will not collapse, or is at the point of collapse (incipient collapse).

---

† This is consistent with the theoretical analyses in Chapter 1. The transverse shear force Q is retained in equations (1.1) and (1.2) but the plastic deformations of the beams in Chapter 1 are controlled by the bending moment M alone (e.g., Figures 1.2 and 1.4).

2.4.3 Upper bound theorem

If the work rate of a system of applied loads during any kinematically admissible collapse of a structure is equated to the corresponding internal energy dissipation rate, then that system of loads will cause collapse, or incipient collapse, of the structure.

2.4.4 Corollaries of collapse theorems

There are a number of corollaries to the plastic collapse theorems and some of the more practically important are now stated for future reference.[2.4, 2.5]

1. The addition (removal) of (weightless) material to (from) a structure without any change in the position of the applied loads cannot result in a lower (higher) collapse load.

2. Increasing (decreasing) the yield strength of the material in any region of a perfectly plastic structure cannot weaken (strengthen) it.

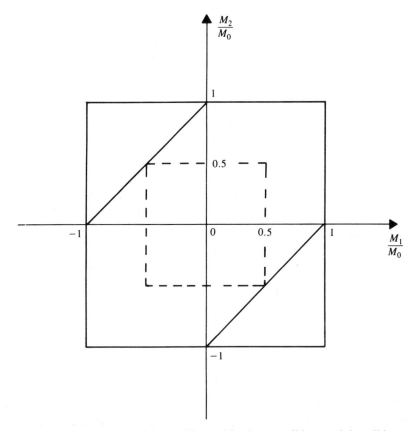

Figure 2.4 Hexagonal yield condition with circumscribing and inscribing square, or Johansen, yield criteria.

3. The exact collapse load associated with a convex yield condition which circumscribes (inscribes) the actual yield condition is an upper (lower) bound on the actual collapse load.

The two-dimensional square yield condition in Figure 2.4 circumscribes the hexagonal, or Tresca, yield criterion, whereas another which is 0.5 times as large would inscribe it. Thus, any calculations using the circumscribing and inscribing square yield criteria would predict upper and lower bounds on the exact collapse load using the hexagonal yield condition, respectively. However, it should be noted that the bounds could be closer than this for a particular problem. For example, if the generalised stresses were to remain within the upper right-hand or lower left-hand quadrants in Figure 2.4, then the circumscribing square and hexagonal yield criteria would predict the same collapse load.
4. Initial stresses or deformations have no effect on collapse provided the displacements remain infinitesimal.

2.5 Basic equations for circular plates

If a thin circular plate of thickness H is subjected to axisymmetrical lateral loads, which produce infinitesimal displacements, then the only non-zero stresses are σ_r, σ_θ and $\sigma_{rz} = \sigma_{zr}$, which are related to the generalised stress resultants

$$M_r = \int_{-H/2}^{H/2} \sigma_r z \, dz, \quad M_\theta = \int_{-H/2}^{H/2} \sigma_\theta z \, dz \quad \text{and} \quad Q_r = \int_{-H/2}^{H/2} \sigma_{rz} \, dz.$$

The moment and lateral equilibrium equations for the element of the plate shown in Figure 2.5 are

$$d(rM_r)/dr - M_\theta - rQ_r = 0 \tag{2.4}$$

and

$$d(rQ_r)/dr + rp = 0, \tag{2.5}$$

respectively.

The radial and circumferential changes of curvature of the middle surfaces are

$$\kappa_r = -d^2w/dr^2 \tag{2.6}$$

and

$$\kappa_\theta = -(1/r) \, dw/dr, \tag{2.7}$$

respectively, where w is the lateral, or transverse, displacement of the middle plane as defined in Figure 2.5.† The in-plane displacements and direct (membrane) strains vanish in the absence of in-plane external loads and/or when the influence of finite-deflections, or geometry changes, is disregarded.

Equations (2.4) to (2.7) are valid for a circular plate made from any material.

---

† Equations (2.4) to (2.7) are consistent according to the principle of virtual velocities. (See equations (A.68), (A.69) and (A.72) without the inertia term and (A.73) of Appendix 4.)

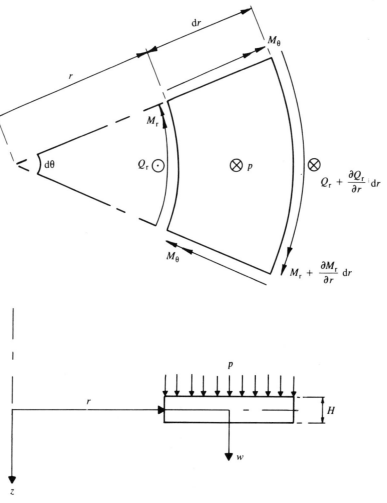

Figure 2.5 Notation for the static transverse axisymmetric loading of a thin circular plate. $\longrightarrow$: bending moment per unit length (right-hand rule). $\bigcirc$: transverse shear force per unit length, with $\cdot$ and $\times$ indicating the head or tail of a direction arrow, respectively.

However, in order to obtain a solution for a particular problem, these equations must be supplemented by the constitutive equations which relate the bending moments in the equilibrium equations to the changes in curvature of the middle plane. It is assumed that plastic flow is controlled by the maximum shear stress or Tresca yield criterion

$$\text{Max.}\{|\sigma_r|, |\sigma_\theta|, |\sigma_r - \sigma_\theta|\} \leqslant \sigma_0, \tag{2.8}‡$$

‡ This notation [2.2] represents six inequalities. The associated six equalities $\sigma_r = \pm\sigma_0$, $\sigma_\theta = \pm\sigma_0$ and $\sigma_r - \sigma_\theta = \pm\sigma_0$ correspond to the six lines in Figure 2.6(a).

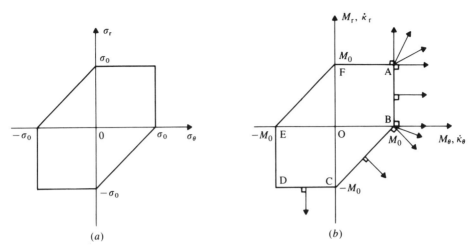

Figure 2.6(a) Tresca yield criterion for a two-dimensional principal stress state. (b) Tresca yield condition for two principal moments with normality rule.

which is illustrated in Figure 2.6(a) (σ_0 is the uniaxial tensile or compressive yield stress). It is now necessary to express this yield condition in terms of the bending moments M_r and M_θ which govern the plastic behaviour of circular plates. Clearly, the above inequality can be multiplied by z and integrated across the plate thickness H to give

$$\text{Max.}\{|M_r|, |M_\theta|, |M_r - M_\theta|\} \leqslant M_0, \tag{2.9}$$

where $M_0 = \sigma_0 H^2/4$ is the plastic collapse moment of a solid cross-section of unit width. Thus, any generalised stresses which lie within the Tresca yield condition illustrated in Figure 2.6(b) are elastic when a plate is made from an elastic, perfectly plastic material with the bilinear moment–curvature relation shown in Figure 1.4.

The lateral, or transverse shear force Q_r in Figure 2.5 does not participate in the yield condition because lateral shear deflections are neglected, as remarked in § 2.3.5. Thus, when plastic flow occurs, the deformation consists of the two components $\dot{\kappa}_r$ and $\dot{\kappa}_\theta$ of the curvature rate vector which is perpendicular to the corresponding portion of the yield surface (or lies within the fan at a corner as indicated in Figure 2.6(b)), according to the normality requirements of plasticity in § 2.3.4.

2.6 Static plastic collapse pressure of circular plates

2.6.1 Introduction

Now consider the simply supported circular plate which is subjected to a uniformly distributed transverse pressure, as shown in Figure 2.7(a). The limit theorems of

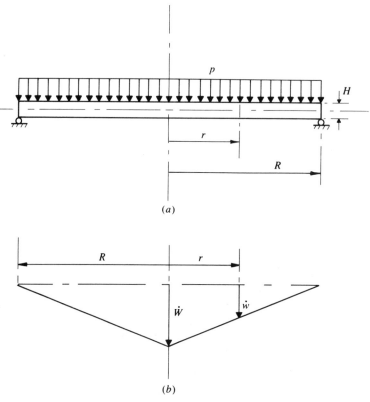

Figure 2.7 (a) Simply supported circular plate. (b) Transverse velocity field at collapse.

plasticity in §§ 2.4.2 and 2.4.3 are used in the next three sections to predict the exact collapse pressure p_c which was first obtained by Hopkins and Prager.[2.6] A partially loaded plate is examined in § 2.6.5, while fully clamped supports are discussed in § 2.6.6.

2.6.2 Lower bound calculation for simple supports

In order to obtain a lower bound (p^l) to the exact collapse pressure (p_c), it is necessary to seek a bending moment distribution which satisfies the equilibrium equations (2.4) and (2.5), nowhere exceeds the yield condition illustrated in Figure 2.6(b), and satisfies the simply supported boundary condition $M_r = 0$ at $r = R$. It appears feasible that the side AB of the yield hexagon might control the behaviour of the plate since this portion offers the possibility that the corner A could be used to give $M_r = M_\theta$ at the plate centre, as demanded from symmetry requirements, while the boundary condition $M_r = 0$ at $r = R$ could be located at the corner B. Thus, let

$$M_\theta = M_0, \qquad 0 \leqslant M_r \leqslant M_0, \qquad (2.10a, b)$$

for $0 \leqslant r \leqslant R$, where $M_0 = \sigma_0 H^2/4$, σ_0 is the uniaxial yield stress and H is the uniform plate thickness.

If equations (2.4) and (2.10a) are substituted into equation (2.5), then

$$\mathrm{d}^2(rM_r)/\mathrm{d}r^2 = -p^\mathrm{l}r$$

or

$$M_r = M_0 - p^\mathrm{l}r^2/6 \tag{2.11}$$

when satisfying the requirement that $M_r = M_\theta = M_0$ at $r = 0$. However, equation (2.11) gives immediately

$$p^\mathrm{l} = 6M_0/R^2 \tag{2.12}$$

since $M_r = 0$ at $r = R$ for a circular plate with a simply supported boundary. Substituting equation (2.12) into equation (2.11) gives the radial bending moment

$$M_r = M_0(1 - r^2/R^2), \tag{2.13}$$

which satisfies $0 \leqslant M_r \leqslant M_0$ for $0 \leqslant r \leqslant R$ and, therefore, lies on the side **AB** of the yield condition in Figure 2.6(b). Thus, equation (2.12) represents a lower bound to the exact collapse pressure p_c because all the requirements necessary for a lower bound calculation in § 2.4.2 have been satisfied.

2.6.3 Upper bound calculation for simple supports

An upper bound p^u to the exact collapse pressure p_c is obtained by equating the external work rate during any kinematically admissible velocity of the plate to the corresponding rate of internal energy dissipation. It appears reasonable from a physical viewpoint to assume that a plate would deform in the kinematically admissible conical velocity mode sketched in Figure 2.7(b) and which can be described in the form

$$\dot{w} = \dot{W}(1 - r/R), \tag{2.14}$$

where $\dot{W}$ is the transverse velocity at the plate centre. Thus, equations (2.6) and (2.7) give

$$\dot{\kappa}_r = 0 \qquad \text{and} \qquad \dot{\kappa}_\theta = \dot{W}/rR \tag{2.15a, b}$$

for $0 < r \leqslant R$, which demand $M_\theta = M_0$, according to the normality requirements associated with the Tresca yield condition illustrated in Figure 2.6(b). If the external work rate is equated to the corresponding internal energy dissipation rate, then

$$\int_0^R p^\mathrm{u}\dot{W}(1 - r/R)2\pi r\,\mathrm{d}r = \int_0^R M_0(\dot{W}/rR)2\pi r\,\mathrm{d}r, \tag{2.16}\dagger$$

or

$$p^\mathrm{u} = 6M_0/R^2. \tag{2.17}$$

---

† It should be noted that the internal energy dissipation is zero at $r = 0$ even though $\dot{\kappa}_r \to \infty$ at $r = 0$. Thus, $M_\theta\dot{\kappa}_\theta$ is the only non-zero term in the version of equation (2.3) for polar coordinates.

2.6.4 Exact collapse pressure for simple supports

Clearly, equations (2.12) and (2.17) show that the collapse pressure

$$p_c = 6M_0/R^2 \tag{2.18}$$

is simultaneously an upper bound and a lower bound and is, therefore, the exact collapse pressure according to the limit theorems of plasticity. Thus, the side AB of the yield condition in Figure 2.6(b) does control the collapse behaviour of the plate, while the incipient velocity profile is conical, as assumed in the theoretical work.

It should be noted that, if the normality requirements associated with the portions of the yield condition which are used in a lower bound calculation in § 2.6.2 are satisfied, then the solution is exact and it is unnecessary to perform the separate upper bound calculation in § 2.6.3. In the present problem, it is evident that equations (2.15a, b) give $\dot{\kappa}_r = 0$ and $\dot{\kappa}_\theta \geq 0$, which are consistent with the portion AB of the yield condition in Figure 2.6(b). Thus, equation (2.12) is the exact static collapse pressure.

This latter theoretical procedure is used in § 4.4.2 and predicts the static collapse pressure given by equation (2.18). (See equation (4.39), page 123.)

2.6.5 Partially loaded case with simple supports

If a uniform pressure is distributed within a central circular zone of radius a, then it is straightforward to extend the foregoing analysis and show that the corresponding exact static collapse pressure is [2.6]

$$p_c = 6M_0/\{a^2(3 - 2a/R)\}, \tag{2.19}$$

which reduces to equation (2.18) when $a = R$. Equation (2.19) may also be used for the collapse of the other extreme case of a central concentrated load. In this case, equation (2.19) may be written in the form $P_c = \pi a^2 p_c$, or

$$P_c = 6\pi M_0/(3 - 2a/R), \tag{2.20a}$$

which predicts a centrally located collapse load

$$P_c = 2\pi M_0 \tag{2.20b}$$

when $a \to 0$. It should be noted that a concentrated load P_c would produce large local lateral shear stresses and shear deformations, the influence of which have not been catered for in the above theoretical analysis.

2.6.6 Fully clamped supports

Hopkins and Prager[2.6] have shown that the exact collapse pressure of a uniformly loaded, fully clamped circular plate is

$$p_c \cong 11.26M_0/R^2. \tag{2.21}$$

2.7 Basic equations for rectangular plates

If a thin rectangular plate of uniform thickness H is subjected to lateral, or transverse loads, which produce infinitesimal displacements, then $N_x = N_y = N_{xy} = N_{yx} = 0$, where the generalised stresses are defined by equations (2.1a–d). In this circumstance, the transverse equilibrium and two moment-equilibrium equations for the element of the rectangular plate in Figure 2.8 are

$$\partial Q_x/\partial x + \partial Q_y/\partial y + p = 0, \tag{2.22}$$

$$\partial M_x/\partial x + \partial M_{xy}/\partial y - Q_x = 0, \tag{2.23}$$

and

$$\partial M_y/\partial y + \partial M_{xy}/\partial x - Q_y = 0, \tag{2.24}$$

respectively.

The transverse shear forces Q_x and Q_y from equations (2.23) and (2.24), respectively, may be differentiated and substituted into equation (2.22) to give the governing equation

$$\partial^2 M_x/\partial x^2 + 2\partial^2 M_{xy}/\partial x\,\partial y + \partial^2 M_y/\partial y^2 + p = 0. \tag{2.25}$$

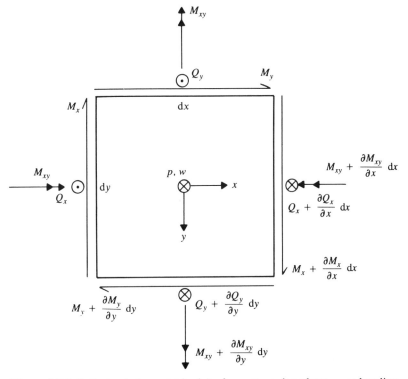

Figure 2.8 Infinitesimal element (dx dy) of a rectangular plate: ⟶: bending moment per unit length (right-hand rule). ⟶: twisting moment per unit length (right-hand rule). ○: transverse shear force per unit length, with · and × indicating the head or tail of the direction arrow, respectively.

It is evident that equations (2.22), (2.23) and (2.25), with $Q_x = Q$, $M_x = M$ and $Q_y = M_y = M_{xy} = 0$, respectively, reduce to equations (1.2), (1.1) and (1.39) on the static plastic behaviour of beams.

It is assumed that a rectangular plate is made from a rigid, perfectly plastic material and that the plastic flow is controlled by the three generalised stresses M_x, M_y and M_{xy}, when the transverse shear forces Q_x and Q_y are taken as reactions. Thus, the plastic flow of a rectangular plate satisfies a three-dimensional yield surface in M_x–M_y–M_{xy} space. However, this yield surface may be represented in a two-dimensional principal moment space, as shown, for example, in Figure 2.4 for the Tresca and square yield criteria.

The principal bending moments M_1 and M_2 in a rectangular plate are related to the moments M_x, M_y and M_{xy}, which are defined in Figure 2.8, as follows[2.7]:

$$M_1 = (M_x + M_y)/2 + \{(M_x - M_y)^2/4 + M_{xy}^2\}^{1/2} \qquad (2.26)$$

and

$$M_2 = (M_x + M_y)/2 - \{(M_x - M_y)^2/4 + M_{xy}^2\}^{1/2}. \qquad (2.27)$$

The form of equations (2.26) and (2.27) is similar to the well-known expressions for the principal stresses in plane stress. In fact, if the stresses on a plane which is inclined at an arbitrary angle to the x-axis are multiplied by the coordinate z and integrated through the plate thickness, then equations (2.26) and (2.27) are obtained when using methods similar to those used to predict the principal stresses in elementary texts.

2.8 Static plastic collapse pressure of rectangular plates

2.8.1 Introduction

The limit theorems of plasticity in §§ 2.4.2 and 2.4.3 are employed in the next two sections to bound the exact static collapse pressure of a simply supported rectangular plate which is subjected to a uniformly distributed transverse pressure, as shown in Figure 2.9. These predictions are discussed in § 2.8.4 and shown to be exact for a square plate which is made from a perfectly plastic material and which obeys the square, or Johansen, yield criterion in Figure 2.4.

The influence of a partial moment restraint around the boundaries of a rectangular plate is discussed in § 2.8.5, while the static collapse behaviour due to a concentrated load is examined in § 2.8.6.

2.8.2 Lower bound calculation for simple supports

Now, the static plastic collapse behaviour of the rectangular plate in Figure 2.9 is symmetric about both the x- and y-axes and, therefore, it is necessary only to examine one quadrant of the plate. Thus, in order to satisfy the symmetry

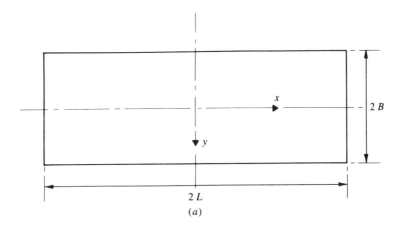

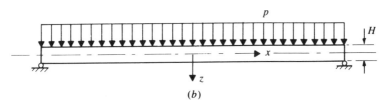

Figure 2.9 Simply supported rectangular plate subjected to a uniformly distributed transverse pressure. (*a*) Plan view. (*b*) Side view.

requirements, the bending moments may be written in the form

$$M_x = M_0 - ax^2,$$

(2.28a)

$$M_y = M_0 - by^2$$

(2.28b)

and

$$M_{xy} = -cxy,$$

(2.28c)

which are the leading terms in the series expansions. Equations (2.28) are acceptable for a lower bound calculation provided the equilibrium equations, boundary conditions and continuity requirements are satisfied without violating the yield criterion.

Equations (2.28a, b) satisfy $M_x = 0$ at $x = \pm L$ and $M_y = 0$ at $y = \pm B$ for simple supports when

$$a = M_0/L^2$$

(2.29a)

and

$$b = M_0/B^2.$$

(2.29b)

Thus, $M_x = M_y = 0$ and $M_{xy} = \mp cLB$ at the plate corners, so that

$$c = M_0/BL$$

(2.29c)

is necessary to avoid a yield violation.† Finally, equations (2.28) and (2.29) give

$$M_x = M_0(1 - x^2/L^2), \tag{2.30a}$$

$$M_y = M_0(1 - y^2/B^2) \tag{2.30b}$$

and

$$M_{xy} = -M_0xy/BL. \tag{2.30c}$$

Now, differentiating the generalised stresses (2.30) and substituting into the governing equilibrium equation (2.25) predicts the pressure

$$p^1 = 2M_0\{1 + L/B + (L/B)^2\}/L^2. \tag{2.31}$$

It is straightforward to show that equations (2.26, 2.27 and 2.30) predict the principal bending moments

$$M_1/M_0 = 1 \tag{2.32a}$$

and

$$M_2/M_0 = 1 - (x^2/L^2 + y^2/B^2), \tag{2.32b}$$

where $-1 \leqslant M_2/M_0 \leqslant 1$. These principal bending moments lie on one side of the square yield condition in Figure 2.4, and the associated generalised stresses given by equations (2.30) are, therefore, statically admissible. Thus, equation (2.31) provides a lower bound to the exact collapse pressure for a rectangular plate made from a rigid, perfectly plastic material which obeys the square, or Johansen, yield condition in Figure 2.4. A square yield curve with $|M_1| = M_0/2$ and $|M_2| = M_0/2$ would inscribe the Tresca yield condition, as shown in Figure 2.4. The corollary (3) in § 2.4.4 shows that the pressure $p^1/2$ is a lower bound to the exact collapse pressure for a rectangular plate which is made from a material controlled by the Tresca yield criterion.

2.8.3 Upper bound calculation for simple supports

Wood[2.8] calculated an upper bound to the static collapse pressure of the rectangular plate in Figure 2.9 with the aid of the kinematically admissible transverse velocity profile illustrated in Figure 2.10. It is evident from Figure 2.11 that this field describes adequately the dominant features of the reinforced concrete plate tested by Sawczuk and Winnicki.[2.9]

The transverse velocity profile may be expressed in the form

$$\dot{w} = \dot{W}(B \tan \phi - x')/B \tan \phi \tag{2.33}$$

in region I and

$$\dot{w} = \dot{W}(B - y)/B \tag{2.34}$$

† Equations (2.26) and (2.27) predict the principal moments $M_1 = M_{xy}$ and $M_2 = -M_{xy}$ when $M_x = M_y = 0$ at the plate corners. Therefore, $|M_{xy}| \leqslant M_0$ is required to avoid a yield violation according to the Johansen, or square, yield condition in Figure 2.4.

in region II, where $\dot{W}$ is the maximum transverse velocity at the plate centre and the in-plane velocities $\dot{u} = \dot{v} = 0$ throughout the entire plate. Thus, regions I and II are rigid, and all plastic flow is concentrated in the plastic hinges located at the boundaries of these regions.

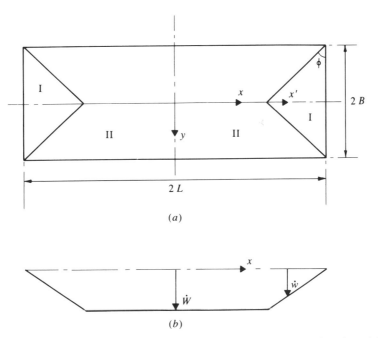

(a)

(b)

Figure 2.10 Static plastic collapse of a simply supported rectangular plate. (a) Plan view of plastic hinge lines. (b) Side view of transverse velocity profile.

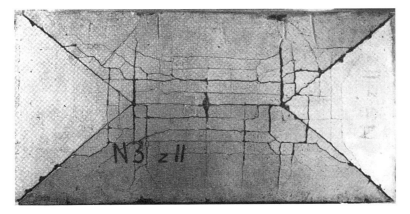

Figure 2.11 Collapse mode of a simply supported reinforced concrete rectangular plate with $L/B = 2$ which was tested by Sawczuk and Winnicki[2.9].

The angular velocities at the outer simply supported boundaries of regions I and II are

$$\dot{\theta}_1 = \dot{W}/B \tan \phi \tag{2.35}$$

and

$$\dot{\theta}_2 = \dot{W}/B, \tag{2.36}$$

respectively, while the changes in angular velocities across the inclined plastic hinges, which lie at the intersections of regions I and II, are

$$\dot{\theta}_3 = \dot{\theta}_1 \cos \phi + \dot{\theta}_2 \sin \phi,$$

or

$$\dot{\theta}_3 = \dot{W}/B \sin \phi. \tag{2.37}$$

The total internal energy dissipation in the plate is

$$\dot{D} = M_0(2L - 2B \tan \phi)2\dot{\theta}_2 + 4M_0B\dot{\theta}_3/\cos \phi,$$

or

$$\dot{D} = 4M_0\dot{W}(L/B + \cot \phi), \tag{2.38}$$

where $M_0 = \sigma_0 H^2/4$.

It is straightforward to show that the external work rate due to a uniformly distributed external pressure p^u is

$$\dot{E} = 4p^u\left[\int_0^{B \tan \phi} \dot{W}(B \tan \phi - x')x' \, dx'/B \tan^2 \phi \right.$$
$$\left. + \int_0^B \dot{W}(B - y)\dot{y} \tan \phi \, dy/B + (L - B \tan \phi) \int_0^B \dot{W}(B - y) \, dy/B \right],$$

which reduces to

$$\dot{E} = 2B^2\dot{W}p^u(L/B - \tan \phi/3). \tag{2.39}$$

An upper bound to the static collapse pressure is obtained by equating equations (2.38) and (2.39), which predicts

$$p^u = 6M_0(1 + \beta/\tan \phi)/\{B^2(3 - \beta \tan \phi)\}, \tag{2.40}$$

where

$$\beta = B/L \tag{2.41}$$

is the aspect ratio of a plate. A minimum upper bound collapse pressure p^u occurs when $\dfrac{\partial p^u}{\partial \tan \phi} = 0$. Thus,

$$\tan^2 \phi = (3 - \beta \tan \phi)/(1 + \beta/\tan \phi),$$

which has a solution

$$\tan \phi = -\beta + \sqrt{3 + \beta^2}. \tag{2.42}$$

If equation (2.42) is substituted into equation (2.40), then

$$p^u = 6M_0/\{B^2(\sqrt{3 + \beta^2} - \beta)^2\}, \tag{2.43}$$

which is an upper bound to the exact collapse pressure.

2.8.4 Comments on the bounds for a simply supported rectangular plate

Now, the lower and upper bounds to the exact collapse pressure of a simply supported rectangular plate are given by equations (2.31) and (2.43), which may be recast in the dimensionless forms

$$p^l/(2M_0/B^2) = 1 + \beta + \beta^2 \tag{2.44}$$

and

$$p^u/(2M_0/B^2) = 3/\{\sqrt{3 + \beta^2} - \beta\}^2, \tag{2.45}$$

respectively. It is evident from Figure 2.12 that the bounds predict the exact static collapse pressure for the two extreme cases $\beta \to 0$ and $\beta = 1$ and are very close for all other values of β.

The case $\beta \to 0$ is associated with a plate having $L \gg B$ and a breadth $2B$, as sketched in Figure 2.13. The corresponding static plastic collapse behaviour is

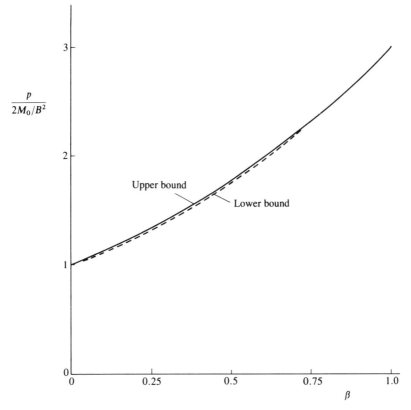

Figure 2.12 Comparison of lower (−−−: equation (2.44)) and upper (———: equation (2.45)) bounds on the static collapse pressure of a simply supported rectangular plate.

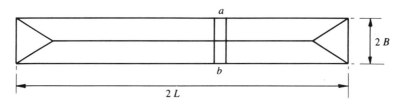

Figure 2.13 The behaviour of a strip ab in a rectangular plate with $\beta \to 0$ is similar to a beam of span $2B$.

similar to the strip ab with a plastic hinge at the mid-span and, therefore, is the same as a simply supported beam with a span $2B$. Equations (2.44) and (2.45) with $\beta \to 0$ predict

$$p^{\mathrm{l}} = p^{\mathrm{u}} = p_{\mathrm{c}} = 2M_0/B^2, \qquad (2.46)$$

which is identical to equation (1.27) for a simply supported beam subjected to a transverse pressure distributed uniformly over a span $2B$.

The particular case $\beta = 1$ (i.e., $B = L$) corresponds to a square plate, and equations (2.44) and (2.45) predict

$$p^{\mathrm{l}} = p^{\mathrm{u}} = p_{\mathrm{c}} = 6M_0/L^2, \qquad (2.47)$$

which agrees with Prager.[2.10] Equation (2.47) also agrees with the exact static collapse pressure according to equation (4.128), which is obtained by solving the governing equations.

2.8.5 Static collapse pressure of rectangular plates with partially restrained supports

Wood[2.8] has also studied the behaviour of a rectangular plate when the outer boundary is restrained with a bending moment m, where $0 \leqslant m \leqslant M_0$. The kinematically admissible transverse velocity field, which is illustrated in Figure 2.10 and described by equations (2.33) and (2.34), may be used to obtain an upper bound to the static plastic collapse pressure for this particular case. Thus, the energy dissipation at the plastic hinges within the plate, which are shown in Figure 2.10, is given by equation (2.38). In addition, however, the energy dissipation around the plate boundaries is

$$\dot{D}_{\mathrm{b}} = m(4B\dot{W}/B \tan \phi + 4L\dot{W}/B), \qquad (2.48)$$

which, when added to equation (2.38), gives the total internal energy dissipation

$$\dot{D}_{\mathrm{t}} = 4M_0\dot{W}(1 + m/M_0)(L/B + 1/\tan \phi). \qquad (2.49)$$

The external work rate due to the transverse pressure is again given by equation (2.39). An upper bound to the exact static collapse pressure is obtained by equating equations (2.39) and (2.49). It may be shown, when minimising this expression, that

$$p^{\mathrm{u}} = 6M_0(1 + m/M_0)/\{B^2(\sqrt{3 + \beta^2} - \beta)^2\} \qquad (2.50)$$

is the smallest upper bound collapse pressure for the assumed collapse profile, where β is defined by equation (2.41) and $\tan \phi$ is given by equation (2.42).

Equation (2.43) for the static plastic collapse pressure of a simply supported rectangular plate is recovered from equation (2.50) with $m = 0$. The other extreme case of fully clamped supports is obtained from equation (2.50) with $m = M_0$, or

$$p^u = 12M_0/\{B^2(\sqrt{3 + \beta^2} - \beta)^2\}, \tag{2.51}$$

which is twice as large as equation (2.43).

A theoretical procedure,[2.11] which is similar to that outlined in § 2.8.2, may be used to construct a statically admissible generalised stress field for a rectangular plate which is fully clamped around the boundary and subjected to a uniformly distributed pressure. This analysis predicts a lower bound to the exact static collapse pressure

$$p^l = 4M_0(1 + \beta^2)/B^2. \tag{2.52}$$

Equations (2.51) and (2.52) with $\beta \to 0$ predict

$$p^l = p^u = p_c = 4M_0/B^2 \tag{2.53}$$

for a fully clamped beam which is subjected to a transverse pressure distributed uniformly across a span $2B$. Equation (2.53), therefore, predicts the same collapse pressure as equation (1.32) for a beam with a span $2B$. However, the bounds separate with increase in β to give

$$p^u = 12M_0/L^2 \tag{2.54a}$$

and

$$p^l = 8M_0/L^2 \tag{2.54b}$$

when $\beta = 1$ for a fully clamped square plate with $B = L$. Fox[2.12] used a numerical scheme to obtain the exact static collapse pressure

$$p_c = 10.71M_0/L^2 \tag{2.54c}$$

for a fully clamped square plate.

2.8.6 Static plastic collapse of a plate due to a concentrated load

Several authors (e.g., Zaid[2.13]) have demonstrated that the magnitude of a static concentrated load necessary to cause plastic collapse of an arbitrarily shaped rigid, perfectly plastic plate with any degree of fixity around the boundary[2.2] is

$$P_c = 2\pi M_0 \tag{2.55}$$

when the material flows plastically according to the Tresca yield condition. Equation (2.55) is the same as equation (2.20b) for the particular case of a simply supported circular plate.

2.9 Basic equations for cylindrical shells

The foundations of shell theory are related closely to those of beam and plate theory and are discussed by Hodge[2.2] and other authors.

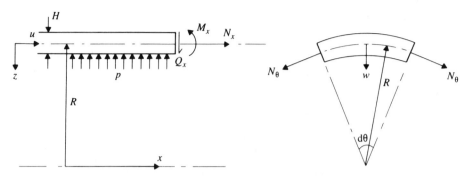

Figure 2.14 Notation for a cylindrical shell.

If the element of a thin-walled cylindrical shell, which is shown in Figure 2.14, is unloaded axially (i.e., $N_x = 0$), then the force and moment equilibrium of a shell element demands

$$R \, dQ_x/dx + N_\theta - Rp = 0 \tag{2.56}$$

and

$$dM_x/dx - Q_x = 0, \tag{2.57}$$

respectively, where p is an axisymmetric pressure distribution.† It is straightforward to show for infinitesimal displacements that the membrane strains

$$\epsilon_x = du/dx \tag{2.58}$$

and

$$\epsilon_\theta = -w/R \tag{2.59}$$

and the curvature change‡

$$\kappa_x = -d^2w/dx^2 \tag{2.60}$$

are consistent with the equilibrium equations (2.56) and (2.57) according to the principle of virtual velocities in Appendix 4.

It is evident that N_θ, M_x and Q_x are the only generalised stresses which appear in the equilibrium equations (2.56) and (2.57). Moreover, the lateral shear force Q_x does not enter into the yield condition for a cylindrical shell when the correspond-

---

† The generalised stresses N_x, N_θ, M_x and Q_x were defined in § 2.2 when θ was identified with the coordinate y. These definitions are adequate for shells with large R/H ratios as discussed by Kraus[2.14] and others.

‡ The circumferential curvature change κ_θ is $1/R - 1/(R - w)$, or $\kappa_\theta \cong -w/R^2$. Thus, the circumferential strain in a cylindrical shell is $e_\theta = \epsilon_\theta + z\kappa_\theta = -(w/R)(1 + z/R)$. However, $z/R \ll 1$ for thin shells, so that $\kappa_\theta \cong 0$.

It is also evident from the general expressions presented by Kraus[2.14] (see also Hodge[2.15] and Calladine[2.16]) that the curvature change κ_θ is zero, and the bending moment M_θ does not appear in the equilibrium equations of an axisymmetrically loaded cylindrical shell which undergoes infinitesimal displacements.

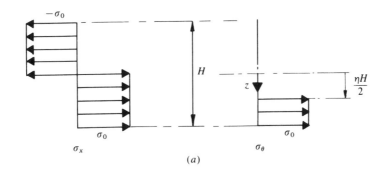

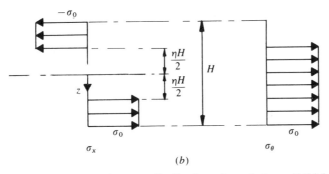

Figure 2.15 Fully plastic stress distributions through the wall thickness H of a cylindrical shell when subjected to an axial bending moment M_x and a circumferential membrane force N_θ.

ing shear strain is neglected. Thus, it is necessary to retain only the two generalised stresses N_θ and M_x when constructing a yield condition for a cylindrical shell which is subjected to an axisymmetric pressure distribution and is unloaded axially. If such a shell is made from a rigid, perfectly plastic material, which flows plastically according to the Tresca yield criterion, then

$$\text{Max.}\{|\sigma_x|, |\sigma_\theta|, |\sigma_x - \sigma_\theta|\} \leqslant \sigma_0, \qquad (2.61)†$$

which is the same as that illustrated in Figure 2.6(a) when the ordinate σ_r is replaced by σ_x.

It is now necessary to express the yield criterion (2.61) in terms of the generalised stresses N_θ and M_x. Consider first the stress distribution across the wall thickness H of a shell which is shown in Figure 2.15(a). This stress distribution has $\sigma_x = -\sigma_0$ and $\sigma_\theta = 0$ for $-H/2 \leqslant z \leqslant 0$, while $\sigma_x = \sigma_0$ and $\sigma_\theta = 0$ for $0 \leqslant z \leqslant \eta H/2$ and $\sigma_x = \sigma_\theta = \sigma_0$ for $\eta H/2 \leqslant z \leqslant H/2$. These combinations of stress satisfy the Tresca yield condition. The stress distribution for σ_x in Figure 2.15(a) gives $N_x = 0$ and $M_x = M_0 = \sigma_0 H^2/4$ for a beam with a solid rectangular cross-section of depth H and

† See also equation (2.8) and footnote.

unit width. The stress distribution for σ_θ in Figure 2.15(a) gives rise to $0 \leqslant N_\theta \leqslant N_0/2$† and $M_\theta > 0$. However, the bending moment due to σ_θ is not considered further, since M_θ does not influence the behaviour of an axisymmetrically loaded cylindrical shell, as noted previously. Thus, a circumferential membrane force which lies within the range $0 \leqslant N_\theta \leqslant N_0/2$ can be associated with a fully plastic axial (longitudinal) bending moment $M_x = M_0$ without violating any of the inequalities in (2.61).† The straight line AB in Figure 2.16 represents this portion of the yield curve.‡

It is evident from Figure 2.15(a) that if $\sigma_\theta = \sigma_0$ extends into the region $z < 0$, then $\sigma_x = -\sigma_0$ and $\sigma_\theta = \sigma_0$, which violates the Tresca yield condition. Thus larger values of N_θ are accompanied by a reduction in the value of M_x, which is necessary for plastic flow. This suggests the stress distribution shown in Figure 2.15(b) for which $M_x = M_0(1 - \eta^2)$ and $N_\theta = N_0(1 + \eta)/2$, or $M_x/M_0 - 4N_\theta/N_0 + 4(N_\theta/N_0)^2 = 0$ when η is eliminated. The corresponding portion of the yield condition is indicated by the curve BC in Figure 2.16. The behaviour in the remaining quadrants of Figure 2.16 can be obtained in a similar manner to give the complete yield curve ABCDEFGH, which has been shown to be exact (e.g., Hodge[2.15]).

Thus, the yield curve in the generalised stress plane $N_\theta - M_x$ is non-linear even though it is based on the Tresca yield criterion which is piecewise linear in the principal stress space. A number of authors (see Hodge[2.2]) have proposed various approximate yield curves in order to avoid some of the mathematical complexities encountered when studying the behaviour of cylindrical shells made from materials which yield according to the Tresca yield criterion. The hexagonal yield curve drawn in Figure 2.16 is one such approximation and is clearly a lower bound to the exact yield condition. It may be shown that another hexagonal yield curve with all dimensions multiplied by 1.125 circumscribes the exact yield condition and is, therefore, an upper bound. The fact that these hexagonal yield criteria provide lower and upper bounds to the exact yield curve follows directly from corollary 3 of the limit theorems of plasticity which is discussed in § 2.4.4.

The theoretical analysis in the following section for the static plastic collapse pressure of a long reinforced cylindrical shell is simplified with the aid of the rectangular yield criterion in Figure 2.17. This is known as a limited-interaction yield curve and circumscribes the exact yield curve, while another 0.75 times as large everywhere inscribes it, as indicated in Figure 2.17.

---

† $N_0 = \sigma_0 H$ is the magnitude of a force necessary to produce plastic flow over the entire wall thickness of a shell. $M_0 = \sigma_0 H^2/4$ is the plastic capacity of the cross-section when subjected to a pure bending moment.

‡ The portion AB in Figure 2.16 may also be obtained, for example, with $\sigma_x = \sigma_0$ and $0 \leqslant \sigma_\theta \leqslant \sigma_0$ for $0 \leqslant z \leqslant H/2$ in Figure 2.15(a).

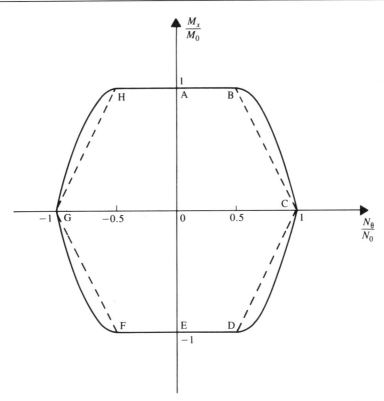

Figure 2.16 Exact (——) and inscribing hexagonal (– – –) yield curves for the combined influence of M_x and N_θ.

2.10 Static collapse pressure of a long reinforced cylindrical shell

Now, consider the particular case of a long cylindrical shell, which is reinforced by equally spaced rigid rings, and is subjected to a uniformly distributed internal pressure p, as shown in Figure 2.18. It is necessary to examine only one bay, which may be considered to be a short cylindrical shell, of length $2L$, fully clamped at both ends. The static plastic collapse pressure is sought for this particular shell when the material is rigid, perfectly plastic and obeys the limited-interaction yield curve indicated in Figure 2.17.

It appears reasonable to explore the possibility that the profile of the shell at collapse is as indicated in Figure 2.19. The shell deforms with a radial displacement which gives rise to $N_\theta = N_0$ for $-L \leqslant x \leqslant L$. Thus, the circumferential plastic hinges which form at $x = -L$ and $x = L$ are located at A in Figure 2.17, while the circumferential plastic hinge which develops at $x = 0$ is located at the corner B. This suggests immediately that the stress profile AB governs the behaviour in the region $0 \leqslant x \leqslant L$, or

$$N_\theta = N_0 \qquad \text{and} \qquad -M_0 \leqslant M_x \leqslant M_0. \qquad (2.62a, b)$$

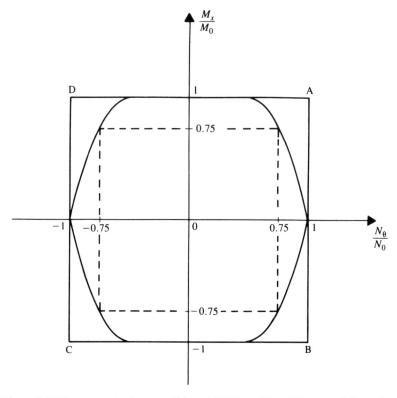

Figure 2.17 Exact (——), circumscribing (ABCD) and inscribing (–––) limited-interaction (rectangular) yield curves for the combined influence of M_x and N_θ.

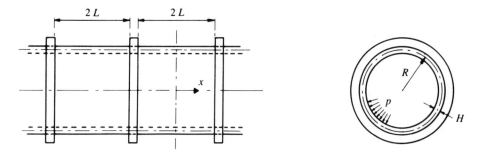

Figure 2.18 Cylindrical shell with rigid reinforcing rings subjected to an internal pressure p.

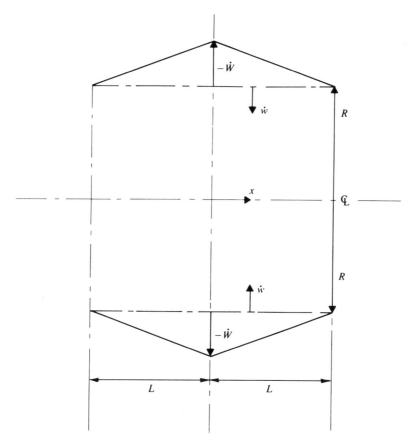

Figure 2.19 Axisymmetric transverse velocity field associated with the static plastic collapse of an internally pressurised cylindrical shell which is fully clamped across a span $2L$.

The portion $-L \leqslant x \leqslant 0$ is not considered henceforth since the problem posed is symmetrical about the vertical plane at $x = 0$ in Figure 2.18.

Equations (2.56) and (2.57) may be combined to give

$$\mathrm{d}^2 M_x / \mathrm{d}x^2 + N_\theta / R = p, \tag{2.63a}$$

which becomes

$$\mathrm{d}^2 M_x / \mathrm{d}x^2 = p - N_0 / R \tag{2.63b}$$

when using equation (2.62a). Thus

$$M_x = (p - N_0/R)x^2/2 + C_1 x - M_0 \tag{2.64}$$

because $M_x = -M_0$ at $x = 0$. Moreover, it is evident from equation (2.57) that $\mathrm{d}M_x/\mathrm{d}x = 0$ at $x = 0$, since $Q_x = 0$ on a plane of symmetry. Therefore,

$$M_x = (p - N_0/R)x^2/2 - M_0. \tag{2.65}$$

Finally, the requirement that equation (2.65) predicts $M_x = M_0$ at the fully clamped support ($x = L$) gives the static plastic collapse pressure

$$p_c = N_0(1 + RH/L^2)/R, \qquad (2.66a)$$

or

$$p_c = N_0(1 + 2/\omega^2)/R, \qquad (2.66b)$$

where

$$\omega^2 = 2L^2/RH. \qquad (2.66c)$$

Equation (2.66b) with $\omega^2 \gg 1$ reduces to

$$p_c = N_0/R \qquad (2.67)$$

for a long cylindrical shell. This static collapse pressure could also be obtained directly from equation (2.56) for a cylindrical shell without any bending (i.e., equation (2.56) with $M_x = Q_x = 0$ gives $N_\theta - Rp = 0$, which yields equation (2.67)).

The bending moment distribution according to equation (2.65) may be recast in the form

$$M_x/M_0 = 2x^2/L^2 - 1 \qquad (2.68)$$

with the aid of equation (2.66a).

It is evident from equation (2.68) that M_x increases monotonically from $M_x = -M_0$ at $x = 0$ to $M_x = M_0$ at $x = L$, as indicated in Figure 2.20. Thus, the above solution is statically admissible and equation (2.66b) is, therefore, a lower bound to

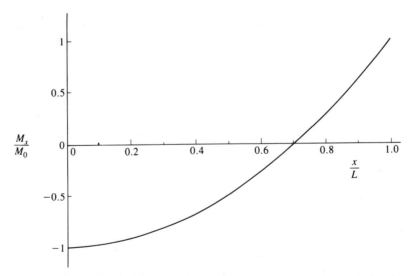

Figure 2.20 Axisymmetric longitudinal bending moment distribution during the static plastic collapse of a fully clamped cylindrical shell according to equation (2.68).

the exact collapse pressure of a cylindrical shell made from a material which obeys the rectangular yield condition in Figure 2.17.

The collapse profile, illustrated in Figure 2.19, may be written as

$$\dot{w} = -\dot{W}(1 - x/L) \tag{2.69}$$

in the region $0 \leqslant x \leqslant L$, when satisfying the boundary condition $\dot{w} = 0$ at $x = L$ and putting $\dot{w} = -\dot{W}$ at $x = 0$. It is evident from equations (2.59) and (2.60) that equation (2.69) gives $\dot{\kappa}_x = 0$ and $\dot{\epsilon}_\theta \geqslant 0$. These generalised strain rates agree with the normality requirement which is associated with the portion AB of the rectangular yield condition in Figure 2.17. A kinematically admissible transverse velocity field (equation (2.69)), therefore, can be associated with the generalised stress profile which was used to construct the above statically admissible solution. Thus, equation (2.66b) is the exact collapse pressure for the cylindrical shell problem which is illustrated in Figure 2.18 when the material flows plastically according to a rectangular yield condition. Finally,

$$p_c = 0.75N_0(1 + 2/\omega^2)/R \tag{2.70}$$

is the exact collapse pressure associated with the inscribing rectangular yield condition drawn in Figure 2.17. Clearly, the exact collapse pressure for a cylindrical shell which is made from a material with plastic yielding controlled by the Tresca yield criterion must lie between the predictions of equations (2.66b) and (2.70) according to corollary 3 of the limit theorems in § 2.4.4.

2.11 Static plastic collapse of a ring-loaded cylindrical shell

Consider a long cylindrical shell which is subjected to a radial ring-load P (per unit circumferential length), as indicated in Figure 2.21(a). In order to simplify the following presentation, it is assumed that the shell material obeys the rectangular

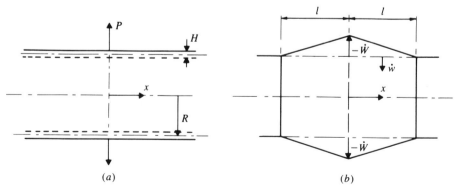

(a)   (b)

Figure 2.21 (a) Long cylindrical shell subjected to an axisymmetric radial ring load P (per unit length). (b) Axisymmetric plastic collapse profile with three circumferential plastic hinges.

yield criterion drawn in Figure 2.17. It appears reasonable to postulate that the kinematically admissible radial velocity profile shown in Figure 2.21(b) develops at incipient collapse of this shell. This profile consists of three circumferential plastic hinges separated by two conical regions of plastic flow. The total external work rate is

$$\dot{E} = 2\pi R P^u \dot{W}, \tag{2.71}$$

while the internal energy dissipated in the three circumferential plastic hinges is

$$\dot{D}_1 = 2\pi R M_0 (\dot{W}/l + 2\dot{W}/l + \dot{W}/l) \tag{2.72}$$

and

$$\dot{D}_2 = 2 \int_0^l N_0 \dot{\epsilon}_\theta 2\pi R \, dx,$$

or

$$\dot{D}_2 = 2 \int_0^l N_0 (\dot{W}/R)(1 - x/l) 2\pi R \, dx \tag{2.73}$$

is the energy dissipated in the conical portions owing to the circumferential membrane force ($N_\theta = N_0$).

An upper bound to the static collapse load is given by $\dot{E} = \dot{D}_1 + \dot{D}_2$ or

$$P^u = 4M_0/l + N_0 l/R, \tag{2.74}$$

the minimum value of which occurs when $l = \sqrt{RH}$.† Thus,

$$P^u = 2N_0(H/R)^{1/2} \tag{2.75}$$

is the minimum upper bound to the exact limit load for the assumed collapse profile and a rectangular yield condition.

Now, it is evident from equations (2.59) and (2.60) that $\dot{\epsilon}_\theta \geqslant 0$ and $\dot{\kappa}_x = 0$ in the conical region $0 \leqslant x \leqslant l$ of the shell illustrated in Figure 2.21(b). The normality requirement of plasticity, therefore, suggests the portion

$$N_\theta = N_1 \qquad \text{and} \qquad -M_1 \leqslant M_x \leqslant M_1 \tag{2.76a, b}$$

of the inscribing, or lower bound, rectangular yield condition shown in Figure 2.22. If equation (2.76a) is substituted into the equilibrium equation (2.63a) with $p = 0$, then

$$d^2 M_x/dx^2 = -N_1/R$$

or

$$M_x = -N_1 x^2/2R + Px/2 - M_1 \tag{2.77}$$

when putting $M_x = -M_1$ and $dM_x/dx = Q_x = P/2$ at $x = 0$.

It can be shown that the bending moment distribution according to equation (2.77) attains a maximum value at $x_1 = PR/2N_1$ so that

$$P^l = (16M_1 N_1/R)^{1/2} \tag{2.78}$$

---

† This may be compared with $2\pi\sqrt{RH}/\{3(1 - v^2)\}^{1/4}$, which is the distance that the disturbance at the end of an axisymmetrically loaded linear elastic cylindrical shell decays to negligible values.

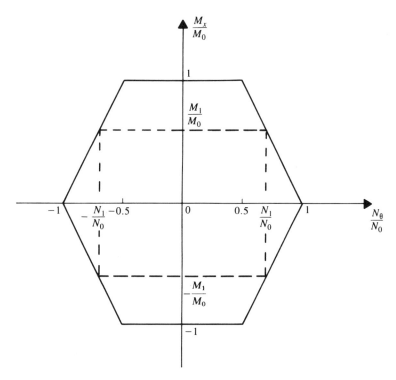

Figure 2.22 Hexagonal and inscribing rectangular yield curves for a ring-loaded cylindrical shell.

if the largest value equals M_1. Equation (2.78), with $N_1 = N_0/2$ and $M_1 = M_0$, predicts the largest possible collapse load associated with generalised stresses which remain on or within the hexagonal yield curve. Thus,

$$P^l = \sqrt{2}N_0(H/R)^{1/2}. \qquad (2.79)$$

Prager[2.1] has shown how the above generalised stress fields may be continued into the region $x \geqslant x_1$ without violating the yield conditions. Equation (2.79) is, therefore, a true lower bound to the exact collapse load. Finally, the exact collapse load P_c of the cylindrical shell shown in Figure 2.21(a) is bounded, viz.,

$$\sqrt{2}N_0(H/R)^{1/2} \leqslant P_c \leqslant 2N_0(H/R)^{1/2} \qquad (2.80)$$

according to equations (2.75) and (2.79).

2.12 Experiments on plates and shells

2.12.1 Static plastic behaviour of a circular plate

Onat and Haythornthwaite[2.17] have subjected simply supported circular plates to concentrated transverse loads and obtained the experimental results which are presented in Figure 2.23. Equation (2.20a) predicts the static plastic collapse load

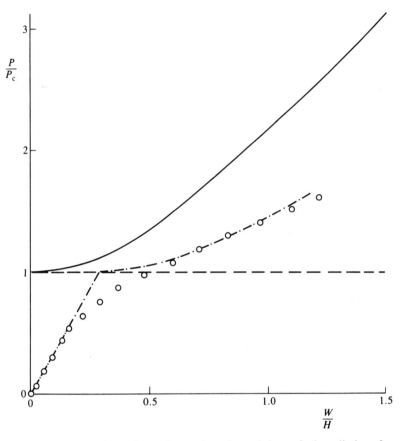

Figure 2.23 Comparison of experimental results and theoretical predictions for a simply supported circular plate which is subjected to a concentrated load at the centre. $\bigcirc$: experimental results on mild steel plates.[2.17] $---$: equation (2.20a) with $a/R = 0.078$; plastic limit load. ———: theoretical predictions[2.17] which include the influence of finite transverse displacements, or geometry changes (see also equations (7.40)). $-\cdot-\cdot-$: approximate theoretical predictions[2.17] which include the influence of finite transverse displacements and elastic effects.

for this particular case. However, it is evident from Figure 2.23 that equation (2.20a) for $P = P_c$ provides reasonable agreement with the experimental results only for dimensionless transverse displacements $W/H \cong 0.5$, i.e., for maximum transverse displacements equal to about one-half the associated plate thickness, approximately. This phenomenon is due primarily to the favourable influence of finite-deflections, or geometry changes, which introduce in-plane, or membrane, forces into a plate during deformation.[2.18] In order to cater for this effect, the equilibrium equations (2.4) and (2.5) must be replaced by others which would be derived in the deformed configuration of a plate. Moreover, membrane strains (ϵ_r

and ϵ_θ) must be considered in addition to the changes in curvature (2.6) and (2.7). However, the limit theorems of plasticity cannot then be used since they were developed for infinitesimal displacements.†

The experimental results and corresponding approximate theoretical predictions‡ of Onat and Haythornthwaite[2.17] for a simply supported circular plate, which is subjected to a centrally located concentrated load, are shown in Figure 2.23. Despite the obvious importance of geometry changes, it is observed that the static collapse load is a useful concept for a designer because it marks an important inflection in the load-displacement characteristics.

2.12.2 Static plastic behaviour of rectangular plates

The static plastic collapse pressures of uniformly loaded rectangular plates are presented in § 2.8. The theoretical predictions for fully clamped rectangular plates according to equation (2.51) in § 2.8.5 are compared with the corresponding experimental results of Hooke and Rawlings[2.19] in Figure 2.24.

It is evident that the theoretical predictions in Figure 2.24 are valid only for maximum transverse displacements with $W/H \cong 0.5$. However, an approximate theoretical procedure,§ which retains the influence of finite transverse displacements, or geometry changes, was developed in reference[2.20] and predicts reasonable estimates of the corresponding experimental results, as shown in Figure 2.24.

2.12.3 Static plastic collapse of a reinforced cylindrical shell

It is shown in § 2.10 that equations (2.66b) and (2.70) bound the exact static collapse pressure of a long cylindrical shell, which is reinforced by equally spaced rigid rings, or

$$0.75N_0(1 + 2/\omega^2)/R \leqslant p_c \leqslant N_0(1 + 2/\omega^2)/R \qquad (2.81)$$

when the material flows plastically according to the limited-interaction or rectangular yield condition in Figure 2.17.

The inequalities (2.81) are compared in Figure 2.25 with the corresponding experimental results* of Augusti and d'Agostino[2.21] and Perrone.[2.22] It is

---

† Simple proofs of the limit theorems for perfectly plastic beams are presented in § 1.3. The statically admissible bending moment distribution $M^s(x)$ in the proof of the lower bound theorem in § 1.3.2(b) satisfies the equilibrium equations (1.1) and (1.2), which do not retain the influence of finite deflections, or geometry changes. Equation (1.13) in § 1.3.3(b) for the upper bound theorem does not cater for the energy dissipated owing to membrane forces which develop because of finite transverse displacements in an axially restrained beam.

‡ See equations (7.40).

§ This theoretical procedure is discussed in § 7.4.5.

* Augusti and d'Agostino[2.21] used two different definitions for the limit pressure due to the difficulty in obtaining its precise value in the experimental tests. Thus, two points are associated with each specimen in Figure 2.25.

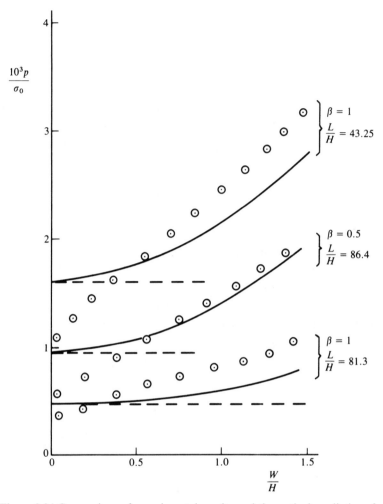

Figure 2.24 Comparison of experimental results and theoretical predictions for rectangular steel plates which are fully clamped around the boundaries and subjected to a uniform transverse pressure. ○: experimental results on mild steel plates.[2.19] ———: equation (2.51); plastic limit load. ——: theoretical predictions[2.20] which include the influence of finite transverse displacements, or geometry changes (see equations (7.57) and (7.59)).

evident that the simple theoretical predictions of equations (2.66b) and (2.70) for a rectangular yield condition provide useful estimates of the experimental results.

Hodge[2.2, 2.15] has examined the problem shown in Figure 2.18 when the shell is made from a material which flows plastically according to the hexagonal yield curve illustrated in Figure 2.16.

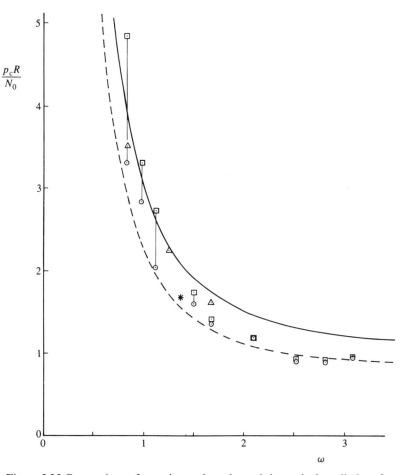

Figure 2.25 Comparison of experimental results and theoretical predictions for cylindrical shells which are fully clamped at the supports and subjected to an internal pressure. $\square$, $\bigcirc$, $\triangle$: experimental results on mild steel shells,[2.21] where $\bigcirc$ and $\square$ are associated with the same specimen.
*: experimental result on a metal shell.[2.22]
    Theoretical predictions for a rectangular yield curve; equations (2.66b) and (2.70) (also inequalities (2.81)):
——: circumscribing yield curve; - - -: inscribing yield curve.

2.12.4 Static plastic collapse of a ring-loaded cylindrical shell

It is shown in § 2.11 that inequality (2.80) bounds the exact collapse load of the cylindrical shell in Figure 2.21(a) when it flows plastically according to the rectangular yield curve in Figure 2.17. In addition, Drucker[2.23] has derived static plastic collapse loads

$$P_c = 1.73 N_0 (H/R)^{1/2} \tag{2.82}$$

and

$$P_c = 1.82N_0(H/R)^{1/2}, \tag{2.83}$$

corresponding, respectively, to the hexagonal and exact (Tresca) yield curves shown in Figure 2.16. Demir and Drucker[2.24] reported that equation (2.83) is valid for shells with $L \geqslant 2.34\sqrt{RH}$, or $\omega \geqslant 3.31$, when ω^2 is defined by equation (2.66c) and $2L$ is the shell length. Eason and Shield[2.25] and Eason[2.26] have examined the theoretical behaviour of rigid-plastic cylindrical shells of any length which are subjected to an axisymmetrical ring loading.

The theoretical predictions of equations (2.75) and (2.83) are plotted in Figure 2.26, together with the theoretical predictions of Eason and Shield[2.25] and Eason[2.26] for shorter shells made from materials which flow plastically according to the circumscribing rectangular and Tresca yield criteria, respectively. It is evident that the experimental results of Demir and Drucker,[2.24] which were obtained on steel and aluminium shells, are close to the corresponding theoretical predictions over the entire range of the dimensionless parameter ω examined in the tests.

2.13 Final remarks

It is the aim of this chapter to provide a brief introduction to the static collapse behaviour of plates and shells which are made from a rigid, perfectly plastic material. The various concepts and methods which are used for the static behaviour are further developed in the subsequent chapters of this book in order to examine the dynamic plastic response.

Many theoretical solutions and experimental test results have been published for the static plastic collapse behaviour of plates and shells with various boundary conditions and external loadings. For example, Hu[2.27] has examined various axisymmetric plate problems, while Mansfield[2.28] has presented some analytical upper bound solutions for the collapse of plates which flow plastically according to the Johansen yield criterion. However, it was noted in § 2.12.1 and § 2.12.2 and illustrated in Figures 2.23 and 2.24 that the the phenomenon of geometry changes must be catered for in the basic equations when the maximum transverse displacement exceeds approximately one-half of the corresponding plate thickness.[2.17,2.20,2.29]

Hodge[2.2] and Olszak and Sawczuk[2.30] have reported a number of theoretical solutions, which have been published, on the behaviour of axisymmetrically loaded, perfectly plastic cylindrical shells. It should be remarked that it is necessary to use a three-dimensional yield surface when a shell is loaded axially (i.e., $N_x \neq 0$).

It is evident from the experimental results of Demir and Drucker,[2.24] which are reported in Figure 2.26, that geometry changes, or finite deflections, do not exercise

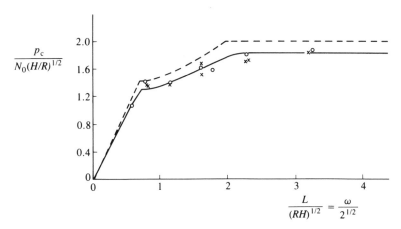

$$\frac{L}{(RH)^{1/2}} = \frac{\omega}{2^{1/2}}$$

Figure 2.26 Comparison of experimental results and theoretical predictions for cylindrical shells subjected to axisymmetric ring loads. $\times$, $\bigcirc$: experimental results on steel and aluminium shells, respectively.[2.24] ———: theoretical predictions for the exact (Tresca) yield curve in Figure 2.16.[2.26] – – – : theoretical predictions for the circumscribing rectangular yield curve in Figure 2.17.[2.25]

an important influence on the behaviour of these shells. However, the theoretical predictions of Duszek[2.31] and Duszek and Sawczuk[2.32] demonstrate convincingly that geometry changes do play an important role in the response of internally pressurised cylindrical shells restrained axially.†

It appears that most of the theoretical work which has been published on rigid, perfectly plastic shells has focused on the behaviour of cylindrical shells. The theoretical analyses of conical, shallow and spherical shells, even when loaded axisymmetrically, are more complicated because the associated yield criteria must involve the presence of at least four generalised stresses, N_θ, N_ϕ, M_θ and M_ϕ.[2.33] Onat and Prager[2.34] derived a four-dimensional yield condition in generalised stress space for axisymmetrically loaded shells of revolution which flow plastically according to the Tresca yield criterion. A number of solutions using either this yield surface, or various simplified yield criteria, are presented in the books by Hodge[2.2, 2.15] and Olszak and Sawczuk.[2.30]

Drucker and Shield[2.35, 2.36] were the first authors to demonstrate how the limit theorems of plasticity might be used to examine the behaviour of pressure vessels. They employed an approximate yield surface in order to simplify the calculations for the static collapse pressure of a cylindrical shell with either torispherical or toriconical ends. Save and Janas[2.37] reported the results of an experimental

† The collapse pressure, which is predicted by the limit theorems of plasticity, still marks an important change in the characteristics of the structural behaviour. Beyond this load, the displacements of the shells increase rapidly compared with those in the elastic range.

investigation which was conducted on mild steel pressure vessels with the same geometry. It was observed that the collapse pressures of the annealed vessels never differ more than 10 per cent from the theoretical predictions of Drucker and Shield,[2.36] while the assumed collapse profiles resemble the actual deformed shapes. However, the carrying capacities of these vessels at large deflections were much higher than the corresponding theoretical static collapse pressures. Nevertheless, the theoretical limit pressure of such vessels is a meaningful concept for design use because it marks the onset of large displacement increments per unit increase of load.

The behaviour of a flush cylindrical nozzle which intersects radially internally pressurised spherical, cylindrical or conical pressure vessels has been examined by Gill,[2.38] Cloud and Rodabaugh[2.39] and in reference 2.40, respectively. The static plastic collapse behaviour of many other pressure vessel and pipeline problems have been examined over the years and an interested reader is referred to Gill,[2.41] Hodge[2.42] and other more recent articles published in the *Transactions of the American Society of Mechanical Engineers*, conference proceedings and various journals on pressure vessels.

PROBLEMS

2.1 Derive the moment and lateral equilibrium equations for the circular plate in Figure 2.5 (i.e., equations (2.4) and (2.5)) and the radial and circumferential curvature changes (equations (2.6) and (2.7)).

Show that equations (2.4) to (2.7) are consistent according to the principle of virtual velocities in Appendix 3. (See Appendix 4 for the corresponding dynamic case.)

2.2 Obtain the static collapse pressure (equation (2.19)) for a simply supported circular plate subjected to a pressure distributed uniformly within a central circular zone of radius a. Assume that the plate is made from a rigid, perfectly plastic material which obeys the Tresca yield condition in Figure 2.6(b).

2.3 The loads acting on a circular plate are idealised as an axisymmetric transverse, or lateral, pressure which varies linearly from a peak value p_0 at the centre to zero at the supports. The circular plate is simply supported around the outer edge of radius R, has a uniform thickness H and is made from a ductile material which may be characterised as perfectly plastic with a uniaxial flow stress σ_0.

Determine the exact magnitude of p_0 which leads to incipient collapse. Show clearly that all the requirements for an exact solution have been satisfied.

2.4 Obtain the exact collapse pressure of a rigid, perfectly plastic annular plate which is simply supported around the outer boundary of radius R and free on the inner boundary with a radius a (see Figure 4.4(a)). The axisymmetric lateral pressure varies linearly from a maximum value at $r = a$ to zero at $r = R$. Assume that the material flows plastically according to the Tresca yield condition in Figure 2.6(b). (The exact static collapse pressure is given by equation (4.7) and the corresponding dynamic case is examined in § 4.3.)

2.5 Assume that the generalised stresses are $0 \leqslant M_\theta \leqslant M_0$ and $M_\theta - M_r = M_0$ in an annular plastic zone of a rigid, perfectly plastic circular plate. If plastic flow of the material is controlled by the Tresca yield condition in Figure 2.6(b), then what is the corresponding velocity profile in the annular zone? Obtain an expression for the rate of internal energy dissipation.

(The behaviour described above occurs in the outer zone of the circular plate in Figure 2.7 with fully clamped supports.[2.6] Lance and Onat[2.43] have etched the surfaces of fully clamped circular plates after loading and found radial lines in a central zone surrounded by an annular region of logarithmic spirals. Sometimes this pattern is found on the bottom of aerosol cans.)

2.6 Obtain the magnitude and distribution of the transverse shear force associated with the collapse load of the uniformly loaded, simply supported circular plate which was examined in §§ 2.6.2 to 2.6.4. Assess the potential importance of the transverse shear forces on the plastic collapse behaviour of a circular plate.

2.7 Derive the equilibrium equations (2.22)–(2.24) for a rectangular plate.

2.8 An upper bound to the static plastic collapse pressure of a simply supported rectangular plate is given by equation (2.40). Obtain equation (2.42) and show that the minimum upper bound is given by equation (2.43).

2.9 Modify the theoretical solution of § 2.8.3 to show that equation (2.51) gives an upper bound to the exact static plastic collapse pressure of a fully clamped rectangular plate.

2.10 Show that the lower bound solution in § 2.8.2 for a simply supported rectangular plate satisfies overall vertical equilibrium.

2.11 Obtain equations (2.56) and (2.57) when using equations (2.58) to (2.60) and the principle of virtual velocities in Appendix 3 for the case with $N_x = 0$.

2.12 Draw the exact yield curve in Figure 2.16 for a cylindrical shell. Use the method which is outlined in § 2.9.

3

Dynamic plastic behaviour of beams

3.1 Introduction

This chapter examines the behaviour of beams when they are subjected to dynamic loads which produce an inelastic response. The external energy is sufficiently large to cause a permanent deformed shape, or visible damage, as shown in Figure 3.1.

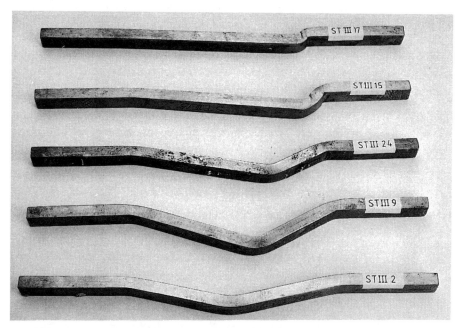

Figure 3.1 Permanent deformed profiles of initially straight fully clamped beams struck by a falling mass at various locations.[3.1]

    The steel beams have $2L = 101.6$ mm, $H = 6.35$ mm, $B = 10.16$ mm, $\sigma_0 = 302$ MN/m^2 and are struck by a 5 kg mass with initial kinetic energies of 99, 214, 137, 240 and 67 N m for specimens STIII 17, 15, 24, 9 and 2, respectively.

Standard static methods of analysis with dynamic magnification factors, for example, are not adequate in many dynamic plastic structural problems, as demonstrated by the static and dynamic tests on idealised buses conducted by Lowe, Al-Hassani and Johnson.[3.2] The roof of the bus shown in Figure 3.2 essentially collapsed as an Euler column when subjected to a static longitudinal load, while the front end was crushed severely under a dynamic longitudinal load, with little damage to the roof elsewhere. An analysis of the bus subjected to static longitudinal loads, therefore, is of little relevance to the design of a crashworthy vehicle.

Studies in this general area are relevant to the design of energy-absorbing systems (e.g., car bumpers and motorway guard rails), earthquake damage of civil engineering structures and many other practical engineering problems.

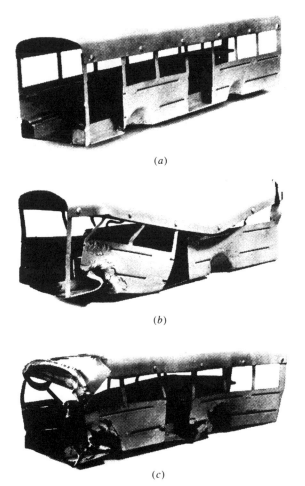

Figure 3.2 Model motor coach.[3.2] (a) Original specimen. (b) Static crumpling. (c) Dynamic crumpling.

The theoretical analysis in Chapter 1 shows that a simply supported beam collapses when subjected to a uniformly distributed static pressure having the limit value $p_c = 2M_0/L^2$, which is given by equation (1.27). A beam, which is made from a rigid, perfectly plastic material, remains rigid for external pressures smaller than p_c, while static equilibrium is not possible for pressures larger than p_c when the influences of material strain hardening and geometry changes, or finite-deflection effects, are ignored. Thus, a beam deforms plastically and inertia forces are generated when an external pressure larger than p_c is applied suddenly. The transverse, or lateral, deflections of a beam would become excessive if this external pressure pulse were to be maintained for a sufficiently long duration. However, if after a finite time the pressure pulse had been removed, or had decayed to small values, then a finite external energy would have been imparted to a beam. In this circumstance, a beam would eventually reach a final or permanent deformed shape when all the external energy was absorbed through plastic deformation.

The same assumptions and simplifications which were found helpful in the previous two chapters for studying the static plastic behaviour of structures are also employed to examine the dynamic plastic response. The influence of material elasticity usually plays a minor role in those problems which are dominated by large plastic strains. Thus, material elasticity is neglected in the dynamic case, which appears to be a reasonable assumption provided the total external dynamic energy is much larger than the total amount of energy that can be absorbed in a wholly elastic fashion. The accuracy of the rigid-plastic method for predicting the response of beams subjected to large dynamic loads is discussed further in § 3.10.

The general procedure for generating theoretical solutions on the dynamic plastic behaviour of beams is similar to that used for the static behaviour in Chapter 1. A kinematically admissible velocity field which describes the motion of a beam is first postulated. Some guidance on the shape of this field is often suggested by the characteristics of the corresponding static collapse profile. The normality requirements of plasticity are then invoked in order to seek the appropriate portion of the yield surface which is associated with the assumed velocity field. It is now usually possible to complete the solution by integrating the governing differential equations and satisfying the initial and boundary conditions. However, as a final step, it is necessary to examine whether or not any yield violations have occurred. If the yield condition is not violated then the assumed velocity profile is correct and the theoretical solution is exact. On the other hand, if yield violations do occur at any time during the entire response, or for certain values of the parameters, then it is necessary to attempt another solution for these cases with an alternate velocity field, the shape of which is usually suggested by the nature of the yield violations. This procedure may be repeated until an exact theoretical solution has been found.

The governing equations for beams are developed in the next section, while the dynamic plastic response of a simply supported beam subjected to a uniformly distributed pressure pulse with a reactangular-shaped pressure–time history is examined in §§ 3.3 and 3.4. The influences of an impulsive velocity loading and fully clamped supports are explored in subsequent sections. A theoretical analysis is presented in § 3.8 for the response of a fully clamped beam which is struck at the mid-span by a mass travelling with an initial velocity. This analysis is modified in § 3.9 for a cantilever beam which is impacted at the tip. Comparisons are also made with experimental results, and § 3.10 contains a discussion on the accuracy of the simplifications and approximations which are introduced in this chapter.

3.2 Governing equations for beams

The dynamic behaviour of the beam in Figure 3.3, for infinitesimal displacements, is governed by the equations

$$Q = \partial M / \partial x, \tag{3.1}$$

$$\partial Q / \partial x = -p + m\, \partial^2 w / \partial t^2 \tag{3.2}$$

and

$$\kappa = -\partial^2 w / \partial x^2, \tag{3.3}$$

which are identical to equations (1.1) to (1.3), except for the inclusion of an inertia term in the lateral equilibrium equation (3.2). (m is mass per unit length of beam, t is time.) In order to simplify the following presentation, the influence of rotatory inertia is not retained in the moment equilibrium equation (3.1), but its influence is discussed in Chapter 6. Moreover, it is assumed that the beam is made from a rigid, perfectly plastic material with a uniaxial yield stress σ_0, as shown in Figure 1.3, and a fully plastic bending moment M_0. (See Figure 1.4.)

Thus, a theoretical solution for a beam must satisfy equations (3.1) and (3.2) in addition to the boundary conditions and initial conditions. The generalised stress, or bending moment M, must remain statically admissible with $-M_0 \leqslant M \leqslant M_0$ and not violate the yield condition. Furthermore, the transverse velocity field must give rise to a generalised strain rate, or curvature rate $\dot{\kappa}$, which satisfies the

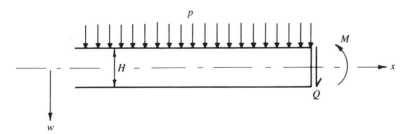

Figure 3.3 Notation for a beam.

normality requirement of plasticity. In other words, the curvature rate vector associated with an active plastic region in a beam must be normal to the yield curve at the corresponding point (i.e., $\dot{\kappa} \geqslant 0$ when $M = M_0$ and $\dot{\kappa} \leqslant 0$ when $M = -M_0$). The theoretical solution is said to be exact if the generalised stress field is statically admissible and the associated transverse velocity field is kinematically admissible.

3.3 Simply supported beam, $p_c \leqslant p_o \leqslant 3p_c$

3.3.1 Introduction

Consider the dynamic response of the simply supported rigid, perfectly plastic beam shown in Figure 3.4(a). The entire span of this particular beam is subjected to the rectangular pressure pulse which is sketched in Figure 3.5 and which may be expressed in the form

$$p = p_0, \qquad 0 \leqslant t \leqslant \tau \tag{3.4}$$

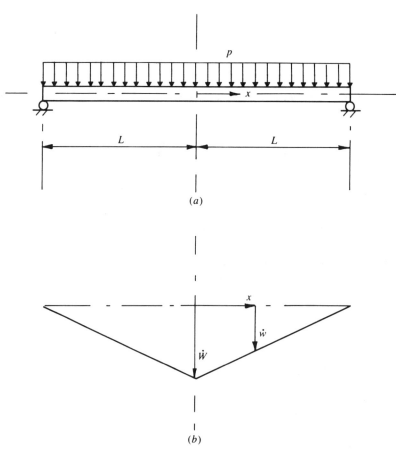

Figure 3.4 (a) Uniformly loaded beam with simple supports. (b) Transverse velocity field.

and

$$p = 0, \qquad t \geqslant \tau. \tag{3.5}$$

It was shown in Chapter 1 that the exact static collapse pressure of this beam is given by equation (1.27), or

$$p_c = 2M_0/L^2, \tag{3.6}$$

with the associated incipient transverse displacement field sketched in Figure 1.7(b). Clearly, a rigid, perfectly plastic beam remains rigid for pressure pulses which satisfy the inequality $0 \leqslant p_0 < p_c$, and accelerates when it is subjected to larger pressures. However, if the pressure is released after a short time, then a beam may reach an equilibrium position (with a deformed profile) when all the external dynamic energy has been expended as plastic work. Equations (3.4) and (3.5) suggest that it is convenient to divide the subsequent analysis into the two parts $0 \leqslant t \leqslant \tau$ and $\tau \leqslant t \leqslant T$, where T is the duration of motion. Moreover, it is necessary to consider only one half of the beam $0 \leqslant x \leqslant L$ owing to its symmetry about the mid-span position $x = 0$.

3.3.2 First phase of motion, $0 \leqslant t \leqslant \tau$

A theoretical solution is sought when using the transverse or lateral velocity field which is sketched in Figure 3.4(b) and which has the same form as the displacement

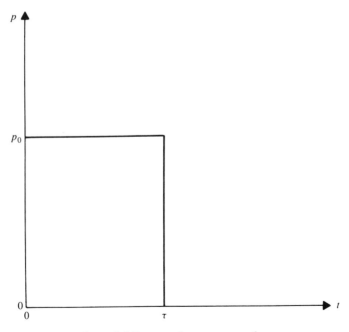

Figure 3.5 Rectangular pressure pulse.

profile associated with the static collapse pressure p_c. Thus,

$$\dot{w} = \dot{W}(1 - x/L), \qquad 0 \leqslant x \leqslant L, \tag{3.7}†$$

where $\dot{W}$ is the lateral velocity at the mid-span. Equations (3.3) and (3.7) give $\dot{\kappa} = 0$ throughout the beam except at $x = 0$, where $\dot{\kappa} \to \infty$. The beam is, therefore, idealised as two rigid arms (i.e., $-M_0 < M < M_0$) which are connected by a central plastic hinge ($M = M_0$).

Now, substituting equation (3.1) into equation (3.2) gives

$$\partial^2 M/\partial x^2 = -p + m\, \partial^2 w/\partial t^2, \tag{3.8}$$

which, when using equations (3.4) and (3.7), becomes

$$\partial^2 M/\partial x^2 = -p_0 + m(1 - x/L)\, \mathrm{d}^2 W/\mathrm{d}t^2. \tag{3.9}$$

Equation (3.9) may be integrated spatially,‡

$$M = -p_0 x^2/2 + m(x^2/2 - x^3/6L)\, \mathrm{d}^2 W/\mathrm{d}t^2 + M_0, \tag{3.10}$$

where the arbitrary constants of integration have been determined from the requirements that $M = M_0$ and $Q = \partial M/\partial x = 0$ at $x = 0$. However, it is necessary to have $M = 0$ at $x = L$ for a simply supported boundary. Thus,

$$\mathrm{d}^2 W/\mathrm{d}t^2 = 3(\eta - 1)M_0/mL^2, \tag{3.11}$$

where

$$\eta = p_0/p_c, \tag{3.12}$$

is the ratio of the magnitude of the dynamic pressure pulse to the corresponding static collapse pressure.§ If equation (3.11) is integrated with respect to time, then

$$W = 3(\eta - 1)M_0 t^2/2mL^2, \tag{3.13}$$

since $W = \dot{W} = 0$ at $t = 0$. The first stage of motion is completed at $t = \tau$,

$$W = 3(\eta - 1)M_0 \tau^2/2mL^2 \tag{3.14}$$

and

$$\dot{W} = 3(\eta - 1)M_0 \tau/mL^2. \tag{3.15}$$

3.3.3 Second phase of motion, $\tau \leqslant t \leqslant T$

The external pressure is removed suddenly at $t = \tau$, so that the beam is unloaded during this stage of motion. However, the beam has a transverse velocity at $t = \tau$ according to equations (3.7) and (3.15) and, therefore, has a finite kinetic energy. Thus, the beam continues to deform for $t \geqslant \tau$ until the remaining kinetic energy is absorbed through plastic energy dissipation at the plastic hinges.

If the velocity field which is indicated in Figure 3.4(b) and described by equation

---

† The notation $(\dot{\ }) = \partial(\ )/\partial t$ is used throughout this chapter.

‡ Equation (3.10) is valid only for beams with a uniform cross-section, i.e., m is assumed to be independent of x.

§ Equation (3.11) gives $\eta = 1$ (i.e., $p_0 = p_c$) when $\ddot{W} = 0$, as expected.

(3.7) is assumed to remain valid during this stage, then it is evident that equation (3.10) remains unchanged except $p_0 = 0$. Thus, equation (3.11) becomes

$$d^2 W/dt^2 = -3M_0/mL^2, \tag{3.16}$$

which may be integrated to give

$$\dot{W} = 3M_0(\eta\tau - t)/mL^2, \tag{3.17a}$$

$$W = 3M_0(2\eta\tau t - t^2 - \eta\tau^2)/2mL^2 \tag{3.17b}$$

when using equations (3.14) and (3.15) as initial conditions ($t = \tau$) for the displacement and velocity profiles during the second stage of motion. The beam reaches its permanent† position when $\dot{W} = 0$, which, according to equation (3.17a), occurs when

$$T = \eta\tau. \tag{3.18}$$

Equations (3.7), (3.17b) and (3.18) predict the final deformed profile

$$w = 3\eta(\eta - 1)M_0\tau^2(1 - x/L)/2mL^2. \tag{3.19}$$

3.3.4 Static admissibility

The theoretical analysis in §§ 3.3.2 and 3.3.3 satisfies the equilibrium equations (3.1) and (3.2), initial conditions and boundary conditions for a simply supported beam subjected to a pressure pulse with a rectangular pressure–time history. However, the bending moment M has only been specified at the supports and the mid-span. It is necessary, therefore, to ensure that the bending moment does not violate the yield condition anywhere in the beam for $0 \leqslant x \leqslant L$ and during both stages of motion $0 \leqslant t \leqslant \tau$ and $\tau \leqslant t \leqslant T$.

Now, equation (3.10) may be written

$$M/M_0 = 1 - \eta(x/L)^2 + (\eta - 1)(3 - x/L)(x/L)^2/2 \tag{3.20}$$

when using equation (3.11), and gives $dM/dx = 0$ at $x = 0$, as expected. Moreover,

$$(L^2/M_0)\,d^2M/dx^2 = \eta - 3 - 3(\eta - 1)x/L, \tag{3.21}$$

which predicts $d^2M/dx^2 \leqslant 0$ at $x = 0$ provided $\eta \leqslant 3$. Thus, no yield violations occur anywhere in the beam during the first stage of motion when $\eta \leqslant 3$ because $d^2M/dx^2 < 0$ over the entire span, as indicated in Figure 3.6(a). The theoretical solution which is presented for the first stage of motion is, therefore, correct provided $1 \leqslant \eta \leqslant 3$. However, a yield violation occurs when $\eta > 3$ since $d^2M/dx^2 > 0$ at $x = 0$, which implies that the bending moment near the mid-span exceeds the plastic limit moment M_0.

It is evident that equation (3.20) with $\eta = 0$ gives

$$M/M_0 = 1 - (3 - x/L)(x/L)^2/2 \tag{3.22}$$

during the second stage of motion $\tau \leqslant t \leqslant T$. Equation (3.22), or equation (3.21),

---

† No elastic unloading occurs for a rigid, perfectly plastic material.

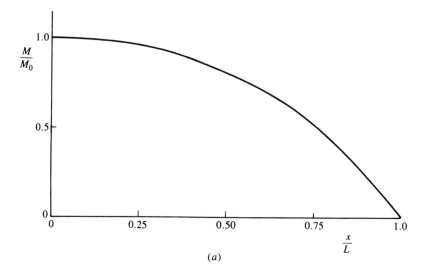

(a)

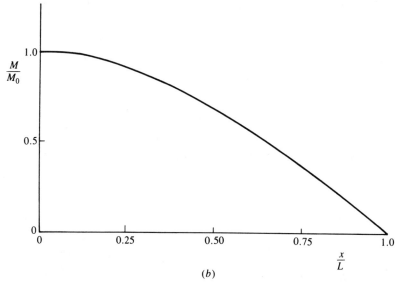

(b)

Figure 3.6 (a) Bending moment distribution ($0 \leqslant x/L \leqslant 1$) during the first phase of motion ($0 \leqslant t \leqslant \tau$) for a simply supported beam subjected to a uniformly distributed rectangular pressure pulse with $\eta = 2$. (b) Bending moment distribution ($0 \leqslant x/L \leqslant 1$) during the second phase of motion ($\tau \leqslant t \leqslant T$) for a simply supported beam subjected to a uniformly distributed rectangular pressure pulse with $1 \leqslant \eta \leqslant 3$.

with $\eta = 0$, gives $d^2M/dx^2 \leqslant 0$, so that no yield violations occur because $M = M_0$ and $dM/dx = 0$ at $x = 0$ and $M = 0$ at $x = L$, as shown in Figure 3.6(b). Thus, the above theoretical solution, which leads to the response time predicted by equation (3.18) with the associated permanent shape given by equation (3.19), is correct provided $1 \leqslant \eta \leqslant 3$, or $p_c \leqslant p_0 \leqslant 3p_c$.†

3.4 Simply supported beam, $p_0 \geqslant 3p_c$

3.4.1 Introduction

It is evident from § 3.3.4 that the theoretical analysis using the transverse velocity profile in Figure 3.4(b) is exact for pressure pulses having $p_c \leqslant p_0 \leqslant 3p_c$ (or $1 \leqslant \eta \leqslant 3$). In order to obtain an exact theoretical solution for pressure pulses with $p_0 \geqslant 3p_c$, it is necessary to seek an alternative transverse velocity field, as discussed in § 3.1.

Equation (3.21) with $x = 0$ is positive when $\eta > 3$ and a local minimum therefore develops in the bending moment (M) distribution at the mid-span, where $M = M_0$. This leads to a yield violation within a central zone and suggests the formation of plastic hinges at $x = \xi_0$ and $x = -\xi_0$ in Figure 3.7(a), where ξ_0 is a function of the load ratio η.‡ It transpires that the plastic hinges remain stationary during the first stage of motion $0 \leqslant t \leqslant \tau$, and propagate inwards towards the mid-span during a second stage of motion $\tau \leqslant t \leqslant T_1$, where T_1 is the time at which the two plastic hinges coalesce at $x = 0$. The final stage of motion is similar to the second phase of motion, which is outlined in § 3.3.3 for the lower pressure case with $1 \leqslant \eta \leqslant 3$.

3.4.2 First phase of motion, $0 \leqslant t \leqslant \tau$

As remarked in § 3.1, the selection of a transverse velocity field is the first step in an analysis for this type of problem. If it transpires that an exact solution is not achieved, then another velocity field must be chosen until the correct one is found.

Now, it is only necessary to consider the portion $0 \leqslant x \leqslant L$ of the beam in Figure 3.4(a) because the behaviour is symmetrical about the mid-span $(x = 0)$. Thus, the transverse velocity profile in Figure 3.7(a) may be written

$$\dot{w} = \dot{W}_1, \qquad 0 \leqslant x \leqslant \xi_0, \tag{3.23a}$$

† The continuity of displacement (w) and velocity $(\partial w/\partial t)$ between the two stages of motion $(t = \tau)$ was imposed on the solution. However, it is observed from equations (3.11) and (3.16) that the acceleration $\partial^2 w/\partial t^2$ is discontinuous at $t = \tau$, which leads to a time discontinuity in the bending moment field, as is evident from equations (3.20) and (3.22). This discontinuity occurs because the pressure is released suddenly at $t = \tau$ and is permissible provided the equilibrium equation (3.8) is satisfied. If $[X]_\tau$ is used to denote $X_{\tau+} - X_\tau$, i.e., the magnitude of the discontinuity in X which occurs at $t = \tau$, then equation (3.8) can be written in the form $[\partial^2 M/\partial x^2]_\tau = [-p]_\tau + m[\partial^2 w/\partial t^2]_\tau$. It can be shown that equations (3.4), (3.5), (3.7), (3.11), (3.16), (3.20) and (3.22) satisfy this requirement.

‡ It is anticipated that ξ_0 is a function of η because the yield violation is more extensive for larger values of η. (See equation (3.30).)

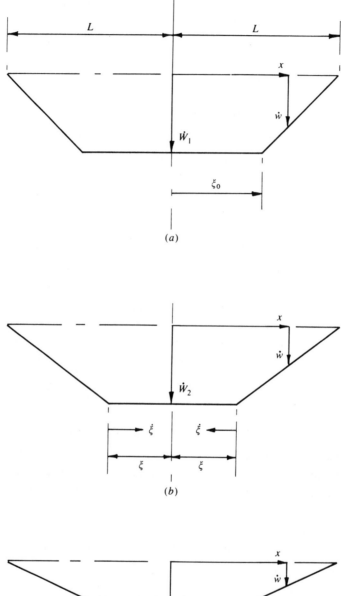

(a)

(b)

(c)

Figure 3.7 Transverse velocity profiles for a simply supported beam subjected to a rectangular pressure pulse with $\eta \geqslant 3$. (a) First phase of motion, $0 \leqslant t \leqslant \tau$. (b) Second phase of motion, $\tau \leqslant t \leqslant T_1$. (c) Third phase of motion, $T_1 \leqslant t \leqslant T$.

and

$$\dot{w} = \dot{W}_1(L - x)/(L - \xi_0), \qquad \xi_0 \leqslant x \leqslant L. \tag{3.23b}$$

Equations (3.3) and (3.23) give $\dot{\kappa} = 0$, except at the plastic hinges $(x = \pm \xi_0)$ where $\dot{\kappa} \to \infty$ and, therefore, $M = M_0$.

It is assumed that the plastic hinge at $x = \xi_0$ in Figure 3.7(a) remains stationary during the first phase of motion when the pressure pulse is constant according to equation (3.4). Equations (3.1), (3.2), (3.4) and (3.23a), therefore, predict

$$\partial^2 M/\partial x^2 = -p_0 + m \, \mathrm{d}^2 W_1/\mathrm{d}t^2, \qquad 0 \leqslant x \leqslant \xi_0,$$

which, when integrated with respect to x, becomes

$$\partial M/\partial x = (-p_0 + m \, \mathrm{d}^2 W_1/\mathrm{d}t^2)x, \tag{3.24}$$

where the constant of integration is zero because $Q = \partial M/\partial x = 0$ at $x = 0$ from considerations of symmetry. However, $\partial M/\partial x = Q = 0$ at the plastic hinge $(x = \xi_0)$ which is satisfied by equation (3.24) when

$$m \, \mathrm{d}^2 W_1/\mathrm{d}t^2 = p_0. \tag{3.25}†$$

Equation (3.24), therefore, reduces to

$$\partial M/\partial x = 0, \qquad 0 \leqslant x \leqslant \xi_0, \tag{3.26}$$

and another integration yields

$$M = M_0, \qquad 0 \leqslant x \leqslant \xi_0, \tag{3.27}$$

where the constant of integration is found from the requirement that $M = M_0$ at $x = \xi_0$.

Equations (3.1), (3.2), (3.4) and (3.23b) give

$$\partial^2 M/\partial x^2 = -p_0 + m\{(L - x)/(L - \xi_0)\} \, \mathrm{d}^2 W_1/\mathrm{d}t^2, \qquad \xi_0 \leqslant x \leqslant L, \tag{3.28}$$

which, integrating, becomes

$$\partial M/\partial x = -p_0 x + p_0(Lx - x^2/2)/(L - \xi_0) + A_1$$

when using equation (3.25) and where the constant of integration is found from the requirement that $Q = \partial M/\partial x = 0$ at $x = \xi_0$, or

$$A_1 = p_0\xi_0 - p_0\xi_0(L - \xi_0/2)/(L - \xi_0).$$

Thus, integrating once more,

$$M = -p_0 x^2/2 + p_0(Lx^2/2 - x^3/6)/(L - \xi_0) + A_1 x + B_1, \qquad \xi_0 \leqslant x \leqslant L, \tag{3.29}$$

with

$$B_1 = p_0 L^2/2 - p_0 L^3/3(L - \xi_0) - p_0\xi_0 L + p_0\xi_0 L(L - \xi_0/2)/(L - \xi_0)$$

in order to satisfy the simply supported boundary condition $M = 0$ at $x = L$.

The bending moment distribution (3.29) must also satisfy the yield condition $M = M_0$ at the plastic hinge $(x = \xi_0)$, or

$$-p_0\xi_0^2/2 + p_0(L\xi_0^2/2 - \xi_0^3/6)/(L - \xi_0) + A_1\xi_0 + B_1 = M_0,$$

† The beam has a constant acceleration throughout this phase.

which may be rearranged to give

$$(1 - \xi_0/L)^2 = 6M_0/p_0L^2,$$

or

$$(1 - \bar{\xi}_0)^2 = 3/\eta, \tag{3.30}$$

where

$$\bar{\xi}_0 = \xi_0/L \tag{3.31}$$

and η is defined by equation (3.12), with p_c given by equation (3.6). Equation (3.30) shows that the extent of the central zone $(0 \leqslant x \leqslant \xi_0)$ in Figure 3.7(a) depends on the magnitude p_0 (or η) of the dynamic pressure pulse. In particular, $\xi_0 = 0$ when $p_0 = 3p_c$, or $\eta = 3$. It is also evident that $\bar{\xi}_0 \to 1$ when $\eta \to \infty$.

The bending moment distribution according to equation (3.29) may now be written in the following form:

$$M/M_0 = \eta[3\bar{\xi}_0(x/L)^2 - (x/L)^3 - 3\bar{\xi}_0^2(x/L) + 1 - 3\bar{\xi}_0 + 3\bar{\xi}_0^2]/3(1 - \bar{\xi}_0),$$
$$\xi_0 \leqslant x \leqslant L. \quad (3.32)$$

Equation (3.25) may be integrated to give

$$dW_1/dt = p_0t/m \tag{3.33a}†$$

and

$$W_1 = p_0t^2/2m, \tag{3.33b}$$

since $w = \dot{w} = 0$ at $t = 0$.

3.4.3 Second phase of motion, $\tau \leqslant t \leqslant T_1$

The pressure pulse is removed at $t = \tau$, as indicated in Figure 3.5 and by equation (3.5), when the transverse velocity and, therefore, the kinetic energy have reached the largest values during the first phase of motion according to equation (3.33a). The transverse velocity profile is again assumed to be given by equations (3.23) except that $\dot{W}_1$ is replaced by $\dot{W}_2$ and the plastic hinge at $x = \xi_0$ in Figure 3.7(a) is allowed to move, as shown in Figure 3.7(b). Thus, it is evident by inspection that equations (3.1), (3.2), (3.5) and (3.23a) lead to equations (3.24) to (3.27) for $0 \leqslant x \leqslant \xi$, but with $p_0 = 0$ and W_1 replaced by W_2. In particular, equation (3.25), with the above modifications, predicts

$$d^2W_2/dt^2 = 0, \tag{3.34}$$

which, integrating with respect to time, yields the velocity

$$dW_2/dt = p_0\tau/m, \tag{3.35}$$

where the constant of integration is selected to ensure that equation (3.35) matches equation (3.33a) at $t = \tau$.

---

† The transverse velocity increases linearly with time and reaches the largest value at $t = \tau$.

The transverse displacement is obtained by a further integration of equation (3.35), or

$$W_2 = p_0 \tau t/m - p_0 \tau^2/2m, \tag{3.36}$$

where the constant of integration ensures that equation (3.33b) is recovered at $t = \tau$.

The governing equation for $\xi \leqslant x \leqslant L$ according to equations (3.1), (3.2), (3.5) and (3.23b), with $\dot{W}_1$ replaced by $\dot{W}_2$, is

$$\partial^2 M/\partial x^2 = m\{(L-x)/(L-\xi)\}\, d^2 W_2/dt^2 + m\{(L-x)/(L-\xi)^2\}\, (dW_2/dt)\, d\xi/dt,$$
$$\xi \leqslant x \leqslant L, \quad (3.37)$$

which, in view of equations (3.34) and (3.35), becomes

$$\partial^2 M/\partial x^2 = p_0 \tau\{(L-x)/(L-\xi)^2\}\, d\xi/dt. \tag{3.38}$$

Now, integrating equation (3.38) with respect to x yields

$$\partial M/\partial x = p_0 \tau (Lx - x^2/2)\dot{\xi}/(L-\xi)^2 + A_2, \tag{3.39}$$

where

$$A_2 = -p_0 \tau \xi(L - \xi/2)\dot{\xi}/(L-\xi)^2 \tag{3.40}$$

in order to satisfy the requirement that $Q = \partial M/\partial x = 0$ at the plastic hinge with $x = \xi$. Integrating equation (3.39) with respect to x gives

$$M = p_0 \tau (Lx^2/2 - x^3/6)\dot{\xi}/(L-\xi)^2 + A_2 x + B_2, \tag{3.41}$$

where

$$B_2 = -p_0 \tau L^3 \dot{\xi}/3(L-\xi)^2 - A_2 L, \tag{3.42}$$

since $M = 0$ at $x = L$ for simple supports. However, $M = M_0$ at the travelling plastic hinge† at $x = \xi$, so that equation (3.41) must satisfy

$$M_0 = p_0 \tau (L\xi^2/2 - \xi^3/6)\dot{\xi}/(L-\xi)^2 + A_2 \xi + B_2,$$

which, with the aid of equations (3.40) and (3.42), can be rearranged in the form

$$M_0 = -p_0 \tau(L - \xi)\dot{\xi}/3, \tag{3.43}$$

which gives the velocity $\dot{\xi}$ of the plastic hinge. This equation may be cast in the form

$$\int_{L\xi_0}^{\xi} (L - \xi)\, d\xi = -\int_{\tau}^{t} (3M_0/p_0 \tau)\, dt$$

to give the position ξ of the travelling plastic hinge for $t \geqslant \tau$,

$$L\xi - \xi^2/2 - L^2\xi_0 + L^2\xi_0^2/2 = 3M_0(\tau - t)/p_0 \tau. \tag{3.44}$$

The two travelling plastic hinges in Figure 3.7(b) coalesce at the mid-span ($x = 0$) when

$$T_1 = p_0 L^2 \tau/6M_0, \tag{3.45a}$$

according to equations (3.30) and (3.44) with $\xi = 0$, or

$$T_1 = \eta\tau/3 \tag{3.45b}$$

when using equations (3.6) and (3.12).

<hr>

† See footnote† on page 77.

It may be shown that equation (3.41) may be rewritten with the aid of equations (3.40), (3.42) and (3.43) in the form

$$M/M_0 = \{2 + 2x/L - x^2/L^2 - 6(\xi/L - \xi^2/2L^2)\}(1 - x/L)/\{2(1 - \xi/L)^3\},$$

$$\xi \leqslant x \leqslant L. \quad (3.46)$$

3.4.4 Third phase of motion, $T_1 \leqslant t \leqslant T$

The analysis for the final phase of motion is similar to that in § 3.3.3 since the beam is unloaded according to equation (3.5) and a plastic hinge remains stationary at the mid-span, as indicated in Figure 3.7(c). Thus, equations (3.1) (3.2), (3.5) and (3.23b) with $\dot{W}_1$ replaced by $\dot{W}_3$ and $\xi_0 = 0$ give

$$\partial^2 M/\partial x^2 = m(1 - x/L)\ddot{W}_3, \quad 0 \leqslant x \leqslant L, \quad (3.47)$$

which, integrating with respect to x, becomes

$$\partial M/\partial x = m(x - x^2/2L)\ddot{W}_3,$$

where the constant of integration is zero because $\partial M/\partial x = Q = 0$ at $x = 0$ from symmetry considerations. The bending moment distribution is obtained by another integration with respect to x, or

$$M = m(x^2/2 - x^3/6L)\ddot{W}_3 + M_0 \quad (3.48)$$

since $M = M_0$ at the plastic hinge which is located at the mid-span ($x = 0$).

Now, equation (3.48) must satisfy the simply supported boundary condition $M = 0$ at $x = L$, which requires

$$\ddot{W}_3 = -3M_0/mL^2. \quad (3.49)\dagger$$

This allows the bending moment distribution according to equation (3.48) to be written as

$$M/M_0 = 1 - (3x^2/L^2 - x^3/L^3)/2, \quad 0 \leqslant x \leqslant L. \quad (3.50)$$

The transverse velocity is found by integrating equation (3.49), or

$$\dot{W}_3 = -3M_0 t/mL^2 + 3p_0\tau/2m,$$

where the constant of integration has been selected to match equation (3.35) at $t = T_1$ when T_1 is defined by equations (3.45). A further integration yields the transverse displacement

$$W_3 = -3M_0 t^2/2mL^2 + 3p_0\tau t/2m - p_0\tau^2/2m - p_0^2 L^2\tau^2/24mM_0, \quad (3.51)$$

where the constant of integration ensures that the displacement is continuous with equation (3.36) when $t = T_1$.

Motion ceases when all the kinetic energy in the beam at the end of the second phase of motion ($t = T_1$) has been absorbed as plastic work in the stationary central plastic hinge during the final phase. This occurs when $\dot{W}_3 = 0$, or

$$T = \eta\tau, \quad (3.52)$$

† The beam has a constant deceleration throughout the final phase of motion.

where η is defined by equation (3.12). Finally, the maximum permanent transverse displacement at the beam centre is given by equation (3.51), with $t = \eta\tau$ according to equation (3.52), or

$$W_f = p_0\tau^2(4\eta - 3)/6m. \tag{3.53}$$

3.4.5 Static admissibility

The bending moment distribution according to equation (3.32) for $\xi_0 \leqslant x \leqslant L$ during the first phase of motion, with $0 \leqslant t \leqslant \tau$, gives $M = M_0$ and $\partial M/\partial x = 0$ at $x = \xi_0$ and $M = 0$ at $x \doteq L$. Moreover, it may be shown that $\partial^2 M/\partial x^2 < 0$ for $\xi_0 < x \leqslant L$ ($\partial^2 M/\partial x^2 = 0$ at $x = \xi_0$ according to equations (3.25) and (3.28)). Thus, the bending moment distribution for $0 \leqslant x \leqslant \xi_0$ and $\xi_0 \leqslant x \leqslant L$ throughout the first phase of motion $0 \leqslant t \leqslant \tau$ satisfies $0 \leqslant M \leqslant M_0$ according to equations (3.27) and (3.32), as indicated in Figure 3.8(a).

Equation (3.46) for the bending moment during the second stage of motion $\tau \leqslant t \leqslant T_1$ gives $M = M_0$ and $\partial M/\partial x = 0$ at $x = \xi$, $M = 0$ at $x = L$ and $\partial^2 M/\partial x^2 < 0$ for $\xi \leqslant x < L$, except that $\partial^2 M/\partial x^2 = 0$ at $x = L$. (See also equation (3.38) with $d\xi/dt < 0$ from equation (3.43).) The bending moment distribution during the second phase of motion is drawn in Figure 3.8(b).†

The bending moment distribution during the third phase of motion $T_1 \leqslant t \leqslant T$ is governed by equation (3.50) which yields $M = M_0$ and $\partial M/\partial x = 0$ at $x = 0$, $M = 0$ at $x = L$ and $\partial^2 M/\partial x^2 < 0$ for $0 \leqslant x < L$, except that $\partial^2 M/\partial x^2 = 0$ at $x = L$. (See also equations (3.47) and (3.49).) This leads to the bending moment distribution shown in Figure 3.8(c).

Thus, the bending moment distribution throughout the beam $0 \leqslant x \leqslant L$ at all times during the response $0 \leqslant t \leqslant T$ satisfies $0 \leqslant M \leqslant M_0$ and does not violate the yield condition. The theoretical solution is, therefore, statically admissible without any limitations on the parameters except $p_0 \geqslant 3p_c$ or $\eta \geqslant 3$.

3.4.6 Kinematic admissibility

Some general expressions are developed initially then specialised later in order to examine the kinematic admissibility of the foregoing theoretical solution.

It is reasonable to postulate, in the absence of transverse shear deformations, that the lateral displacement (w) is continuous everywhere in a beam.‡ Thus, $w_1 = w_2$, or

$$[w]_\xi = 0, \tag{3.54}$$

---

† Note that M is continuous across a travelling plastic hinge, or $[M]_\xi = 0$ where the notation $[\ \ ]_\xi$ is defined in § 3.4.6. Differentiating with respect to time gives $[\partial M/\partial t]_\xi + \dot{\xi}[\partial M/\partial x]_\xi = 0$ which is satisfied since $M = M_0$ and $\partial M/\partial x = 0$ at $x = \xi$ for all time t.

‡ Lee and Symonds[3.3] have shown that the displacement (w) and velocity ($\dot{w}$) must be continuous functions of the coordinate x at a travelling plastic hinge.

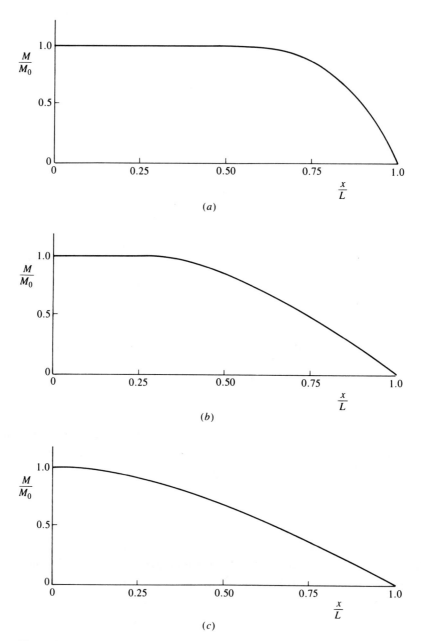

Figure 3.8 Bending moment distributions for a simply supported beam with $\eta \geqslant 3$. (a) First phase of motion, $0 \leqslant t \leqslant \tau$, with $\eta = 12$. (b) Second phase of motion, $\tau \leqslant t \leqslant T_1$, when $\xi/L = 0.25$ and $\eta \geqslant 3$. (c) Third phase of motion, $T_1 \leqslant t \leqslant T$, with $\eta \geqslant 3$.

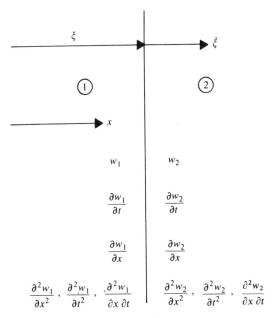

Figure 3.9 Two adjacent regions ① and ② separated by an interface at $x = \xi$ which is travelling with a velocity $\dot{\xi}$.

where w_1 and w_2 refer to the two adjacent regions shown in Figure 3.9 and $[X]_\xi$ is used to denote the difference between the values of a quantity X on either side of the interface (e.g., plastic hinge) at $x = \xi$. Now, differentiating equation (3.54) with respect to time gives

$$\frac{d}{dt}[w]_\xi = \left[\frac{\partial w}{\partial t}\right]_\xi + \left[\frac{\partial w}{\partial x}\frac{\partial x}{\partial t}\right]_\xi,$$

or

$$\left[\frac{\partial w}{\partial t}\right]_\xi + \dot{\xi}\left[\frac{\partial w}{\partial x}\right]_\xi = 0 \tag{3.55}$$

since the time derivative of equation (3.54) following an interface, which travels with a velocity $\dot{\xi}$, is zero. Moreover, it is postulated that the slope $\partial w/\partial x$ is continuous at a travelling plastic hinge, i.e.,

$$\left[\frac{\partial w}{\partial x}\right]_\xi = 0 \tag{3.56}\dagger$$

† Symonds, Ting and Robinson[3.4] show that the continuity of $\partial w/\partial x$ maintains finite bending strains and angular velocities at a moving plastic hinge. However, this restriction is not necessary at a stationary plastic hinge.

so that equation (3.55) becomes

$$\left[\frac{\partial w}{\partial t}\right]_\xi = 0. \tag{3.57}†$$

The derivative of equation (3.57) following the interface at $x = \xi$ gives

$$\frac{d}{dt}\left[\frac{\partial w}{\partial t}\right]_\xi = \left[\frac{\partial^2 w}{\partial t^2}\right]_\xi + \left[\frac{\partial^2 w}{\partial x\, \partial t}\frac{\partial x}{\partial t}\right]_\xi,$$

or

$$\left[\frac{\partial^2 w}{\partial t^2}\right]_\xi + \dot\xi\left[\frac{\partial^2 w}{\partial x\, \partial t}\right]_\xi = 0. \tag{3.58}$$

Similarly, equation (3.56) requires

$$\left[\frac{\partial^2 w}{\partial x\, \partial t}\right]_\xi + \dot\xi\left[\frac{\partial^2 w}{\partial x^2}\right]_\xi = 0. \tag{3.59}$$

It is observed that at a stationary plastic hinge or interface (i.e., $\dot\xi = 0$), equation (3.55) requires the satisfaction of equation (3.57) but not equation (3.56). Thus, changes of slope are allowed to develop at stationary plastic hinges but not at travelling ones.

In summary, the slope and transverse velocity must be continuous across a travelling plastic hinge in a beam according to equations (3.56) and (3.57), respectively. Any changes in acceleration and curvature across a travelling plastic hinge, or interface, are related through equations (3.58) and (3.59) to the corresponding change in $\partial^2 w/\partial x\, \partial t$ and the plastic hinge speed $\dot\xi$.

Now, returning to the theoretical analysis in §§ 3.4.2 to 3.4.4 for the problem in Figure 3.4(a), the transverse velocity distribution in the first phase of motion according to equations (3.23a, b) is continuous at the stationary plastic hinge, or $[\dot w] = 0$ at $x = \xi_0$. Equation (3.55), therefore, is satisfied when $\dot\xi = 0$. Moreover, $[\ddot w] = 0$ at $x = \xi_0$, so that equation (3.58) with $\dot\xi = 0$ is also satisfied. Equation (3.59) need not be satisfied by equations (3.23a, b) because it is based on equation (3.56), which is not required for a stationary plastic hinge.

The transverse velocity profile for the second phase of motion in § 3.4.3 is again given by equations (3.23a, b), except that $\dot W_1$ is replaced by $\dot W_2$ and the plastic hinge at $x = \xi_0$ in Figure 3.7(a) is allowed to move, as indicated in Figure 3.7(b). Clearly, equation (3.57) is satisfied since the transverse velocity profile is continuous at the travelling plastic hinge. Now, the transverse displacement accumulated during the second phase of motion at x is

$$w_2 = \int_\tau^{t^*} \dot W_2\, dt + \int_{t*}^t \dot W_2(L - x)\, dt/(L - \xi), \qquad \xi \leqslant x \leqslant \xi_0, \tag{3.60}$$

---

† Alternatively, if equation (3.57) is assumed and substituted into equation (3.55), then equation (3.56) follows.

where t^* is the time that the travelling hinge reaches x. Clearly, $dt = d\xi/\dot{\xi}$, which, using equation (3.43), allows equation (3.60) to be written in the form

$$w_2 = -\int_{L\xi_0}^{x} \dot{W}_2 p_0 \tau(L - \xi) \, d\xi/3M_0 - \int_{x}^{\xi} \dot{W}_2 p_0 \tau(L - x) \, d\xi/3M_0,$$

or

$$w_2 = -\dot{W}_2 p_0 \tau(x^2/2 - L^2\xi_0 + L^2\xi_0^2/2 + L\xi - \xi x)/3M_0, \qquad \xi \leqslant x \leqslant \xi_0$$

$$(3.61)\dagger$$

which gives $\partial w_2/\partial x = 0$ at $x = \xi^+$. Equation (3.23a) with $\dot{W}_2$ for $\dot{W}_1$ also indicates that $\partial w_2/\partial x = 0$ at $x = \xi^-$, where ξ^- is the side of the plastic hinge which lies in the central region. Thus, equation (3.56) is satisfied.

Equations (3.23a, b) give $[\ddot{w}_2]_\xi = \dot{W}_2 \dot{\xi}/(L - \xi)$ and $[\partial \dot{w}_2/\partial x]_\xi = -\dot{W}_2/(L - \xi)$ and, therefore, equation (3.58) with $\dot{\xi} \neq 0$ is satisfied. Equation (3.61) predicts that $\partial^2 w_2/\partial x^2 = -\dot{W}_2 p_0 \tau/3M_0$, whereas equation (3.23a) gives $\partial^2 w_2/\partial x^2 = 0$. Thus, $[\partial^2 w_2/\partial x^2]_\xi = -\dot{W}_2 p_0 \tau/3M_0$, and equation (3.59) is satisfied when using equation (3.43).

The plastic hinge remains stationary at the mid-span of the beam during the third and final phase of motion, as indicated in Figure 3.7(c). The transverse velocity profile is, therefore, given by equation (3.23b), with $\xi_0 = 0$ and $\dot{W}_1$ replaced by $\dot{W}_3$. Thus, the kinematic relations (3.55) and (3.58), with $\dot{\xi} = 0$, are satisfied due to symmetry of the velocity field.

It is evident from the foregoing calculations that the conditions across all the stationary and travelling plastic hinges during the three phases of motion in the theoretical analysis presented in §§ 3.4.2 to 3.4.4 satisfy the required kinematic relations. Furthermore, the transverse displacement and velocity profiles satisfy the boundary and initial conditions for the problem illustrated in Figures 3.4(a) and are continuous at $t = \tau$ and $t = T_1$ between the three phases of motion. Thus, the theoretical analysis in §§ 3.4.2 and 3.4.4 is kinematically admissible and, therefore, exact, since it was shown to be statically admissible in § 3.4.5.

The theoretical predictions presented in §§ 3.3.2, 3.3.3 and 3.4.2 to 3.4.4 were first obtained by Symonds[3.5] but using a method of analysis which is similar to that outlined later in §§ 3.5.2 and 3.5.3 for the impulsive loading case. The influence of various external pressure–time characteristics was also studied by Symonds.[3.5]

3.5 Simply supported beam loaded impulsively

3.5.1 Introduction

It is evident from Figure 3.10 that the maximum permanent transverse displacement of the simply supported beam illustrated in Figure 3.4(a) is, from a practical

---

† $\dot{W}_2$ is constant during the second phase of motion according to equation (3.35).

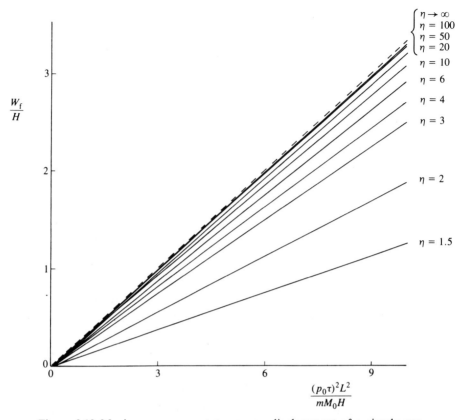

Figure 3.10 Maximum permanent transverse displacement of a simply supported beam (with unit width) subjected to a rectangular pressure pulse having
various values of the dimensionless pressure ratio η.

viewpoint, insensitive to the value of the dimensionless pressure ratio when $\eta > 20$,
approximately. External pressure loadings having a finite impulse with an infinitely
large magnitude ($\eta \to \infty$) and an infinitesimally short duration ($\tau \to 0$) (Dirac delta
function) are known as impulsive. In other words, a beam of unit breadth acquires
instantaneously a uniform transverse velocity V_0, where, to conserve linear
momentum,†

$$(2mL)V_0 = (p_0 2L)\tau,$$

or

$$V_0 = p_0\tau/m. \tag{3.62}$$

In this circumstance, equations (3.52) and (3.53) predict

$$T = mV_0/p_c \tag{3.63a}$$

† Newton's second law requires $F = d(mv)/dt$, or $F\,dt = d(mv)$. Thus, $\int F\,dt = $ change in linear
momentum, where $\int F\,dt$ is impulse.

and

$$W_{\rm f} = mV_0^2 L^2/3M_0 \qquad\qquad (3.63b)$$

for the duration of response and maximum permanent transverse displacement, respectively.

The approximation of an impulse loading usually simplifies a theoretical analysis and is an acceptable idealisation for many practical problems. Thus, the impulsive loading of a simply supported beam is examined in the following sections, but by using an alternative method which employs momentum and energy conservation principles.

3.5.2 First phase of motion, $0 \leqslant t \leqslant T_1$

Initially, a discontinuity in velocity occurs at $x = L$ since the support remains fixed spatially (i.e., $w = \dot{w} = 0$ at $x = L$ for all t). It appears reasonable to postulate that plastic hinges would develop at the supports when $t = 0$ then travel inwards with a speed $\dot{\xi}$ towards the beam centre, as indicated at a time t by the velocity profile in Figure 3.7(b). The central portion $0 \leqslant x \leqslant \xi$ remains rigid since it continues to travel at the initial velocity V_0 until disturbed by the inwards moving plastic hinge. This velocity field can be written in the form

$$\dot{w} = V_0, \qquad 0 \leqslant x \leqslant \xi \qquad\qquad (3.64)$$

and

$$\dot{w} = V_0(L - x)/(L - \xi), \qquad \xi \leqslant x \leqslant L. \qquad\qquad (3.65)$$

Now, the conservation of angular momentum of one half of the beam about a support demands

$$\int_0^L mV_0(L - x)\, {\rm d}x = \int_0^\xi mV_0(L - x)\, {\rm d}x + \int_\xi^L mV_0(L - x)^2\, {\rm d}x/(L - \xi) + M_0 t \qquad\qquad (3.66)$$

since the entire beam is rigid† except at the travelling plastic hinge with a limit moment M_0. Equation (3.66) gives

$$t = mV_0(L - \xi)^2/6M_0, \qquad\qquad (3.67)$$

which is the time required for a plastic hinge to travel inwards from a support to the location $x = \xi$.

The first stage of motion is completed at $t = T_1$ when the two travelling hinges coalesce at $x = 0$. Equation (3.67) predicts

$$T_1 = mV_0 L^2/6M_0 \qquad\qquad (3.68)$$

when $\xi = 0$. Thus, the corresponding lateral displacement at $x = 0$ is $W_1 = V_0 T_1$, or

$$W_1 = mV_0^2 L^2/6M_0. \qquad\qquad (3.69)$$

† Equations (3.3), (3.64) and (3.65) give $\dot{\kappa} = 0$.

3.5.3 Final phase of motion, $T_1 \leqslant t \leqslant T$

Now, at the end of the first stage of motion ($t = T_1$), the beam has a linear velocity profile with a peak value V_0 and, therefore, possesses a kinetic energy $mV_0^2 L/3$ which remains to be dissipated during subsequent motion. It appears reasonable to assume that this kinetic energy will be dissipated in a plastic hinge which remains stationary at $x = 0$ during the second stage of motion, as shown in Figure 3.7(c). In this circumstance, the conservation of energy requires that

$$mV_0^2 L/3 = 2M_0\theta_2, \tag{3.70}$$

where $2\theta_2$ is the angular change of the central plastic hinge. The displacement acquired by the beam centre during the second stage of motion is $W_2 = L\theta_2$, or

$$W_2 = mV_0^2 L^2/6M_0. \tag{3.71}$$

Finally, the total permanent lateral displacement at the beam centre is $W_f = W_1 + W_2$ which agrees with equation (3.63b). This expression may be written in the dimensionless form

$$W_f/H = \lambda/3, \tag{3.72}$$

where

$$\lambda = mV_0^2 L^2/M_0 H \tag{3.73}$$

is a non-dimensionalised form of the initial kinetic energy and H is the beam depth.

It is interesting to observe that the contributions to the central lateral displacement of the beam are identical during both phases of motion (i.e., $W_1 = W_2$) even though two-thirds of the initial kinetic energy is dissipated during the first phase of motion and only one-third during the final phase. However, two travelling plastic hinges are present in the beam during the first phase of motion, as shown in Figure 3.7(b), while only one stationary plastic hinge develops at the beam centre during the second phase.

The foregoing theoretical predictions must remain tentative until the solution has been shown to be both statically (i.e., no yield violations) and kinematically (i.e., no geometrical violations) admissible.

3.5.4 Static admissibility

Now, no external forces act on the beam, so that the only loading is due to inertia ($m\ddot{w}$), which is indicated in Figure 3.11(a) for the first phase of motion and which is obtained by differentiating the corresponding velocity profile sketched in Figure 3.7(b), or described by equations (3.64) and (3.65). The lateral shear force (Q) distribution in Figure 3.11(b) is obtained directly from the inertia loading distribution and equation (3.2) (i.e., $Q = \int m\ddot{w}\, dx + c_1$). Finally, the bending moment (M) distribution in Figure 3.11(c) is determined using equation (3.1) (i.e.,

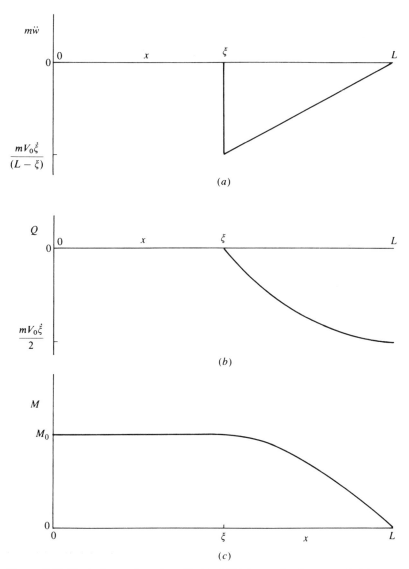

Figure 3.11 First phase of motion ($0 \leqslant t \leqslant T_1$) for a simply supported beam loaded impulsively. (a) Distribution of inertia loading ($\ddot{\xi} \leqslant 0$). (b) Distribution of transverse shear force ($\dot{\xi} \leqslant 0$). (c) Bending moment distribution.

$M = \int Q \, dx + c_2$). Similarly, the inertia loading, shear force and bending moment distributions sketched in Figure 3.12 are obtained during the second phase of motion.

It is evident from these distributions that the theoretical solution in §§ 3.5.2 and 3.5.3 is statically admissible throughout both phases of motion. An interested reader who does the calculations for static admissibility will find that the response

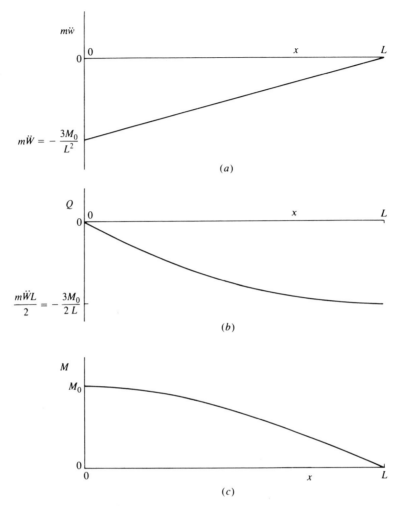

Figure 3.12 Second phase of motion ($T_1 \leqslant t \leqslant T$) for a simply supported beam loaded impulsively. (a) Distribution of inertia loading ($\ddot{W} \leqslant 0$). (b) Distribution of transverse shear force ($\ddot{W} \leqslant 0$). (c) Bending moment distribution.

duration† is the same as equation (3.63a), while the permanent deformed shape in Figure 3.13 is

$$w_f/H = \lambda(1 - x/L)(2 + x/L)/6, \qquad (3.74)$$

where λ is defined by equation (3.73).

3.5.5. Kinematic admissibility

It is straightforward to show that the displacement and velocity profiles in Figure 3.7(b, c), which are also associated with the theoretical solution for an impulsively

---

† The response duration is not predicted by the calculations in § 3.5.3.

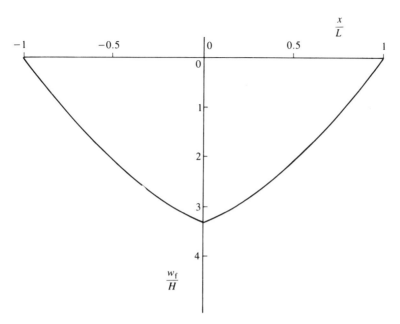

Figure 3.13 Permanent transverse displacement profile for an impulsively loaded simply supported beam according to equation (3.74) with $\lambda = 10$.

loaded simply supported beam, satisfy equations (3.56) to (3.59) and are, therefore, kinematically admissible.

It is evident that the theoretical solution in §§ 3.5.2 and 3.5.3 is both statically and kinematically admissible and is, therefore, the correct one. Thus, the exact duration of response is predicted by equation (3.63a), while the exact permanent deformed shape is given by equation (3.74).

The theoretical prediction for the maximum permanent transverse displacement at the mid-span according to equation (3.72), or equation (3.74), with $x = 0$, is compared in Figure 3.10 with the theoretical analyses in §§ 3.3 and 3.4 for various values of the dimensionless pressure ratio η.

3.6 Fully clamped beam, $\bar{p}_{\mathrm{c}} \leqslant p_0 \leqslant 3\bar{p}_{\mathrm{c}}$

3.6.1 Introduction

Consider a beam which is fully clamped at both ends, as shown in Figure 3.14(a), and subjected to the rectangular pressure pulse in Figure 3.5 which is described by equations (3.4) and (3.5). This beam collapses under a uniformly distributed static pressure

$$\bar{p}_{\mathrm{c}} = 4M_0/L^2 \tag{3.75}$$

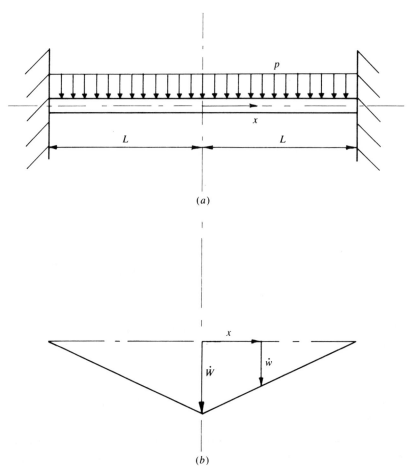

Figure 3.14 (a) Uniformly loaded beam with fully clamped supports. (b) Transverse velocity profile.

according to equation (1.32) and with the associated incipient transverse displacement field in Figure 1.7(b).

3.6.2 First phase of motion, $0 \leqslant t \leqslant \tau$

If the theoretical procedure described in § 3.3.2 is again followed for the present case with the velocity field in Figure 3.14(b), then equation (3.10) remains unchanged. However, a fully clamped support requires $M = -M_0$ at $x = L$ so that equation (3.11) is replaced by

$$\mathrm{d}^2 W/\mathrm{d}t^2 = 6(\bar{\eta} - 1)M_0/mL^2,$$ (3.76a)

where

$$\bar{\eta} = p_0/\bar{p}_{\mathrm{c}}.$$ (3.76b)

Equation (3.76a) is identical to equation (3.11) but with M_0 and η replaced by $2M_0$ and $\bar{\eta}$, respectively. Thus, equations (3.14) and (3.15) with M_0 and η replaced by $2M_0$ and $\bar{\eta}$, respectively, give the transverse displacement and velocity at the end of the first phase of motion for the fully clamped beam in Figure 3.14(a).

3.6.3 Final phase of motion, $\tau \leqslant t \leqslant T$

It is evident that equation (3.16), with M_0 replaced by $2M_0$, controls this phase of motion for a beam with fully clamped supports. Thus, the response duration is

$$T = \bar{\eta}\tau \tag{3.77}$$

with an associated final deformed profile

$$w_f = 3\bar{\eta}(\bar{\eta} - 1)M_0\tau^2(1 - x/L)/mL^2, \tag{3.78}$$

where $\bar{\eta}$ is defined by equation (3.76b).

3.6.4 Static and kinematic admissibility

It is left as an exercise for the reader to show that the foregoing theoretical solution is kinematically admissible and statically admissible provided $\bar{\eta} \leqslant 3$.

3.7 Fully clamped beam, $p_0 \geqslant 3\bar{p}_c$

It was observed in § 3.6.4 that the theoretical solution in §§ 3.6.2 and 3.6.3 is statically admissible provided $\bar{\eta} \leqslant 3$. Thus, following the theoretical details in § 3.4 for a beam with simple supports, it is assumed that the transverse velocity profile shown in Figure 3.7 and described by equations (3.23) also governs the response of a fully clamped beam subjected to a dynamic pressure pulse having $\bar{\eta} \geqslant 3$.

If a procedure similar to that in § 3.6 is followed, then it may be shown that the response duration for the present case is

$$T = \bar{\eta}\tau, \tag{3.79}$$

with an associated maximum permanent transverse displacement at the beam centre

$$W_f = p_0\tau^2(4\bar{\eta} - 3)/6m. \tag{3.80}$$

Equations (3.79) and (3.80) for the impulsive loading case described by equation (3.62) predict

$$T = mV_0L^2/4M_0 \tag{3.81}$$

and

$$W_f/H = \lambda/6, \tag{3.82}$$

respectively, where λ is defined by equation (3.73). Again it is left for a reader to demonstrate that these theoretical solutions are kinematically admissible and statically admissible provided $\bar{\eta} \geqslant 3$.

It is evident from equations (3.63a) and (3.72) and equations (3.81) and (3.82) that a fully clamped beam acquires one-half the lateral displacement in one-half the time of a similar simply supported beam which is subjected to the same impulsive velocity distributed uniformly over the entire span.

The theoretical predictions of equations (3.79) and (3.80), for the response duration and maximum permanent transverse displacement for a beam with fully clamped supports and subjected to a rectangular pressure pulse, were first obtained by Symonds.[3.5]

3.8 Impact of a mass on a fully clamped beam

3.8.1 Introduction

This section examines the dynamic plastic response of the fully clamped beam of length $2L$ shown in Figure 3.15(a) when struck at the mid-span by a mass M travelling with an initial velocity V_0. The mid-span of the beam travels with a velocity V_0 at the instant of impact, while the remainder of the beam is stationary. Therefore, to maintain dynamic equilibrium, a disturbance propagates away from the mid-span, while the striker is assumed to remain in contact with the beam. In fact, two distinct phases of motion occur.

A plastic hinge develops under the impact point at $t = 0$ and two plastic hinges propagate the disturbance away from the mid-span towards the supports and into the undeformed portions of the beam during the first phase of motion, as shown in Figure 3.15(b). The plastic hinges remain stationary at the supports and the mid-span during the final phase of motion, as indicated in Figure 3.15(c), until the beam and striker come to rest, when all the initial kinetic energy of the striking mass $MV_0^2/2$ is dissipated plastically.

3.8.2 First phase of motion $0 \leqslant t \leqslant t_1$

The transverse velocity profile in Figure 3.15(b) can be written for the right-hand portion of the beam $0 \leqslant x \leqslant L$ as

$$\dot{w} = \dot{W}(1 - x/\xi), \qquad 0 \leqslant x \leqslant \xi \qquad (3.83a)$$

and

$$\dot{w} = 0, \qquad \xi \leqslant x \leqslant L, \qquad (3.83b)$$

where ξ is the time-dependent location of the travelling plastic hinge.

Vertical equilibrium for the central portion of the beam between the two travelling hinges demands that

$$M\ddot{W} + 2\int_0^\xi m\ddot{w}\,\mathrm{d}x = 0 \qquad (3.84)$$

since the transverse shear force Q is zero at the travelling plastic hinges ($x = \pm\xi$) where the maximum bending moment develops in the beam (see equation (3.1)).

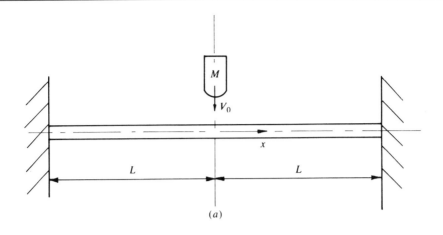

(a)

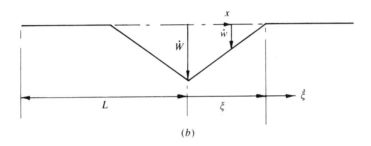

(b)

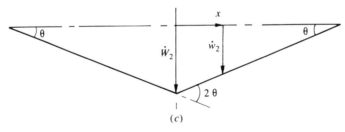

(c)

Figure 3.15 (a) Impact of a fully clamped beam with a mass M travelling with a velocity V_0. (b) Transverse velocity field during the first phase of motion $(0 \leqslant t \leqslant t_1)$. (c) Transverse velocity field during the second phase of motion $(t_1 \leqslant t \leqslant T)$.

Substituting equation (3.83a) into equation (3.84) gives

$$M\ddot{W} + 2m \int_0^\xi \{\ddot{W}(1 - x/\xi) + \dot{W}x\dot{\xi}/\xi^2\}\, \mathrm{d}x = 0$$

or

$$M\ddot{W} + m(\ddot{W}\xi + \dot{W}\dot{\xi}) = 0. \tag{3.85}$$

Now, $d(\dot{W}\xi)/dt = \ddot{W}\xi + \dot{W}\dot{\xi}$, so integrating equation (3.85) with respect to time yields

$$M\dot{W} + m\dot{W}\xi = MV_0, \tag{3.86}$$

where the constant of integration satisfies the initial conditions $\dot{W} = V_0$ and $\xi = 0$ at $t = 0$. Equation (3.86) may be rearranged in the form

$$\dot{W} = V_0/(1 + m\xi/M). \tag{3.87}$$

Now, considering moment equilibrium of the portion $0 \leqslant x \leqslant \xi$ with plastic hinges at $x = 0$ and $x = \xi$ and taking moments about the mid-span gives

$$2M_0 - \int_0^\xi m\ddot{w}x \, dx = 0, \tag{3.88}$$

since $M = M_0$ at $x = 0$ and $M = -M_0$ and $Q = 0$ at $x = \xi$. Substituting the time derivative of equation (3.83a) into equation (3.88) yields

$$\int_0^\xi m\{\ddot{W}(1 - x/\xi) + \dot{W}\dot{\xi}x/\xi^2\}x \, dx = 2M_0,$$

or

$$m(\ddot{W}\xi^2/6 + \dot{W}\dot{\xi}\xi/3) = 2M_0,$$

which can be written in the form

$$d(\dot{W}\xi^2)/dt = 12M_0/m. \tag{3.89}$$

Integrating equation (3.89) with respect to time and using the initial condition $\xi = 0$ when $t = 0$ finally gives the location–time characteristic of the travelling hinge

$$t = m\dot{W}\xi^2/12M_0, \tag{3.90}$$

or, when using equation (3.87),

$$t = mMV_0\xi^2/\{12M_0(M + m\xi)\}, \tag{3.91}$$

which, differentiating with respect to time, predicts the velocity of the travelling hinge

$$\dot{\xi} = 12M_0(M + m\xi)^2/\{mMV_0\xi(2M + m\xi)\}. \tag{3.92}$$

It is evident from Figure 3.15(b) that the transverse displacement at any position x in the beam is zero until the time $t(x)$ when the travelling plastic hinge reaches x. Thus, the transverse displacement at a position x for a time $t \geqslant t(x)$ is

$$w = \int_{t(x)}^t \dot{w} \, dt, \tag{3.93}$$

where $\dot{w}$ is given by equation (3.83a) and the time $t(x)$ is predicted by equation (3.91) with $\xi = x$. This equation may be written

$$w = \int_x^\xi \dot{w} \, d\xi/\dot{\xi} \tag{3.94}$$

since $\dot{\xi} = d\xi/dt$. Now, substituting equations (3.83a), (3.87) and (3.92) into equation (3.94) gives

$$w = \int_x^\xi \frac{V_0(1 - x/\xi)mMV_0\xi(2M + m\xi)\,d\xi}{(1 + m\xi/M)12M_0(M + m\xi)^2}$$

which, integrating† and rearranging, becomes

$$w = \frac{M^2V_0^2}{24mM_0}\left\{\frac{1 + \beta}{(1 + \alpha)^2} - \frac{(1 + 2\beta)}{(1 + \beta)} + \frac{2\beta}{(1 + \alpha)} + 2\log_e\left(\frac{1 + \alpha}{1 + \beta}\right)\right\}, \quad (3.95)$$

where

$$\alpha = m\xi/M \quad \text{and} \quad \beta = mx/M. \quad (3.96a, b)$$

This phase of motion is completed when the travelling hinges reach the supports which occurs when $\xi = L$ and

$$t_1 = mMV_0L^2/\{12M_0(M + mL)\}, \quad (3.97)$$

according to equation (3.91). However, it is evident from equation (3.87) that the transverse velocity is not zero at $t = t_1$. Therefore the total kinetic energy in the mass M and the beam when using equations (3.83a) and (3.87) is

$$MV_0^2(1 + 2mL/3M)/\{2(1 + mL/M)^2\}, \quad (3.98)$$

which remains to be dissipated during a second phase of motion.

3.8.3 Second phase of motion, $t_1 \leqslant t \leqslant T$

It is assumed that the kinetic energy remaining in the beam at the end of the first phase of motion according to equation (3.98) is dissipated in the stationary plastic hinges which are located at both supports and at the beam centre during the second phase of motion, as shown in Figure 3.15(c). The transverse displacement profile associated with this deformation pattern is

$$w_2 = (L - x)\theta, \quad (3.99)$$

where θ and 2θ are the angular rotations accumulated during the second phase of motion alone at the support and central hinges, respectively. A straightforward energy balance predicts that

$$4M_0\theta = MV_0^2(1 + 2mL/3M)/\{2(1 + mL/M)^2\}, \quad (3.100)$$

where the left- and right-hand sides of equation (3.100) are the energy absorbed in the three plastic hinges and the remaining kinetic energy according to equation (3.98), respectively. Thus, the transverse displacement acquired during the second phase of motion according to equations (3.99) and (3.100) is

$$w_2 = MV_0^2L(1 + 2mL/3M)(1 - x/L)/\{8M_0(1 + mL/M)^2\}. \quad (3.101)$$

---

† The integral may be recast into several standard forms which are the same as cases 29, 32 and 35 on page 413 of reference 3.6.

The final permanent transverse displacement profile for the problem in Figure 3.15(a) is obtained by adding equation (3.95), with $\xi = L$, and equation (3.101), or

$$w_f = \frac{M^2 V_0^2}{24mM_0} \left\{ \frac{\bar{\alpha} - \beta}{(1 + \bar{\alpha})(1 + \beta)} + 2 \log_e \left(\frac{1 + \bar{\alpha}}{1 + \beta} \right) \right\}, \qquad 0 \leqslant \beta \leqslant \bar{\alpha}, \quad (3.102)$$

where

$$\bar{\alpha} = mL/M \tag{3.103}$$

and β is defined by equation (3.96b).

3.8.4 Special case. Heavy strikers

Equation (3.87) at the end of the first phase of motion (i.e., $\xi = L$) predicts that $\dot{W} \cong V_0$ when $M/mL \gg 1$. Moreover, the kinetic energy to be dissipated during the second phase of motion is $MV_0^2/2$, approximately, according to equation (3.98) with $M/mL \gg 1$. These observations show that the first phase of motion plays an unimportant role for heavy strikers. Thus, energy conservation for the second phase of motion, with the velocity profile shown in Figure 3.15(c), requires

$$4M_0 \theta = MV_0^2/2, \tag{3.104}$$

where θ is now the final total angular rotation at a support. The final permanent transverse displacement is

$$w_f = (L - x)\theta$$

or

$$w_f = MV_0^2 L(1 - x/L)/8M_0. \tag{3.105}$$

It is left to the reader to show that equation (3.105) can be recovered from equation (3.102) when $\bar{\alpha} \to 0$ and $\beta \to 0$.

3.8.5 Special case. Light strikers

In this particular case, $mL/M \gg 1$, or $M/mL \to 0$, and it is observed that $\dot{W} \cong 0$ at the end of the first phase of motion according to equation (3.87) with $\xi = L$. Thus, the kinetic energy of the beam and mass is zero at the end of the first phase of motion, which is confirmed by equation (3.98). The movement of the beam, therefore, ceases at the end of the first phase of motion with no kinetic energy available to cause plastic deformation at the stationary hinges which normally form at the supports during a second phase of motion.

Equation (3.102) may be written in the form

$$w_f = \frac{MV_0^2 L}{24M_0} \left\{ \frac{1 + \beta}{\bar{\alpha}(1 + \bar{\alpha})^2} - \frac{(1 + 2\beta)}{\bar{\alpha}(1 + \beta)} + \frac{2\beta}{\bar{\alpha}(1 + \bar{\alpha})} + \frac{2}{\bar{\alpha}} \log_e \left(\frac{1 + \bar{\alpha}}{1 + \beta} \right) \right.$$

$$\left. + (3 + 2\bar{\alpha})(1 - \beta/\bar{\alpha})/(1 + \bar{\alpha})^2 \right\}$$

which, with $\bar{\alpha} \gg 1$, reduces to

$$w_{\mathrm{f}} \cong \frac{MV_0^2 L}{12M_0\bar{\alpha}} \log_e\left(\frac{\bar{\alpha}}{1+\beta}\right)$$

or

$$w_{\mathrm{f}} = \frac{M^2 V_0^2}{12mM_0} \log_e\left(\frac{mL/M}{1+mx/M}\right). \tag{3.106}$$

It is evident that the permanent transverse displacement profile is logarithmic for light strikers (equation (3.106)) which differs markedly from the linear profile for heavy strikers according to equation (3.105).

3.8.6 Kinematic admissibility

The kinematic conditions according to equations (3.56) to (3.59) must be satisfied by equations (3.83) for the velocity fields during the first phase of motion. Equation (3.99) for the displacement field during the final phase of motion must also satisfy equations (3.56) to (3.59).

Now, differentiating equation (3.95) for the first phase of motion with respect to $\beta = mx/M$, gives

$$\frac{\partial w}{\partial \beta} = \frac{M^2 V_0^2}{24mM_0}\left\{\frac{1}{(1+\alpha)^2} - \frac{2}{1+\beta} - \frac{1}{(1+\beta)^2} + \frac{2}{1+\alpha}\right\}, \tag{3.107}$$

which predicts that $\partial w/\partial\beta = 0$ when $\alpha = \beta$ (or $x = \xi^-$). Moreover, on the other side of the travelling hinge ($x = \xi^+$), $\partial w/\partial\beta = 0$ according to equation (3.83b). Thus, equation (3.56) is satisfied, while equations (3.83a, b) clearly satisfy equation (3.57).

Equation (3.83a) predicts that $\partial^2 w/\partial t^2 = \dot{W}\dot{\xi}/\xi$ at the travelling plastic hinge located at $x = \xi^-$, which, in view of equation (3.83b), gives

$$[\partial^2 w/\partial t^2]_\xi = -\dot{W}\dot{\xi}/\xi. \tag{3.108}$$

Differentiation of equation (3.107) with respect to time yields

$$\partial^2 w/\partial x\, \partial t = -mV_0^2 \dot{\xi}(2+\alpha)/\{12M_0(1+\alpha)^3\}$$

at $x = \xi^-$, which, together with equation (3.83b), indicates that

$$[\partial^2 w/\partial x\, \partial t]_\xi = mV_0^2 \dot{\xi}(2+\alpha)/\{12M_0(1+\alpha)^3\}. \tag{3.109}$$

Finally, it may be shown that equations (3.108) and (3.109), together with equations (3.87) and (3.92), satisfy the kinematic requirement demanded by equation (3.58).

Now, differentiating equation (3.107) with respect to β yields

$$\partial^2 w/\partial \beta^2 = M^2 V_0^2(2+\alpha)/\{12mM_0(1+\alpha)^3\} \tag{3.110}$$

when $x = \xi^-$, while $\partial^2 w/\partial \beta^2 = 0$ at $x = \xi^+$ according to equation (3.83b). Thus, equations (3.109) and (3.110) satisfy equation (3.59) when noting that $\partial^2 w/\partial x^2 = (m/M)^2\, \partial^2 w/\partial \beta^2$.

In addition to the above observations, the transverse displacement and velocity fields satisfy the initial and boundary conditions and are, therefore, kinematically admissible throughout the first phase of motion.

It is left to a reader to demonstrate that equation (3.99) for the transverse displacement profile with stationary plastic hinges is also kinematically admissible during the final phase of motion.

3.8.7 Static admissibility

It is necessary to demonstrate that the theoretical solutions in §§ 3.8.2 and 3.8.3 do not violate the yield condition and are, therefore, statically admissible.

Equations (3.1) and (3.2) predict the governing equation $\partial^2 M/\partial x^2 = m \, \partial^2 w/\partial t^2$ which when substituting equations (3.83a) and (3.87) becomes

$$\partial^2 M/\partial x^2 = m^2 V_0 \dot{\xi}(2\alpha\beta + \beta - \alpha^2)/\{M\alpha^2(1 + \alpha)^2\}, \qquad 0 \leqslant \beta \leqslant \alpha, \qquad (3.111)$$

where α and β are defined by equations (3.96a, b), respectively. Integrating equation (3.111) twice with respect to x and using the boundary conditions $Q = \partial M/\partial x = 0$ at $x = \xi$ and $M = M_0$ at $x = 0$ leads to the bending moment distribution

$$M/M_0 = (4\alpha\beta^3 + 2\beta^3 - 6\alpha^2\beta^2 - 6\alpha^2\beta + 2\alpha^3 + \alpha^4)/\{\alpha^3(2 + \alpha)\}, \qquad 0 \leqslant \beta \leqslant \alpha. \tag{3.112}$$

This bending moment distribution is plotted in Figure 3.16(a) and has $M = M_0$, $\partial M/\partial x < 0$ and $\partial^2 M/\partial x^2 < 0$ at $x = 0$ and $M = -M_0$, $\partial M/\partial x = 0$ and $\partial^2 M/\partial x^2 > 0$ at $x = \xi$. Moreover, equation (3.111) has only one root so that the bending moment nowhere exceeds the plastic limit moment for $0 \leqslant x \leqslant \xi$.

Equations (3.1), (3.2) and (3.83b) for the first phase of motion with $\xi \leqslant x \leqslant L$ gives $\partial^2 M/\partial x^2 = 0$, which predicts $M = -M_0$ since $Q = \partial M/\partial x = 0$ and $M = -M_0$ at $x = \xi$. Thus, the theoretical solution is statically admissible for $0 \leqslant x \leqslant L$ throughout the first phase of motion with $0 \leqslant t \leqslant t_1$, as indicated in Figure 3.16(a).

A simple energy method of analysis was used in § 3.8.3 to predict the final results at the end of the second phase of motion. Unfortunately, it is more difficult to show that the theoretical solution is statically admissible during the second phase of motion.

Equations (3.1), (3.2) and (3.99) may be used to obtain the bending moment distribution

$$M/M_0 = 1 - 2(3\bar{\alpha} - \beta)\beta^2/\{\bar{\alpha}^2(3 + 2\bar{\alpha})\} - 6\beta/\{\bar{\alpha}(3 + 2\bar{\alpha})\}, \qquad 0 \leqslant \beta \leqslant \bar{\alpha}, \tag{3.113}$$

where β and $\bar{\alpha}$ are defined by equations (3.96b) and (3.103), respectively. The transverse shear force at $x = 0$ was taken as $Q = ML\ddot{\theta}/2$, where $L\ddot{\theta}$ is the acceleration of the mass, and used as a boundary condition together with $M = M_0$ at $x = 0$ and $M = -M_0$ at $x = L$. Equation (3.113) predicts that $\partial M/\partial x < 0$ at

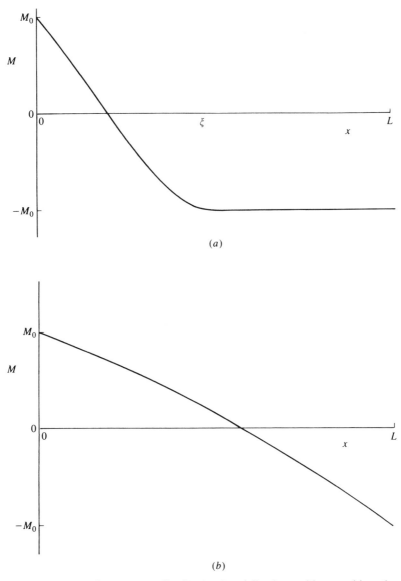

Figure 3.16 Bending moment distribution in a fully clamped beam subjected to a mass impact at the mid-span. (a) First phase of motion, $0 \leqslant t \leqslant t_1$. (b) Second phase of motion, $t_1 \leqslant t \leqslant T$.

$x = 0$ and $x = L$, while $\partial^2 M/\partial x^2 \leqslant 0$ for $0 \leqslant x \leqslant L$. No yield violations occur, therefore, in the beam during the second phase of motion $t_1 \leqslant t \leqslant T$, as indicated in Figure 3.16(b).

The theoretical solution in §§ 3.8.2 and 3.8.3 for the particular beam in Figure 3.15(a) is, therefore, exact since it is both statically admissible and kinematically admissible, as demonstrated in § 3.8.6. Figure 3.17 shows the variation of the dimensionless maximum permanent transverse displacement at the mid-span with non-dimensional impact energy for various values of the mass ratio $\bar{\alpha}$ defined by equation (3.103). A comparison is also made with the theoretical predictions for the heavy and light strikers.

Parkes[3.7] has examined the beam problem in Figure 3.15(a) but with the mass striking at any position in the span. It is left to an interested reader to show that the theoretical predictions obtained by Parkes,[3.7] when specialised to the central impact case, are identical to the corresponding results in §§ 3.8.2 to 3.8.5.

Parkes[3.7] also conducted experimental tests on beams made from several different metals and struck by moving masses. The beams were held by devices which prevented rotation at the supports, but permitted axial movement. Parkes observed that the general features of the experimental results agreed with the corresponding theoretical predictions. The two halves of the beams struck by heavy strikers remained straight with plastic hinges forming at the supports and the impact point, as predicted by equation (3.105) in § 3.8.4 and indicated in Figure 3.15(c) for the second phase of motion. However, the two halves of the beam did not remain straight when struck by light strikers, as predicted by equation (3.106) in § 3.8.5. In order to achieve good agreement between the theoretical predictions and experimental results for the maximum transverse displacements, Parkes[3.7]

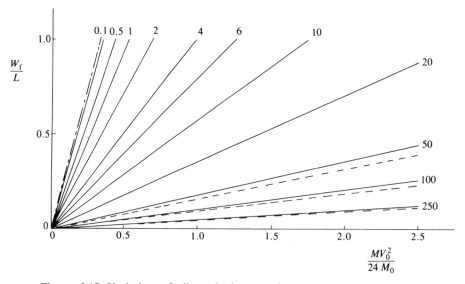

Figure 3.17 Variation of dimensionless maximum permanent transverse displacement at the mid-span of the beam in Figure 3.15(a) with dimensionless impact energy. ——: equation (3.102) with $\beta = 0$. ————: equation (3.105) with $x = 0$ for heavy strikers ($\bar{\alpha} \to 0$). ———: equation (3.106) with $x = 0$ for light strikers ($\bar{\alpha} \gg 1$). Numbers: value of $\bar{\alpha} = mL/M$.

found that it was necessary to consider the influence of material strain rate sensitivity, a phenomenon which is discussed in Chapter 8.

3.9 Impact of a cantilever beam

3.9.1 Introduction

The dynamic response of a cantilever beam subjected to large dynamic loads, which produce an inelastic material behaviour, has been examined theoretically by many authors using rigid plastic methods of analysis. In particular, Parkes[3.8] studied the behaviour of a cantilever beam of length L which was struck at the tip by a mass G travelling with an initial velocity V_0, as shown in Figure 3.18(a).

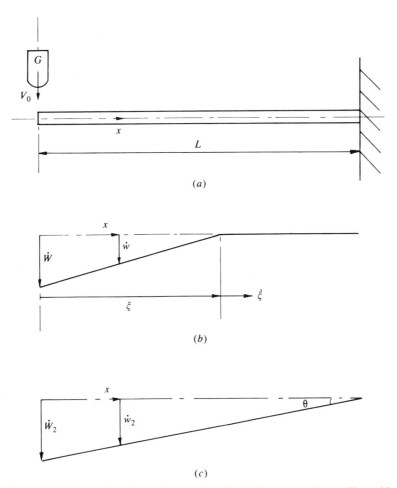

Figure 3.18 (a) cantilever beam impacted at the tip by a mass G travelling with a velocity V_0. (b) Transverse velocity during the first phase of motion, $0 \leqslant t \leqslant T_1$. (c) Transverse velocity during the second phase of motion, $T_1 \leqslant t \leqslant T_2$.

Now, when a mass G strikes the tip of the cantilever beam, a disturbance develops immediately underneath the mass which propagates by means of a plastic hinge into the undeformed region of the beam, as shown in Figure 3.18(b). The travelling plastic hinge eventually reaches the base of the cantilever at the end of the first phase of motion. The residual kinetic energy in the beam and mass is then dissipated at the plastic hinge which remains stationary at the support throughout a second phase of motion, as illustrated in Figure 3.18(c).

It transpires that the phases of motion in Figures 3.18(b) and 3.18(c) are respectively similar to those in Figures 3.15(b) and 3.15(c) for one-half of a beam $(0 \leqslant x \leqslant L)$ which is fully clamped at both supports and struck by a mass at the mid-span. Thus, the theoretical analysis for the particular problem in Figure 3.18(a) is obtained in the following sections by analogy with the theoretical analysis presented in § 3.8.

3.9.2 First phase of motion, $0 \leqslant t \leqslant T_1$

Now, the transverse velocity pattern for the beam in Figure 3.18(b) is similar to that for the right-hand part of the beam in Figure 3.15(b). Thus, it is evident from the theoretical analysis § 3.8.2 that equations (3.83a, b) remain valid, while equation (3.84) is also applicable to the present problem, provided G is substituted for $M/2$. Equations (3.85) to (3.87), with M replaced by $2G$, control, therefore, the behaviour of the cantilever beam in Figure 3.18(a). Moreover, if the cantilever beam is made from a rigid-plastic material with a limit moment M_p, then equation (3.88) is again obtained, provided $2M_0$ is replaced by M_p, since the bending moment is zero at the tip ($M = 0$ at $x = 0$).

If similar arguments are followed throughout the remainder of § 3.8.2, it is evident that equations (3.89) to (3.97) also describe the behaviour of the cantilever beam in Figure 3.18(a) during the first phase of motion when $M = 2G$ and $2M_0 = M_p$. Thus, the transverse displacement according to equation (3.95) is, therefore,

$$w = \frac{G^2 V_0^2}{3mM_p} \left\{ \frac{1+\beta}{(1+\alpha)^2} - \frac{(1+2\beta)}{(1+\beta)} + \frac{2\beta}{(1+\alpha)} + 2\log_e\left(\frac{1+\alpha}{1+\beta}\right) \right\},$$

(3.114)

where

$$\alpha = m\xi/2G \quad \text{and} \quad \beta = mx/2G. \tag{3.115a, b}$$

The first phase of motion is completed when the travelling plastic hinge reaches the support, or

$$T_1 = mGV_0L^2/\{3M_p(2G + mL)\} \tag{3.116}$$

by analogy with equation (3.97), while the kinetic energy remaining in the cantilever beam is

$$GV_0^2(1 + mL/3G)/\{2(1 + mL/2G)^2\}. \tag{3.117}$$

3.9.3 Second phase of motion, $T_1 \leqslant t \leqslant T_2$

The transverse velocity pattern in Figure 3.18(c) for the second phase of motion is similar to that in Figure 3.15(c) for the portion $0 \leqslant x \leqslant L$ of a fully clamped beam struck by a mass. Therefore, equation (3.99) remains unchanged, while an energy balance using equation (3.117) gives

$$M_p \theta = G V_0^2 (1 + mL/3G)/\{2(1 + mL/2G)^2\}, \qquad (3.118)$$

which also may be obtained from equation (3.100) with the same substitutions used in § 3.9.2 (i.e., M and $2M_0$ replaced by $2G$ and M_p, respectively).

The final, or permanent, transverse displacement profile by analogy with equation (3.102) is

$$w_f = \frac{G^2 V_0^2}{3mM_p} \left\{ \frac{1+\beta}{(1+\bar{\alpha})^2} - \frac{(1+2\beta)}{(1+\beta)} + \frac{2\beta}{(1+\bar{\alpha})} + 2\log_e \left(\frac{1+\alpha}{1+\beta} \right) \right.$$

$$\left. + \bar{\alpha}(3 + 2\bar{\alpha})(1 - \beta/\bar{\alpha})/(1+\bar{\alpha})^2 \right\}, \qquad 0 \leqslant \beta \leqslant \bar{\alpha}, \quad (3.119a)$$

or

$$\frac{6M_p w_f}{GLV_0^2} = \frac{1 - x/L}{(1+\bar{\alpha})(1+\beta)} + \frac{2}{\bar{\alpha}} \log_e \left(\frac{1+\bar{\alpha}}{1+\beta} \right), \qquad (3.119b)$$

where

$$\bar{\alpha} = mL/2G \qquad (3.120)$$

and β is given by equation (3.115b).

It is left as an exercise to an interested reader to show that this theoretical solution is statically and kinematically admissible and, therefore, exact.

Equation (3.119) for the special case of heavy strikers (i.e., $G/mL \gg 1$, or $\bar{\alpha} \to 0$) reduces to

$$w_f = G V_0^2 L (1 - x/L)/2M_p, \qquad (3.121)$$

while light strikers ($G/mL \ll 1$ or $\bar{\alpha} \gg 1$) give rise to a permanent transverse displacement

$$w_f = \frac{2G^2 V_0^2}{3mM_p} \log_e \left(\frac{mL/2G}{1 + mx/2G} \right). \qquad (3.122)$$

Equations (3.121) and (3.122) may also be obtained respectively from equations (3.105) and (3.106), with $M = 2G$ and $2M_0 = M_p$.

Figure 3.19 shows a comparison between the exact permanent transverse displacement profiles, which are predicted by equation (3.119b) for various mass ratios ($\bar{\alpha}$), and the approximations of equations (3.121) and (3.122) for heavy and light strikers, respectively.

3.9.4 Comparison with experimental results

Parkes[3.8] used a test arrangement, similar to that indicated in Figure 3.18(a), to obtain some experimental data for cantilever beams which were made from mild

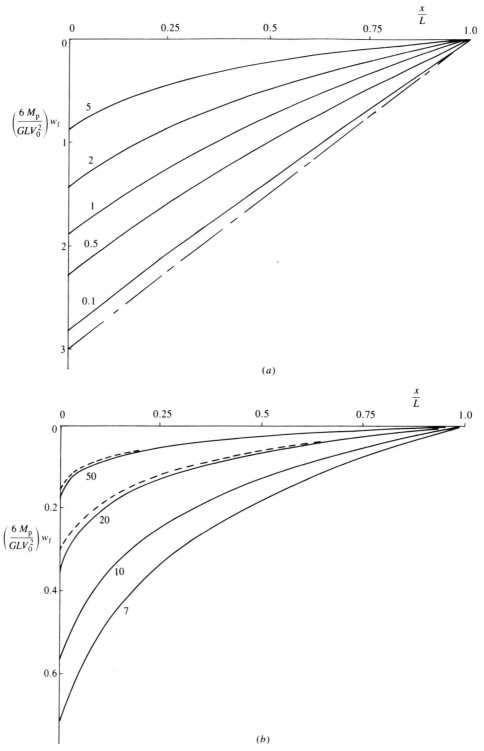

Figure 3.19 Permanent transverse displacement profile for the cantilever beam in Figure 3.18(a): ——: equation (3.119) with $\bar{\alpha}$ indicated. (a) ————: equation (3.121) for heavy strikers ($\bar{\alpha} \to 0$). (b) ———: equation (3.122) for light strikers ($\bar{\alpha} \gg 1$).

steel and other metals. It was observed that gravitational effects could exercise an important influence on the dynamic response of some test specimens.

Now, the plastic hinge at the base of a cantilever beam deforming with the transverse velocity field in Figure 3.18(c) must absorb the change of potential energy of a striker (GgW_f) in addition to the initial kinetic energy ($GV_0^2/2$). The change of potential energy of a beam was considered negligible for heavy strikers with $\bar{\alpha} = mL/2G \to 0$. Thus, energy conservation requires

$$GV_0^2/2 + GgW_f = M_p(W_f/L), \qquad (3.123)$$

where W_f/L is the total change of angle across a stationary plastic hinge at the cantilever support. Equation (3.123) may be rearranged in the form

$$W_f = GV_0^2L/\{2(M_p - GgL)\}, \qquad (3.124)$$

which reduces to equation (3.121) with $x = 0$ when gravitational effects are ignored.

A comparison of the theoretical predictions of equations (3.121) and (3.124) in Figure 3.20 reveals the increase in maximum permanent transverse displacements

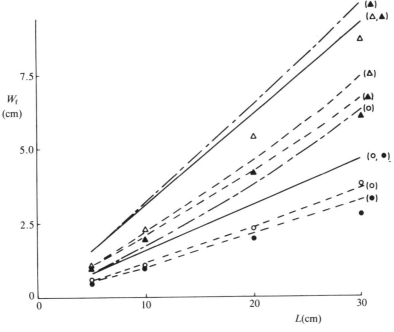

Figure 3.20 Comparison of experimental results and theoretical predictions for maximum permanent transverse displacements of mild steel cantilever beams struck by heavy strikers. $\bigcirc$, $\bullet$, $\triangle$, $\blacktriangle$: experimental results.[3.8] ———: equation (3.121). — · —: equation (3.124) (including a correction for gravitational effects). – – –: theoretical predictions of Parkes[3.8] which take account of material strain rate sensitivity and gravitational effects (equation (3.124) with a correction for material strain rate sensitivity).

due to gravitational effects for the mild steel test specimens examined by Parkes.[3.8] It is evident that the maximum permanent transverse displacements predicted by these equations are larger than the corresponding experimental values. However, mild steel is a strain-rate-sensitive material, a phenomenon which is discussed in Chapter 8. Parkes[3.8] compensated for this effect by using an enhanced plastic bending moment which is estimated from the average strain rate in a cantilever beam during a test. Reasonable agreement is then found in Figure 3.20 between the experimental results and the theoretical predictions of equation (3.124) with an enhanced value for the plastic bending moment (M_p).

3.10 Final remarks

The results in Figure 3.20 indicate that the theoretical rigid-plastic procedure developed in this chapter is capable of predicting reasonable agreement with experimental test results, at least for the impact loading of cantilever beams.

Florence and Firth[3.9] conducted an experimental investigation into the behaviour of metal beams which were subjected to a uniformly distributed impulsive velocity field across the entire span. The specimens were free to move axially at the supports and were either simply supported or clamped. The high-speed photographs taken during the response of these beams lend some credence to the concept of a travelling plastic hinge, which was used in the theoretical solutions outlined earlier in this chapter.† The theoretical analysis in § 3.7 for an impulsively loaded beam with clamped supports predicts the location–time history of the travelling plastic hinge,

$$L - \xi = (12M_0 t/mV_0)^{1/2}, \tag{3.125}$$

according to equation (3.67), with M_0 replaced by $2M_0$. A comparison between the predictions of equation (3.125) and the corresponding experimental values recorded by Florence and Firth,[3.9] using a high-speed camera, are shown in Figure 3.21.

Parkes[3.7, 3.8] found it necessary to cater for the influence of material strain rate sensitivity in order to improve the agreement between his theoretical predictions and the test results recorded on mild steel cantilever and fully clamped beams. This important phenomenon is discussed further in Chapter 8. However, there are several other assumptions incorporated into the basic rigid-plastic method. For example, the influence of transverse shear effects and finite-deflections, or geometry changes, are neglected. Nevertheless, they are sometimes sufficiently important in certain practical cases to warrant exploration in Chapter 6 and 7, respectively.

The attractive simplicity of rigid-plastic analyses for both static and dynamic

---

† Recently, Symonds and Fleming[3.10] have used a numerical scheme to examine the travelling hinge phase of a rigid-plastic beam loaded dynamically.

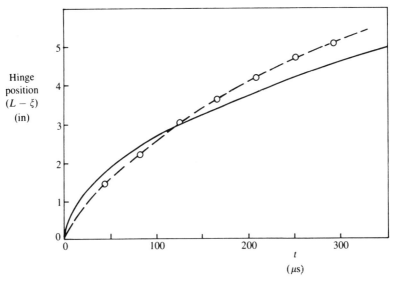

Figure 3.21 Comparison between equation (3.125) and the temporal location of a travelling plastic hinge observed by Florence and Firth[3.9] in an impulsively loaded clamped aluminium 2024-T4 beam. ——: equation (3.125). --○--: experimental results.[3.9]

structural problems is due largely to the neglect of material elasticity. In fact, only one exact elastic-plastic theoretical solution exists within the entire class of beam structures loaded dynamically. Duwez, Clark and Bohnenblust[3.11] considered the rather special case of an infinitely long beam which is impacted at mid-span with a load which imparts a constant velocity.† It is indicated by Symonds, Ting and Robinson[3.4] that Bohnenblust's problem, when repeated for a rigid, perfectly plastic material, is simplified considerably and, under certain circumstances, can give reasonable agreement with the corresponding exact elastic, perfectly plastic solution. When the central impact velocity is 6 times the maximum impact velocity which can be tolerated in a wholly elastic manner, then the angular change of a rigid perfectly plastic beam under the impacted point is 13.6 per cent higher, while this difference is less than 2 per cent when the velocity is 60 times larger.

Generally speaking, elastic effects are not important when the external dynamic energy (e.g., kinetic energy K_e) is significantly larger than the maximum amount of strain energy (S_e) which may be absorbed in a wholly elastic manner. This requirement may be expressed as an energy ratio

$$E_r = K_e/S_e \gg 1. \tag{3.126}$$

† Conroy[3.12] has modified slightly Bohnenblust's method in order to examine the influence of a special form of dynamic loading at one end of a semi-infinite beam.

The elastic strain energy in a beam with a volume v is

$$S = \int_v \frac{\sigma_x^2 \, \mathrm{d}v}{2E},$$ (3.127)

where E is Young's modulus. Thus, the maximum possible amount of elastic strain energy which a beam can absorb occurs when $\sigma_x = \sigma_0$ throughout the entire volume, i.e.,

$$S_e = \sigma_0^2 v / 2E.$$ (3.128)

Equation (3.128) is undoubtedly an overestimate of the maximum amount of energy which a perfectly plastic beam can absorb in a wholly elastic manner because local plastic deformations would occur at smaller values of elastic strain energy.

The initial kinetic energy imparted to a volume v of a beam which is subjected to a uniformly distributed impulsive velocity V_0 is†

$$K_e = \rho v V_0^2 / 2.$$ (3.129)

The energy ratio now becomes

$$E_r = \rho E V_0^2 / \sigma_0^2 \gg 1$$ (3.130)

according to equation (3.126).

The initial kinetic energy imparted to a beam which is struck by a mass G travelling with an initial velocity V_0, as shown, for example, in Figures 3.15(a) and 3.18(a), is

$$K_e = G V_0^2 / 2.$$ (3.131)

Thus, equations (3.126), (3.128) and (3.131) predict

$$E_r = E G V_0^2 / \sigma_0^2 2LBH \gg 1$$ (3.132)

and

$$E_r = E G V_0^2 / \sigma_0^2 LBH \gg 1$$ (3.133)

since $v = 2LBH$ and $v = LBH$ when the fully clamped and cantilever beams in Figures 3.15(a) and 3.18(a), respectively, have solid rectangular cross-sections.

Symonds[3.13] has examined the dynamic behaviour of impulsively loaded simple one-degree-of-freedom, spring–mass models with elastic, perfectly plastic or rigid, perfectly plastic springs. He found that elastic effects may be disregarded when the initial kinetic energy is much larger than the maximum possible strain energy which can be absorbed elastically (i.e., large energy ratios). The rigid, perfectly plastic model overestimates the permanent displacement by less than 10 per cent when the energy ratio is larger than 10, approximately. However, if a pressure pulse is maintained constant for a duration τ, then released, as shown in Figure 3.5, it is

† $m = \rho \times$ cross-sectional area of beam (BH), where ρ is density of material and B and H are the respective breadth and depth of a beam with a rectangular cross-section.

observed that the difference between the predictions of these two models is a function of τ/T, where T is the fundamental elastic period. As the ratio τ/T increases, the accuracy of a rigid, perfectly plastic analysis deteriorates† when compared to the predictions of an elastic, perfectly plastic analysis. Thus, when $\tau/T = 1/2\pi$, the energy ratio must be larger than about 20 in order to achieve an error of less than 10 per cent. However, if $\tau/T = 0.01$, then the error is less than 10 per cent when the energy ratio is larger than eight.

It appears, therefore, that elastic effects may be disregarded when the energy ratio is larger than about 10, provided the duration of a pressure pulse is sufficiently short compared with the corresponding natural elastic period, or

$$\tau/T \ll 1. \tag{3.134}$$

The fundamental period of elastic vibration (T) is

$$T = \frac{8L^2}{\pi}\left(\frac{m}{EI}\right)^{1/2} \tag{3.135}$$

and

$$T = \frac{8\pi L^2}{(4.73)^2}\left(\frac{m}{EI}\right)^{1/2} \tag{3.136}$$

for uniform beams which are simply supported and fully clamped, respectively.

These observations on the behaviour of a one-dimensional model and those discussed earlier with regard to Bohnenblust's analysis for an elastic-plastic beam, provide an estimate of the probable accuracy and range of validity‡ of rigid-plastic solutions for other structural problems.

Symonds[3.5] has employed the rigid-plastic procedure in order to examine the response of beams subjected to triangular or exponentially decaying pressure pulses. Some results are also presented by Symonds[3.5] for simply supported and fully clamped beams which are acted on by a lateral concentrated impact load at the mid-span. A number of other articles have been written on the theoretical behaviour of rigid-plastic beams subjected to various dynamic loads. Some of these studies are reviewed by Symonds,[3.13] Goldsmith[3.16] and Johnson,[3.17] while more recent work on beams is reported, for example, in references 3.10 and 3.18 to 3.24.

In this chapter the theoretical predictions were developed for beams with a fully plastic bending moment M_0 or M_p. The analyses are, therefore, valid for beams

---

† Symonds and Frye[3.14] have examined recently various dynamic pulse loadings having a range of τ/T values up to $\tau/T \simeq 10$. They find that the two predictions, for a pressure pulse with a finite rise time, are very close at regular intervals, while the rigid-plastic prediction has negative (unconservative) errors in between, relative to the elastic-plastic solution.

‡ It has been reported by Bodner and Symonds[3.15] and Florence and Firth[3.9] that elementary rigid-plastic procedures provide reasonable agreement with experimental impulsive loading tests on cantilever and unrestrained beams with energy ratios as small as three.

with a cross-section having any shape which is symmetric about the vertical plane containing w and x in Figure 3.3. However, dynamic loadings may cause distortion or buckling of certain cross-sections. This phenomenon is ignored here, since the original cross-sectional shape of a beam is maintained. Wegener and Martin[3.25] have examined the distortion of a simply supported square tube subjected to a uniformly distributed impulsive velocity. They find that the local distortion of the cross-section was almost finished before the global deformations commenced. This observation allowed the uncoupling of these two phenomena and the development of a relatively simple numerical scheme.

PROBLEMS

3.1 Use equation (3.43) to calculate the variation of speed of the plastic bending hinge during the second phase of motion (Figure 3.7(b)) for the rigid, perfectly plastic beam in Figure 3.4(a). Assume that the beam is simply supported across a span of 200 mm and is subjected to a uniformly distributed pressure pulse (Figure 3.5) with a dimensionless magnitude $\eta = 10$ and a duration $\tau = 100$ μs.

Discuss your calculations and contrast them with the longitudinal elastic wave speeds $((E/\rho)^{1/2})$ of 5150 m/s and 5100 m/s for mild steel and aluminium, respectively.

3.2 Use the results in § 3.4 to obtain an expression for the permanent transverse displacement profile of the simply supported beam in Figure 3.4(a).

3.3 The fully clamped beam in Figure 3.14(a) is made from a rigid, perfectly plastic material and is subjected to a uniformly distributed pressure pulse having the rectangular-shaped pressure–time characteristics in Figure 3.5. Obtain the maximum permanent transverse displacement when $\bar{\eta} \leqslant 3$, where $\bar{\eta}$ is defined by equation (3.76b).

Show that your theoretical solution is statically and kinematically admissible.

3.4 Repeat Problem 3.3 for the case when $\bar{\eta} \geqslant 3$.

3.5 Obtain the transverse displacement which is acquired during the second phase of motion for the fully clamped beam illustrated in Figure 3.15(a). Use the equations of equilibrium rather than the energy balance in § 3.8.3.

3.6 Prove that the maximum permanent transverse displacement of a fully clamped beam struck by a heavy mass (i.e., equation (3.105) with $x = 0$) is obtained from equation (3.102), with $\bar{\alpha} \to 0$ and $\beta \to 0$, where $\bar{\alpha}$ and β are defined by equations (3.103) and (3.96b), respectively.

3.7 The tip of an initially straight beam, which is built-in at one end and free at the other, is hit by a mass G travelling with an initial velocity V_0. If the beam of total length L is made from a rigid, perfectly plastic material with a plastic limit moment M_0, prove that the final deflection w at any distance x from the tip is

$$w = \frac{GV_0^2 L}{2M_0}\left(1 - \frac{x}{L}\right),$$

when the mass G is very much larger than the mass of the beam.

3.8 (a) A straight bar of mass M_1 travels with a uniform velocity V_0. The ends of the bar then hit a ballistic pendulum which has a mass M_2 and is suspended on wires of length l from the ceiling. The beam travels in the same plane as the centre of gravity of the ballistic pendulum and hits both supports simultaneously. Neglecting any rebound and using conservation principles, show that the maximum angular displacement of the pendulum θ_m is

$$\theta_m = 2\sin^{-1}\left\{\frac{V_0 M_1}{2M_2\sqrt{gl}}\right\},$$

when $M_1 \ll M_2$ and the bar and pendulum are considered to be rigid bodies.

(b) A rigid, perfectly plastic beam of length $2L$ and mass m per unit length travels with a velocity V_0 which is measured by means of a ballistic pendulum. Prove that the maximum permanent transverse deflection of the beam is

$$\frac{M_2^2 gl}{3mM_0}\sin^2\left(\frac{\theta_m}{2}\right)$$

when it is simply supported, and M_0 is the plastic limit moment for the beam cross-section.

3.9 A series of dynamic tests were conducted on cantilever beams in connection with the shock design of some shipboard equipment. The cantilever beams of length L, thickness H and breadth B could be considered to be fixed spatially at one end and free at the other. A mass G was attached securely to the free end and subjected to a pressure pulse which may be taken as equivalent to an impulsive velocity of magnitude V_0.

The cantilever beam as a composite cross-section with thin upper and lower sheets of thickness t ($t \ll H$), enclosing a core which supports only transverse shear forces and no bending moments. The two thin sheets are made from a ductile material which is perfectly plastic and flows at a uniaxial tensile stress σ_0.

Neglect the influence of any material elastic effects and (a) estimate the maximum transverse displacement when the mass of the cantilever beam is negligible compared with the tip mass G; (b) estimate the maximum strain and strain rate in the cantilever beam when it is observed that the deformation is largely concentrated within a region of length $2H$ at the fixed end; (c) estimate the response duration.

4

Dynamic plastic behaviour of plates

4.1 Introduction

The general method which was presented in the previous chapter for examining the dynamic plastic behaviour of beams is developed further in this chapter to study the dynamic plastic response of plates. It is assumed that an initially flat plate is made from a rigid, perfectly plastic material and is subjected to an external dynamic load which produces a permanent inelastic deformation, as illustrated in Figure 4.1 for the particular case of a rectangular plate.

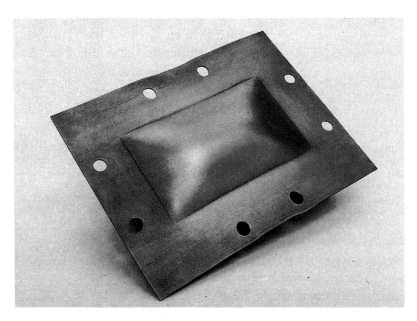

Figure 4.1 Permanent deformed profile of an initially flat fully clamped rectangular plate subjected to an impulsive velocity loading.[4.1]

A kinematically admissible transverse velocity field is selected as the first step in the analyses for beams which were presented in the previous chapter. This field must be capable of satisfying the initial displacement and velocity requirements and boundary conditions for a beam (e.g., equation (3.7) in § 3.3.2). The change in curvature (κ) of a beam was then calculated using equation (3.3) in § 3.2. A region of a beam with $\dot{\kappa} = 0$ is rigid. Otherwise, a plastic hinge forms with $M = M_0$ when $\dot{\kappa} > 0$ and $M = -M_0$ when $\dot{\kappa} < 0$. It is then possible to integrate the equilibrium equation (3.2) and evaluate the constants of integration from the symmetry requirements and the boundary conditions for the bending moment (M) and the transverse shear force (Q).

Once a theoretical solution has been found it is then necessary to examine whether or not it is statically admissible. This is undertaken by exploring whether the yield condition is satisfied within the interior of a beam (i.e., $-M_0 \leqslant M \leqslant M_0$). An enquiry sometimes reveals restrictions in the validity of the theoretical solution, as found, for example, by equation (3.21) in § 3.3.4, which gives a limitation on the magnitude of a dynamic pressure acting on a simply supported beam.

The general theoretical procedure described above for the response of beams is also used in this chapter to examine the dynamic plastic behaviour of circular and square plates.

The governing equations for circular plates are presented in the following section and are used to study a simply supported annular plate subjected to a linearly distributed rectangular pressure pulse in § 4.3. A simply supported circular plate is examined in §§ 4.4 and 4.5 and the impulsively loaded, fully clamped case in § 4.6. The governing equations for rectangular plates are developed in § 4.7 and the particular solution of a simply supported square plate subjected to a rectangular pressure pulse is studied in §§ 4.8 and 4.9. The chapter ends with some final remarks on other work in § 4.10.

4.2 Governing equations for circular plates

The dynamic axisymmetric behaviour of the element of a circular plate, which is drawn in Figure 4.2, is governed by the equations

$$\partial(rM_r)/\partial r - M_\theta - rQ_r = 0, \tag{4.1}$$

$$\partial(rQ_r)/\partial r + rp - \mu r\, \partial^2 w/\partial t^2 = 0, \tag{4.2}$$

$$\kappa_r = -\partial^2 w/\partial r^2 \tag{4.3}$$

and

$$\kappa_\theta = -(\partial w/\partial r)/r, \tag{4.4}$$

which are identical to equations (2.4) to (2.7) except for the inclusion of the transverse inertia term in the lateral equilibrium equation (4.2). (μ is mass per unit area and t is time.) Equations (4.1) and (4.2) govern the behaviour of plates which

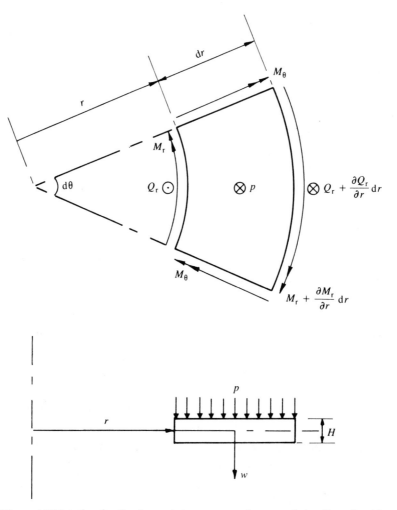

Figure 4.2 Notation for the dynamic transverse axisymmetric loading of a thin circular plate.

undergo infinitesimal displacements. However, the influence of finite lateral displacements, or geometry changes, is explored in Chapter 7. Moreover, the influence of rotatory inertia is not retained in the moment equilibrium equation (4.1) but its influence is discussed in Chapter 6.

It is assumed that a circular plate is made from a rigid, perfectly plastic material, the plastic flow of which is controlled by the Tresca yield criterion which is drawn in $M_r - M_\theta$ space in Figure 4.3. The transverse shear force Q_r is taken as a reaction and is assumed not to influence plastic yielding. However, the effect of the transverse shear force on the plastic flow of the material is examined later in Chapter 6.

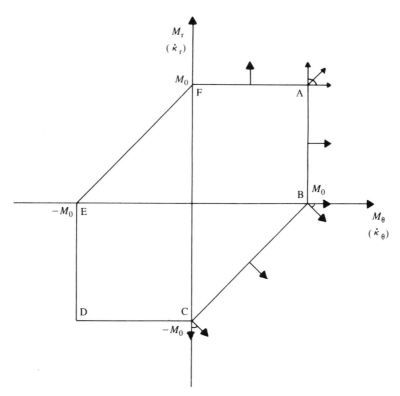

Figure 4.3 Tresca yield criterion for a rigid, perfectly plastic circular plate with plastic yielding controlled by radial and circumferential bending moments. The arrows indicate the normality requirements of plasticity for the associated radial and circumferential curvature rates.

The general procedure which is employed in order to analyse the dynamic plastic behaviour of circular plates is similar to that outlined earlier for beams.

4.3 Annular plate loaded dynamically

4.3.1 Introduction

The general theoretical procedure for the dynamic plastic response of plates, which is described in §§ 4.1 and 4.2, is used to study the simple annular plate problem shown in Figure 4.4(a). The plate is simply supported around the outer edge and is made from a rigid, perfectly plastic material which obeys the Tresca yield criterion in $M_r - M_\theta$ space shown in Figure 4.3. It is subjected to the rectangular pressure pulse illustrated in Figure 4.5, which may be written

$$p = p_0(R - r)/(R - a), \qquad a \leqslant r \leqslant R, \tag{4.5}$$

when $0 \leqslant t \leqslant \tau$, and

$$p = 0, \qquad a \leqslant r \leqslant R, \tag{4.6}$$

for $t \geqslant \tau$.

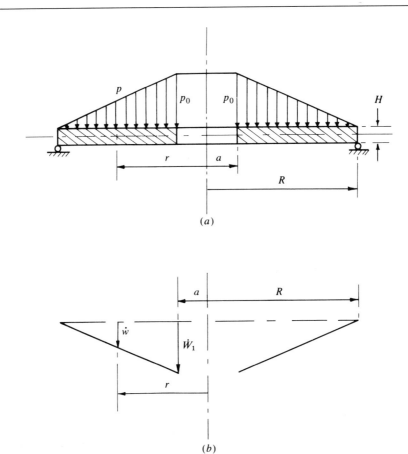

Figure 4.4 (a) Simply supported annular plate subjected to an axisymmetric pressure pulse with a peak value p_0 at the inner boundary $(r = a)$. (b) Transverse velocity profile which is associated with the static collapse pressure p_c given by equation (4.7) for the annular plate loaded as shown in (a).

Now, it may be shown, when using the theoretical methods in Chapter 2,† that the annular plate in Figure 4.4(a) collapses when the magnitude p_0 of the linearly distributed static pressure described by equation (4.5) with $\tau \to \infty$ reaches

$$p_c = 12M_0/\{R^2(1 - \alpha)(1 + 3\alpha)\}, \tag{4.7}$$

where

$$\alpha = a/R \tag{4.8}$$

and M_0 is the static fully plastic bending moment (per unit length) for the plate cross-section. The transverse velocity profile associated with this collapse pressure is illustrated in Figure 4.4(b).

† See Problem 2.4.

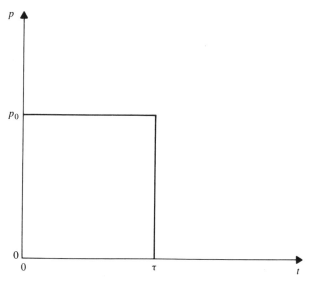

Figure 4.5 Rectangular pressure pulse.

It turns out that the dynamic plastic response of the annular plate consists of two phases of motion. These phases correspond with the durations of the external pressure pulses described by equations (4.5) and (4.6) which are valid for $0 \leqslant t \leqslant \tau$ and $\tau \leqslant t \leqslant T$, respectively, where T is the response duration.

4.3.2 First phase of motion, $0 \leqslant t \leqslant \tau$

It is necessary to select a kinematically admissible transverse velocity profile as the first step in the theoretical solution for the dynamic case. Thus, it is postulated that

$$\dot{w} = \dot{W}_1 (R - r)/(R - a), \qquad a \leqslant r \leqslant R, \tag{4.9}$$

which has the same form as the corresponding static plastic collapse profile in Figure 4.4(b), where $(\dot{\ }) = \partial(\ )/\partial t$.

Equations (4.3), (4.4) and (4.9) predict that the curvature rates $\dot{\kappa}_r = 0$ and $\dot{\kappa}_\theta = \dot{W}_1/\{r(R - a)\} \geqslant 0$. The normality requirement of plasticity (Drucker's postulate), therefore, suggests that the associated generalised stress field lies on the side AB of the Tresca yield criterion in Figure 4.3, where

$$M_\theta = M_0, \qquad 0 \leqslant M_r \leqslant M_0. \tag{4.10a, b}$$

Equation (4.1), with the aid of equation (4.10a), predicts the transverse shear force

$$rQ_r = \partial(rM_r)/\partial r - M_0, \tag{4.11}$$

which may be substituted into equation (4.2), together with equations (4.5) and (4.9), to give the governing equation

$$\partial^2(rM_r)/\partial r^2 = r(\mu \ddot{W}_1 - p_0)(R - r)/(R - a). \tag{4.12}$$

Now, integrating equation (4.12) twice with respect to r gives

$$M_r = (\mu \ddot{W}_1 - p_0)r^2(2R - r)/\{12(R - a)\} + A + B/r, \qquad (4.13)$$

where the constants† of integration

$$A = (\mu \ddot{W}_1 - p_0)R^2(\alpha^3 - \alpha^2 - \alpha - 1)/\{12(1 - \alpha)\} \qquad (4.14a)$$

and

$$B = (\mu \ddot{W}_1 - p_0)R^3\alpha(1 + \alpha - \alpha^2)/\{12(1 - \alpha)\} \qquad (4.14b)$$

are evaluated from the boundary conditions $M_r = 0$ at both $r = a$ and $r = R$. In addition, equation (4.13) must also satisfy the requirement that the transverse shear force $Q_r = 0$ at $r = a$ on the unloaded inner boundary of the annular plate. Thus, equations (4.11) and (4.13) predict that

$$(\mu \ddot{W}_1 - p_0)(Ra/2 - a^2/3)/(R - a) + A/a - M_0/a = 0,$$

which, using equation (4.14a), may be rearranged in the form

$$\mu \ddot{W}_1 = p_0 - p_c, \qquad (4.15)$$

where p_c is the peak value of the linearly distributed static plastic collapse pressure according to equation (4.7).

It is evident from equation (4.15) that the plate accelerates throughout the first phase of motion when $p_0 > p_c$. Moreover, the static solution with $\ddot{W}_1 = 0$ is recovered when $p_0 = p_c$ as expected.

Now, integrating equation (4.15) twice with respect to time gives

$$\mu W_1 = (p_0 - p_c)t^2/2, \qquad 0 \leqslant t \leqslant \tau, \qquad (4.16)$$

where the two contants of integration are zero because $W_1 = \dot{W}_1 = 0$ when motion commences at $t = 0$. This phase of motion is completed at $t = \tau$, according to equation (4.5), and Figure 4.5, when, according to equation (4.16), the annular plate has an associated peak transverse displacement

$$W_1 = (p_0 - p_c)\tau^2/2\mu \qquad (4.17a)$$

and an associated maximum transverse velocity

$$\dot{W}_1 = (p_0 - p_c)\tau/\mu. \qquad (4.17b)$$

It is evident from equation (4.17b) that the inner edge ($r = a$) of the annular plate in Figure 4.4(a) has a finite transverse velocity when the rectangular pressure pulse is removed at $t = \tau$. Thus, the annular plate has a total kinetic energy

$$K_e = \int_a^R \left\{ \frac{(p_0 - p_c)(R - r)\tau}{\mu(R - a)} \right\}^2 \mu\pi r \, dr$$

according to equations (4.9) and (4.17b), or

$$K_e = \pi(p_0 - p_c)^2 R^2 \tau^2 (1 - \alpha)(1 + 3\alpha)/12\mu, \qquad (4.18)$$

which remains to be dissipated during a second phase of motion.

† $\mu \ddot{W}_1 - p_0 = -p_c$ according to equation (4.15).

4.3.3 Second phase of motion, $\tau \leqslant t \leqslant T$

Now, it is again assumed that the transverse velocity profile for the annular plate has the form described by equation (4.9) and shown in Figure 4.4(b), or

$$\dot{w} = \dot{W}_2(R - r)/(R - a), \qquad a \leqslant r \leqslant R. \tag{4.19}$$

Thus, equations (4.3), (4.4) and (4.19) again predict that $\dot{\kappa}_r = 0$ and $\dot{\kappa}_\theta \geqslant 0$, so that the generalised stresses lie on the side AB of the yield criterion in Figure 4.3 and, therefore, equations (4.10a, b) remain valid.

It is evident that equations (4.6), (4.10a) and (4.19), when substituted into equations (4.1) and (4.2), lead to equation (4.12) but with $p_0 = 0$ and W_1 replaced by W_2. Equations (4.13) and (4.14), with $p_0 = 0$ and W_1 replaced by W_2, then follow, and equation (4.15) is replaced by

$$\mu \ddot{W}_2 = -p_c. \tag{4.20}†$$

Integrating equation (4.20) with respect to time predicts the transverse velocity

$$\mu \dot{W}_2 = -p_c t + C, \tag{4.21a}$$

while another integration gives the transverse displacement

$$\mu W_2 = -p_c t^2/2 + Ct + D. \tag{4.21b}$$

The transverse displacement and velocity of an annular plate at the start of the second phase of motion must be continuous with the corresponding values at the end of the first phase of motion, when $t = \tau$. Thus, equations (4.17) and (4.21) lead to

$$C = p_0 \tau \tag{4.22a}$$

and

$$D = -p_0 \tau^2/2. \tag{4.22b}$$

Motion of an annular plate ceases and reaches its final, or permanent, position when $\dot{W}_2 = 0$, which occurs when

$$T = \eta \tau, \tag{4.23}$$

according to equations (4.21a) and (4.22a), where

$$\eta = p_0/p_c \tag{4.24}$$

is a dynamic load factor. Equations (4.21b), (4.22) and (4.23) give the permanent transverse displacement profile

$$w_f = 6M_0\eta(\eta - 1)\tau^2(1 - r/R)/\{\mu R^2(1 - \alpha)^2(1 + 3\alpha)\}, \tag{4.25}$$

when using equations (4.7) and (4.19).

It is left as an exercise to the reader to show that the plastic energy absorbed by the plate during this phase of motion, i.e.,

$$\int_\tau^T \int_a^R (M_\theta \dot{\kappa}_\theta + M_r \dot{\kappa}_r)2\pi r \, dr \, dt,$$

---

† The plate decelerates throughout the unloaded second phase of motion.

equals the kinetic energy remaining in the plate at the end of the first phase of motion.

4.3.4 Static admissibility

The theoretical analysis presented in §§ 4.3.2 and 4.3.3 satisfies the equilibrium equations (4.1) and (4.2) for the annular plate problem illustrated in Figure 4.4(a). It satisfies the boundary conditions at the inner and outer edges of the plate, and $M_\theta = M_0$ for $a \leqslant r \leqslant R$, according to equation (4.10a) for both phases of motion. However, the normality requirement demands that the generalised stress field lies on the side AB of the yield criterion in Figure 4.3, with $0 \leqslant M_r \leqslant M_0$. Thus, in order to show that the theoretical solution is statically admissible, it is necessary to demonstrate that $0 \leqslant M_r \leqslant M_0$ for $a \leqslant r \leqslant R$.

It is straightforward to show that $\partial M_r/\partial r > 0$ and $\partial^2 M_r/\partial r^2 < 0$ at $r = a$, and $\partial M_r/\partial r < 0$ and $\partial^2 M_r/\partial r^2 > 0$ at $r = R$ for the radial bending moment distribution during both phases of motion, as indicated in Figure 4.6.† It is possible to demonstrate analytically after a considerable amount of algebra that $0 \leqslant M_r \leqslant M_0$.

Equations (4.13), (4.14a) and (4.14b) all contain the term $\mu\ddot{W}_1 - p_0$, which equals $-p_c$, during the first phase of motion, according to equation (4.15). Moreover, the same term, which is replaced by $\mu\ddot{W}_2$ during the second phase of motion, also equals $-p_c$, according to equation (4.20). Furthermore, the term $\mu\ddot{W}_1 - p_0$ reduces to $-p_c$ for static external pressures, with $\ddot{W}_1 = 0$ and $p_0 = p_c$. It is evident, therefore, that the radial bending moment distributions for the first and second phases of motion are identical, time-independent, and the same for a static pressure with the linearly varying distribution shown in Figure 4.4(a). In fact, it is also clear that the magnitude and distribution of the radial bending moment is independent of the dynamic pressure ratio η defined by equation (4.24) and the duration (τ) of the rectangular pressure pulse shown in Figure 4.5.

The radial bending moment distributions have been calculated for annular plates with $\alpha = 0.1$, 0.5 and 0.9, and are presented in Figure 4.6. It is evident that $0 \leqslant M_r \leqslant M_0$ and, therefore, the theoretical solution in §§ 4.3.2 and 4.3.3 is statically admissible, as noted above.

The transverse velocity fields (4.9) and (4.19) for the first and second phases of motion, respectively, contain neither stationary nor travelling plastic hinges. Moreover, the boundary conditions $w = \dot{w} = 0$ at $r = R$ are satisfied. Furthermore, the transverse velocity and the associated transverse displacement profiles satisfy the

---

† It is evident from the numerical calculations that $\partial^2 M_r/\partial r^2 > 0$ at $r = R$, though it is not readily apparent from the curves in Figure 4.6.

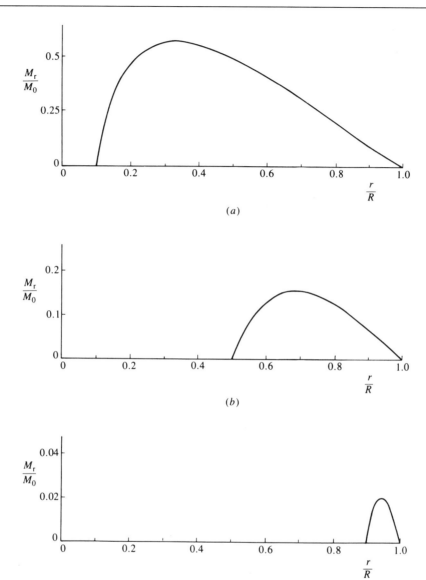

Figure 4.6 Radial bending moment distribution during the first and second phases of motion for the annular plate in Figure 4.4(a). (a) $\alpha = 0.1$. (b) $\alpha = 0.5$. (c) $\alpha = 0.9$.

initial conditions and continuity requirements between the two phases of motion. Thus, the theoretical analysis in §§ 4.3.2 and 4.3.3 is both kinematically and statically admissible and is, therefore, an exact rigid, perfectly plastic solution for the annular plate problem in Figure 4.4(a).

4.3.5 Impulsive loading

If the rectangular pressure pulse in Figure 4.5 has a large magnitude (i.e., $\eta = p_0/p_c \gg 1$) and a short duration ($\tau \to 0$), then it is often simpler to idealise it as an impulsive velocity loading, with the total impulse equalling the change in linear momentum at the instant motion commences. Thus,

$$\int_a^R p_0\tau 2\pi r\, dr(R-r)/(R-a) = \int_a^R \mu V_0 2\pi r\, dr(R-r)/(R-a),$$

or

$$p_0\tau = \mu V_0, \tag{4.26}$$

where V_0 is the magnitude of the linearly distributed initial impulsive velocity at $r = a$.

Now, equation (4.25), with $\eta \gg 1$, reduces to

$$w_f = 6M_0\eta^2\tau^2(1 - r/R)/\{\mu R^2(1 - \alpha)^2(1 + 3\alpha)\}, \tag{4.27}$$

or

$$w_f = \mu V_0^2 R^2(1 + 3\alpha)(1 - r/R)/24M_0 \tag{4.28}$$

when using $\eta\tau = \mu V_0/p_c$ from equations (4.24) and (4.26) and equation (4.7) for p_c.

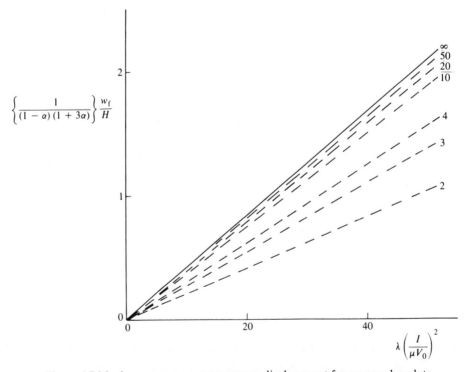

Figure 4.7 Maximum permanent transverse displacement for an annular plate loaded as shown in Figure 4.4(a). ———: impulsive loading, equation (4.29). ---: rectangular pressure pulse; equation (4.31), with η given by numbers on lines.

Equation (4.28) may be recast in the dimensionless form

$$w_f/H = \lambda(1 + 3\alpha)(1 - r/R)/24, \tag{4.29}$$

where

$$\lambda = \mu V_0^2 R^2/M_0 H \tag{4.30}$$

is a non-dimensional initial kinetic energy and H is the plate thickness. Equation (4.25) may also be written in the form

$$w_f/H = \lambda(I/\mu V_0)^2(1 - 1/\eta)(1 + 3\alpha)(1 - r/R)/24, \tag{4.31}$$

where

$$I = p_0\tau, \tag{4.32}$$

and reduces to equation (4.29) when using equation (4.26) and $\eta \to \infty$.

It is evident from the theoretical results in Figure 4.7 that the theoretical predictions of equation (4.31), with $\eta > 50$, are only slightly smaller than the impulsive loading case with $\eta \to \infty$ according to equation (4.29).

4.4 Simply supported circular plate loaded dynamically, $p_c \leqslant p_0 \leqslant 2p_c$

4.4.1 Introduction

A study is made in this section of the dynamic plastic response of the circular plate in Figure 4.8(a) which is simply supported around the outer boundary and subjected to a uniformly distributed dynamic pressure pulse with the rectangular pressure–time history shown in Figure 4.5. Hopkins and Prager[4.2] examined this particular problem in 1954 and used the general theoretical procedure which is outlined in the previous section for an annular plate made from a rigid, perfectly plastic material.

The exact static collapse pressure for a uniformly loaded, simply supported circular plate is

$$p_c = 6M_0/R^2 \tag{4.33}$$

according to equation (2.18). The associated static collapse profile has the form shown in Figure 2.7(b) and Figure 4.8(b). A circular plate accelerates if it is subjected to pressures larger than p_c, but eventually absorbs the external dynamic energy and reaches an equilibrium position (i.e., permanent deformed shape) if the pressure pulse is removed after a short time.

It transpires that, for pressure pulses lying within the range $p_c \leqslant p_0 \leqslant 2p_c$, the dynamic response may be divided, for convenience, into the two separate phases of motion $0 \leqslant t \leqslant \tau$ and $\tau \leqslant t \leqslant T$, where τ is the pulse duration and T is the duration of response.

4.4.2 First phase of motion, $0 \leqslant t \leqslant \tau$

A circular plate is subjected, during this phase of motion, to a uniformly distributed pressure

$$p = p_0, \qquad 0 \leqslant r \leqslant R, \tag{4.34}$$

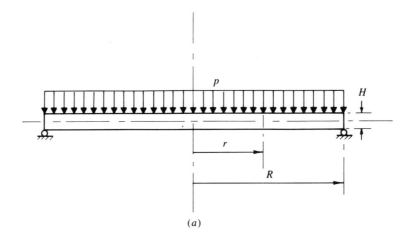

(a)

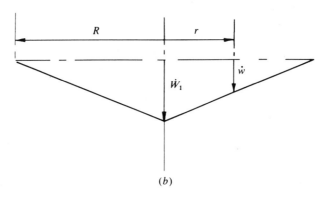

(b)

Figure 4.8 (a) Simply supported circular plate subjected to an axisymmetric pressure pulse distributed uniformly over the plate. (b) Transverse velocity profile which is associated with the static collapse pressure p_c given by equation (4.33) for the circular plate loaded as shown in (a).

which is assumed to be larger than the corresponding static collapse pressure (p_c) according to equation (4.33).

It appears reasonable to postulate that a plate deforms with the velocity field sketched in Figure 4.8(b) which, as noted in § 4.4.1, has the same general characteristics as the corresponding static collapse profile. Thus,

$$\dot{w} = \dot{W}_1(1 - r/R), \qquad 0 \leqslant r \leqslant R, \tag{4.35}$$

where $\dot{W}_1$ is the transverse velocity at the centre of a plate.

Now, equations (4.3), (4.4) and (4.35) predict that $\dot{\kappa}_r = 0$ and $\dot{\kappa}_\theta = \dot{W}_1/rR \geqslant 0$ which are associated with the side AB of the yield condition in Figure 4.3 according to the normality requirements of plasticity. Thus,

$$M_\theta = M_0, \qquad 0 \leqslant M_r \leqslant M_0. \tag{4.36a, b}$$

Equations (4.1) and (4.36a) again give equation (4.11) for the transverse shear force which may be substituted with equations (4.34) and (4.35) into equation (4.2) for the governing equation

$$\partial^2(rM_r)/\partial r^2 = -rp_0 + \mu r\ddot{W}_1(1 - r/R). \tag{4.37}$$

Integrating twice with respect to r gives the radial bending moment

$$M_r = -p_0 r^2/6 + \mu\ddot{W}_1(r^2/6 - r^3/12R) + M_0 \tag{4.38}$$

when eliminating the constants of integration with the aid of the boundary conditions $M_r = M_0$ and $Q_r = 0$ at the plate centre ($r = 0$). However, equation (4.38) must satisfy the simply supported boundary condition ($M_r = 0$ at $r = R$) which, therefore, requires

$$\mu\ddot{W}_1 = 2p_0 - 12M_0/R^2, \tag{4.39}\dagger$$

or

$$W_1 = (p_0 - p_c)t^2/\mu \tag{4.40}$$

when integrating twice with respect to time and satisfying the initial conditions $W_1 = \dot{W}_1 = 0$ at $t = 0$ and where p_c is defined by equation (4.33).

The transverse displacement and velocity at the end of the first phase of motion are

$$w = (p_0 - p_c)\tau^2(1 - r/R)/\mu \tag{4.41a}$$

and

$$\dot{w} = 2(p_0 - p_c)\tau(1 - r/R)/\mu, \tag{4.41b}$$

respectively. Equation (4.41b) leads to a kinetic energy

$$K_e = \pi R^2(p_0 - p_c)^2\tau^2/3\mu. \tag{4.42}$$

which is dissipated plastically during a second phase of motion.

4.4.3 Second phase of motion, $\tau \leqslant t \leqslant T$

The circular plate is unloaded during this phase of motion and, therefore,

$$p = 0, \qquad 0 \leqslant r \leqslant R, \tag{4.43}$$

but plastic deformation continues in order to dissipate the kinetic energy present in the plate at $t = \tau$ according to equation (4.42). It is assumed that the transverse velocity profile during this phase of motion has the same form as equation (4.35) for the first phase, or

$$\dot{w} = \dot{W}_2(1 - r/R), \qquad 0 \leqslant r \leqslant R. \tag{4.44}$$

Equations (4.3), (4.4) and (4.44) predict that $\dot{\kappa}_r = 0$ and $\dot{\kappa}_\theta \geqslant 0$ and, therefore, again lead to equations (4.36a, b) through the normality requirement of plasticity for the yield condition in Figure 4.3.

---

† The acceleration of the plate is constant during the first phase of motion since p_0 is constant for $0 \leqslant t \leqslant \tau$, as shown in Figure 4.5. If $\ddot{W}_1 = 0$ for static pressures, then $p_0 = p_c$, where p_c is given by equation (4.33).

Now, it may be shown, when substituting equations (4.11), (4.36a), (4.43) and (4.44) into equation (4.2), that equation (4.37) is again obtained, except that $\ddot{W}_1$ is replaced by $\ddot{W}_2$ and $p_0 = 0$. Thus, integrating twice with respect to r and satisfying the boundary conditions $M_r = M_0$ and $Q_r = 0$ at $r = 0$ and $M_r = 0$ at $r = R$ leads to

$$\mu\ddot{W}_2 = -12M_0/R^2, \tag{4.45}†$$

which may be obtained from equation (4.39) with the substitutions noted above.

A straightforward integration of equation (4.45) with respect to time predicts the transverse displacement at the plate centre

$$W_2 = p_0\tau(2t - \tau)/\mu - p_c t^2/\mu \tag{4.46}$$

when eliminating the two constants of integration by ensuring continuity with the transverse displacement and velocity at the end of the first phase of motion ($t = \tau$). Equation (4.46) gives $\dot{W}_2 = 0$ when $t = T$, where

$$T = \eta\tau \tag{4.47}$$

is the total duration of response and η is defined by equation (4.24) when p_c is defined by equation (4.33). The associated permanently deformed transverse profile is

$$w_f = \eta(\eta - 1)p_c\tau^2(1 - r/R)/\mu \tag{4.48}$$

according to equations (4.44), (4.46) and (4.47).

4.4.4 Static admissibility

The tangential bending moment M_θ equals the fully plastic value M_0 throughout both phases of motion in §§ 4.4.2 and 4.4.3 according to the normality requirements of plasticity and equation (4.36a). However, it is necessary to demonstrate that $0 \leqslant M_r \leqslant M_0$ in order to satisfy inequality (4.36b).

Now, equation (4.38) for the radial bending moment during the first phase of motion may be written

$$M_r/M_0 = 1 + (\eta - 2)r^2/R^2 - (\eta - 1)r^3/R^3, \qquad 0 \leqslant r \leqslant R. \tag{4.49}$$

when using equations (4.24), (4.33) and (4.39) and is plotted in Figure 4.9(a). Thus, $\partial M_r/\partial r = 0$ at $r = 0$ and

$$\partial^2 M_r/\partial r^2 = 2(\eta - 2)M_0/R^2 \tag{4.50}$$

at $r = 0$. In order to avoid a yield violation near $r = 0$ in the bending moment distribution in Figure 4.9(a), where $M_r = M_0$, it is necessary to ensure that $\partial^2 M_r/\partial r^2 \leqslant 0$. However, equation (4.50) shows that a yield violation develops near the plate centre when $\eta > 2$. Therefore, the bending moment distribution according to equation (4.49) is statically admissible during the first phase of motion when $\eta \leqslant 2$, in which circumstance $\partial M_r/\partial r = 0$ at $r = 0$ and $\partial^2 M_r/\partial r^2 \leqslant 0$ for $0 \leqslant r \leqslant R$.

---

† The plate has a constant deceleration throughout the second phase.

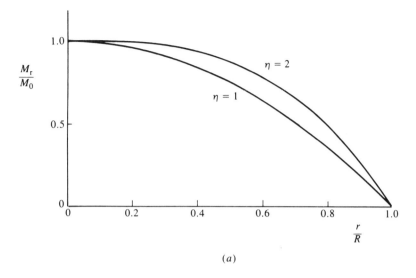

(a)

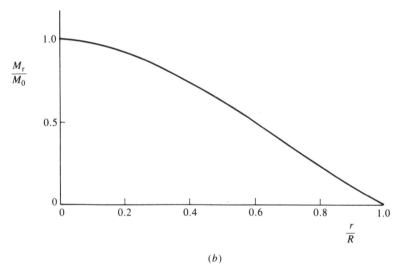

(b)

Figure 4.9 Radial bending moment distributions for the circular plate in Figure 4.8(a) with $\eta \leqslant 2$. (a) First phase of motion for $\eta = 1$ and $\eta = 2$. (b) Second phase of motion for $1 \leqslant \eta \leqslant 2$.

The radial bending moment distribution for the second phase of motion from equation (4.38), with $p_0 = 0$, and $\ddot{W}_1$ replaced by $\ddot{W}_2$ from equation (4.45), is

$$M_r/M_0 = 1 - (2 - r/R)r^2/R^2, \qquad 0 \leqslant r \leqslant R, \tag{4.51}$$

and is plotted in Figure 4.9(b). This bending moment distribution is independent of η and time and is statically admissible, since the only maximum occurs at $r = 0$ (i.e., $\partial M_r/\partial r = 0$ at $r = 0$) and $\partial^2 M_r/\partial r^2 < 0$ at $r = 0$.

The theoretical analysis in §§ 4.4.2 and 4.4.3 is statically admissible provided $1 \leqslant \eta \leqslant 2$, while the associated transverse velocity fields (4.35) and (4.44) are kinematically admissible. Thus, the solution is exact throughout the entire response of a rigid, perfectly plastic circular plate. An analysis is presented in the next section for circular plates having rectangular pressure pulses with $\eta > 2$.

4.5 Simply supported circular plate loaded dynamically, $p_0 \geqslant 2p_c$

4.5.1 Introduction

It was observed in the previous section that equation (4.50) predicts $\partial^2 M_r/\partial r^2 > 0$ at $r = 0$ during the first phase of motion when $\eta > 2$. This leads to a yield violation near the plate centre, since $M_r = M_0$ and $\partial M_r/\partial r = 0$ at $r = 0$. In order to overcome this difficulty for larger pressure pulses, it is assumed in this section that the yielding of a central area of the plate is controlled by the corner A ($M_r = M_\theta = M_0$) of the yield condition in Figure 4.3. Thus, a circular plate is divided into two concentric regions, with the inner region $0 \leqslant r \leqslant \xi_0$ governed by the corner A of the yield condition in Figure 4.3, while plastic flow in the outer annular zone ($\xi_0 \leqslant r \leqslant R$) is controlled by the side AB.

Hopkins and Prager[4.2] found it necessary to consider three stages of motion. A stationary circular plastic hinge develops at a radius ξ_0 during the first stage of motion of duration τ. At $t = \tau$, when the pressure pulse is removed, this plastic hinge starts to move inwards towards the centre of the plate. The final stage of motion is similar to the second stage of motion for the low pressure case ($1 \leqslant \eta \leqslant 2$) which was examined in § 4.4.3.†

4.5.2 First phase of motion, $0 \leqslant t \leqslant \tau$

It is assumed that a stationary circular plastic hinge develops in a circular plate at $r = \xi_0$ while the pressure pulse in Figure 4.5 is active. The associated transverse velocity field is shown in Figure 4.10(a) and may be written in the form

$$\dot{w} = \dot{W}_1, \qquad 0 \leqslant r \leqslant \xi_0, \tag{4.52}$$

and

$$\dot{w} = \dot{W}_1(R - r)/(R - \xi_0), \qquad \xi_0 \leqslant r \leqslant R. \tag{4.53}$$

Equations (4.3) and (4.4), therefore, predict that $\dot{\kappa}_r = \dot{\kappa}_\theta = 0$ in the inner circular zone $0 \leqslant r \leqslant \xi_0$, while $\dot{\kappa}_r = 0$ and $\dot{\kappa}_\theta \geqslant 0$ in the outer annular zone $\xi_0 \leqslant r \leqslant R$. Moreover, it is evident from Figure 4.10(a) that $\dot{\kappa}_r \to \infty$ and $\dot{\kappa}_\theta \geqslant 0$ at the plastic hinge ($r = \xi_0$). These curvature rates are consistent with the regions A and AB of

---

† This sequence of events is similar to that described in § 3.4.1 for a simply supported beam subjected to a rectangular pressure pulse.

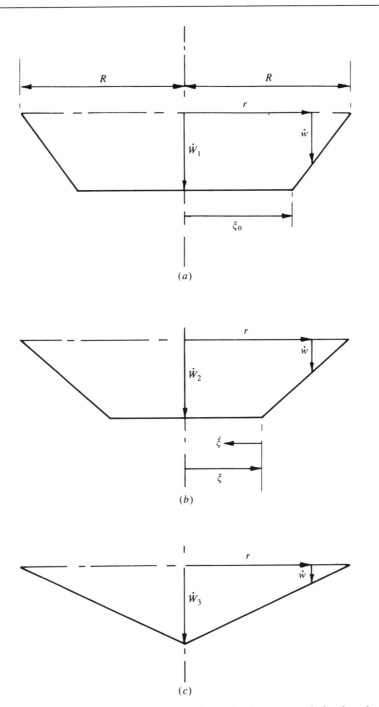

Figure 4.10 Transverse velocity profiles for a simply supported circular plate subjected to a rectangular pressure pulse with $\eta \geqslant 2$. (a) First phase of motion, $0 \leqslant t \leqslant \tau$. (b) Second phase of motion, $\tau \leqslant t \leqslant T_1$. (c) Third phase of motion, $T_1 \leqslant t \leqslant T$.

the Tresca yield condition in Figure 4.3 according to the normality requirements of plasticity. Thus,

$$M_\theta = M_0, \qquad M_r = M_0, \qquad 0 \leqslant r \leqslant \xi_0, \qquad (4.54a, b)$$

and

$$M_\theta = M_0, \qquad 0 \leqslant M_r \leqslant M_0, \qquad \text{when } \xi_0 \leqslant r \leqslant R. \qquad (4.55a, b)$$

Now, consider the central circular zone $0 \leqslant r \leqslant \xi_0$ which is governed by

$$\partial^2(rM_r)/\partial r^2 = -rp_0 + \mu r \ddot{W}_1 \qquad (4.56)$$

according to equations (4.1), (4.2), (4.34), (4.52) and (4.54a). Integrating twice with respect to r gives

$$M_r = (\mu \ddot{W}_1 - p_0)r^2/6 + M_0, \qquad (4.57)$$

where the two constants of integration have been evaluated from the requirements that M_r remains finite at $r = 0$ and $Q_r = 0$ at $r = 0$ for symmetry.† It is evident that $M_r = M_0$ for $0 \leqslant r \leqslant \xi_0$ when

$$\mu \ddot{W}_1 = p_0. \qquad (4.58)‡$$

Thus,

$$\dot{W}_1 = p_0 t/\mu \qquad (4.59a)$$

and

$$W_1 = p_0 t^2/2\mu, \qquad (4.59b)$$

since $W_1 = \dot{W}_1 = 0$ when motion commences at $t = 0$.

Equations (4.1), (4.2), (4.34), (4.53) and (4.55a) lead to the governing equation for the outer annular region

$$\partial^2(rM_r)/\partial r^2 = -rp_0 + \mu r \ddot{W}_1(R - r)/(R - \xi_0), \qquad \xi_0 \leqslant r \leqslant R, \qquad (4.60)$$

which, integrating twice with respect to r, predicts the radial bending moment

$$M_r = -p_0 r^2/6 + \mu \ddot{W}_1(Rr^2/6 - r^3/12)/(R - \xi_0) + A_1 + B_1/r, \qquad (4.61)$$

where

$$A_1 = p_0(R^3 - R^2\xi_0 - R\xi_0^2 - \xi_0^3)/\{12(R - \xi_0)\} - M_0\xi_0/(R - \xi_0) \qquad (4.62a)$$

and

$$B_1 = M_0 R\xi_0/(R - \xi_0) - p_0 R\xi_0(R^2 - R\xi_0 - \xi_0^2)/\{12(R - \xi_0)\} \qquad (4.62b)$$

are evaluated, with the aid of equation (4.58), from the simply supported boundary condition $M_r = 0$ at $r = R$ and the continuity requirement $M_r = M_0$ at $r = \xi_0$. In view of equations (4.54a) and (4.55a), equation (4.11) for the transverse shear force Q_r is valid for both zones $0 \leqslant r \leqslant \xi_0$ and $\xi_0 \leqslant r \leqslant R$ of a circular plate, and must be continuous across the stationary plastic hinge at $r = \xi_0$. It is evident from equations (4.11) and (4.54) that $Q_r = 0$ for $0 \leqslant r \leqslant \xi_0$. Thus, equation (4.11), with

---

† In addition, $M_r = M_\theta = M_0$ at the centre of a circular plate loaded axisymmetrically.

‡ The plate has a constant acceleration during the first phase of motion.

equation (4.61) for $\xi_0 \leqslant r \leqslant R$, must also equal zero at $r = \xi_0$, which is satisfied when

$$p_0/p_c = 2R^3/\{(R + \xi_0)(R - \xi_0)^2\}, \tag{4.63}$$

where p_c is the static plastic collapse pressure defined by equation (4.33).

Equation (4.63) predicts the position of the stationary circular plastic hinge (ξ_0) in terms of the dynamic pressure ratio ($\eta = p_0/p_c$). It is evident that $\xi_0 = 0$ when $p_0 = 2p_c$, which agrees with the analysis in § 4.4. Moreover, the stationary plastic hinge forms at the supports for intense pressure loadings since $\xi_0 \to R$ when $p_0 \gg p_c$, or $\eta \to \infty$.

4.5.3 Second phase of motion, $\tau \leqslant t \leqslant T_1$

Once the rectangular pressure pulse is released at $t = \tau$, then the circular plastic hinge at $r = \xi_0$ commences to move inwards towards the plate centre, as indicated in Figure 4.10(b). The associated transverse velocity profile may be expressed in the form

$$\dot{w} = \dot{W}_2, \qquad 0 \leqslant r \leqslant \xi, \tag{4.64a}$$

and

$$\dot{w} = \dot{W}_2(R - r)/(R - \xi), \qquad \xi \leqslant r \leqslant R, \tag{4.64b}$$

where ξ is the time-dependent position of the travelling plastic hinge. Again the normality requirements of plasticity lead to equations (4.54) and (4.55) which, to account for a travelling plastic hinge, are written

$$M_\theta = M_0, \qquad M_r = M_0, \qquad 0 \leqslant r \leqslant \xi, \tag{4.65a, b}$$

and

$$M_\theta = M_0, \qquad 0 \leqslant M_r \leqslant M_0, \qquad \xi \leqslant r \leqslant R. \tag{4.66a, b}$$

The governing equation for the central circular zone is

$$\partial^2(rM_r)/\partial r^2 = \mu r \ddot{W}_2, \qquad 0 \leqslant r \leqslant \xi, \tag{4.67}$$

from equations (4.1), (4.2), (4.64a), (4.65a) and $p = 0$. Integrating twice with respect to r gives

$$M_r = M_0 + \mu r^2 \ddot{W}_2/6$$

when evaluating the constants of integration in order to satisfy the requirements that $Q_r = 0$ and $M_r = M_0$ at $r = 0$. However, reconciliation with equation (4.65b) demands that

$$\mu \ddot{W}_2 = 0, \tag{4.68}$$

which, integrating with respect to time, gives

$$\dot{W}_2 = p_0\tau/\mu \tag{4.69a}$$

and

$$W_2 = p_0\tau t/\mu - p_0\tau^2/2\mu, \tag{4.69b}$$

where the constants of integration have been evaluated to ensure that the transverse velocity and displacement are continuous with equations (4.59) at $t = \tau$.

Now, the governing equations (4.1) and (4.2) for the outer annular region $\xi \leqslant r \leqslant R$ of the circular plate may be written with the aid of equations (4.64b) and (4.66a) in the form

$$\partial^2(rM_r)/\partial r^2 = \mu r\{\ddot{W}_2(R - r)/(R - \xi) + \dot{W}_2\xi(R - r)/(R - \xi)^2\}$$

or

$$\partial^2(rM_r)/\partial r^2 = p_0\tau\dot{\xi}(Rr - r^2)/(R - \xi)^2 \qquad (4.70)$$

when using equations (4.68) and (4.69a). Integrating twice with respect to r yields

$$M_r = p_0\tau\dot{\xi}(2Rr^2 - r^3)/\{12(R - \xi)^2\} + A_2 + B_2/r, \qquad (4.71)$$

where, in order to satisfy the boundary condition $M_r = 0$ at $r = R$ and the continuity requirement $M_r = M_0$ at $r = \xi$,

$$A_2 = -\xi M_0/(R - \xi) - p_0\tau\dot{\xi}(R^3 + R^2\xi + R\xi^2 - \xi^3)/\{12(R - \xi)^2\} \qquad (4.72a)$$

and

$$B_2 = \xi R M_0/(R - \xi) + p_0\tau\dot{\xi}\xi R(R^2 + R\xi - \xi^2)/\{12(R - \xi)^2\}. \qquad (4.72b)$$

In addition, the transverse shear force Q_r must be continuous across the travelling plastic hinge. $Q_r = 0$ throughout the central circular zone $0 \leqslant r \leqslant \xi$, so that equations (4.1), (4.66a) and (4.71) for the annular zone $\xi \leqslant r \leqslant R$ give $Q_r = 0$ at $r = \xi$ when

$$\dot{\xi} = -2p_c R^3/\{p_0\tau(R^2 + 2\xi R - 3\xi^2)\}, \qquad (4.73)$$

where p_c is defined by equation (4.33).

The velocity of the travelling circular plastic hinge is given by equation (4.73), which, integrating with respect to time, predicts the hinge position

$$(\xi/R)^3 - (\xi/R)^2 - \xi/R = 2p_c t/p_0\tau - 1 \qquad (4.74)$$

when using equation (4.63) and satisfying the initial condition $\xi = \xi_0$ at $t = \tau$. The second phase of motion is complete at $t = T_1$ when the travelling plastic hinge reaches the plate centre (i.e., $\xi = 0$). In this circumstance, equation (4.74) with $\xi = 0$ predicts that

$$T_1 = p_0\tau/2p_c. \qquad (4.75)$$

4.5.4 Third phase of motion, $T_1 \leqslant t \leqslant T$

The transverse velocity at the plate centre remains constant throughout the second phase of motion according to equation (4.69a). Thus, the maximum transverse velocity at the end of the second phase of motion is the same as that at the end of the first phase. A third phase of motion, therefore, is necessary to absorb the kinetic energy remaining in the plate.

The travelling plastic hinge reaches the plate centre at the end of the second phase of motion. It is assumed that the shape of the transverse velocity profile then remains unchanged during the final phase of motion, as shown in Figure 4.10(c), or

$$\dot{w} = \dot{W}_3(1 - r/R), \qquad 0 \leqslant r \leqslant R. \qquad (4.76)$$

The theoretical analysis for this phase of motion is similar to that in § 4.4.3. Thus,

$$\mu \ddot{W}_3 = -2p_c \qquad (4.77)$$

according to equation (4.45), with $\ddot{W}_2$ replaced by $\ddot{W}_3$, where p_c is the corresponding static collapse pressure defined by equation (4.33). Now, integrating equation (4.77) with respect to time gives

$$\mu \dot{W}_3 = 2p_c(p_0\tau/p_c - t) \qquad (4.78)$$

when ensuring continuity with equations (4.69a) and (4.75) at the end of the second phase of motion. A further integration yields

$$W_3 = 2p_0\tau t/\mu - p_c t^2/\mu - p_0^2\tau^2/4\mu p_c - p_0\tau^2/2\mu, \qquad (4.79)$$

where the constant of integration is found by matching with equation (4.69b) when $t = T_1$ and T_1 is defined by equation (4.75).

It is evident from equation (4.78) that motion finally ceases at $t = T$, where

$$T = \eta\tau, \qquad (4.80)$$

which is identical to equation (4.47) for the low-pressure case $1 \leqslant \eta \leqslant 2$ and equals $2T_1$, where T_1 is defined by equation (4.75). The associated maximum permanent transverse displacement (at the plate centre) from equation (4.79) is

$$W_f = p_0\tau^2(3p_0/2p_c - 1)/2\mu. \qquad (4.81)$$

4.5.5 Static admissibility

It is evident from equations (4.54a), (4.55a), (4.65a), (4.66a) and § 4.5.4 that $M_\theta = M_0$ for $0 \leqslant r \leqslant R$ throughout the three phases of motion. Thus, it is only necessary to examine the static admissibility of the radial bending moment M_r.

During the first phase of motion, the radial bending moment is $M_r = M_0$ within the central circular zone ($0 \leqslant r \leqslant \xi_0$) according to equation (4.54b), and from equations (4.58), (4.61) and (4.62) is

$$M_r/M_0 = 1 - (r/R - \xi_0/R)^3(r/R + \xi_0/R)/\{(1 - \xi_0/R)^3(1 + \xi_0/R)(r/R)\} \quad (4.82)$$

in the outer annular zone, $\xi_0 \leqslant r \leqslant R$. Equation (4.82) predicts that $\{\partial(M_r/M_0)\}/\{\partial(r/R)\} \leqslant 0$ for $\xi_0 \leqslant r \leqslant R$ so that inequality (4.55b) is satisfied, as indicated in Figure 4.11(a).

Now, equation (4.65b) indicates that $M_r = M_0$ within the central circular zone $0 \leqslant r \leqslant \xi$ during the second phase of motion in § 4.5.3. Equation (4.71) for the annular zone $\xi \leqslant r \leqslant R$ may be rearranged with the aid of equation (4.73) in the form

$$M_r/M_0 = 1 + R(r - \xi)^2(r^2 - 2Rr + 2\xi r - 4R\xi + 3\xi^2)/\{r(R - \xi)^3(R + 3\xi)\}. \qquad (4.83)$$

It transpires that $\{\partial(M_r/M_0)\}/\{\partial(r/R)\} \leqslant 0$ for $\xi \leqslant r \leqslant R$ so that $0 \leqslant M_r \leqslant M_0$ and inequality (4.66b) is satisfied, as shown in Figure 4.11(b).

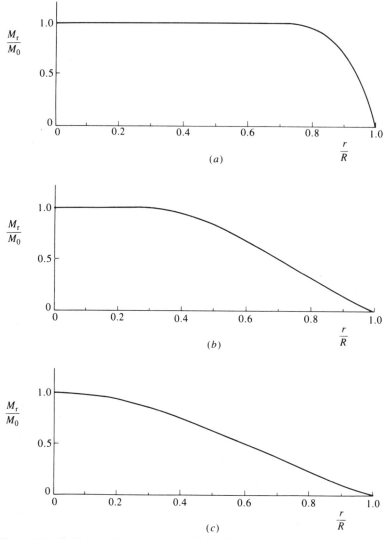

Figure 4.11 Radial bending moment distribution during the response of the simply supported circular plate in Figure 4.8(a). (a) First phase of motion $(0 \leqslant t \leqslant \tau)$ when $\eta = 13$ $(\xi_0 = 0.6991)$. (b) Second phase of motion $(\tau \leqslant t \leqslant T_1)$ when $t = 0.637T_1 = 0.3185\eta\tau$ and $\xi = 0.3$. (c) Third phase of motion $(T_1 \leqslant t \leqslant T)$.

The radial bending moment for the final phase of motion is

$$M_r/M_0 = 1 - (2 - r/R)(r/R)^2, \qquad 0 \leqslant r \leqslant R, \qquad (4.84)$$

according to §§ 4.4.3 and 4.5.4 and equations (4.33) and (4.77). Equation (4.84) predicts that $\{\partial(M_r/M_0)\}/\{\partial(r/R)\} \leqslant 0$ for $0 \leqslant r \leqslant R$ and, therefore, $0 \leqslant M_r \leqslant M_0$, as indicated in Figure 4.11(c).

The radial bending moment M_r is continuous across a travelling plastic hinge, or

$$[M_r]_\xi = 0. \tag{4.85}$$

Thus, differentiating with respect to time gives

$$[\partial M_r/\partial t]_\xi + \dot{\xi}[\partial M_r/\partial r]_\xi = 0 \tag{4.86}$$

since the time derivative of equation (4.85) is zero for a travelling hinge, or interface, which propagates with a velocity $\dot{\xi}$. The theoretical solution in § 4.5.3 has $M_r = M_0$ and $\partial M_r/\partial r = 0$ at the travelling plastic hinge so that equations (4.85) and (4.86) are automatically satisfied.

The foregoing comments reveal that the theoretical solution which is developed in §§ 4.5.2 to 4.5.4 is statically admissible throughout the entire plate during the whole response.

4.5.6 Kinematic admissibility

Some general expressions are developed initially for an axisymmetrically loaded circular plate, then specialised later in order to examine the kinematic admissibility of the foregoing theoretical solution.

It turns out that the theoretical development in § 3.4.6 for the dynamic response of beams remains valid for circular plates undergoing an axisymmetric response, provided x is replaced by r. Thus, equations (3.54) to (3.59) in § 3.4.6, with x replaced by r, require

$$[w]_\xi = 0, \tag{4.87a}$$

$$[\dot{w}]_\xi + \dot{\xi}[\partial w/\partial r]_\xi = 0, \tag{4.87b}$$

$$[\partial w/\partial r]_\xi = 0, \tag{4.87c}$$

$$[\dot{w}]_\xi = 0, \tag{4.87d}$$

$$[\ddot{w}]_\xi + \dot{\xi}[\partial \dot{w}/\partial r]_\xi = 0 \tag{4.87e}$$

and

$$[\partial \dot{w}/\partial r]_\xi + \dot{\xi}[\partial^2 w/\partial r^2]_\xi = 0. \tag{4.87f}$$

Now, returning to the theoretical solution in §§ 4.5.2 to 4.5.4, the continuity requirements (4.87a) and (4.87d) have been used in the analyses for the three phases of motion. The plastic hinges are stationary in the first and third phases of motion and equations (4.87b) and (4.87e) are satisfied, therefore, while equations (4.87c) and (4.87f) are not necessary. Thus, the kinematic requirements are satisfied in the first and third phases of motion. Only the second phase of motion with a travelling plastic hinge, which is developed in § 4.5.3, remains to be examined further. In fact, only equations (4.87c), (4.87e) and (4.87f) require checking, since equation (4.87b) is satisfied in view of equations (4.87c) and (4.87d).

Equations (4.64) give $[\ddot{w}]_\xi = \dot{W}_2\dot{\xi}/(R - \xi)$ and $[\partial \dot{w}/\partial r]_\xi = -\dot{W}_2/(R - \xi)$ and, therefore, equation (4.87e) is satisfied.

Equations (4.52) and (4.64a) show that equation (4.69b) gives the transverse displacement within the central circular zone during the second phase of motion,

$$w = p_0 \tau t/\mu - p_0 \tau^2/2\mu, \qquad 0 \leqslant r \leqslant \xi. \tag{4.88}$$

The transverse displacement in the annular zone $\xi \leqslant r \leqslant \xi_0$ is

$$w = p_0 \tau^2/2\mu + \int_\tau^{t(r)} \dot{W}_2 \, dt + \int_{t(r)}^t \dot{W}_2 (R - r) \, dt/(R - \xi) \tag{4.89}$$

according to equations (4.52), (4.59b), (4.64a) and (4.64b). However, $\dot{\xi} = d\xi/dt$, or $dt = d\xi/\dot{\xi}$, where $\dot{\xi}$ is given by equation (4.73), which, with equation (4.69a), allows equation (4.89) to be written as

$$w = p_0 \tau^2/2\mu + (p_0 \tau/\mu) \int_{\xi_0}^r d\xi/\dot{\xi} + (p_0 \tau/\mu) \int_r^\xi (R - r) \, d\xi/\{\dot{\xi}(R - \xi)\}. \tag{4.90}$$

Equations (4.88) and (4.90) predict that $[\partial w/\partial r]_\xi = 0$ and $[\partial^2 w/\partial r^2]_\xi = -(p_0 \tau)^2 (R + 3\xi)/2\mu p_c R^3$. Thus, equation (4.87c) is satisfied. It is straightforward to show that equation (4.87f) is also satisfied during the second phase of motion.

The foregoing remarks have demonstrated that the theoretical solution in §§ 4.5.2 to 4.5.4 is kinematically admissible and is, therefore, exact within the spirit of plasticity theory since it is also statically admissible, as shown in § 4.5.5.

4.5.7 Impulsive loading

A rectangular pressure pulse is known as an impulsive loading when $p_0/p_c \gg 1$ (i.e., $\eta \to \infty$) and $\tau \to 0$. In this circumstance, conservation of linear momentum at $t = 0$ demands

$$\int_0^R p_0 \tau 2\pi r \, dr = \int_0^R \mu V_0 2\pi r \, dr$$

or

$$p_0 \tau = \mu V_0, \tag{4.91}$$

where V_0 is the magnitude of an initial impulsive velocity distributed uniformly over the entire plate $0 \leqslant r \leqslant R$.

The first phase of motion in § 4.5.2 no longer exists when $\tau \to 0$ and the theoretical analysis for the impulsive loading problem consists only of the two phases of motion examined in § 4.5.3 and § 4.5.4. It is evident from equation (4.63) with $\eta \to \infty$ that the circular plastic hinge forms initially at the outer boundary (i.e., $\xi_0 = R$). The transverse velocity profiles for the two phases of motion are shown, therefore, in Figures 4.10(b, c), except that the plastic hinge originates at the outer support when $t = 0$. Wang[4.3] has examined this particular problem, but the results reported here are obtained from the theoretical analysis in § 4.5.3 and § 4.5.4.

The duration of response from equations (4.33), (4.80) and (4.91) for impulsive loading is

$$T = \mu V_0 R^2/6M_0. \tag{4.92}$$

Equation (4.81) with $p_0/p_c \gg 1$ may be written

$$W_f \cong 3p_0^2\tau^2/4\mu p_c,$$

which, using equations (4.33) and (4.91), becomes

$$W_f/H = \lambda/8, \tag{4.93a}$$

where

$$\lambda = \mu V_0^2 R^2/M_0 H \tag{4.93b}$$

is the same as equation (4.30) and is a dimensionless initial kinetic energy, and H is the plate thickness.

The theoretical predictions of equation (4.93a) are compared in Figure 4.12 with

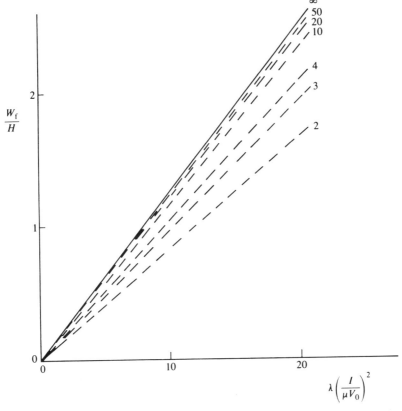

Figure 4.12 Maximum permanent transverse displacement of a simply supported circular plate loaded as shown in Figures 4.5 and 4.8(a). ——: impulsive loading; equation (4.93a). – – –: rectangular pressure pulse; equation (4.94a) with η given by numbers on lines.

equation (4.81) for a rectangular pressure pulse which may be written

$$W_f/H = \lambda(I/\mu V_0)^2(1 - 2/3\eta)/8,$$ (4.94a)

where

$$I = p_0\tau$$ (4.94b)

is the pulse per unit surface area of the plate.

It is left as an exercise to a reader to show that the analysis in § 4.5.3 for impulsive loads predicts a transverse displacement

$$w_1/H = \lambda\{2 - (r/R)^2 - (r/R)^3\}/24$$ (4.95a)

at the end of the first phase of motion, while the transverse displacement accumulated during the second and final phase of motion in § 4.5.4 is

$$w_2/H = \lambda(1 - r/R)/24.$$ (4.95b)

Thus, the final transverse displacement profile for an impulsive velocity may be written

$$w_f/H = \lambda\{3 - r/R - (r/R)^2 - (r/R)^3\}/24,$$ (4.96)

which is drawn in Figure 4.13.

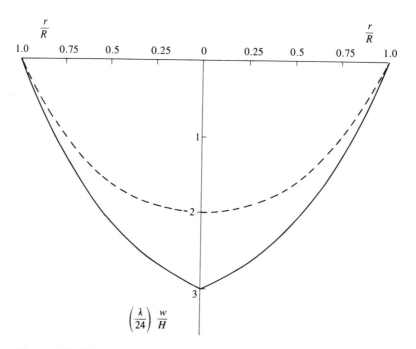

Figure 4.13 Axisymmetric transverse displacement profile for an impulsively loaded simply supported circular plate. ---: equation (4.95a); transverse displacement at end of first phase of motion. ——: equation (4.96); permanent profile.

4.6 Fully clamped circular plate loaded impulsively

4.6.1 Introduction

Florence[4.4] has studied the behaviour of a rigid, perfectly plastic circular plate which is fully clamped around an outer boundary and subjected to rectangular pressure pulse, while Wang and Hopkins[4.5] have examined the corresponding impulsive velocity case shown in Figure 4.14(a). The general theoretical procedure

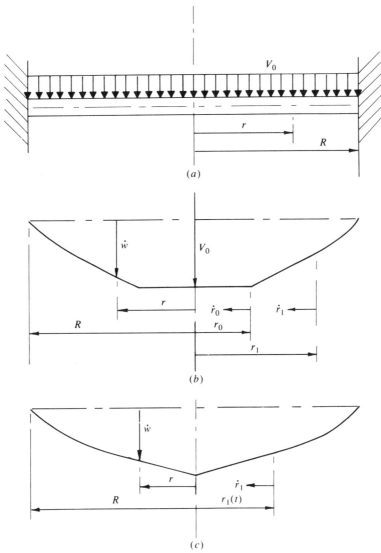

Figure 4.14 Fully clamped circular plate with outer radius R subjected to a uniformly distributed impulsive velocity V_0. (a) Notation. (b) Transverse velocity field during the first phase of motion ($0 \leqslant t \leqslant T_1$). (c) Transverse velocity field during the second phase of motion ($T_1 \leqslant t \leqslant T$).

used by Wang and Hopkins[4.5] is similar to that employed in § 4.5, except that the change of boundary condition leads to differential equations which must be integrated numerically. Wang and Hopkins[4.5] found that the response consisted of two phases of motion, which are now described briefly.

4.6.2 First phase of motion, $0 \leqslant t \leqslant T_1$

Wang and Hopkins[4.5] observed that the response of the circular plate in Figure 4.14(a) is controlled by three different plastic zones during the first phase of motion. The plastic flow in the central zone of the plate, $0 \leqslant r \leqslant r_0(t)$, is controlled by the corner A of the Tresca yield criterion in Figure 4.3. The central zone of the circular plate travels throughout this phase with a transverse velocity equal to the initial impulsive velocity V_0, as shown in Figure 4.14(b). This central area of the plate is surrounded by two annular zones, $r_0(t) \leqslant r \leqslant r_1(t)$ and $r_1(t) \leqslant r \leqslant R$, which are governed by the regimes AB and BC in Figure 4.3, respectively, with C located at the fully clamped support to allow for the development of a circular plastic hinge.

The plastic behaviour of the whole plate is controlled initially by the corner A in Figure 4.3 which, therefore, requires that $r_0(0) = r_1(0) = R$. As motion proceeds, both $r_0(t)$ and $r_1(t)$ decrease monotonically with $\dot{r}_0(t) \cong 2.87\dot{r}_1(t)$ until $r_0(T_1) = 0$, when the central plastic zone vanishes. This occurs at a time

$$T_1 \cong 0.57\mu V_0 R^2/12M_0 \tag{4.97}$$

when the first phase of motion is complete.

4.6.3 Second phase of motion, $T_1 \leqslant t \leqslant T$

At the end of the first phase of motion $r_0(T_1) = 0$, but $r_1(T_1) \neq 0$. Thus, a circular plate is divided into two zones, $0 \leqslant r \leqslant r_1(T_1)$ and $r_1(T_1) \leqslant r \leqslant R$, which are governed, respectively, by the regimes AB and BC of the Tresca yield criterion in Figure 4.3. It is assumed that the behaviour during the second phase of motion is controlled by the same two regimes, which are separated by a time-dependent boundary $r_1(t)$. The associated transverse velocity profile is shown in Figure 4.14(c).

Wang and Hopkins[4.5] found that $r_1(t)$ remains almost stationary during this phase of motion and that the transverse velocity at the plate centre decreases gradually until motion ceases. The duration of the second phase of motion alone is

$$T_2 \cong 0.51\mu V_0 R^2/12M_0, \tag{4.98}$$

which is slightly shorter than T_1. Thus, the total response duration is

$$T \cong 1.08\mu V_0 R^2/12M_0, \tag{4.99}$$

which is almost one-half of equation (4.92) for an impulsively loaded circular plate with simple supports around an outer boundary.

The maximum permanent transverse displacement, which occurs at the plate centre, is

$$W_f/H \cong 0.84\lambda/12,$$ (4.100)

when λ is defined by equation (4.93b). It is evident from a comparison of equations (4.93a) and (4.100) that a change of boundary conditions from simply supported to fully clamped leads to maximum permanent transverse displacements which are almost one-half as large.

4.7 Governing equations for rectangular plates

The dynamic behaviour of the element of a rectangular plate, which is drawn in Figure 4.15, is governed by the equations

$$\partial Q_x/\partial x + \partial Q_y/\partial y + p = \mu\ddot{w},$$ (4.101)

$$\partial M_x/\partial x + \partial M_{xy}/\partial y - Q_x = 0,$$ (4.102)

$$\partial M_y/\partial y + \partial M_{xy}/\partial x - Q_y = 0,$$ (4.103)

$$\kappa_x = -\partial^2 w/\partial x^2,$$ (4.104)

$$\kappa_y = -\partial^2 w/\partial y^2$$ (4.105)

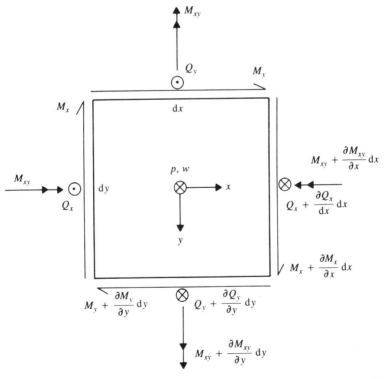

Figure 4.15 Infinitesimal element ($dx\,dy$) of a rectangular plate. ——→: bending moments per unit length. —→→: twisting moments per unit length. ○: transverse shear forces per unit length with · and × indicating head or tail of direction arrow.

and

$$\kappa_{xy} = -\partial^2 w/\partial x\,\partial y. \tag{4.106}$$

Equations (4.101) to (4.103) are identical to equations (2.22) to (2.24), except for the inclusion of the transverse inertia term in the lateral, or transverse, equilibrium equation (4.101).† (μ is mass per unit area and $(\dot{\ }) = \partial(\ )/\partial t$, where t is time.)

The principal bending moments M_1 and M_2 in a rectangular plate are related to the moments M_x, M_y and M_{xy}, which are defined in Figure 4.15, as follows[4.6]:

$$M_1 = (M_x + M_y)/2 + \{(M_x - M_y)^2 + 4M_{xy}^2\}^{1/2}/2 \tag{4.107}$$

and

$$M_2 = (M_x + M_y)/2 - \{(M_x - M_y)^2 + 4M_{xy}^2\}^{1/2}/2. \tag{4.108}$$

It is assumed that a rectangular plate is made from a rigid, perfectly plastic material and that the plastic flow is controlled by the three generalised stresses M_x, M_y and M_{xy} when the transverse shear forces Q_x and Q_y are taken as reactions. Thus, the plastic flow of a rectangular plate satisfies a three-dimensional yield surface in M_x–M_y–M_{xy} space. However, one of the two-dimensional yield criteria in Figure 4.16 governs plastic flow when the generalised stresses are arranged in the principal form given by equations (4.107) and (4.108).

Cox and Morland[4.7] appear to have obtained the only exact theoretical solution for the dynamic plastic response of a rectangular plate. They used the general theoretical procedure outlined for the dynamic plastic response of circular plates in §§ 4.3 to 4.5 and examined the dynamic plastic behaviour of a simply supported square plate which is subjected to a rectangular pressure pulse distributed uniformly over the entire surface area.

4.8 Simply supported square plate loaded dynamically, $p_{\mathrm{c}} \leqslant p_0 \leqslant 2p_{\mathrm{c}}$

4.8.1 Introduction

The only theoretical solution which has been published on the dynamic plastic response of rectangular plates was obtained by Cox and Morland.[4.7] They examined the particular case of a square plate which is simply supported around the boundary, as shown in Figure 4.17, and subjected to the rectangular pressure pulse in Figure 4.5. In order to simplify the analysis, Cox and Morland[4.7] found it necessary to employ the simple square, or Johansen, yield criterion in Figure 4.16. It transpires that the dynamic response then consists of two phases of motion, $0 \leqslant t \leqslant \tau$ and $\tau \leqslant \tau \leqslant T$, which are now examined in the following two sections.

---

† Equations (4.101), (4.102) and (4.104) with $Q_x = Q$, $M_x = M$ and $Q_y = M_y = M_{xy} = 0$ reduce to equations (3.1) to (3.3) for the dynamic plastic behaviour of beams.

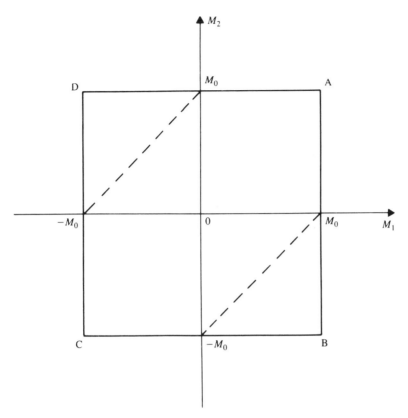

Figure 4.16 Yield criteria for a rigid, perfectly plastic rectangular plate with plastic yielding controlled by the principal moments M_1 and M_2. ——: Johansen. ---: Tresca.

4.8.2 First phase of motion, $0 \leqslant t \leqslant \tau$

The static collapse pressure for a simply supported square plate is

$$p_c = 6M_0/L^2, \qquad (4.109)$$

according to equation (2.47), where $2L$ is the length of each side, as indicated in Figure 4.17. Transverse motion commences when $p_0 \geqslant p_c$ and it is assumed that the transverse velocity field has the same form as the static plastic collapse profile in Figure 4.17(c), which is associated with equation (4.109). Thus,

$$\dot{w} = \dot{W}_1(1 - z), \qquad 0 \leqslant z \leqslant 1, \qquad (4.110)$$

where

$$z = (x + y)/\sqrt{2}L, \qquad (4.111)$$

as indicated in Figure 4.17(b). The coordinates x and y are measured along the diagonals of a square plate where the plastic hinges develop. It is only necessary to consider one quadrant of a plate since the response is symmetric about the x- and y-axes.

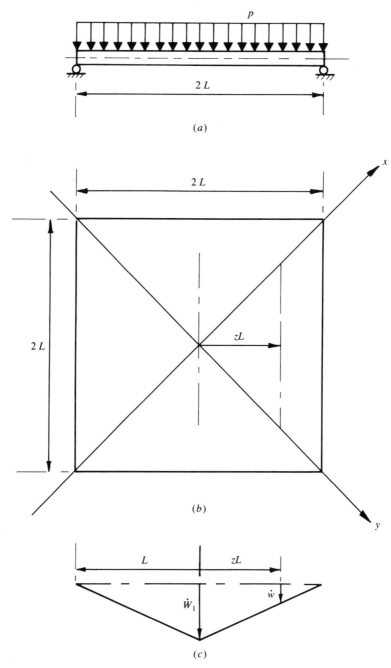

Figure 4.17 Simply supported square plate subjected to a uniformly distributed pressure pulse. (*a*) Side view of a square plate. (*b*) Plan view of a square plate. (*c*) Transverse velocity profile.

If the transverse shear forces are eliminated from equation (4.101) using equations (4.102) and (4.103), then

$$\partial^2 M_x/\partial x^2 + 2\partial^2 M_{xy}/\partial x\, \partial y + \partial^2 M_y/\partial y^2 = \mu\ddot{w} - p. \tag{4.112}$$

Now,

$$p = p_0, \quad 0 \leqslant t \leqslant \tau, \tag{4.113}$$

so that equation (4.112) becomes

$$\partial^2 M_x/\partial x^2 + 2\partial^2 M_{xy}/\partial x\, \partial y + \partial^2 M_y/\partial y^2 = \mu\ddot{W}_1(1-z) - p_0 \tag{4.114}$$

when using equation (4.110).

It is evident that

$$M_x = M_y = M_0 \quad \text{and} \quad M_{xy} = 0 \tag{4.115a–c}$$

from considerations of symmetry at the plate centre. This suggests that the plastic flow of the material in the centre of the plate is governed by regime A of the Johansen yield condition in Figure 4.16. The development of plastic hinge lines along the x- and y-axes in Figure 4.17 requires

$$M_y = M_0 \quad \text{when} \quad y = 0 \quad \text{and} \quad 0 \leqslant x \leqslant \sqrt{2}L \tag{4.116}$$

and

$$M_x = M_0 \quad \text{when} \quad x = 0 \quad \text{and} \quad 0 \leqslant y \leqslant \sqrt{2}L, \tag{4.117}$$

respectively. Thus, plastic flow of the material in the hinges could be controlled by the regime AB of the Johansen yield condition in Figure 4.16, or

$$M_1 = (M_x + M_y)/2 + \{(M_x - M_y)^2 + 4M_{xy}^2\}^{1/2}/2 = M_0 \tag{4.118}$$

and

$$-M_0 \leqslant M_2 = (M_x + M_y)/2 - \{(M_x - M_y)^2 + 4M_{xy}^2\}^{1/2}/2 \leqslant M_0 \tag{4.119}$$

when using equations (4.107) and (4.108).

Cox and Morland[4.7] simplified the theoretical analysis by choosing the elementary series

$$M_x = M_0 + x^2 f_1(z), \tag{4.120}$$
$$M_y = M_0 + y^2 f_1(z) \tag{4.121}$$

and

$$M_{xy} = xy f_1(z), \tag{4.122}$$

which satisfy equations (4.115) to (4.117) and where $f_1(z)$ is an arbitrary function which remains to be determined. Equations (4.120) to (4.122) may be substituted into equation (4.114) to rewrite the governing equation in the form

$$z^2\, \partial^2 f_1/\partial z^2 + 6z\, \partial f_1/\partial z + 6f_1 = \mu\ddot{W}_1(1-z) - p_0, \tag{4.123}$$

the general solution of which is

$$f_1 = (\mu\ddot{W}_1 - p_0)/6 - \mu\ddot{W}_1 z/12 + C_1/z^2 + C_2/z^3, \tag{4.124}$$

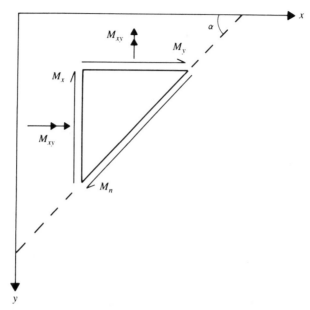

Figure 4.18 Bending moment M_n per unit length along a plane inclined at an angle α to the x-axis. ———►: bending moments per unit length. ———►: twisting moments per unit length.

where C_1 and C_2 are arbitrary constants of integration. However,

$$C_1 = C_2 = 0 \tag{4.125}$$

because the bending moments given by equations (4.120) and (4.121) must satisfy equation (4.115) at the plate centre.†

The bending moment M_n across any plane which is inclined at an angle α, as shown in Figure 4.18, is

$$M_n = M_x \sin^2 \alpha + 2M_{xy} \sin \alpha \cos \alpha + M_y \cos^2 \alpha. \tag{4.126}‡$$

Therefore, the simply supported boundary condition at $z = 1$ requires

$$(M_x + M_y)/2 + M_{xy} = 0, \qquad z = 1, \tag{4.127}$$

according to equation (4.126) with $\alpha = 45°$. Thus, substituting equations (4.120) to (4.122), (4.124) and (4.125) with $z = 1$ into equation (4.127) gives

$$\mu \ddot{W}_1 = 2(p_0 - p_c) \tag{4.128}§$$

---

† For example, taking $f_1 = C_1/z^2 + C_2/z^3$ from equation (4.124) and substituting into equation (4.120) gives $M_x = M_0 + x^2 2L^2 C_1/(x + y)^2 + x^2 2\sqrt{2}L^3 C_2/(x + y)^3$ when using equation (4.111). Thus, $M_x = M_0 + 2L^2 C_1 + 2\sqrt{2}L^3 C_2/x$ when $y = 0$. Clearly, $C_1 = C_2 = 0$ in order to satisfy equation (4.115a).

‡ This expression is similar to that developed in elementary texts for the stress normal to a plane which is inclined at an arbitrary angle α. See also reference 4.6.

§ The plate accelerates throughout the first phase of motion $0 \leqslant t \leqslant \tau$ when $p_0 > p_c$ and the static solution with $\ddot{W}_1 = 0$ is recovered when $p_0 = p_c$.

when using equations (4.109) and (4.111). Integrating this equation twice with respect to time and using equation (4.110) gives the transverse displacement

$$w = (p_0 - p_c)t^2(1 - z)/\mu, \qquad 0 \leqslant z \leqslant 1, \tag{4.129}$$

since $w = \dot{w} = 0$ when $t = 0$. Therefore, the transverse velocity at the end of the first phase of motion is

$$\dot{w} = 2(p_0 - p_c)\tau(1 - z)/\mu, \qquad 0 \leqslant z \leqslant 1. \tag{4.130}$$

4.8.3 Second phase of motion, $\tau \leqslant t \leqslant T$

Equation (4.130) shows that the transverse velocity of the plate is finite at the end of the first phase of motion ($t = \tau$) when the rectangular pressure pulse is released. The corresponding kinetic energy in the square plate is

$$K_e = 4(p_0 - p_c)^2\tau^2 L^2/3\mu. \tag{4.131}$$

A second phase of motion is, therefore, necessary to absorb this energy plastically.

It is again assumed that the form of the transverse velocity profile remains the same as the static collapse profile shown in Figure 4.17(c), or

$$\dot{w} = \dot{W}_2(1 - z), \qquad 0 \leqslant z \leqslant 1, \tag{4.132}$$

which, with

$$p = 0, \qquad 0 \leqslant z \leqslant 1 \qquad \text{and} \qquad \tau \leqslant t \leqslant T, \tag{4.133}$$

allows the governing equation (4.112) to be written

$$\partial^2 M_x/\partial x^2 + 2\partial^2 M_{xy}/\partial x\,\partial y + \partial^2 M_y/\partial y^2 = \mu\ddot{W}_2(1 - z). \tag{4.134}$$

The expressions (4.120) to (4.122) with $f_1(z)$ replaced by $f_2(z)$ remain valid for the bending moments during the second phase of motion, since equations (4.115) to (4.119) must again be satisfied. Therefore, a differential equation for $f_2(z)$, which is similar to equation (4.123) but with $\ddot{W}_1$ replaced by $\ddot{W}_2$ and $p_0 = 0$, yields

$$f_2(z) = \mu\ddot{W}_2/6 - \mu\ddot{W}_2 z/12, \tag{4.135}$$

which is obtained from equations (4.124) and (4.125), with the above changes. Thus, the simply supported boundary condition (4.127) at $z = 1$ yields

$$\mu\ddot{W}_2 = -2p_c, \tag{4.136}$$

which may also be obtained from equation (4.128), with the changes noted above.

Now, integrating equation (4.136) with respect to time gives

$$\dot{W}_2 = 2(p_0\tau - p_c t)/\mu \tag{4.137}$$

when matching with equation (4.130) at $t = \tau$. Another integration with respect to time yields

$$W_2 = -p_c t^2/\mu + 2p_0\tau t/\mu - p_0\tau^2/\mu, \tag{4.138}$$

where the constant of integration has been found by ensuring displacement continuity with equation (4.129) at $t = \tau$.

The permanent deformed profile of a square plate is reached when all the kinetic energy of equation (4.131) is absorbed, which occurs when $\dot{W}_2 = 0$, or

$$T = \eta\tau \tag{4.139}$$

from equation (4.137), where

$$\eta = p_0/p_c \tag{4.140}$$

and p_c is defined by equation (4.109). The associated permanent transverse displacement from equations (4.138) and (4.139) is

$$w_f = p_c\tau^2\eta(\eta - 1)(1 - z)/\mu, \qquad 0 \leqslant z \leqslant 1. \tag{4.141}$$

It is interesting to note that the duration of response (equation (4.139)) and maximum permanent transverse displacement (equation (4.141)) of a square plate are identical to equations (4.47) and (4.48) for a simply supported circular plate having $L = R$. In other words, the major response characteristics for a square plate may be obtained from a simply supported circular plate with a radius which just inscribes the square form.

4.8.4 Static admissibility

It has not been demonstrated whether the theoretical solution satisfies equation (4.118) or inequality (4.119). However, substituting equations (4.120) to (4.122), together with equations (4.124), (4.125) and (4.128) into equation (4.118), gives

$$M_1 = M_0, \qquad 0 \leqslant z \leqslant 1, \tag{4.142}$$

so that equation (4.118) is satisfied during the first phase of motion. Similarly, using equations (4.120) to (4.122), inequality (4.119) becomes

$$-M_0 \leqslant M_0 + (x^2 + y^2)f_1(z) \leqslant M_0, \qquad 0 \leqslant z \leqslant 1, \tag{4.143}†$$

which, with the aid of equations (4.124), (4.125) and (4.128), may be written in the form

$$-2 \leqslant (x^2 + y^2)\{\eta - 2 - (\eta - 1)z\}/L^2 \leqslant 0. \tag{4.144}$$

The right-hand inequality requires

$$\eta \leqslant 2 \tag{4.145}$$

in order to avoid a yield violation at the plate centre. The left-hand inequality of (4.144) becomes an equality at the plate corners, but does not lead to any restrictions on η.†

It may be shown that equation (4.142) and inequality (4.144) with $\eta = 0$ apply to the second phase of motion in § 4.8.3. Thus, the theoretical solution is statically admissible throughout both phases of motion provided inequality (4.145) is satisfied, i.e., $1 \leqslant \eta \leqslant 2$.

---

† It is evident from equations (4.118), (4.119) and Figure 4.16 that $-M_0$ on the left-hand side of (4.143) would be replaced by 0 for the Tresca yield criterion. However, it may be shown that $M_1 = M_0$ and $M_2 = -M_0$ at the corners of a square plate. Thus, the theoretical analysis is only valid for the Johansen yield criterion.

4.9 Simply supported square plate loaded dynamically, $p_0 \geqslant 2p_c$

4.9.1 Introduction

It is evident from inequality (4.145) that the theoretical analysis in § 4.8 for the simply supported rigid, perfectly plastic square plate in Figure 4.17 is statically admissible, and, therefore, exact, provided $p_c \leqslant p_0 \leqslant 2p_c$. For pressure pulses with larger magnitudes, the right-hand side of inequality (4.144) is first violated near the plate centre, since $\{\eta - 2 - (\eta - 1)z\} \geqslant 0$ at $z = 0$ when $\eta \geqslant 2$. This indicates that a yield violation develops near the plate centre which, therefore, requires a modification of the transverse velocity field, as also found in § 4.5 for a simply supported circular plate.

It transpires that the dynamic response may be divided into three distinct phases, which are examined in the following sections.

4.9.2 First phase of motion, $0 \leqslant t \leqslant \tau$

The yield violation which occurs at the plate centre when $\eta \geqslant 2$ suggests that a central zone develops with the plastic flow controlled by the corner A of the yield condition in Figure 4.16. Thus,

$$M_1 = M_0 \quad \text{and} \quad M_2 = M_0, \qquad 0 \leqslant z \leqslant \xi_0, \qquad \text{(4.146a, b)}$$

while any plastic flow in the remainder of the plate is controlled by the side AB, or

$$M_1 = M_0, \qquad -M_0 \leqslant M_2 \leqslant M_0, \qquad \xi_0 \leqslant z \leqslant 1. \qquad \text{(4.147a, b)}$$

It is assumed that the dynamic response during the first phase of motion, while the rectangular pressure pulse shown in Figure 4.5 is active, is governed by the transverse velocity profile

$$\dot{w} = \dot{W}_1, \qquad 0 \leqslant z \leqslant \xi_0, \qquad \text{(4.148)}$$

and

$$\dot{w} = \dot{W}_1(1 - z)/(1 - \xi_0), \qquad \xi_0 \leqslant z \leqslant 1, \qquad \text{(4.149)}$$

as shown in Figure 4.19(a), where z is defined by equation (4.111) and ξ_0 is time-independent. Equations (4.148) and (4.149) reduce to equation (4.110) when $p_0 \leqslant 2p_c$ since, in this case, $\xi_0 = 0$.†

The governing differential equation (4.112), with equations (4.113) and (4.148), simplifies to

$$\mu \ddot{W}_1 = p_0, \qquad 0 \leqslant z \leqslant \xi_0, \qquad \text{(4.150)}$$

for the central zone, since it follows from equations (4.107), (4.108) and (4.146a, b) that

$$M_x = M_0, \qquad M_y = M_0, \qquad M_{xy} = 0, \qquad 0 \leqslant z \leqslant \xi_0. \qquad \text{(4.151a–c)}$$

---

† Equation (4.158) gives $\xi_0 = 0$ when $\eta = 2$.

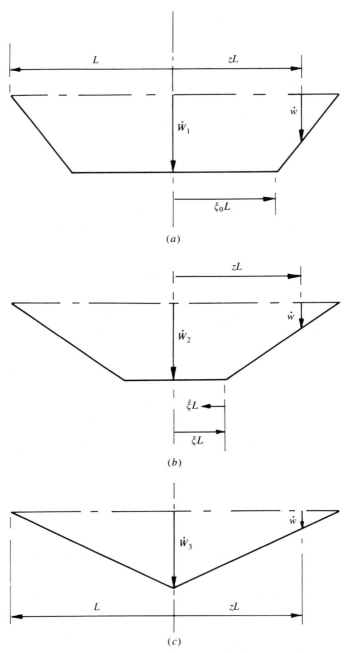

Figure 4.19 Transverse velocity profiles for a simply supported square plate subjected to a rectangular pressure pulse with $\eta \geqslant 2$. (a) First phase of motion, $0 \leqslant t \leqslant \tau$. (b) Second phase of motion, $\tau \leqslant t \leqslant T_1$. (c) Third phase of motion, $T_1 \leqslant t \leqslant T$.

Integrating equation (4.150) twice with respect to time yields

$$\dot{W}_1 = p_0 t / \mu \tag{4.152}$$

and

$$W_1 = p_0 t^2 / 2\mu \tag{4.153}$$

when satisfying the initial conditions $w = \dot{w} = 0$ at $t = 0$.

The bending moments in the outer zone of the plate $\xi_0 \leqslant z \leqslant 1$, are again assumed to be given by equations (4.120) to (4.122) but with $f_1(z)$ replaced by $g_1(z)$. Substituting these expressions into equation (4.112), together with equations (4.113) and (4.149), gives

$$z^2 \, \partial^2 g_1 / \partial z^2 + 6z \, \partial g_1 / \partial z + 6g_1 = \mu \ddot{W}_1 (1 - z) / (1 - \xi_0) - p_0, \tag{4.154}$$

the general solution for which is

$$g_1 = -p_0/6 + p_0(2 - z)/\{12(1 - \xi_0)\} + C_3/z^2 + C_4/z^3 \tag{4.155}$$

when using equation (4.150). It is necessary to ensure that Q_x, Q_y, M_x, M_y and M_{xy} are continuous across the plastic hinge located at $z = \xi_0$. It may be shown that equations (4.102) and (4.103) and equations (4.120) to (4.122), with $f_1(z)$ replaced by g_1 from equation (4.155), are continuous at $z = \xi_0$, with equations (4.151a–c) and with $Q_x = Q_y = 0$ in the central zone, when

$$C_3 = -p_0 \xi_0^3 / \{6(1 - \xi_0)\} \tag{4.156}$$

and

$$C_4 = p_0 \xi_0^4 / \{12(1 - \xi_0)\}. \tag{4.157}$$

In addition, equation (4.127) with $z = 1$ must be satisfied for a square plate with simple supports. Thus, substituting equations (4.120) to (4.122), with $f_1(z)$ replaced by g_1, leads to the relation

$$(1 - \xi_0)^2 (1 + \xi_0) = 2/\eta \tag{4.158}$$

for the location of the plastic hinge, where η is the pressure ratio which is defined by equation (4.140). Equation (4.158) predicts that $\xi_0 = 0$ when $\eta = 2$, as expected from § 4.8, and $\xi_0 \to 1$ when $\eta \to \infty$ for impulsive loadings.

4.9.3 Second phase of motion, $\tau \leqslant t \leqslant T_1$

The square plate accelerates throughout the first phase of motion, according to equation (4.150), while the rectangular pressure pulse is active and has a finite transverse velocity, according to equations (4.148), (4.149) and (4.152) at $t = \tau$. Thus, motion continues during a second phase of motion and it is assumed, by analogy with the theoretical solution for a rigid, perfectly plastic circular plate in § 4.5, that the plastic hinge at $z = \xi_0$ in Figure 4.19(a) is now active and propagates inwards towards the plate centre. In this circumstance,

$$\dot{w} = \dot{W}_2, \qquad 0 \leqslant z \leqslant \xi, \tag{4.159}$$

and
$$\dot{w} = \dot{W}_2(1 - z)/(1 - \xi), \qquad \xi \leqslant z \leqslant 1, \tag{4.160}$$
as indicated in Figure 4.19(b).

Equations (4.151a–c) still control the behaviour in the central zone so that the governing equation (4.112), with equation (4.159) and $p = 0$, predicts
$$\mu \ddot{W}_2 = 0, \qquad 0 \leqslant z \leqslant \xi, \tag{4.161}$$
which may be integrated twice with respect to time to give
$$\dot{W}_2 = p_0 \tau / \mu \tag{4.162}$$
and
$$W_2 = p_0 \tau t / \mu - p_0 \tau^2 / 2\mu \tag{4.163}$$
when matching, respectively, with equations (4.152) and (4.153) for the transverse velocity and displacement at the end of the first phase of motion.

The bending moments in the outer zone of the plate, $\xi \leqslant z \leqslant 1$, are given by equations (4.120) to (4.122) but with f_1 replaced by g_2. Thus, substituting these expressions into equation (4.112), together with equation (4.160) and $p = 0$, requires
$$z^2 \, \partial^2 g_2/\partial z^2 + 6z \, \partial g_2/\partial z + 6g_2 = \mu \ddot{W}_2(1 - z)/(1 - \xi) + \mu \dot{W}_2 \dot{\xi}(1 - z)/(1 - \xi)^2, \tag{4.164}$$
the general solution for which is
$$g_2 = p_0 \tau \dot{\xi}(2 - z)/\{12(1 - \xi)^2\} + C_5/z^2 + C_6/z^3 \tag{4.165}$$
when using equations (4.161) and (4.162). Continuity of M_x, M_y, M_{xy}, Q_x and Q_y at the travelling plastic hinge ($z = \xi$) is satisfied when
$$C_5 = p_0 \tau \dot{\xi} \xi^2 (2\xi - 3)/\{6(1 - \xi)^2\} \tag{4.166}$$
and
$$C_6 = p_0 \tau \dot{\xi} \xi^3 (4 - 3\xi)/\{12(1 - \xi)^2\}, \tag{4.167}$$
while the simply supported boundary condition demands
$$(1 - \xi)(1 + 3\xi)\dot{\xi} = -2/\eta\tau \tag{4.168}$$
according to equations (4.109), (4.127) and (4.140). A time integration of equation (4.168) for the hinge propagation speed predicts that
$$(1 - \xi)^2(1 + \xi) = 2t/\eta\tau \tag{4.169}$$
when satisfying the initial condition represented by equation (4.158) at $t = \tau$.

The second phase of motion is completed when $\xi = 0$ which, according to equation (4.169), occurs when
$$T_1 = \eta\tau/2. \tag{4.170}$$
The transverse displacement and velocity at the end of the second phase of motion may be obtained from equations (4.159), (4.160), (4.162), (4.163) and (4.170). In particular, the maximum transverse displacement at the plate centre is
$$W_2 = p_0 \tau^2 (\eta - 1)/2\mu. \tag{4.171}$$

4.9.4 Third phase of motion, $T_1 \leqslant t \leqslant T$

It is evident from equation (4.162) that a square plate possesses a transverse velocity and, therefore, a kinetic energy at $t = T_1$ which must be absorbed during further motion. It is assumed that the transverse velocity profile at $t = T_1$, which is given by equation (4.160) with a vanishing central plastic zone (i.e., $\xi = 0$), persists throughout the final phase of motion. Thus,

$$\dot{w} = \dot{W}_3(1 - z), \qquad 0 \leqslant z \leqslant 1, \tag{4.172}$$

as shown in Figure 4.19(c).

The analysis for this phase of motion is similar to that presented in § 4.8.3 for the second phase of motion when $p_c \leqslant p_0 \leqslant 2p_c$, but with $f_2(z)$ replaced by $g_3(z)$. Equation (4.136), therefore, predicts

$$\mu \ddot{W}_3 = -2p_c, \qquad T_1 \leqslant t \leqslant T, \tag{4.173}$$

which may be integrated to give

$$\dot{W}_3 = 2p_c(\eta \tau - t)/\mu \tag{4.174}$$

and

$$W_3 = p_c\{(2\eta\tau - t)t - \eta^2\tau^2/4 - \eta\tau^2/2\}/\mu \tag{4.175}$$

when matching with equations (4.162) and (4.163) at $t = T_1$, where η and T_1 are defined by equations (4.140) and (4.170), respectively.

A plate reaches its permanent position at $t = T$ when $\dot{W}_3 = 0$, or

$$T = \eta \tau \tag{4.176}$$

which equals $2T_1$, and the associated maximum permanent transverse displacement is

$$W_f = \eta p_c \tau^2(3\eta - 2)/4\mu. \tag{4.177}$$

4.9.5 Static and kinematic admissibility

It may be shown that the principal bending moments M_1 and M_2 during the three phases of motion in the foregoing theoretical analysis do not violate the Johansen yield condition in Figure 4.16 for rectangular pressure pulses with $p_0 \geqslant 2p_c$.[†] Despite the fact that $M_n = 0$ along the simply supported boundaries of the plate according to equation (4.127), it may be shown that $M_1 = M_0$ and $M_2 = -M_0$ at the plate corners.[‡] Thus, the Johansen rather than the Tresca yield criterion controls plastic flow. In addition, the continuity requirements $[Q_x] = [Q_y]$ $= [M_x] = [M_y] = [M_{xy}] = 0$ are satisfied at the plastic hinges which are located at $z = \xi_0$, $z = \xi$ and $z = 0$ during the first, second and third phases of motion, respectively.

---

† It appears that analytical methods cannot be used to show that the condition $-M_0 \leqslant M_2 \leqslant M_0$ from equation (4.119) for $\xi \leqslant z \leqslant 1$ is satisfied during the second phase of motion except at $z = \xi$ and $z = 1$. However, Cox and Morland[(4.7)] state that the theoretical solution is statically admissible.

‡ See footnote to equation (4.143).

The theoretical solution also satisfies the kinematic requirements. In particular, equations (4.87) in § 4.5.6 remain valid for the travelling plastic hinges at $z = \xi$ in a square plate, provided r is replaced by z.

It is concluded that the theoretical analysis in § 4.9.2 to § 4.9.4 for a rigid, perfectly plastic square plate is statically and kinematically admissible and, therefore, exact within the spirit and limitations of the general method.

4.9.6 Impulsive loading

As noted already in § 4.5.7 for circular plates, impulsive loading is associated with the condition

$$\mu V_0 = p_0 \tau, \qquad (4.178)$$

where V_0 is the magnitude of the impulsive velocity which is distributed uniformly over the plate. The theoretical analysis in § 4.9.2 to § 4.9.4, with $p_0 \gg p_c$ (i.e., $\eta \to \infty$)

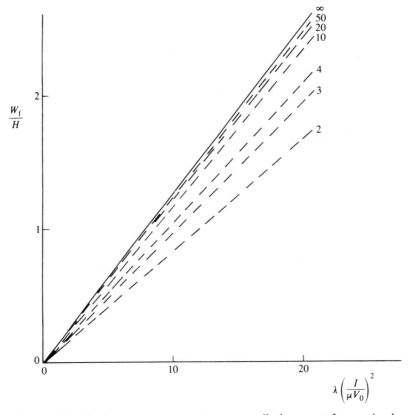

Figure 4.20 Maximum permanent transverse displacement for a simply supported square plate loaded as shown in Figures 4.5 and 4.17(a). ———: impulsive loading; equation (4.180). ———: rectangular pressure pulse; equation (4.182) with η given by numbers on lines.

and $\tau \to 0$, reduces to the impulsive loading case. In this circumstance, the first phase of motion no longer exists and the plastic hinge develops initially at $z = \xi_0$ $= 1$ according to equation (4.158). Thus, equations (4.176) and (4.177) predict respectively the response duration

$$T = \mu V_0/p_c \tag{4.179}$$

and the associated maximum permanent transverse displacement

$$W_f/H = \lambda/8, \tag{4.180}$$

where

$$\lambda = \mu V_0^2 L^2/M_0 H. \tag{4.181}$$

Equation (4.177) for a square plate subjected to a rectangular pressure pulse may be recast in the dimensionless form

$$W_f/H = \lambda(I/\mu V_0)^2(1 - 2/3\eta)/8, \tag{4.182}$$

where

$$I = p_0\tau \tag{4.183}$$

and is compared in Figure 4.20 with the theoretical predictions of equation (4.180) for impulsive loading. It is evident that, when a square plate has a side length $2L$, which is equal to the diameter $2R$ of a circular plate, then equations (4.180), (4.182) and Figure 4.20 for square plates are identical to equations (4.93a), (4.94a) and Figure 4.12 for circular plates, respectively.

4.10 Final remarks

The response of several rigid, perfectly plastic, annular, circular and square plates subjected to dynamic loads has been examined in this chapter. However, many other exact and approximate theoretical solutions have been published, and brief literature surveys may be found in references 4.8 and 4.9. Only those articles which employ the theoretical methods outlined in this chapter are now discussed in this section. In fact, it is fairly straightforward to use the theoretical methods for plates with a variety of other loadings and boundary conditions, although an exact theoretical solution has not been published for a dynamically loaded rectangular plate.

Shapiro[4.10] examined the dynamic behaviour of an annular plate which is fully clamped around the inner edge and free at the outer edge. The outer edge is subjected to a constant axisymmetric velocity, which is removed suddenly after a specified time. Florence[4.11] studied a similar annular plate, except that the outer edge was subjected to a transverse impulse rather than a constant velocity being maintained for a short time. Perrone[4.12] investigated the dynamic plastic response of an annular plate which was simply supported around the outer boundary and free at the inner edge, and subjected to an axisymmetric impulsive velocity which was distributed linearly over the entire upper surface of a plate with a peak value at the inner radius and zero at the outer boundary. This problem was also examined in reference 4.13 and in § 4.3.5 here, while the associated pressure case (see Figure 4.4(a)) in § 4.3 was studied in reference 4.14. Annular plates with either simply supported or fully clamped outer supports and subjected to a uniformly distributed impulse have been examined in reference 4.15.

Many articles have been written on the dynamic plastic response of circular plates, in addition to those already cited in § 4.4 to § 4.6. For example, both Florence[4.16] and Conroy[4.17] have examined the response of a circular plate subjected to a dynamic load with a rectangular pressure–time history (see Figure 4.5) which is distributed uniformly within a central circular zone. Florence[4.16] studied the fully clamped case, while Conroy[4.17] considered simple supports, and both authors assumed that the plastic flow of the circular plate was controlled by the Tresca yield criterion shown in Figure 4.3. More recently, Florence[4.18] demonstrated that his previous analysis[4.16] is simplified when the plastic flow of the material is governed by the Johansen yield criterion in Figure 4.16 which is often used for reinforced concrete slabs.

Perzyna[4.19] explored the influence of the external pressure pulse shape on the dynamic plastic response of a simply supported circular plate, and obtained solutions for blast type loads[4.20] having

$$\int_0^t p(t) \, dt \geqslant t p(t), \tag{4.184}$$

which he distinguished from impact type loads with

$$\int_0^t p(t) \, dt \leqslant t p(t). \tag{4.185}$$

Youngdahl[4.21] has also studied the influence of pulse shape on the dynamic plastic response of an axisymmetrically loaded simply supported circular plate. He observed that the maximum permanent transverse displacement depends significantly upon the pulse shape. However, Youngdahl[4.21] found that the strong dependence on pulse shape could be practically eliminated by introducing an effective load

$$P_e = I/2t_c, \tag{4.186}$$

where

$$I = \int_{t_y}^{t_f} P(t) \, dt \tag{4.187}$$

is the total impulse, $P(t)$ is the external load, t_y and t_f are the times when plastic deformation starts and finishes and where t_c is the centroid of the pulse given by

$$I t_c = \int_{t_y}^{t_f} (t - t_y) P(t) \, dt. \tag{4.188}$$

For the purpose of estimating the response time t_f, which is not known *a priori*, Youngdahl[4.21] replaced equation (4.187) by the simpler expression

$$I \cong P_y(t_f - t_y), \tag{4.189}$$

where P_y is the static plastic collapse load for a circular plate.

The correlation parameters for the effective load (P_e), the total impulse (I) and the mean time (t_c), which are defined by equations (4.186) to (4.188), respectively, contain integrals of the external loading and are, therefore, insensitive to small perturbations in the pulse shape. This accounts for the essential collapse onto a single curve of the theoretical predictions for simply supported circular plates subjected to different pulse shapes. These observations are encouraging from a practical viewpoint because it is often difficult to record accurately the pressure–time histories of dynamic loadings and to model the actual dynamic loads in laboratory tests.

Krajcinovic[4.22,4.23] has undertaken a similar study to that reported in reference 4.21 for the dynamic plastic response of circular plates which are fully clamped around the outside edge. A numerical scheme was used to examine the influence of external pressures having arbitrary pressure–time histories and distributed uniformly either over the entire plate[4.22] or within a central region.[4.23] Again, Krajcinovic observed that the final transverse displacements are sensitive to the pulse shape.† It transpires that the correlation parameters defined by equations (4.186) to (4.188) lead to collapse of the theoretical predictions onto almost one curve for pressure pulses with rectangular, linear decay, triangular and parabolic shapes.

Cox and Morland[4.7] appear to be the only authors who have obtained exact theoretical solutions for the dynamic plastic behaviour of non-axisymmetric plates. In addition to the theoretical solutions for a simply supported square plate presented in § 4.8 and § 4.9, Cox and Morland[4.7] have examined a regular n-sided polygonal plate either with simple supports or an edge bending moment. The permissible edge bending moment ranges from zero for simple supports when $n = 4$ for a square plate to fully clamped supports when $n \to \infty$ for a built-in circular plate. Hopkins[4.25] has derived the governing equations for plates, including further consideration of the kinematic requirements in § 4.5.6, but no exact theoretical solutions have been published using these equations for specific non-axisymmetric problems.

Many experimental studies have demonstrated that the influence of finite-deflections, or geometry changes, leads to significant changes in the dynamic plastic

---

† Stronge[4.24] has examined the requirements for the most efficient pressure–time histories which maximise the maximum permanent transverse displacements of simply supported beams loaded uniformly over the entire span. No similar studies appear to have been published for plates.

response of initially flat circular and rectangular plates. Thus, this important phenomenon is incorporated into theoretical rigid, perfectly plastic analyses in Chapter 7, wherein comparison are made with experimental results. Furthermore, the strain rate sensitive behaviour of materials exercises an important effect on the dynamic plastic response of strain rate sensitive structures, as discussed in Chapter 8. As far as the author is aware, the influence of material elasticity and pulse duration have not been examined for plates, so that guidelines similar to those developed for beams in § 3.4.10 are also used for plates.

PROBLEMS

4.1 Confirm that the energy absorbed through plastic deformations during the second phase of motion for the annular plate in § 4.3.3 equals the kinetic energy at the end of the first phase of motion (equation (4.18)).

4.2 Examine the static admissibility for the annular plate problem in §§ 4.3.2 and 4.3.3 when $\alpha \to 0$.

4.3 Obtain the permanent transverse displacement profile (equation (4.96)) for an impulsively loaded, simply supported circular plate which is made from a rigid, perfectly plastic material.

4.4 (a) Describe clearly (but do not derive) the theoretical analysis for a simply supported circular plate of radius R and thickness H when subjected to an axisymmetric uniformly distributed pressure pulse having a rectangular pressure–time history of magnitude p_0 and duration τ. The plate is made from a ductile and perfectly plastic material with a flow stress σ_0 and a mass per unit surface area μ.
(b) A simply supported circular diaphragm is used as a safety valve and remains intact for pressure pulses which do not produce a maximum transverse displacement larger than B. Calculate the maximum impulsive velocity which the diaphragm can tolerate without rupture and discuss the assumptions you have used.

Note that, when $p_0/p_c > 2$ (p_c is the associated static collapse pressure) then

$$W_2 = p_0\tau(t - \tau/2)/\mu \qquad \text{for} \quad \tau \leqslant t \leqslant T_1,$$

with

$$T_1 = p_0\tau/2p_c, \qquad T = p_0\tau/p_c \quad \text{and} \quad W_f = p_0\tau^2(3p_0/2p_c - 1)/2\mu,$$

where $p_c = 6M_0/R^2$, T is the response duration and W_f is the maximum final transverse displacement.

(c) A strain gauge is located at the centre of the unloaded side of the simply supported circular diaphragm in order to activate safety systems in the event of failure. Derive the physical limitation in the operation of this system if the strain gauge recorder is insensitive to elastic stress waves. How would you remove this operating limitation?

4.5 Consider the simply supported square plate which is examined in § 4.8, and show that the generalised stress field at a corner of the plate satisfies the Johansen yield criterion but not the Tresca yield criterion.

4.6 (a) Evaluate the effective load (P_e) and the mean time (t_e) from equations (4.186) to (4.188) for the rectangular pressure pulse in Figure 4.5. (b) Show that the response durations are predicted exactly from equation (4.189) for the simply supported circular and square plates examined in this chapter. Show that equation (4.189) is almost satisfied for the fully clamped circular plate in § 4.6.3 when using equation (4.91) and equation (2.21). (c) Recast equations (4.48), (4.81), (4.141) and (4.177) for the permanent transverse displacements of simply supported circular and square plates in terms of P_e and I^2, where $I = p_0\tau$ is the total pulse (per unit area). Use these equations to predict the corresponding maximum permanent transverse displacements for pressure–time histories with a linear decay from an initial peak value p_0 to zero pressure at a time t_0 as shown in Figure 5.23.

5

Dynamic plastic behaviour of shells

5.1 Introduction

The two previous chapters have examined the response of beams and plates when made from rigid, perfectly plastic materials and subjected to large dynamic loads. This chapter employs similar methods of analysis to study the dynamic stable response of shells.

Shells are thin-walled structural members having either one non-zero curvature (e.g., cylindrical and conical shells) or two non-zero curvatures (e.g., spherical and toroidal shells). They are used throughout engineering for storage (e.g., gas storage tanks), transportation (e.g., pipelines and railway tank cars) and for protection purposes (e.g., crash helmets), and are vital components of submersibles, offshore platforms, chemical plant and many other applications.

In some cases, the dynamic response occurs during normal operation as, for example, in an energy-absorbing cushion made of thin-walled shells which is positioned underneath an object dropped from an aircraft. However, it is also necessary to provide protection of some shell structures against unexpected loads, as might occur, for example, when a turbine disc bursts and produces energetic missiles which threaten shells containing hazardous substances.

It is evident that shell structures may have a variety of shapes and are subject to a wide range of internal and external dynamic loads. However, the structural response studied in this chapter is stable, while dynamic plastic buckling, or unstable behaviour, is examined in Chapter 10.

The shell structures of interest in this chapter are thin-walled and are subjected to dynamic loads which produce large plastic deformations. Thus, the general theoretical procedure for beams and plates, which is outlined in the two previous chapters, is again employed.

The first step in a theoretical analysis of a particular shell is to postulate the velocity field which is associated with the dynamic load. Usually, the velocity field associated with the corresponding static loading problem is suitable for small dynamic loads. The generalised strain rates are then calculated and used, together with the normality requirements of plasticity, to locate the active portion of the yield condition which controls plastic flow. This information, together with the equilibrium equations and boundary conditions, permits integration of the governing equations to predict the permanent deformed profile and other response characteristics.

An exact theoretical solution must satisfy the initial, boundary and kinematical conditions on the displacement and velocity fields. It is usually quite straightforward to check these conditions, which are invariably satisfied. However, enquiries into the static admissibility are often more difficult, as found in previous chapters, and may lead to important restrictions on the theoretical solution. In other words, the theoretical solution might be statically admissible only within a certain range of the dynamic loads (e.g., inequality (4.145)), or for restricted values of the geometrical parameters (e.g., equation (5.43) in this chapter). In this circumstance, a new velocity profile is required and the whole process must be repeated until an exact theoretical solution is achieved. The selection of a revised velocity profile is guided by the nature of the static inadmissibility, or yield violations, of the theoretical solutions. A yield violation suggests the formation of a plastic hinge or the development of a plastic zone.

The basic equations which govern the dynamic plastic behaviour of a cylindrical shell are presented in the next section. The cylindrical shell problems examined in this chapter are axisymmetric, so that the equilibrium equations are developed for an axisymmetric response which remains infinitesimal. Some simple yield criteria are introduced, but a thorough review is beyond the scope of this chapter. However, Hodge[5.1] has examined the various yield criteria for the static loading of shells and these may be used for the dynamic behaviour when neglecting the influence of material strain rate sensitivity, which is introduced in Chapter 8.

The dynamic behaviour of long unreinforced and reinforced cylindrical shells are examined in §§ 5.3 and 5.4, respectively, while the dynamic response of a short cylindrical shell is studied in § 5.5. All three shells are subjected to rectangular-shaped pressure pulses. The dynamic spherically symmetric behaviour of an elastic, perfectly plastic complete spherical shell is explored in § 5.6 and compared with the theoretical predictions of a simpler rigid, perfectly plastic analysis. These comparisons allow certain observations to be made on the importance of elastic effects and the accuracy of a rigid plastic analysis. Shallow and spherical shells are examined in §§ 5.7 and 5.8, respectively, while § 5.9 introduces a simplified method for studying the influence of complicated pressure pulses.

5.2 Governing equations for cylindrical shells

The dynamic axisymmetric behaviour of the element of a circular cylindrical shell, which is drawn in Figure 5.1, is governed by the equations

$$\partial M_x / \partial x - Q_x = 0, \tag{5.1}$$

$$R\, \partial Q_x / \partial x + N_\theta - Rp - \mu R\, \partial^2 w / \partial t^2 = 0, \tag{5.2}$$

$$\epsilon_x = \partial u / \partial x, \tag{5.3}$$

$$\epsilon_\theta = -w/R \tag{5.4}$$

and

$$\kappa_x = -\partial^2 w / \partial x^2, \tag{5.5}$$

which are identical to equations (2.56) to (2.60), except for the inclusion of the transverse inertia term in the lateral or transverse equilibrium equation (5.2). (μ is the mass per unit area of middle surface and t is time.) The cylindrical shell is not loaded axially (i.e., $N_x = 0$) and the influence of rotatory inertia is not retained in equation (5.1), but is discussed briefly in Chapter 6.

It is evident that N_θ, M_x and Q_x are the only generalised stresses which appear in the equilibrium equations (5.1) and (5.2). Moreover, the lateral, or transverse, shear force Q_x does not enter into the yield condition for a cylindrical shell when the associated shear strain is neglected. Thus, it is necessary to retain only the two generalised stresses N_θ and M_x when constructing a yield criterion for a cylindrical shell which is subjected to an axisymmetric pressure distribution and is unloaded axially. The exact and two approximate square and hexagonal yield curves in M_x–N_θ space are drawn in Figure 5.2.

The general theoretical procedure which is employed in order to analyse the dynamic plastic behaviour of cylindrical shells is similar to that outlined in the two earlier chapters on beams and plates.

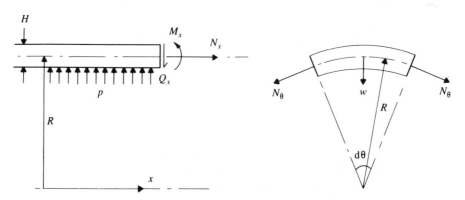

Figure 5.1 Notation for the dynamic axisymmetric loading of a cylindrical shell with a mean radius R and wall thickness H.

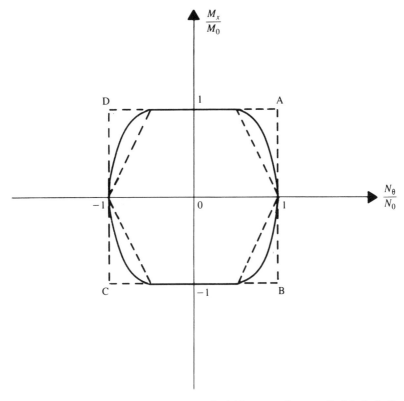

Figure 5.2 Exact, square and hexagonal yield curves for a cylindrical shell.
——: exact yield curve. ---: approximate yield curves.

5.3 Long cylindrical shell

5.3.1 Introduction

The general theoretical procedure which is developed in the two previous chapters is now used to obtain the dynamic plastic response of an infinitely long cylindrical shell. The shell is subjected to an axisymmetric internal pressure pulse with the rectangular-shaped pressure-time history shown in Figure 5.3 and described by

$$p = p_0, \qquad 0 \leqslant t \leqslant \tau \qquad (5.6)$$

and

$$p = 0, \qquad t \geqslant \tau. \qquad (5.7)$$

The cylindrical shell is made from a rigid, perfectly plastic material. Moreover, the response of the shell is axisymmetric with no axial variation, which implies that $M_x = Q_x = 0$. Thus, yielding of the cylindrical shell is controlled by

$$N_\theta = N_0, \qquad (5.8)$$

where N_0 is the fully plastic membrane force for the shell cross-section.

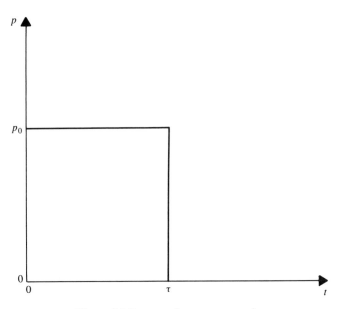

Figure 5.3 Rectangular pressure pulse.

Now, it may be shown when using the theoretical methods in Chapter 2 that the cylindrical shell collapses when the magnitude p_0 of the internal pressure with $\tau \to \infty$ reaches the static collapse pressure

$$p_c = N_0/R, \tag{5.9}$$

which is predicted by equation (2.67).

It turns out that the dynamic plastic response of a long cylindrical shell, with $p_0 \geqslant p_c$, consists of two phases of motion. These phases correspond with the durations of the external pressure pulses described by equations (5.6) and (5.7) which are valid for $0 \leqslant t \leqslant \tau$ and $\tau \leqslant t \leqslant T$, respectively, where T is the response duration.

5.3.2 First phase of motion, $0 \leqslant t \leqslant \tau$

As noted in § 5.3.1, $Q_x = M_x = 0$, so that equation (5.1) is automatically satisfied and equation (5.2) becomes

$$\mu \, \partial^2 w/\partial t^2 = N_0/R - p_0 \tag{5.10}†$$

when using equations (5.6) and (5.8). The right-hand side of equation (5.10) is constant, while w is independent of x and θ, so that a straightforward integration leads to

$$w = (N_0/R - p_0)t^2/2\mu, \tag{5.11}$$

† It is evident that equation (5.9) is recovered for static pressures since $p_0 = p_c = N_0/R$ when $\partial^2 w/\partial t^2 = 0$.

since the initial conditions at $t = 0$ demand $w = \dot{w} = 0$.† At the end of the first phase of motion $t = \tau$, and the radial displacement and velocity are

$$w = (N_0/R - p_0)\tau^2/2\mu \tag{5.12}$$

and

$$\dot{w} = (N_0/R - p_0)\tau/\mu, \tag{5.13}†$$

respectively.

5.3.3 Second phase of motion, $\tau \leqslant t \leqslant T$

The radial velocity increases linearly with time throughout the first phase of motion until the greatest value given by equation (5.13) is reached at $t = \tau$. Thus, although the cylindrical shell is unloaded according to equation (5.7) when $t \geqslant \tau$, radial motion must continue during a second phase of motion until the kinetic energy at $t = \tau$, associated with the radial velocity given by equation (5.13), is dissipated plastically.

Equation (5.2) now becomes

$$\mu \, \partial^2 w/\partial t^2 = N_0/R \tag{5.14}$$

by analogy with equation (5.10) and using equation (5.7). Integrating twice with respect to time and matching the radial displacement and velocity at $t = \tau$ with equations (5.12) and (5.13), respectively, gives

$$w = N_0 t^2/2\mu R - p_0\tau t/\mu + p_0\tau^2/2\mu. \tag{5.15}$$

Motion continues until $\partial w/\partial t = 0$, which occurs when

$$T = \eta\tau, \tag{5.16}$$

where

$$\eta = p_0/p_c \tag{5.17}$$

is the ratio of the dynamic pressure pulse (p_0) to the corresponding static collapse pressure (p_c) which is defined by equation (5.9). The maximum radial displacement from equations (5.15) and (5.16) is, therefore,

$$w_f = p_c\tau^2\eta(1 - \eta)/2\mu. \tag{5.18}‡$$

This displacement is also the final or permanent radial displacement, since no elastic unloading occurs in a cylindrical shell which is made from a rigid, perfectly plastic material.

5.3.4 Comments

The foregoing theoretical solution is kinematically admissible for $0 \leqslant t \leqslant T$ because the uniform radially outwards displacement is independent of the axial coordinate x

---

† $(\dot{\ }) = \partial(\ )/\partial t$.

‡ This is negative because the dynamic load acts so as to increase the diameter, whereas the displacement w is defined positive towards the shell axis, as shown in Figure 5.1.

and the circumferential coordinate θ, and the ends of the cylindrical shell are free. It is evident from equations (5.3) to (5.5) that $\dot{\epsilon}_x = 0$, $\dot{\epsilon}_\theta \geqslant 0$ and $\dot{\kappa}_x = 0$, which, through the normality requirement of plasticity, are consistent with $N_x = 0$, $N_\theta \geqslant 0$ and $M_x = 0$. Moreover, the theoretical solution is statically admissible since $N_\theta = N_0$ throughout both phases of motion and $N_x = M_x = Q_x = 0$. The theoretical solution is, therefore, exact according to plasticity theory. It is evident that the theoretical analysis is exact for a cylindrical shell having any axial length, including the particular case of a ring.

Thus, the impulsive loading case, with $\eta \gg 1$, may be obtained from the theoretical analysis in §§ 5.3.2 and 5.3.3. If $\tau \to 0$ and $\eta \to \infty$, then Newton's law ($\int F \, dt$ = change in linear momentum) gives

$$p_0 \tau = \mu V_0, \qquad (5.19)$$

where V_0 is the initial radially outwards impulsive velocity. In this circumstance, equation (5.16), for the response duration, predicts

$$T = \mu V_0/p_c, \qquad (5.20)$$

and equation (5.18) gives an associated permanent radial displacement

$$w_f/H = -\lambda/8, \qquad (5.21)$$

where

$$\lambda = 4\mu V_0^2 R/N_0 H \qquad (5.22)$$

is a non-dimensional form of the initial kinetic energy which becomes

$$\lambda = \mu V_0^2 R/M_0 \qquad (5.23)$$

for a cylindrical shell with a homogeneous solid wall of thickness H.

5.3.5 Discussion on energies

The total pulse imparted to the cylindrical shell in §§ 5.3.2 and 5.3.3 is

$$I = p_0 \tau \qquad (5.24)$$

per unit area. On the other hand, the total energy of the internal pressure pulse equals

$$E_T = -\int_0^{w(\tau)} p_0 \, dw$$

per unit area, when $w(\tau)$ is the radial deflection at $t = \tau$ according to equation (5.12). Thus, $E_T = -p_0 w(\tau)$ since p_0 is constant for $0 \leqslant t \leqslant \tau$, or

$$E_T = I^2(1 - 1/\eta)/2\mu. \qquad (5.25)$$

Equation (5.25), with $\eta \to \infty$ for impulsive loading, and equations (5.19) and (5.24) predict $E_T = I^2/2\mu = \mu V_0^2/2$, which is the initial kinetic energy per unit area.

The total energy dissipated in the shell due to the circumferential membrane force $N_\theta = N_0$ is

$$D_T = -\int_0^{w_f} N_\theta \, dw/R$$

per unit area, where w_f is the final radial displacement according to equation (5.18). Thus, $D_T = -N_0 w_f/R$ since $N_\theta = N_0$ for $0 \leqslant t \leqslant T$, or

$$D_T = I^2(1 - 1/\eta)/2\mu \tag{5.26}$$

which is identical to equation (5.25), as expected. A similar calculation yields the energy dissipated through circumferential membrane effects during the first phase of motion in § 5.3.2 as

$$D_1 = I^2(1 - 1/\eta)/2\mu\eta \tag{5.27}$$

and

$$D_2 = I^2(1 - 1/\eta)^2/2\mu \tag{5.28}$$

during the second phase of motion in § 5.3.3. It may be shown that $D_1 + D_2 = D_T = E_T$.

It is evident that the total kinetic energy per unit area of the cylindrical shell at the end of the first phase of motion ($t = \tau$) is $K_T = \mu\{\dot{w}(\tau)\}^2/2$ which, using equation (5.13), becomes

$$K_T = I^2(1 - 1/\eta)^2/2\mu, \tag{5.29}$$

which is $1 - 1/\eta$ times the total external energy (E_T) imparted to the cylindrical shell during the first phase of motion. A comparison of equation (5.29) with

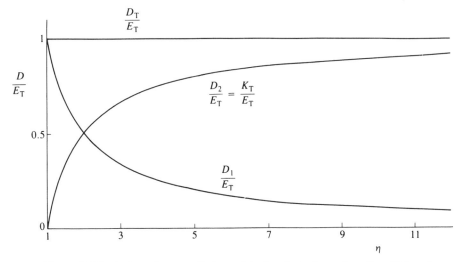

Figure 5.4 Variation of energy dissipated during first phase of motion (D_1) and second phase of motion (D_2) for a long cylindrical shell with dimensionless magnitude of pressure pulse η. D_T = total energy absorbed by shell per unit area. E_T = total energy of pressure pulse per unit area. K_T = kinetic energy per unit area of shell at end of first phase of motion ($t = \tau$).

equation (5.28) reveals that $K_T = D_2$, as expected, while a further check shows that $D_1 + K_T = E_T = D_T$.

Now, it is observed from equations (5.27) and (5.28) that $D_1/D_2 = 1/(\eta - 1)$, while equations (5.25) and (5.29) give $K_T/E_T = 1 - 1/\eta$. Thus, if $\eta \to 1$ then $D_1/D_2 \to \infty$ (i.e., motion is confined to the first phase of motion) and $K_T/E_T \to 0$ (i.e., static behaviour). On the other hand, if $\eta \to \infty$ for impulsive loading, then $D_1/D_2 \to 0$ (i.e., most of the energy is dissipated during the second phase) and $K_T/E_T \to 1$, which implies that all of the internal pressure pulse is converted to kinetic energy. Finally, it is evident from the results in Figure 5.4 that $D_2 \geqslant D_1$ and $K_T/E_T \geqslant \frac{1}{2}$ when $\eta \geqslant 2$.

5.4 Long reinforced cylindrical shell

5.4.1 Introduction

Now consider the particular case of a long cylindrical shell, which is reinforced by equally spaced rigid reinforcing rings, as shown in Figure 5.5. The response of this arrangement is sought when it is subjected to an axisymmetric internal dynamic pressure pulse with the characteristics which are described by equations (5.6) and (5.7), and indicated in Figure 5.3. If the internal pressure pulse does not vary axially, then it is necessary to examine only one bay, which is equivalent to a fully clamped cylindrical shell of length $2L$.

The static collapse pressure of a fully clamped cylindrical shell is given by equation (2.336), or

$$p_c = (1 + 2/\omega^2)N_0/R, \qquad (5.30)\dagger$$

where

$$\omega^2 = 2L^2/RH. \qquad (5.31)$$

Equation (5.30) is developed for a material which obeys the square yield criterion drawn in Figure 5.2.

5.4.2. First phase of motion, $0 \leqslant t \leqslant \tau$

It is assumed that the radial velocity field for a cylindrical shell subjected to a dynamic rectangular pressure pulse is similar to the kinematically admissible displacement profile which is associated with the static collapse pressure. This velocity field is indicated in Figure 5.6 and may be described in the form

$$\dot{w} = -\dot{W}(1 - x/L), \qquad 0 \leqslant x \leqslant L. \qquad (5.32)\ddagger$$

---

† When comparing this result with equation (5.9), which is the static collapse pressure of an infinitely long cylindrical shell, or short shell with unsupported ends, it is evident that the term $2N_0/\omega^2 R$ gives a measure of the increase in strength due to a fully clamped support. If each bay of a clamped shell is very long (i.e., $\omega^2 \gg 1$) then equation (5.30) reduces to equation (5.9).

‡ The cylindrical shell is symmetrical about a plane at $x = 0$ so it is necessary to examine only the portion $0 \leqslant x \leqslant L$.

Equations (5.4), (5.5) and (5.32) give $\dot{\varepsilon}_\theta \geqslant 0$ and $\dot{\kappa}_x = 0$ (except at the plastic circumferential hinges located at $x = 0$ and $x = L$) and, therefore, demand

$$N_\theta = N_0 \qquad \text{and} \qquad -M_0 \leqslant M_x \leqslant M_0 \qquad \text{(5.33a, b)}$$

according to the normality requirements of plasticity for the square yield criterion in Figure 5.2.

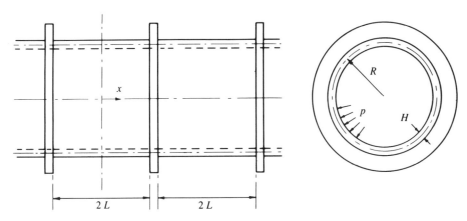

Figure 5.5 Cylindrical shell with rigid reinforcing rings.

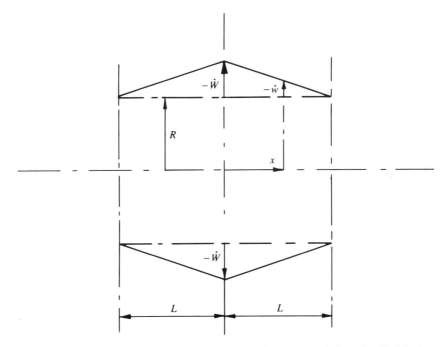

Figure 5.6 Axisymmetric radial velocity field for a long reinforced cylindrical shell.

If equations (5.1), (5.6), (5.32) and (5.33a) are substituted into equation (5.2), then

$$R\, \partial^2 M_x/\partial x^2 = Rp_0 - N_0 - \mu R \ddot{W}(1 - x/L), \qquad (5.34)$$

or

$$M_x = (Rp_0 - N_0)x^2/2R - \mu \ddot{W}(x^2/2 - x^3/6L) - M_0 \qquad (5.35)$$

since, at $x = 0$, $M_x = -M_0$ for the formation of a circumferential plastic hinge and $Q_x = \partial M_x/\partial x = 0$ to satisfy the symmetry requirements. However, the fully clamped support condition demands $M_x = M_0$ at $x = L$, which, when using equation (5.35), gives

$$(\mu L^2/3)\, \mathrm{d}^2 W/\mathrm{d}t^2 = (Rp_0 - N_0)L^2/2R - 2M_0. \qquad (5.36)\dagger$$

Integrating this equation twice with respect to time and using the initial conditions $W = \dot{W} = 0$ at $t = 0$ yields

$$W = 3p_c(\eta - 1)t^2/4\mu, \qquad (5.37)$$

where η and p_c are defined by equations (5.17) and (5.30), respectively.

5.4.3 Second phase of motion, $\tau \leqslant t \leqslant T$

It is now necessary to consider a second phase of motion with $t \geqslant \tau$, since the shell has a finite radial velocity at $t = \tau$ according to the time derivative of equation (5.37). If the radial velocity field is again described by equation (5.32) and substituted into equation (5.2) with equations (5.1), (5.7) and (5.33a), then equation (5.34) is obtained but with $p_0 = 0$. Thus, equations (5.35) and (5.36), with $p_0 = 0$, then follow, and

$$\mathrm{d}^2 W/\mathrm{d}t^2 = -3p_c/2\mu \qquad (5.38)$$

when using equation (5.30) for p_c.

Now, integrating equation (5.38) twice with respect to time and matching the radial velocity and displacement with the corresponding values from equation (5.37) with $t = \tau$ gives

$$W = -3p_c(t^2 - 2\eta\tau t + \eta\tau^2)/4\mu. \qquad (5.39)$$

Finally, the cylindrical shell reaches its permanent position at $t = T$, where

$$T = \eta\tau \qquad (5.40)$$

is obtained from equation (5.39) when $\dot{W} = 0$, while the corresponding deformed profile is

$$w_f = 3p_c\tau^2\eta(1 - \eta)(1 - x/L)/4\mu. \qquad (5.41)$$

5.4.4 Kinematic and static admissibility

It is evident that the foregoing theoretical solution is kinematically admissible, while it is also statically admissible provided M_x satisfies the inequality

---

† $\mathrm{d}^2 W/\mathrm{d}t^2 = 0$ for static loads, in which circumstance equation (5.36) reduces to $p_0 = N_0(1 + RH/L^2)/R$ which equals p_c according to equations (5.30) and (5.31).

$- M_0 \leqslant M_x \leqslant M_0$ given by inequality (5.33b). It turns out that no yield violations occur during the first stage of motion $0 \leqslant t \leqslant \tau$ provided $\partial^2 M_x/\partial x^2 \geqslant 0$ at $x = 0$, or, when using equations (5.30), (5.35) and (5.36),

$$\partial^2 M_x/\partial x^2 = (N_0/R)\{(1 + 2/\omega^2)(3 - \eta)/2 - 1\} \geqslant 0,$$

which requires

$$\eta \leqslant \frac{6 + \omega^2}{2 + \omega^2}. \tag{5.42}$$

However, in order to avoid a yield violation during the second stage of motion, $\tau \leqslant t \leqslant T$, it transpires that the most critical requirement is to maintain $\partial M_x/\partial x \geqslant 0$ at $x = L$, which is satisfied when

$$\omega^2 \leqslant 6. \tag{5.43}$$

Thus, the assumed velocity profile (equation (5.32)) is correct and the theoretical predictions for the response duration (equation (5.40)) and permanent deformed profile (equation (5.41)) are exact for cylindrical shells with dimensions which satisfy inequality (5.43) and which are subjected to pressure pulses having a magnitude which satisfies inequality (5.42). Hodge[5.2] has extended the foregoing theoretical analysis in order to study the behaviour of cylindrical shells with dimensions $\omega^2 > 6$ and subjected to pressure pulses which are larger than those permitted by inequality (5.42).

5.5 Fully clamped short cylindrical shell

5.5.1 Introduction

The theoretical analysis in § 5.4 focused on a long cylindrical shell which is reinforced periodically with rigid circular rings located at a distance $2L$ apart, as shown in Figure 5.5. It was found necessary to examine only one bay, which is idealised as a fully clamped cylindrical shell of axial length $2L$ subjected to an internal dynamic pressure having the rectangular pressure–time history in Figure 5.3. It transpired that a theoretical analysis using the radial velocity profile given by equation (5.32) and the associated side AB of the square yield condition in Figure 5.2 is valid only provided inequalities (5.42) and (5.43) are satisfied. The theoretical analysis in this section relaxes inequality (5.42) in order to examine the behaviour of cylindrical shells subjected to large pressure pulses. However, the restriction on the spacing of the stiffening rings, which is imposed by inequality (5.43), is retained.

Now, $\partial^2 M_x/\partial x^2 = 0$ at $x = 0$ when equality (5.42) is satisfied. Thus, rectangular pressure pulses having values of η which violate inequality (5.42) cause a yield violation at $x = 0$. This suggests that a plastic zone, with $M_x = - M_0$, develops in the central region of the cylindrical shell. It turns out that the dynamic response consists of three phases of motion $0 \leqslant t \leqslant \tau$, $\tau \leqslant t \leqslant t_1$ and $t_1 \leqslant t \leqslant T$, where τ is the duration of the rectangular pressure pulse in Figure 5.3, t_1 is the time when two travelling circumferential plastic hinges coalesce at $x = 0$ and T is the response duration.

5.5.2 First phase of motion, $0 \leqslant t \leqslant \tau$

In view of the foregoing comments, it is assumed that the radial velocity field is

$$\dot{w} = -\dot{W}_1 \qquad \text{for} \qquad 0 \leqslant x \leqslant \xi_0 \tag{5.44a}$$

and

$$\dot{w} = -\dot{W}_1(L - x)/(L - \xi_0) \qquad \text{when} \qquad \xi_0 \leqslant x \leqslant L, \tag{5.44b}$$

where ξ_0 is the time-independent position of the circumferential plastic hinge, as indicated in Figure 5.7(a).†

Equations (5.4), (5.5) and (5.44) predict that $\dot{\epsilon}_\theta \geqslant 0$ for $0 \leqslant x \leqslant L$, and $\dot{\kappa}_x = 0$ for $0 \leqslant x < \xi_0$ and $\xi_0 < x \leqslant L$, while $\dot{\kappa}_x \to -\infty$ and $\dot{\kappa}_x \to \infty$ at the stationary circumferential plastic hinges located at $x = \xi_0$ and $x = L$.‡ Thus, the normality requirements associated with the square yield condition in Figure 5.2 suggest that

$$N_\theta = N_0, \qquad M_x = -M_0 \qquad \text{for} \quad 0 \leqslant x \leqslant \xi_0 \tag{5.45a,b}$$

and

$$N_\theta = N_0, \qquad -M_0 \leqslant M_x \leqslant M_0 \qquad \text{when} \quad \xi_0 \leqslant x \leqslant L. \tag{5.46a,b}$$

Now, substituting equations (5.1), (5.6), (5.44a), (5.45a) and (5.45b) into equation (5.2) yields

$$\mu R \ddot{W}_1 = R p_0 - N_0 \tag{5.47)§}$$

for the central zone $0 \leqslant x \leqslant \xi_0$. Equation (5.47) may be integrated twice with respect to time to obtain the radial displacement

$$W_1 = (R p_0 - N_0) t^2 / 2\mu R \tag{5.48}$$

when satisfying the initial conditions which require $W_1 = \dot{W}_1 = 0$ when $t = 0$.

If equations (5.1), (5.6), (5.44b), (5.46a) and (5.47) are substituted into equation (5.2), then

$$R \, \partial^2 M_x / \partial x^2 = R p_0 - N_0 - (R p_0 - N_0)(L - x)/(L - \xi_0), \tag{5.49}$$

or

$$M_x = (p_0 - N_0/R)(x - \xi_0)^3 / 6(L - \xi_0) - M_0, \qquad \xi_0 \leqslant x \leqslant L, \tag{5.50}$$

when integrating twice with respect to x, and satisfying the conditions $M_x = -M_0$ and $Q_x = \partial M_x / \partial x = 0$ at the circumferential plastic hinge at $x = \xi_0$. Equation

---

† The response of a cylindrical shell is symmetric about the vertical plane located at $x = 0$. Thus, it is only necessary to examine the portion $0 \leqslant x \leqslant L$.

‡ It is assumed that $\dot{u} = 0$, which gives $\dot{\epsilon}_x = 0$ and, since $N_x = 0$, neither $\dot{u}$ nor N_x enters into the theoretical analysis.

§ $p_0 > N_0/R$ according to equation (5.30), so that the acceleration of the cylindrical shell is constant throughout this phase of motion, with a constant pressure pulse.

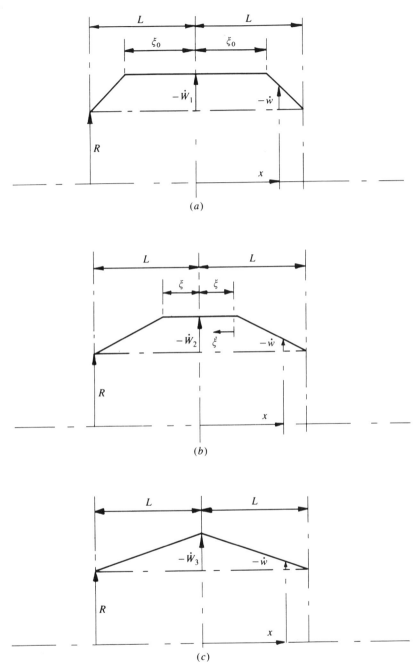

Figure 5.7 Axisymmetric radial velocity field for a fully clamped short cylindrical shell. (a) First phase of motion, $0 \leqslant t \leqslant \tau$, (b) Second phase of motion, $\tau \leqslant t \leqslant t_1$. (c) Third phase of motion, $t_1 \leqslant t \leqslant T$.

(5.50) must also satisfy the clamped boundary condition $M_x = M_0$ at $x = L$, or

$$(L - \xi_0)^2 = 12M_0/(p_0 - N_0/R), \tag{5.51}\dagger$$

which gives the position of the circumferential plastic hinge and allows equation (5.50) to be written in the form

$$M_x/M_0 = 2\{(x - \xi_0)/(L - \xi_0)\}^3 - 1, \qquad \xi_0 \leqslant x \leqslant L. \tag{5.52}$$

Equation (5.51) may also be written

$$(1 - \xi_0/L)^2 = 6/\{\eta(2 + \omega^2) - \omega^2\} \tag{5.53}$$

when using equations (5.17), (5.30) and (5.31). Equality (5.42) is achieved when $\eta = (6 + \omega^2)/(2 + \omega^2)$ which, when substituted into equation (5.53), predicts $\xi_0 = 0$, as expected.

5.5.3 Second phase of motion, $\tau \leqslant t \leqslant t_1$

It is evident from the time derivative of equation (5.48) that the radial velocity increases linearly with time throughout the first phase of motion. Thus, the shell reaches its greatest velocity at $t = \tau$ during the first phase and, therefore, continues to deform after the pressure pulse is released.

If the radial velocity field given by equations (5.44) was used for this phase of motion, then equations (5.47) and (5.49) to (5.51) would again be obtained, but with $p_0 = 0$. However, equation (5.51), with $p_0 = 0$, now predicts a different value for ξ_0, which suggests that the circumferential plastic hinge travels along the shell during this phase of motion. Thus, it is assumed that

$$\dot{w} = -\dot{W}_2 \qquad \text{for} \quad 0 \leqslant x \leqslant \xi \tag{5.54a}$$

and

$$\dot{w} = -\dot{W}_2(L - x)/(L - \xi) \qquad \text{for} \quad \xi \leqslant x \leqslant L, \tag{5.54b}$$

where ξ is now the time-dependent axial position of the circumferential plastic hinge, as indicated in Figure 5.7(b).

The normality requirements of plasticity for the square yield condition in Figure 5.2 are similar to those in § 5.5.2 and suggest the regimes

$$N_\theta = N_0, \qquad M_x = -M_0 \qquad \text{for} \quad 0 \leqslant x \leqslant \xi, \tag{5.55a, b}$$

and

$$N_\theta = N_0, \qquad -M_0 \leqslant M_x \leqslant M_0 \qquad \text{for} \quad \xi \leqslant x \leqslant L. \tag{5.56a, b}$$

Thus, equations (5.1), (5.2), (5.7), (5.54a), (5.55a) and (5.55b) predict that

$$\mu R \ddot{W}_2 = -N_0 \tag{5.57}$$

governs the behaviour in the central plastic zone $0 \leqslant x \leqslant \xi$. Now integrating equation (5.57) twice with respect to time gives

$$W_2 = -N_0 t^2/2\mu R + p_0\tau(t - \tau/2)/\mu \tag{5.58}$$

---

† This equation shows that the position ξ_0 of the circumferential plastic hinge remains constant, as assumed in the analysis.

when obtaining the two constants of integration by matching the radial displacement and radial velocity at $t = \tau$ with those obtained from equation (5.48) at the end of the first phase of motion.

The radial acceleration from equation (5.54b) for the zone $\xi \leqslant x \leqslant L$ is

$$\ddot{w} = -\partial\{\dot{W}_2(L - x)/(L - \xi)\}/\partial t, \qquad (5.59)$$

where ξ is the time-dependent axial position of the plastic hinge, as indicated in Figure 5.7(b). Equations (5.1), (5.2), (5.7), (5.56a) and (5.59) lead to the governing equation

$$\partial^2 M_x/\partial x^2 = -N_0/R - \mu(L - x)\,\partial\{\dot{W}_2/(L - \xi)\}/\partial t, \qquad \xi \leqslant x \leqslant L, \quad (5.60)$$

from which the axial bending moment is

$$M_x = -M_0 - N_0(x - \xi)^2/2R - \mu(x - \xi)^2(3L - x - 2\xi)\,\partial\{\dot{W}_2/6(L - \xi)\}/\partial t \tag{5.61}$$

when integrating twice with respect to x and satisfying the conditions at ξ for the plastic hinge (i.e., $M_x = -M_0$ and $Q_x = \partial M_x/\partial x = 0$).

Now the term

$$\partial\{\dot{W}_2/6(L - \xi)\}/\partial t = \ddot{W}_2/6(L - \xi) + \dot{W}_2\dot{\xi}/6(L - \xi)^2 \tag{5.62}$$

which, when using equation (5.57) and the derivative of equation (5.58), becomes

$$\partial\{\dot{W}_2/6(L - \xi)\}/\partial t = -N_0/6\mu R(L - \xi) + \dot{\xi}(p_0\tau - N_0t/R)/6\mu(L - \xi)^2. \tag{5.63}$$

The temporal axial location (ξ) of the axisymmetric circumferential plastic hinge in equation (5.61) is unknown. However, the cylindrical shell is fully clamped at both ends so that equation (5.61) must satisfy $M_x = M_0$ at $x = L$, or

$$-M_0 - N_0(L - \xi)^2/2R - \mu(L - \xi)^3\,\partial\{\dot{W}_2/3(L - \xi)\}/\partial t = M_0 \tag{5.64}$$

which, using equations (5.17), (5.30), (5.31) and (5.63), predicts the velocity of the travelling plastic hinge

$$\dot{\xi} = \frac{L\{6 + \omega^2(1 - \xi/L)^2\}}{2(1 - \xi/L)\{\omega^2 t - \eta(2 + \omega^2)\tau\}}. \tag{5.65}$$

This first-order differential equation may be integrated using standard methods to predict

$$(1 - \xi/L)^2 = 6t/\{\eta(2 + \omega^2)\tau - \omega^2 t\}, \tag{5.66}$$

since $\xi = \xi_0$ at $t = \tau$, where ξ_0 is defined by equation (5.53).

The second phase of motion is completed at $t = t_1$ when $\xi = 0$, or

$$t_1 = \eta(2 + \omega^2)\tau/(6 + \omega^2) \tag{5.67}$$

according to equation (5.66).

5.5.4 Third phase of motion, $t_1 \leqslant t \leqslant T$

Now, the axisymmetric circumferential plastic hinge reaches the mid-span ($\xi = x = 0$) of a cylindrical shell when $t = t_1$. However, the cylindrical shell has a kinetic energy

$$K_3 = 72p_0^2\tau^2\pi RL/\{\mu(6 + \omega^2)^2\} \tag{5.68}$$

which is associated with the maximum velocity given by the time derivative of equation (5.58) at $t = t_1$. This kinetic energy is dissipated as plastic work during a third and final phase of motion which is taken to have a transverse velocity profile with a form similar to equation (5.54b), with $\zeta = 0$, i.e.,

$$\dot{w} = -\dot{W}_3(L - x)/L, \qquad 0 \leqslant x \leqslant L. \tag{5.69}$$

In this circumstance, equations (5.4) and (5.5), together with the normality rule, suggest the portion

$$N = N_0, \qquad -M_0 \leqslant M_x \leqslant M_0, \qquad 0 \leqslant x \leqslant L, \tag{5.70a, b}$$

of the yield condition in Figure 5.2. Thus, equations (5.1), (5.2), (5.7), (5.69) and (5.70a) give

$$\partial^2 M_x/\partial x^2 = -N_0/R - \mu\dot{W}_3(L - x)/L, \tag{5.71}$$

or

$$M_x = -N_0 x^2/2R - \mu\dot{W}_3(Lx^2/2 - x^3/6)/L - M_0 \tag{5.72}$$

when satisfying $Q_x = \partial M_x/\partial x = 0$ at $x = 0$ and $M_x = -M_0$ at $x = 0$.

Equation (5.72), with $M_x = M_0$ at $x = L$, predicts

$$\mu\ddot{W}_3 = -3M_0(2 + \omega^2)/L^2 \tag{5.73}$$

which, integrating with respect to time, gives

$$\mu\dot{W}_3 = -3M_0(2 + \omega^2)t/L^2 + 3p_0\tau/2 \tag{5.74}$$

when ensuring continuity with the velocity at the end of the second phase ($t = t_1$) from equation (5.58). Motion therefore ceases when

$$T = \eta\tau, \tag{5.75}$$

where η is defined by equation (5.17) and p_c by equation (5.30). Another integration of equation (5.74) predicts the radial displacement at the mid-span

$$\mu W_3/p_c = -3t^2/4 + 3\eta\tau t/2 - \eta\tau^2/2 - \eta^2\tau^2(2 + \omega^2)/4(6 + \omega^2) \tag{5.76}$$

when evaluating the constant of integration to ensure that equation (5.58) is recovered at $t = t_1$. Thus, the maximum permanent radial displacement from equations (5.69) and (5.76) is

$$W_f = -p_0\tau^2\{\eta(8 + \omega^2) - (6 + \omega^2)\}/\{2\mu(6 + \omega^2)\}. \tag{5.77}$$

5.5.5 Static admissibility

Equations (5.45) and (5.46) show that the theoretical solution in § 5.5.2 is statically admissible throughout the first phase of motion, provided inequality (5.46b) is satisfied. Now, equation (5.50) may be written

$$M_x/M_0 = 2\{(x - \xi_0)/(L - \xi_0)\}^3 - 1, \qquad \xi_0 \leqslant x \leqslant L, \tag{5.78}$$

with the aid of equation (5.51), and is statically admissible as indicated in Figure 5.8(a), since $\partial^2 M_x/\partial x^2 \geqslant 0$ for $\xi_0 \leqslant x \leqslant L$ and $\partial M_x/\partial x = 0$ at $x = \xi_0$.

The inequality (5.56b) must be satisfied for static admissibility during the second phase of motion examined in § 5.5.3. Equations (5.61), (5.63) and (5.65) predict

$$\frac{M_x}{M_0} = -1 - \omega^2 \left(\frac{x}{L} - \frac{\xi}{L}\right)^2 \left[1 - \frac{\left(3 - \frac{x}{L} - \frac{2\xi}{L}\right)}{3\left(1 - \frac{\xi}{L}\right)} \left\{ 1 + \left\{ \frac{6 + \omega^2(1 - \xi/L)^2}{2\omega^2(1 - \xi/L)^2} \right\} \right\} \right],$$

$$\xi \leqslant x \leqslant L. \qquad (5.79)$$

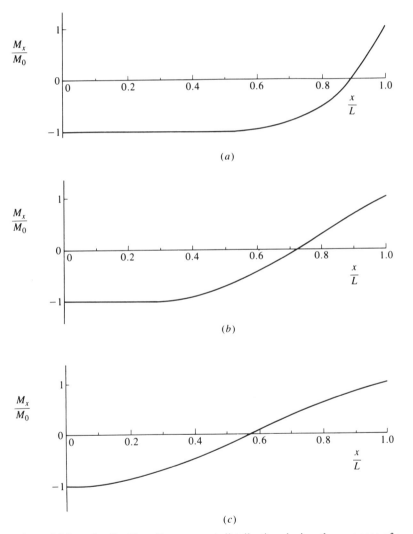

(a)

(b)

(c)

Figure 5.8 Longitudinal bending moment distribution during the response of a fully clamped short cylindrical shell with $\omega^2 = 3$. (a) First phase of motion $(0 \leqslant t \leqslant \tau)$ with $\eta = 5$ $(\xi_0/L = 0.4778)$. (b) Second phase of motion $(\tau \leqslant t \leqslant t_1)$ when $\xi/L = 0.3$. (c) Final phase of motion $(t_1 \leqslant t \leqslant T)$.

This satisfies inequality (5.56b) since $M_x = -M_0$ and $\partial M_x/\partial x = 0$ at $x = \xi$, $M_x = M_0$ at $x = L$ and $\partial^2 M_x/\partial x^2 \leqslant 0$ at $x = L$, while $\partial^2 M_x/\partial x^2 \geqslant 0$ at $x = \xi$ and $\partial M_x/\partial x \geqslant 0$ at $x = L$ provided inequality (5.43) is satisfied. It may be shown that $\partial^2 M_x/\partial x^2$ has only one zero value, so that M_x has the characteristic shape indicated in Figure 5.8(b) and is, therefore, statically admissible.

Finally, it is straightforward to show that, during the third phase of motion in § 5.5.4, equations (5.72) and (5.73) give

$$M_x/M_0 = \{6 + \omega^2 - (2 + \omega^2)(x/L)\}(x/L)^2/2 - 1, \qquad 0 \leqslant x \leqslant L, \quad (5.80)$$

which satisfies inequality (5.70b) and is, therefore, statically admissible, as indicated in Figure 5.8(c).

The theoretical solution in §§ 5.5.2 to 5.5.4 for a dynamically loaded cylindrical shell is statically admissible throughout the entire response for the rectangular pressure pulse in Figure 5.3 having any magnitude p_0 (or η). However, the dimensions of the cylindrical shell must still satisfy inequality (5.43), i.e., $\omega^2 \leqslant 6$.

5.5.6 Kinematic admissibility

The plastic hinges in the theoretical solution in §§ 5.5.2 to 5.5.4 for a dynamically loaded cylindrical shell are circumferential and axisymmetric with $M_x = \pm M_0$. Thus, it is evident that equations (3.54) to (3.59) in § 3.4.6, for the kinematic requirements associated with plastic hinges in beams, remain identical for axisymmetric circumferential plastic hinges in a cylindrical shell, where w is now interpreted as the radial displacement.

The normality requirements of plasticity are satisfied throughout all three phases of motion, together with the initial and support conditions for radial displacement and radial velocity. Stationary plastic hinges with $[M_x] = [Q_x] = 0$ develop during the first and third phases of motion, so that the theoretical solutions in §§ 5.5.2 and 5.5.4 are kinematically admissible. It is left as an exercise to an interested reader to show that the theoretical solution in § 5.5.3 for the second phase of motion with travelling plastic hinges is also kinematically admissible.†

5.5.7 Impulsive loading

In order to remove the restrictions imposed by inequality (5.42), a theoretical solution was developed in §§ 5.5.2 to 5.5.4 using the transverse velocity profile with a travelling plastic hinge, as indicated in Figure 5.7. In fact, this theoretical solution is valid for any value of $\eta \geqslant (6 + \omega^2)/(2 + \omega^2)$, although the inequality (5.43)

---

† The radial displacement profiles during all phases of motion may be obtained analytically after some algebra (see Hodge[5.2]).

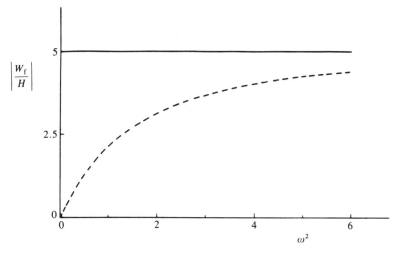

Figure 5.9 Comparison of the dimensionless magnitude of the maximum permanent radial displacement for an impulsively loaded long cylindrical shell (——: equation (5.21)) and an impulsively loaded fully clamped short cylindrical shell (– – –: equation (5.82)) when $\lambda = 40$.

($\omega^2 \leqslant 6$) for short shells remains. In particular, the equations in §§ 5.5.2 to 5.5.4 may be used to predict the response of an impulsively loaded cylindrical shell with $\eta \gg 1$ and $\tau \to 0$. Thus, the response duration according to equations (5.19) and (5.75) is

$$T = \mu V_0/p_c, \tag{5.81}$$

where p_c is the static collapse pressure defined by equation (5.30). The maximum permanent radial displacement from equation (5.77) and equations (5.19), (5.23) and (5.30) is

$$W_f/H = -(\lambda/8)[\omega^2(8 + \omega^2)/\{(2 + \omega^2)(6 + \omega^2)\}]. \tag{5.82}$$

The strengthening influence of fully clamped supports on cylindrical shells, with $\omega^2 \leqslant 6$, is found by comparing the theoretical predictions of equations (5.21) and (5.82), as shown in Figure 5.9.

Equation (5.77) may be written

$$W_f/H = -(\lambda/8)[\omega^2\{8 + \omega^2 - (6 + \omega^2)/\eta\}/\{(2 + \omega^2)(6 + \omega^2)\}], \tag{5.83}$$

where

$$\lambda = (p_0\tau)^2R/\mu M_0 \tag{5.84}$$

which reduces to equation (5.23) when using equation (5.19) for impulsive velocity loading. The maximum permanent radial displacements predicted by equations (5.82) and (5.83) are compared in Figure 5.10.

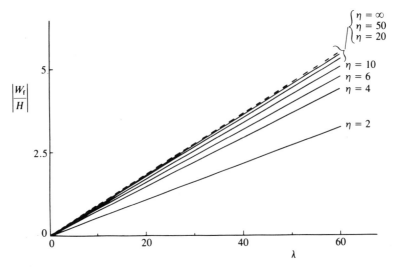

Figure 5.10 Variation of the dimensionless maximum permanent radial displacement of a short cylindrical shell ($\omega^2 = 3$) with dimensionless pulse (λ given by equation (5.84)) and dimensionless magnitude of the rectangular pressure pulse (η). ---: equation (5.82) and equation (5.83) with $\eta \to \infty$. ——: equation (5.83).

5.6 Elastic, perfectly plastic spherical shell subjected to a spherically symmetric dynamic pressure

5.6.1 Introduction

The spherically symmetric response of a complete spherical shell with a thin wall is sufficiently simple to allow a departure to be made from the usual approximation of a rigid, perfectly plastic material which is employed in the previous chapters. An exploration is made in this section into the importance of material elastic effects, together with an examination of the accuracy of the rigid plastic method.

A complete spherical shell is subjected to a spherically symmetric internal dynamic pressure pulse of the rectangular form shown in Figure 5.3 and described by equations (5.6) and (5.7). The elastic response is examined in § 5.6.3, while the dynamic elastic, perfectly plastic and rigid, perfectly plastic cases are considered in §§ 5.6.4 and 5.6.5, respectively. Some comparisons are made in § 5.6.6 between the various theoretical solutions, which leads to observations on the validity of an analysis using a rigid, perfectly plastic material.

5.6.2 Governing equations

An element of a spherical shell which is undergoing a spherically symmetric response is shown in Figure 5.11. The in-plane and radial shear forces and all

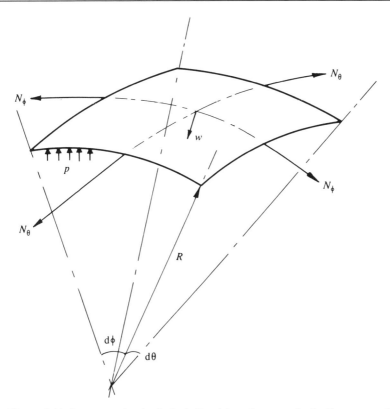

Figure 5.11 Segment of spherical shell subjected to a spherically symmetric internal pressure.

bending moments are zero for a spherically symmetric response.† Moreover, $N_\theta = N_\phi = N$. Thus, radial equilibrium of the shell element in Figure 5.11 demands

$$\mu \, d^2w/dt^2 - 2N/R + p = 0, \tag{5.85}$$

where μ is the mass per unit surface area of the shell.

The biaxial membrane strains in the spherical shell are $\epsilon_\theta = \epsilon_\phi = \epsilon$, where

$$\epsilon = -w/R, \tag{5.86}$$

when w is the radially inwards displacement shown in Figure 5.11.

---

† The radius R increases to $R - w$ (note that w is negative for a radially outwards displacement) and, therefore, the initial curvature $1/R$ of a spherical shell decreases to $1/(R - w)$, giving a change in curvature $1/(R - w) - 1/R \cong w/R^2$. This gives rise to a maximum strain $Hw/2R^2$, where H is the wall thickness of a shell. However, this strain is $H/2R$ times the membrane strain according to equation (5.86) which is, therefore, negligible for thin-walled shells with $H/2R \ll 1$.

5.6.3 Elastic response

5.6.3(a) First phase of motion, $0 \leqslant t \leqslant \tau$

Now, for a linear elastic material, it is assumed that

$$N = EH\epsilon/(1 - v) \tag{5.87}\dagger$$

which, using equation (5.86), becomes

$$N = -EHw/(1 - v)R. \tag{5.88}$$

Therefore, equation (5.85) may be written

$$\mu \, d^2w/dt^2 + 2EHw/(1 - v)R^2 = -p_0, \qquad 0 \leqslant t \leqslant \tau, \tag{5.89}$$

when using equations (5.6) and (5.88), or

$$d^2w/dt^2 + a^2w = -b, \qquad 0 \leqslant t \leqslant \tau, \tag{5.90}\ddagger$$

where

$$a^2 = 2EH/\mu(1 - v)R^2 \qquad \text{and} \qquad b = p_0/\mu. \tag{5.91a, b}$$

Equation (5.90) has the solution

$$w = b\{\cos(at) - 1\}/a^2 \tag{5.92}$$

when satisfying the initial conditions $w = \dot{w} = 0$ at $t = 0$.

5.6.3(b) Second phase of motion, $\tau \leqslant t \leqslant T$

In this phase of motion, $p = 0$ according to equation (5.7) for a rectangular pressure pulse, so that equation (5.90) becomes

$$d^2w/dt^2 + a^2w = 0, \qquad \tau \leqslant t \leqslant T, \tag{5.93}\S$$

which now governs the dynamic behaviour. Equation (5.93) has the solution

$$w = b[\{1 - \cos(a\tau)\} \cos(at) - \sin(a\tau) \sin(at)]/a^2 \tag{5.94}$$

when matching the radial displacement and radial velocity with equation (5.92) and the time derivative of equation (5.92) at $t = \tau$, respectively.

The maximum radial displacement is reached at $t = T$ when $\dot{w} = 0$, or

$$\tan(aT) = -\sin(a\tau)/\{1 - \cos(a\tau)\} \tag{5.95}$$

which, when substituted in equation (5.94), predicts the maximum radial displacement

$$w_m = -\sqrt{2}b\{1 - \cos(a\tau)\}^{1/2}/a^2. \tag{5.96}$$

5.6.3(c) Third phase of motion, $t \geqslant T$

The spherical shell commences to unload elastically at $t = T$ and the response is, therefore, governed by equations (5.93) and (5.94) which remain valid for $t \geqslant T$. Thus, a shell continues to vibrate elastically, as indicated in Figure 5.12.

---

† $\epsilon_\theta = (\sigma_\theta - v\sigma_\phi)/E$ according to Hooke's law for a biaxial stress state, or $\sigma = E\epsilon/(1 - v)$ when $\sigma_\theta = \sigma_\phi$ $= \sigma$. Thus, $\sigma H = N = EH\epsilon/(1 - v)$ for a shell with a solid homogeneous cross-section of thickness H.

‡ This differential equation is similar to equation (10.11) in § 10.2.3. It is evident from equations (5.90) and (5.92) that the shell wall is accelerating radially outwards throughout this phase of motion.

§ The shell wall is decelerating during this phase of motion.

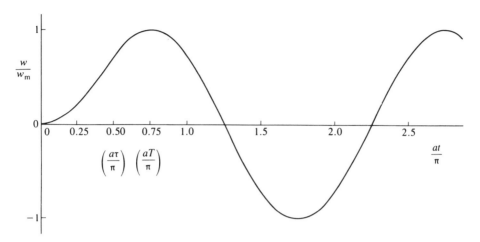

Figure 5.12 Dynamic response of an elastic spherical shell subjected to a rectangular pressure pulse, with the pressure–time history in Figure 5.3, according to equations (5.92) and (5.94) and $a\tau = \pi/2(\gamma_0 = \tau/T = 0.25)$. The maximum radial displacement (w_m) is predicted by equation (5.96).

5.6.3(d) Discussion of energies

The kinetic energy in a spherical shell at $t = \tau$, when the rectangular pressure pulse is released, equals $K_1 = \mu(4\pi R^2)\dot{w}^2(\tau)/2$, where $\dot{w}(\tau)$ is given by equation (5.92) with $t = \tau$. Thus,

$$K_1 = \{\pi(1 - v)p_0^2 R^4/EH\} \sin^2(a\tau). \tag{5.97}$$

The total elastic strain energy associated with the biaxial membrane forces in a spherical shell at $t = \tau$ is $S_1 = (2N\epsilon)(4\pi R^2)/2$, or

$$S_1 = \{\pi(1 - v)p_0^2 R^4/EH\}\{\cos(a\tau) - 1\}^2 \tag{5.98}$$

when using equations (5.86), (5.87) and (5.92).

Now, the rectangular pressure pulse imparts a total energy

$$E_T = -\int_0^\tau p_0 \dot{w}(4\pi R^2)\, \mathrm{d}t$$

to a spherical shell, which, using the derivative of equation (5.92), becomes

$$E_T = 2\pi(1 - v)p_0^2 R^4\{1 - \cos(a\tau)\}/EH. \tag{5.99}$$

It is straightforward to show that $K_1 + S_1 = E_T$, as required for energy conservation.

5.6.3(e) Impulsive loading

Now, for impulsive loading, $p_0 \to \infty$ and $\tau \to 0$, so that conservation of linear momentum at $t = 0$ requires

$$\mu V_0 = p_0\tau, \tag{5.100}$$

where V_0 is the initial radially outwards impulsive velocity.

Clearly, the first phase of motion in § 5.6.3(a) vanishes when $\tau \to 0$, while equation (5.95) gives $\tan(aT) \to -\infty$, or $aT = \pi/2$, which predicts a response duration

$$T = \pi R \{\mu(1 - v)/8EH\}^{1/2}. \tag{5.101}$$

Moreover, equation (5.96) gives a maximum radial displacement $w_m = -b\tau/a$, or

$$w_m/H = -V_0 R \{\mu(1 - v)/2EH^3\}^{1/2}. \tag{5.102}$$

5.6.4 Elastic, perfectly plastic response

5.6.4(a) Introduction

The theoretical analysis which is outlined in § 5.6.3 for a linear elastic material is valid provided the membrane forces in a spherical shell remain within the elastic range, or

$$N \leqslant N_0, \tag{5.103}†$$

where $N_0 = \sigma_0 H$ for a shell with a solid homogeneous cross-section of thickness H. Thus, equation (5.88) shows that a spherical shell remains elastic provided

$$-w \leqslant RN_0(1 - v)/EH \tag{5.104}$$

since $w < 0$ for internal pressures, as indicated in Figure 5.11.

Now, it is assumed that inequality (5.104) is satisfied throughout the first phase of motion in § 5.6.3(a) and that it is just violated at a time $t_1 \geqslant \tau$ during the second phase of motion in § 5.6.3(b). The details of this analysis are presented in the following sections.

5.6.4(b) First phase of motion, $0 \leqslant t \leqslant \tau$

The analysis of this phase of motion is identical to that outlined in § 5.6.3(a).

5.6.4(c) Second phase of motion, $\tau \leqslant t \leqslant t_1$

The governing equation (5.93) and the radial displacement given by equation (5.94) remain valid during this phase of motion until the time t_1 when

$$w(t_1) = -RN_0(1 - v)/EH \tag{5.105}‡$$

according to equality (5.104). Thus, equations (5.94) and (5.105) give

$$\{1 - \cos(a\tau)\} \cos(at_1) - \sin(a\tau) \sin(at_1) = -2N_0/p_0 R, \tag{5.106}§$$

the solution of which predicts the time t_1 when the shell material first yields plastically.

---

† The biaxial membrane forces in a spherical shell undergoing a spherically symmetric response may be written $N_\theta = N_\phi = N$. Thus, plastic flow occurs when $N = N_0$ according to both the Tresca and von Mises yield criteria.

‡ It is assumed that the spherical shell is sufficiently thin for the entire cross-section to become plastic instantaneously at $t = t_1$.

§ See equations (5.164) and (5.165) for alternative forms of equation (5.106).

5.6.4(d) Third phase of motion, $t_1 \leqslant t \leqslant T_1$

If $t_1 \leqslant T$, where T is defined by equation (5.95), then radial motion continues for a perfectly plastic spherical shell with the constant membrane forces

$$N_\theta = N_\phi = N_0. \tag{5.107}$$

In this circumstance, equations (5.7) and (5.85) yield

$$d^2w/dt^2 = 2N_0/\mu R \tag{5.108}$$

or

$$w = N_0 t^2/\mu R + A_3 t + B_3, \tag{5.109}$$

where the constants of integration A_3 and B_3 are obtained by matching with equation (5.105) at $t = t_1$ and the derivative of equation (5.94) at $t = t_1$, which give

$$A_3 = -b[\{1 - \cos(a\tau)\} \sin(at_1) + \sin(a\tau)\cos(at_1)]/a - 2N_0 t_1/\mu R \tag{5.110a}$$

and

$$B_3 = -RN_0(1 - v)/EH - N_0 t_1^2/\mu R - A_3 t_1. \tag{5.110b}$$

Radial motion continues until $t = T_1$, when $\dot{w} = 0$, where

$$T_1 = -\mu R A_3/2N_0, \tag{5.111}$$

which gives a maximum radial displacement

$$w_{\mathrm{m}} = B_3 - \mu R A_3^2/4N_0. \tag{5.112}$$

5.6.4(e) Fourth phase of motion, $t \geqslant T_1$

It is assumed that a shell unloads elastically at $t = T_1$ and continues to vibrate elastically during this phase of motion. Now, during unloading, the membrane force N varies, as indicated in Figure 5.13, or

$$N = N_0 + EH(w_{\mathrm{m}} - w)/(1 - v)R \tag{5.113}$$

when using equation (5.87), since the change in elastic strain is $\epsilon = -(w_{\mathrm{m}} - w)/R$

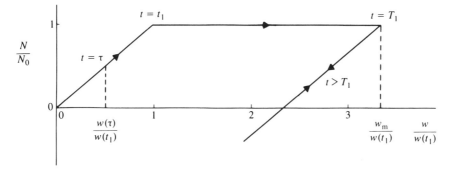

Figure 5.13 Variation of the dimensionless membrane force (N/N_0) during the dynamic response of an elastic-plastic spherical shell according to equations (5.88), (5.107) and (5.113) when using equation (5.105).

during unloading. Thus, the equilibrium equation (5.85), together with equation (5.7), becomes

$$\mu \, d^2w/dt^2 - 2\{N_0 + EH(w_m - w)/(1 - v)R\}/R = 0$$

or

$$d^2w/dt^2 + a^2w = c, \qquad (5.114)$$

where

$$c = 2\{N_0 + EHw_m/(1 - v)R\}/\mu R \qquad (5.115)$$

and a^2 and w_m are defined by equations (5.91a) and (5.112), respectively.

The differential equation (5.114) predicts the radial displacement

$$w = (w_m - c/a^2)\{\cos(aT_1)\cos(at) + \sin(aT_1)\sin(at)\} + c/a^2, \qquad t \geqslant T_1, \qquad (5.116)$$

when satisfying the initial conditions $w = w_m$ and $\dot{w} = 0$ at $t = T_1$. The minimum radial displacement w^* shown in Figure 5.14 occurs at $t = T_2$ when $\dot{w} = 0$, or

$$T_2 = T_1 + \pi/a, \qquad (5.117)$$

and the associated radial displacement is

$$w^* = -w_m + 2c/a^2. \qquad (5.118)$$

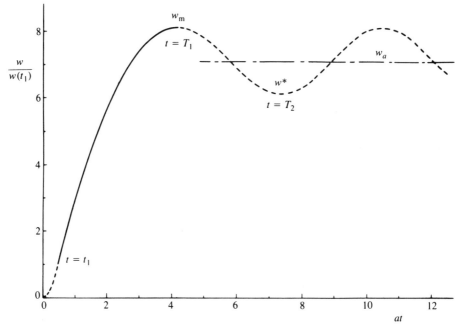

Figure 5.14 Dimensionless radial displacement of an elastic, perfectly plastic spherical shell subjected to a rectangular pressure pulse of the form shown in Figure 5.3 with $a\tau = \pi/8$ ($\gamma_0 = \tau/T = \frac{1}{16}$) and $p_0R/2N_0 = \eta = 10$. ---: elastic response. ——: plastic response. —·—: average permanent radial displacement (w_a) for $t \geqslant T_1$.

The average final radial displacement w_a is taken as $(w_m + w^*)/2$, or

$$w_a = c/a^2. \tag{5.119}\dagger$$

5.6.4(f) Discussion of energies

Equations (5.97) to (5.99) for K_1, S_1 and E_T remain valid for the kinetic, strain and total external energies at $t = \tau$, respectively.

The elastic strain energy at $t = t_1$, when plastic yielding first occurs, is $(4\pi R^2)(2N)(-w(t_1)/R)/2$, which, using equation (5.88), becomes

$$S_2 = 4\pi EHw^2(t_1)/(1 - v), \tag{5.120}$$

where $w(t_1)$ is defined by equation (5.105). The associated kinetic energy at $t = t_1$ is

$$K_2 = 2\pi\mu R^2\dot{w}^2(t_1), \tag{5.121}$$

where $\dot{w}(t_1)$ is found from the time derivative of equations (5.94) or (5.109) evaluated at $t = t_1$. It is straightforward to show that $K_2 + S_2 = E_T$ as required for the conservation of energy.

The plastic energy dissipated during the third phase of motion $t_1 \leqslant t \leqslant T_1$ in § 5.6.4(d) equals $2N_0(4\pi R^2)\{-w_m + w(t_1)\}/R$, or

$$D_3 = 8\pi N_0 R\{w(t_1) - w_m\}, \tag{5.122}$$

where w_m is defined by equation (5.112). It may be shown that $D_3 = K_2$ as expected.

Now, the ratio between the elastic and total energies may be expressed in the dimensionless form

$$\alpha = S_2/E_T, \tag{5.123}$$

which, using equations (5.99) and (5.120), gives

$$\alpha = 1/\{2\eta^2(1 - \cos(2\pi\gamma_0))\} \tag{5.124}$$

when

$$\gamma_0 = a\tau/2\pi \tag{5.125}$$

and

$$\eta = p_0/p_c \tag{5.126}$$

is a dimensionless ratio between the magnitude of the dynamic pressure pulse and the static collapse pressure

$$p_c = 2N_0/R \tag{5.127}\ddagger$$

for a perfectly plastic complete spherical shell. The period of vibration for a linear elastic spherical shell is $T = 2\pi/a$, so that $\gamma_0 = a\tau/2\pi = \tau/T$ is a dimensionless ratio between the duration of the rectangular pressure pulse and the natural elastic period.

† It may be shown that $w_m - w^* = 2w(t_1)$ when using equations (5.91a), (5.115) and (5.118). Thus, $w_m - w_a = w(t_1)$ which, substituting into equation (5.113), predicts that $N = 0$ when $w = w_a$.
‡ See footnote to equation (5.141).

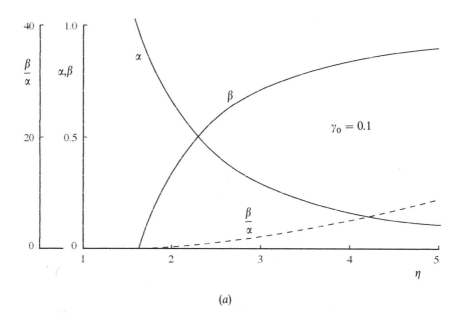

(a)

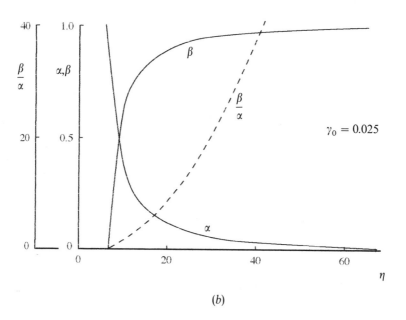

(b)

Figure 5.15 Energy ratios for an elastic, perfectly plastic spherical shell subjected to a rectangular pressure pulse with a dimensionless magnitude $\eta = p_0/p_c$ and dimensionless duration $\gamma_0 = \tau/T$. ——— α: equations (5.123) and (5.124). ——— β: equations (5.128) and (5.129). $---$: β/α. (a) $\gamma_0 = 0.10$. (b) $\gamma_0 = 0.025$. (c) $\gamma_0 = 0.01$.

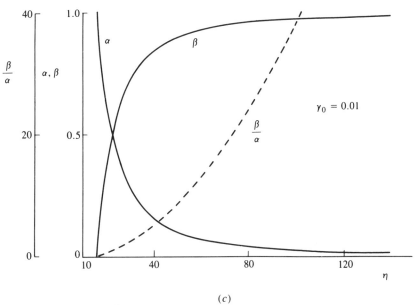

(c)

Figure 5.15—contd.

The ratio between the plastic energy absorbed and the total external energy is

$$\beta = D_3/E_T \tag{5.128}$$

or

$$\beta = 1 - \alpha, \tag{5.129}$$

while the ratio of the plastic to elastic energies is $\beta/\alpha = 1/\alpha - 1$.

The variation of the dimensionless energies α, β and β/α, with the dimensionless load η, is shown in Figure 5.15 for several values of the dimensionless time γ_0.

5.6.4(g) Impulsive loading

The foregoing analysis remains valid for an impulsive velocity loading which satisfies equation (5.100), with $p_0 \to \infty$ and $\tau \to 0$. The first phase of motion in § 5.6.4(b) with a duration τ therefore vanishes, while the duration of the second phase in § 5.6.4(c), which now becomes the first phase for an impulsive velocity loading, is given by equation (5.106), with $\tau \to 0$ and $p_0 \to \infty$, or

$$\sin(at_1) = -aw(t_1)/V_0, \tag{5.130}†$$

where $w(t_1)$ is defined by equation (5.105). Equation (5.130) may be recast in the form

$$\sin(2\pi\gamma_1) = \lambda^{-1/2}, \tag{5.131}†$$

† The analysis for a elastic, perfectly plastic spherical shell has been developed for pressure pulses which give rise to $t_1 \geqslant \tau$, as noted in § 5.6.4(a). Thus, t_1 is finite according to equation (5.130) when $\tau \to 0$ but $t_1 \to 0$ for $\lambda \gg 1$.

where, from § 5.6.4(f),

$$\gamma_1 = at_1/2\pi = t_1/T \tag{5.132}$$

and T is the natural period for a linear elastic spherical shell, and where

$$\lambda = \mu EHV_0^2/2(1 - v)N_0^2 \tag{5.133}$$

is a dimensionless initial kinetic energy.

The second and third phases of motion follow from § 5.6.4(d) and § 5.6.4(e), with $\tau \to 0$ and $p_0 \to \infty$. Equation (5.110a), therefore, reduces to

$$A_3/V_0 = -\cos(2\pi\gamma_1) - 2\pi\gamma_1\lambda^{-1/2}, \tag{5.134}$$

while equation (5.110b) becomes

$$B_3/w(t_1) = 1 - (2\pi\gamma_1)^2/2 - 2\pi\gamma_1\lambda^{1/2}\cos(2\pi\gamma_1). \tag{5.135}$$

Thus,

$$T_1 = t_1 + \lambda^{1/2}(T/2\pi)\cos(2\pi\gamma_1) \tag{5.136}$$

and

$$w_m/w(t_1) = 1 + (\lambda/2)\cos^2(2\pi\gamma_1). \tag{5.137}$$

Finally, it may be shown with the aid of equation (5.131) that

$$w_m/w(t_1) = (\lambda + 1)/2 \tag{5.138}$$

and

$$w_a/w(t_1) = \{w_m - w(t_1)\}/w(t_1)$$

or

$$w_a/w(t_1) = (\lambda - 1)/2. \tag{5.139}$$

5.6.5 Rigid, perfectly plastic response

5.6.5(a) Introduction

The wholly elastic dynamic response of a complete spherical shell is examined in § 5.6.3, while the behaviour for a shell made from an elastic, perfectly plastic material is presented in § 5.6.4 for sufficiently large dynamic pressures which cause plastic yielding of the material. A simple theoretical solution is developed in this section for a shell made from a rigid, perfectly plastic material which is valid for severe dynamic loadings when elastic effects are not important.

5.6.5(b) First phase of motion, $0 \leqslant t \leqslant \tau$

A complete spherical shell is subjected to a spherically symmetric internal pressure with the rectangular pressure–time history shown in Figure 5.3 and described by equations (5.6) and (5.7). The biaxial membrane forces are equal, or $N_\theta = N_\phi = N$, where

$$N = N_0 \tag{5.140}$$

is the fully plastic membrane force. Thus, the equilibrium equation (5.85) becomes

$$\mu \, d^2w/dt^2 = 2N_0/R - p_0 \qquad (5.141)\dagger$$

or

$$w = (N_0/R - p_0/2)t^2/\mu \qquad (5.142)$$

when integrating and satisfying the initial conditions $w = \dot{w} = 0$ at $t = 0$.

The internal pressure is released at $t = \tau$, and equation (5.142) for the associated radial displacement may be written in the dimensionless form

$$w(\tau)/w(t_1) = (\eta - 1)(2\pi\gamma_0)^2/2, \qquad (5.143)$$

while the corresponding radial velocity is

$$\dot{w}(\tau)/V_0 = (1 - \eta)(2\pi\gamma_0)\lambda^{-1/2} \qquad (5.144)$$

when using equations (5.105), (5.125) to (5.127) and (5.133).

5.6.5(c) Second phase of motion, $\tau \leqslant t \leqslant T$

The internal pressure is zero during this phase of motion according to equation (5.7), but the shell has a finite radial velocity at $t = \tau$ which is given by equation (5.144). Thus, a second phase of motion is required and the governing equation (5.141) now becomes

$$d^2w/dt^2 = 2N_0/\mu R, \qquad (5.145)$$

which predicts

$$\dot{w} = 2\pi V_0(\gamma - \eta\gamma_0)\lambda^{-1/2}, \qquad \gamma \geqslant \gamma_0, \qquad (5.146)$$

and

$$w/w(t_1) = \pi\eta(2\pi\gamma_0)(2\gamma - \gamma_0) - (2\pi\gamma)^2/2, \qquad \gamma \geqslant \gamma_0, \qquad (5.147)$$

when matching with equations (5.143) and (5.144) at $t = \tau$ and where

$$\gamma = at/2\pi = t/T. \qquad (5.148)$$

Radial motion finally ceases at $t = T$ when $\dot{w} = 0$, or

$$\gamma_2 = \eta\gamma_0, \qquad (5.149)$$

where

$$\gamma_2 = aT/2\pi = T/T. \qquad (5.150)$$

The associated dimensionless permanent radial displacement is

$$w_p/w(t_1) = \eta(\eta - 1)(2\pi\gamma_0)^2/2. \qquad (5.151)$$

5.6.5(d) Discussion of energies

The total external energy applied by a rectangular pressure pulse to a rigid, perfectly plastic spherical shell is $E_T = -\int_0^\tau p_0\dot{w}(4\pi R^2) \, dt$, which, using the time

---

† $d^2w/dt^2 = 0$ for static behaviour and, therefore, $p_0 = p_c = 2N_0/R$, which is the static collapse pressure given by equation (5.127).

derivative of equation (5.142), becomes

$$E_T = 4\pi(1 - v)R^2 N_0^2 (2\pi\gamma_0)^2 \eta(\eta - 1)/EH. \tag{5.152}$$

All of this energy is, of course, dissipated plastically by the biaxial membrane force field. However, it is of interest to observe that the elastic strain energy S_2 in a complete elastic, perfectly plastic spherical shell, when plastic yielding first commences, is given by equation (5.120). This equation may be recast in the form

$$S_2 = 4\pi(1 - v)R^2 N_0^2/EH \tag{5.153}$$

with the aid of equation (5.105). Thus, the energy ratio

$$E_r = E_T/S_2 \tag{5.154}$$

becomes

$$E_r = \eta(\eta - 1)(2\pi\gamma_0)^2, \tag{5.155a}$$

or

$$E_r = 2w_p/w(t_1) \tag{5.155b}$$

when using equation (5.151).

If $E_r < 1$, then the response of an elastic, perfectly plastic spherical shell is wholly elastic and, therefore, a rigid perfectly plastic analysis would be entirely inappropriate. On the other hand, if $E_r \gg 1$, then elastic effects should be unimportant. Further discussion on the validity of the rigid plastic analysis developed in this section is found in § 5.6.6.

5.6.5(e) Impulsive loading

If a rigid, perfectly plastic complete spherical shell is subjected to a spherically symmetric radially outwards impulsive velocity V_0, then equation (5.145) governs the entire response with the initial conditions $\dot{w} = -V_0$ and $w = 0$ at $t = 0$. Thus, integrating equation (5.145) and satisfying the initial conditions gives the radial displacement

$$w = N_0 t^2/\mu R - V_0 t. \tag{5.156}$$

Motion ceases when $\dot{w} = 0$ at $t = T$, where

$$T = \mu V_0 R/2N_0, \tag{5.157}$$

which may be rewritten

$$2\pi\gamma_2 = \lambda^{1/2} \tag{5.158}$$

with the aid of equations (5.91a), (5.133) and (5.150). The associated permanent radial displacement from equations (5.156) and (5.157) is

$$w_p = -\mu R V_0^2/4N_0 \tag{5.159}$$

or

$$w_p/w(t_1) = \lambda/2 \tag{5.160}$$

when using equations (5.105) and (5.133).

It may be shown for impulsive loading that equations (5.158) and (5.160) may be recovered from equations (5.149) and (5.151), respectively.

5.6.6 Discussion

5.6.6(a) Introduction

The theoretical analyses in §§ 5.6.3 to 5.6.5 for a complete spherical shell which is subjected to an internal axisymmetric dynamic pressure pulse are fairly straightforward and, therefore, may be used to assess the accuracy of a rigid, perfectly plastic solution which neglects the influence of material elasticity. The analysis for an elastic, perfectly plastic material in § 5.6.4 is taken as exact, while the rigid, perfectly plastic solution in § 5.6.5 is approximate.

5.6.6(b) Energy ratios

A sufficiently large energy ratio is often taken to show the unimportance of material elasticity in a particular dynamic problem and to justify the use of a rigid, perfectly plastic analysis. This energy ratio (E_r) may be defined as the total external energy input (E_T) compared with the amount of elastic strain energy (S_2) required to produce initial yielding in a complete spherical shell made from an elastic, perfectly plastic material. Thus, E_r is given by equation (5.154), or

$$E_r = 2\eta^2\{1 - \cos(2\pi\gamma_0)\}, \tag{5.161}$$

from equation (5.124), for a complete spherical shell made from an elastic, perfectly plastic material, and

$$E_r = \eta(\eta - 1)(2\pi\gamma_0)^2, \tag{5.162}$$

from equations (5.154) and (5.155a), for a rigid, perfectly plastic material.

The theoretical results in Figure 5.16 give the energy ratios E_r for various values of γ_0 and η. The differences between the predictions of equations (5.161) and (5.162) reflect the different radial velocities of the spherical shell throughout the duration of the pressure pulse. This leads to different values of E_T, as shown by a comparison of equations (5.99) and (5.152) for the elastic, perfectly plastic and rigid, perfectly plastic cases, respectively. If $\gamma_0 > 0.25$, then an elastic, perfectly plastic shell unloads during the phase $0 \leqslant t \leqslant \tau$ when the pressure pulse is acting (see equation (5.92)), while a rigid, perfectly plastic shell does not (see equation (5.142)). Clearly, a particular spherical shell, with $E_r < 1$, remains wholly elastic throughout the entire response. On the other hand, both equation (5.161) and equation (5.162) predict $E_r \cong \eta^2(2\pi\gamma_0)^2$ when $\eta \gg 1$ and $\gamma_0 \to 0$ for an impulsive velocity loading. The partition of energy in an elastic, perfectly plastic shell is shown in Figures 5.15(a)–(c).

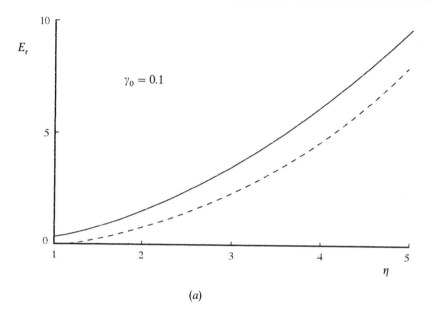

(a)

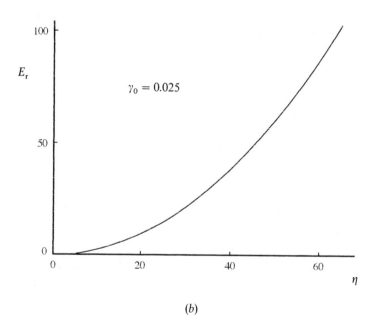

(b)

Figure 5.16 Comparison of energy ratios $(E_r = E_T/S_2)$ for elastic, perfectly plastic (——) and rigid, perfectly plastic (– – –) spherical shells for various $\eta = p_0/p_c$ and $\gamma_0 = \tau/T$. ——: equation (5.161) for an elastic, perfectly plastic material. – – –: equation (5.162) for a rigid, perfectly plastic material. (a) $\gamma_0 = 0.10$. (b) $\gamma_0 = 0.025$, (c) $\gamma_0 = 0.01$. (——: equations (5.161) and (5.162 in (b) and (c)).

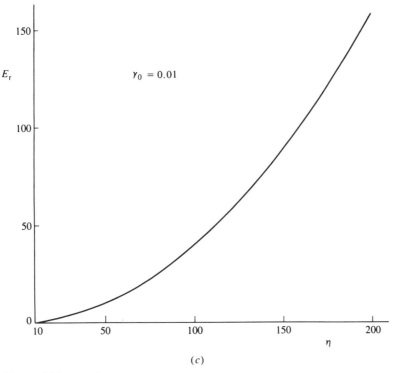

(c)

Figure 5.16—contd.

5.6.6(c) Radial displacement ratios

Equation (5.112) for the maximum radial displacement of a complete spherical shell, which is made from an elastic, perfectly plastic material, may be written in the form

$$w_m/w(t_1) = 1 + \eta^2[\{1 - \cos(2\pi\gamma_0)\} \sin(2\pi\gamma_1) + \sin(2\pi\gamma_0) \cos(2\pi\gamma_1)]^2/2 \tag{5.163}$$

when using equations (5.110a, b) together with equation (5.105) and the definitions given by equations (5.125) to (5.127) and (5.132). Equation (5.163) is a ratio between the maximum radial displacement and the radial displacement when plastic yielding first commences in a complete spherical shell. The dimensionless time γ_1 is found from the transcendental equation (5.106), or

$$\cos(2\pi\gamma_1) = \cos\{2\pi(\gamma_0 - \gamma_1)\} - 1/\eta, \tag{5.164}$$

which may also be written in the form

$$2\pi\gamma_1 = \pi\gamma_0 + \sin^{-1}[1/\{2\eta \sin(\pi\gamma_0)\}] \tag{5.165}$$

when using standard trigonometric identities.

Equation (5.163) is plotted against the dimensionless pressure η in Figure 5.17

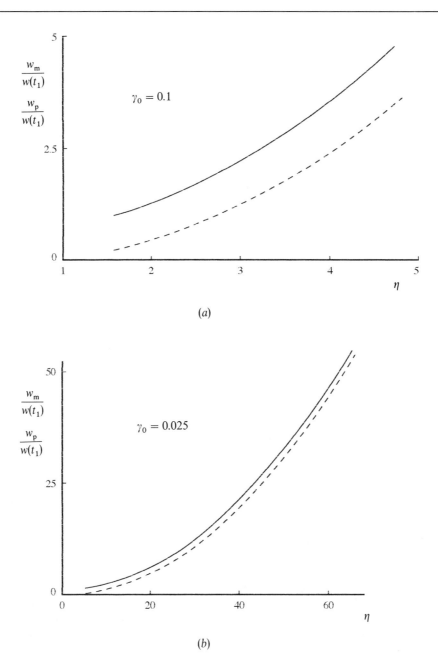

Figure 5.17 Comparison of dimensionless radial displacements for elastic, perfectly plastic and rigid, perfectly plastic spherical shells for various $\eta = p_0/p_c$ and $\gamma_0 = \tau/T$. ——: equation (5.163) for an elastic, perfectly plastic material. – – –: equation (5.151) for a rigid, perfectly plastic material (indistinguishable from equation (5.163) in (c)). (a) $\gamma_0 = 0.10$. (b) $\gamma_0 = 0.025$. (c) $\gamma_0 = 0.01$.

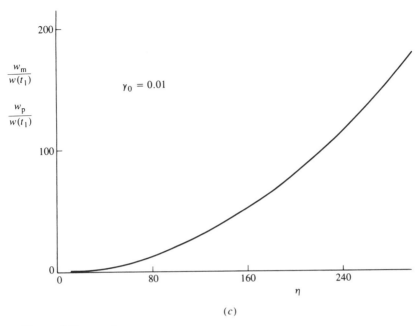

Figure 5.17—*contd.*

for several values of the dimensionless pressure duration γ_0. The dimensionless average final radial displacement is

$$w_a/w(t_1) = \eta^2[\{1 - \cos(2\pi\gamma_0)\} \sin(2\pi\gamma_1) + \sin(2\pi\gamma_0) \cos(2\pi\gamma_1)]^2/2, \quad (5.166)$$

according to equation (5.119) for a spherical shell, which is made from an elastic, perfectly plastic material. For purposes of comparison, the permanent radial displacement, according to equation (5.151) for a complete spherical shell, which is made from a rigid, perfectly plastic material, is also presented in Figure 5.17; this is also virtually indistinguishable from equation (5.163) when $\gamma_0 = 0.01$.

5.6.6(d) General remarks

It is evident from the comparisons made between the various theoretical predictions in Figures 5.16 and 5.17 that the rigid, perfectly plastic analysis in § 5.6.5 gives reasonable predictions for the maximum radial displacement of a complete spherical shell when $\gamma_0 = \tau/T \leqslant 0.25$, provided the energy ratio (E_r) is not too small.

The accuracy is explored further in Figure 5.18 for smaller energy ratios. It appears from Figure 5.18(a) that the permanent radial displacements ($w_p/w(t_1)$) of a rigid, perfectly plastic shell (equation (5.151)) are less than 10 per cent smaller than the average final radial displacement ($w_a/w(t_1)$) of an elastic, perfectly plastic

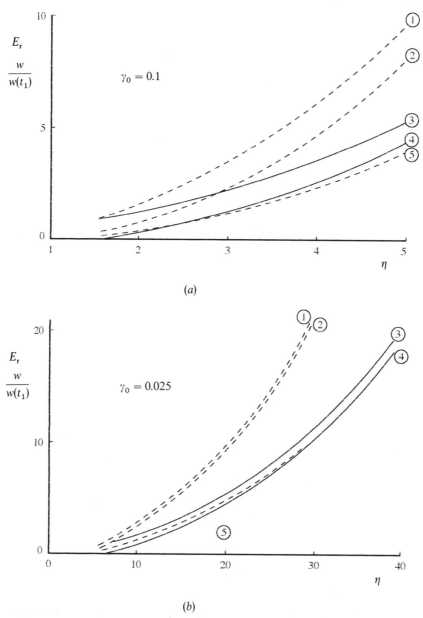

(a)

(b)

Figure 5.18 Comparison of dimensionless radial displacements and energy ratios for elastic, perfectly plastic and rigid, perfectly plastic spherical shells for various values of η and γ_0. ①: Energy ratio (E_r) according to equation (5.161) for an elastic, perfectly plastic material. ②: Energy ratio (E_r) according to equation (5.162) for a rigid, perfectly plastic material. ③: $w_m/w(t_1)$ from equation (5.163) for an elastic, perfectly plastic material. ④: $w_a/w(t_1)$ from equation (5.166) for an elastic, perfectly plastic material. ⑤: $w_p/w(t_1)$ from equation (5.151) for a rigid, perfectly plastic material. (a) $\gamma_0 = 0.10$. (b) $\gamma_0 = 0.025$. (c) $\gamma_0 = 0.01$.

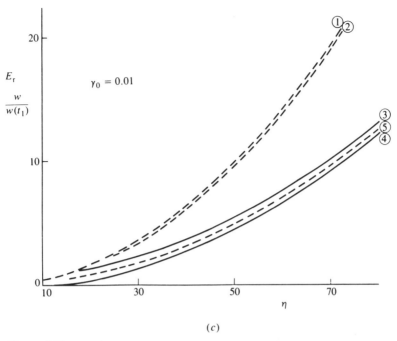

(c)

Figure 5.18—contd.

shell when $\gamma_0 = 0.10$ and $E_r > 4$, approximately. On the other hand, $w_p/w(t_1)$ is slightly larger than $w_a/w(t_1)$ in Figure 5.18(c) for $\gamma_0 = 0.01$, but the difference is less than 10 per cent when $E_r > 8$, approximately, which agrees exactly with the observation of Symonds for a one-degree-of-freedom spring-mass model in § 3.10.

The radial displacement–time history is plotted in Figure 5.14 for an elastic, perfectly plastic spherical shell with $\gamma_0 = \frac{1}{16}$ and $\eta = 10$. The corresponding energy ratio is 15.22, according to equation (5.161).† Thus, the theoretical results in Figure 5.18(a) for the larger value of $\gamma_0 = 0.1$ suggest that a rigid, perfectly plastic analysis, with $\gamma_0 = 0.0625$, might predict reasonable theoretical predictions. Equation (5.151), therefore, predicts $w_p/w(t_1) = 6.94$, which is only 2.39 per cent smaller than the corresponding elastic, perfectly plastic value of $w_a/w(t_1) = 7.11$ in Figure 5.14. The maximum, or permanent, radial displacement of a rigid, perfectly plastic spherical shell is reached at a dimensionless time given by equation (5.149), which, for the parameters in Figure 5.14, predicts $aT = 3.93$. This is 7.1 per cent smaller than the dimensionless time $aT_1 = 4.23$ when the maximum radial displacement $w_m/w(t_1)$ is reached in Figure 5.14.

† Equation (5.123) gives $\alpha = 1/E_r = 0.066$ and, therefore, $\beta = 0.934$ according to equation (5.129). This shows that 93.4 per cent of the external energy is dissipated as plastic work.

The maximum elastic strain energy may be estimated as $S_e = \sigma_0^2(4\pi R^2 H)/2E$ according to the simplified calculations suggested by equation (3.128) in § 3.10. Thus, the energy ratio $E_r = E_T/S_e$ becomes

$$E_r = 2(1 - v)\eta(\eta - 1)(2\pi\gamma_0)^2 \qquad (5.167)†$$

when using E_T given by equation (5.152) for a rigid, perfectly plastic spherical shell. This is $2(1 - v)$ times larger than equation (5.155a) and is, therefore, 40 per cent larger when $v = 0.3$.

It may be shown that the theoretical solutions in §§ 5.6.4 and 5.6.5 for sufficiently thin complete spherical shells which are made from elastic, perfectly plastic and rigid, perfectly plastic materials, respectively, are both statically and kinematically admissible and are, therefore, exact. However, it has been assumed in the analysis in § 5.6.4 that $\tau \leqslant t_1$ so that longer pressure pulses have not been examined.

Baker[5.3] has examined the elastic, linear strain hardening response of a thin-walled spherical shell subjected to a spherically symmetric internal transient pressure pulse having a linear decay. A numerical scheme was also employed to explore both thinning of the shell thickness and large deflection effects. Duffey[5.4] investigated the influence of a linear material strain rate sensitivity law on the response of a complete spherical shell which is loaded impulsively.

5.7 Shallow shells

5.7.1 Introduction

The dynamic response of a complete spherical shell was examined thoroughly in § 5.6 and used in order to explore the accuracy of the rigid plastic method. This analysis was simplified by assuming that the shell was sufficiently thin for only membrane effects to be important. In addition, the response was spherically symmetric. However, many practical spherical shell problems are not spherically symmetric and the influence of bending moments is important. Thus, even for an axisymmetric response, it is necessary to retain four generalised stresses (N_θ, N_ϕ, M_θ and M_ϕ), which complicates a theoretical analysis because a four-dimensional yield condition controls plastic flow. Many approximate yield criteria have been developed to simplify the dynamic plastic analysis of shells and some are discussed by Hodge.[5.1]

A shallow shell approximation has proved valuable for simplifying theoretical calculations for the elastic behaviour of axisymmetric shells.[5.5] Thus, this approximation, together with a simplified yield condition, is used in the following sections to examine the dynamic plastic response of a simply supported shallow shell subjected to an exponentially decaying pressure pulse.

---

† Equation (5.167) with $\eta \gg 1$ reduces to equation (3.130) for an impulsive velocity.

5.7.2 Basic equations

It may be shown[5.1, 5.6] that the strain and curvature rates for the element of a shallow shell in Figure 5.19 are

$$\dot{\epsilon}_\theta = (\dot{v} - z'\dot{w})/r, \qquad (5.168a)$$

$$\dot{\epsilon}_\phi = \dot{v}' - z''\dot{w}, \qquad (5.168b)$$

$$\dot{\kappa}_\theta = -(\dot{w}' + z''\dot{v})/r \qquad (5.168c)$$

and

$$\dot{\kappa}_\phi = (r\dot{\kappa}_\theta)' \qquad (5.168d)$$

provided the shallow shell approximation

$$(z')^2_{max} \ll 1 \qquad (5.168e)$$

is satisfied and where the normal (w) and meridional (v) displacements are defined in Figure 5.19 and

$$(\dot{\ }) = \partial(\ )/\partial t \qquad \text{and} \qquad (\ )' = \partial(\ )/\partial r. \qquad (5.169a, b)$$

The principle of virtual velocities[5.7] in Appendix 3 may be used to obtain the consistent set of equilibrium equations[5.1, 5.6]

$$rz''N_\phi + z'N_\theta + (rQ_\phi)' - rp = r\mu\ddot{w}, \qquad (5.170a)$$

$$(rN_\phi)' - N_\theta - rz''Q_\phi = r\mu\ddot{v} \qquad (5.170b)$$

and

$$rQ_\phi = (rM_\phi)' - M_\theta, \qquad (5.170c)$$

when normal loading (p) only is considered, provided rotatory inertia is neglected, and where the various quantities are defined in Figure 5.19.

Onat and Prager[5.8] derived a four-dimensional yield surface for a rotationally symmetric shell made from a rigid, perfectly plastic material which obeys the Tresca yield condition. However, as noted in § 5.7.1, it is usually necessary to simplify the yield surface[5.1] in order to make this general class of problems tractable unless wholly numerical schemes are to be employed. The uncoupled diamond yield condition illustrated in Figure 5.20 is used in this section to obtain the dynamic response of a shallow shell.

Yield surfaces which are 2 times and 0.618 times as large as that drawn in Figure 5.20 would circumscribe and inscribe the yield surface derived by Onat and Prager.[5.8] However, if a theoretical solution to a particular problem utilised only a portion of the uncoupled diamond yield condition, then the upper and lower bounds might be made closer.

5.7.3 Dynamic behaviour of a simply supported shallow shell

5.7.3(a) Introduction

Now,

$$z = Z(r/R_0)^n \qquad (5.171)$$

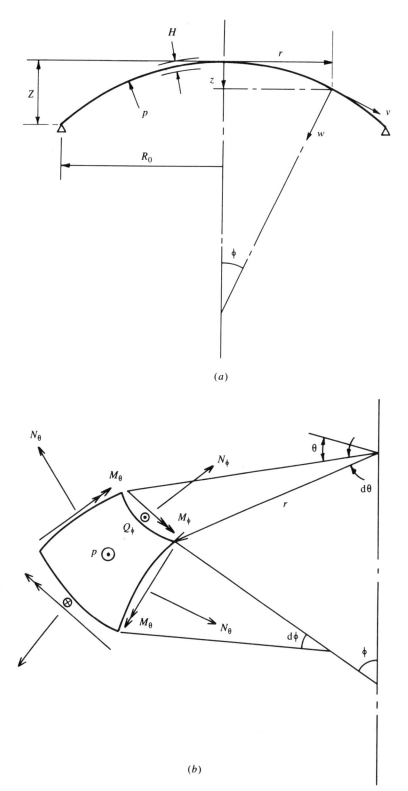

Figure 5.19 (a) Notation for a shallow shell. (b) Stress resultants for a shallow shell.

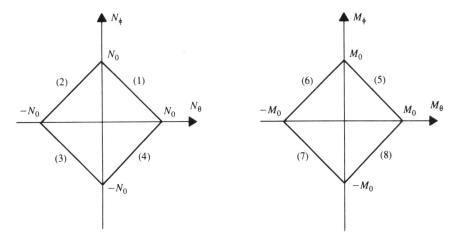

Figure 5.20 Uncoupled diamond yield condition.

for a shallow shell of degree n, where Z is the total depth of a shell, as shown in Figure 5.19. We restrict our attention to shallow shells of second degree ($n = 2$) which may be shallow spherical, paraboloidal, ellipsoidal or hyperboloidal caps. The response of a shell is sought when subjected to a uniformly distributed external pressure

$$p = -p_0\, \mathrm{e}^{-t/T_0}, \tag{5.172}$$

where p_0 is the initial magnitude and T_0 is the time when the magnitude of the pressure pulse has decayed to p_0/e.

5.7.3(b) Small dynamic pressures

It is assumed that a shallow shell deforms with the velocity field

$$\dot{v} = 0 \tag{5.173a}$$

and

$$\dot{w} = \dot{w}_0\{1 - (r/R_0)^2\}, \tag{5.173b}$$

where $\dot{w}_0$ is the normal velocity at $r = 0$. Equations (5.168a) and (5.171), with $n = 2$, predict, therefore, that $\dot{\epsilon}_\theta = -2Z\dot{w}_0\{1 - (r/R_0)^2\}/R_0^2$, or $\dot{\epsilon}_\theta \leqslant 0$ for an external pressure pulse, which produces $\dot{w}_0 \geqslant 0$. Similarly, $\dot{\epsilon}_\phi = \dot{\epsilon}_\theta$, $\dot{\kappa}_\theta = 2\dot{w}_0/R_0^2 \geqslant 0$ and $\dot{\kappa}_\phi = \dot{\kappa}_\theta$. Thus the normality requirements of plasticity suggest the portion (3)–(5) of the yield condition in Figure 5.20 for which

$$N_\theta + N_\phi = -N_0 \qquad \text{and} \qquad M_\theta + M_\phi = M_0, \tag{5.174a, b}$$

where N_0 and M_0 are the fully plastic membrane force and bending moment for the shell cross-section, respectively.

Now, substituting equations (5.171) with $n = 2$, (5.172), (5.173b) and (5.174a) into the equilibrium equation (5.170a) and integrating gives

$$rQ_\phi = \mu\ddot{w}_0(r^2/2 - r^4/4R_0^2) - (r^2/2)p_0\,e^{-t/T_0} + r^2ZN_0/R_0^2 \qquad (5.175)$$

which, substituting into equation (5.170c) with equation (5.174b), yields

$$M_\phi = M_0/2 + \mu\ddot{w}_0(r^2/8 - r^4/24R_0^2) - (r^2/8)p_0\,e^{-t/T_0} + r^2ZN_0/4R_0^2. \qquad (5.176)$$

However, $M_\phi = 0$ for simple supports at the boundary with $r = R_0$, which is satisfied by equation (5.176) when

$$\mu\ddot{w}_0 = 3(p_0\,e^{-t/T_0} - p_s)/2, \qquad (5.177)$$

where

$$p_s = 4M_0/R_0^2 + 2ZN_0/R_0^2 \qquad (5.178)$$

is the corresponding static collapse pressure[5.9] since equation (5.177), with $t = 0$ and no inertia forces ($\mu\ddot{w}_0 = 0$), gives $p_0 = p_s$.

It is now possible to integrate the equilibrium equation (5.170b) to predict the meridional membrane force

$$N_\phi = -N_0/2 + (r^2Z/8R_0^2)\{(p_0\,e^{-t/T_0} - p_s)(1 - r^2/R_0^2) - 8M_0/R_0^2\}. \qquad (5.179)$$

Thus, the four generalised stresses in the uncoupled diamond yield condition in Figure 5.20 may be obtained from equations (5.174a, b), (5.176) and (5.179).

Now, integrating equation (5.177) with respect to time,

$$\mu\dot{w}_0 = -3(p_0T_0\,e^{-t/T_0} + p_st)/2 + 3p_0T_0/2, \qquad (5.180)$$

where the constant of integration has been obtained from the requirement that $\dot{w}_0 = 0$ at $t = 0$. Motion ceases when $t = T$, where

$$p_0T_0(1 - e^{-T/T_0}) - p_sT = 0. \qquad (5.181)$$

A further integration of equation (5.180) yields the normal displacement

$$\mu w_0 = 3p_0T_0^2(e^{-t/T_0} - 1)/2 - 3p_st^2/4 + 3p_0T_0t/2. \qquad (5.182)$$

Thus, the two components of the permanent displacement profile are

$$\mu v_p = 0 \qquad (5.183a)$$

and

$$\mu w_p = 3T\{2T_0(p_0 - p_s) - p_sT\}\{1 - (r/R_0)^2\}/4 \qquad (5.183b)$$

according to equations (5.173a, b), (5.181) and (5.182).

5.7.3(c) Kinematic and static admissibility

The theoretical solution in § 5.7.3(b) does not have any travelling plastic hinges so it is a straightforward exercise to demonstrate that it is kinematically admissible. The analysis is also statically admissible if the generalised stresses predicted by equations (5.174a, b), (5.176) and (5.179) remain on the portion (3)–(5) of the uncoupled diamond yield surface in Figure 5.20. It was shown in reference 5.6 that the solution is statically admissible provided

$$p_0 - p_s - p_sT/T_0 + 2N_0H/R_0^2 \geqslant 0 \qquad (5.184)$$

and

$$p_0 - p_s \leqslant 4(1.5 + \sqrt{2})N_0 H/R_0^2. \tag{5.185}$$

Despite the relative simplicity of the foregoing analysis, it was not possible, using analytical techniques, to extend it in reference 5.6 to cater for pressure pulses larger than the limits of inequalities (5.184) and (5.185). It appears likely, therefore, that a numerical method of analysis is required.

5.7.3(d) General comments

Theoretical predictions for a uniformly distributed exponential pressure pulse (equation (5.172)) have also been reported[5.6] for shallow shells made from materials governed by the two-moment limited interaction and uncoupled square yield conditions in Figure 5.21. It transpires that the theoretical analysis reported in § 5.7.3(b), using the uncoupled diamond yield surface in Figure 5.20, leads to the simplest analysis. Thus, a numerical scheme is required to predict the exponential pressure loading of shallow shells with large pressure pulses which exceed the restrictions of inequalities (5.184) and (5.185), since further simplification of the yield condition is probably unrealistic.

It may be shown that the two yield surfaces in Figure 5.21(a) and (b) completely circumscribe the Tresca yield surface of Onat and Prager,[5.8] while others respectively 0.618 and 0.309 times as large would inscribe it. As already remarked in § 5.7.2, yield surfaces which are 2 times and 0.618 times as large as the uncoupled diamond yield surface in Figure 5.20 would circumscribe and inscribe the Tresca yield surface,[5.8] respectively.

The corollaries of the limit theorems of plasticity, which are introduced in § 2.4.4, show that yield surfaces which inscribe and circumscribe the exact yield surface respectively provide lower and upper bounds to the exact static collapse load of a perfectly plastic structure. Unfortunately, a similar theorem is not available for perfectly plastic structures which are loaded dynamically. However, Hodge and Paul[5.10] have compared the dynamic response of simply supported cylindrical shells which are made from materials which flow plastically according to two different yield criteria. The theoretical predictions of Hodge and Paul[5.10] show that the error involved when employing an approximate yield condition is small provided the static collapse pressure for the approximate yield condition is adjusted to equal the corresponding exact static value.

It is demonstrated in reference 5.6 that the suggestions of Hodge and Paul[5.10] are satisfied when taking the analysis for the two-moment limited interaction yield surface as exact. In this circumstance, the predictions for the maximum permanent normal displacement for a shell obeying the uncoupled diamond yield condition in Figure 5.20, but with the static collapse pressure p_s (equation (5.178)) replaced by

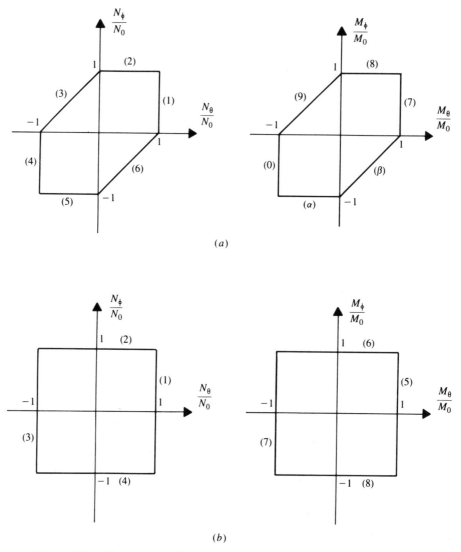

Figure 5.21 (a) Two-moment limited interaction yield condition. (b) Uncoupled square yield condition.

the corresponding value for the yield surface in Figure 5.21(a), gave good agreement with the exact value. Moreover, the analysis in § 5.7.3(b) for the uncoupled diamond yield condition in Figure 5.20, when made to either inscribe or circumscribe the two-moment limited interaction yield condition in Figure 5.21(a), respectively, gave upper and lower bounds to the maximum permanent deflections, which were predicted according to the exact yield surface (two-moment limited interaction yield surface).

5.8 Some comments on spherical shells

The dynamic response of a complete spherical shell, which is subjected to a spherically symmetric pressure pulse with the rectangular shaped pressure–time history in Figure 5.3, was examined in § 5.6 for elastic, perfectly plastic and rigid, perfectly plastic materials. A simply supported rigid, perfectly plastic shallow shell, which may have a shallow spherical shape, was studied in § 5.7 for a pressure pulse having an exponentially decaying pressure–time history. The presence of supports around the boundary of this shell introduced bending moments and a transverse shear force as well as membrane forces, even though the geometry and dynamic pressure loading were both axisymmetric throughout the response. Despite the simplifications introduced by a shallow shell approximation, it was not possible to obtain a theoretical solution for exponential pressure pulses with peak values which are larger than inequalities (5.184) and (5.185), even when using an approximate yield surface.

Sankaranarayanan[5.11] appears to have been the first author to examine the dynamic plastic response of a spherical cap. The two-moment limited interaction yield surface in Figure 5.21(a) was employed for a cap which was simply supported around an outer boundary and subjected to an exponentially decaying pressure pulse. It transpires that the theoretical solution must satisfy an inequality which contains the magnitude of the peak pressure and which is particularly restrictive for total subtended angles larger than $20°$, approximately.

Sankaranarayanan[5.12] also examined the influence of the rectangular pressure pulse loading shown in Figure 5.3. He demonstrated that the actual details of the pressure–time history are important, at least when the peak value of the pressure pulse is only slightly larger than the corresponding static collapse pressure.

The dynamic plastic behaviour of simply supported deep spherical shells was also examined in reference 5.6. The spherical shells were made from rigid, perfectly plastic materials controlled by the yield criteria in Figures 5.20 and 5.21 and subjected to a uniformly distributed exponentially decaying pressure pulse. An attempt was made to extend the validity of Sankaranarayanan's[5.11] theoretical solution for the two-moment limited interaction yield condition in Figure 5.21(a) but was abandoned because of its complexity. The peak values of the pressure pulses in the theoretical solutions, using the two yield criteria in Figures 5.20 and 5.21(b), were restricted in order to satisfy the static admissibility requirements.

The difficulty in achieving an exact theoretical solution for the practically important problem of a spherical shell, which is supported around an outer boundary, motivated the development of an approximate method of analysis.[5.13] This particular procedure, which was developed for an arbitrarily shaped shell, was specialised in reference 5.14 for the particular case of a spherical shell which was fully clamped around the outer boundary and subjected to a uniformly distributed impulsive velocity. It turns out that reasonable agreement was obtained with some experimental results[5.15] which were recorded on aluminium hemispherical shells, with $R/H \leqslant 14.7$, approximately.

5.9 Influence of pressure pulse characteristics

The general theoretical procedure in § 5.3 to § 5.5 has been used to study the dynamic plastic response of cylindrical shells with several boundary conditions and subjected to axisymmetric pressure distributions having various idealised pressure–time characteristics.[5.2, 5.10, 5.16–5.21] Hodge[5.16] examined the response of an infinitely long cylindrical shell with equally spaced rigid reinforcing rings when subjected to uniform axisymmetric pressure pulses having rectangular, triangular and exponentially decaying pressure–time histories. Even for pulses having the same impulse and peak value, Hodge[5.16] found that the blast characteristics have a profound effect on the final deformation for peak loads slightly larger than the associated static collapse pressure, but that the percentage effect decreases for larger pressures.

It was mentioned in § 4.10 that Youngdahl[5.22] has examined the influence of pulse shape on the dynamic plastic response of an axisymmetrically loaded simply supported circular plate. He observed that the strong dependence on pulse shape could be practically eliminated by introducing a correlation parameter for the effective load (P_e) and a mean time (t_c), which are defined by equations (4.186) and (4.188) in § 4.10, respectively.

Youngdahl[5.22] has also used his method to examine the dynamic plastic behaviour of an infinitely long cylindrical shell which is reinforced by equally spaced rigid reinforcing rings and subjected to a time-dependent, uniformly distributed radial pressure. This particular problem is shown in Figure 5.5 and was studied in § 5.4 and § 5.5 for the rectangular pressure pulse illustrated in Figure 5.3. Thus, it is possible to use the theoretical results in § 5.4 and § 5.5 to predict the response of a cylindrical shell (with $\omega^2 \leqslant 6$), which is subjected to a pressure pulse having any pressure–time history. However, it is necessary to recast the results in § 5.4 and § 5.5 in terms of the correlation parameters introduced by Youngdahl.[5.22]

The rectangular pressure pulse described by equations (5.6) and (5.7) and shown in Figure 5.3 immediately begins to deform a rigid plastic material, and, therefore, it is evident that the total impulse (per unit area) $I = p_0\tau$ (i.e., $t_y = 0$ in equation (4.187)). Equation (4.188) gives the centroid of the pressure pulse $t_c = \tau/2$ and, therefore, the effective pressure is $p_e = p_0$ according to equation (4.186). This allows equations (5.41) and (5.77) for the maximum permanent radial deformations of fully clamped cylindrical shells with $\omega^2 \leqslant 6$, and subjected to rectangular pressure pulses to be written,

$$W_f = 3I^2(p_c/p_e - 1)/4\mu p_c \qquad \text{for} \quad \eta \leqslant (6 + \omega^2)/(2 + \omega^2) \qquad (5.186)$$

and

$$W_f = I^2\{p_c/p_e - (8 + \omega^2)/(6 + \omega^2)\}/2\mu p_c \qquad \text{for} \quad \eta \geqslant (6 + \omega^2)/(2 + \omega^2). \qquad (5.187)$$

Equations (5.186) and (5.187), with $\omega^2 = 5$, are plotted in Figure 5.22.

The observations of Youngdahl[5.22] suggest that equations (5.186) and (5.187), or Figure 5.22, would also predict the response for a similar cylindrical shell which is subjected to any pulse shape. For example, consider the uniformly distributed linearly decaying pressure pulse

$$p = p_0(1 - t/t_0), \qquad 0 \leqslant t \leqslant t_0, \qquad (5.188a)$$

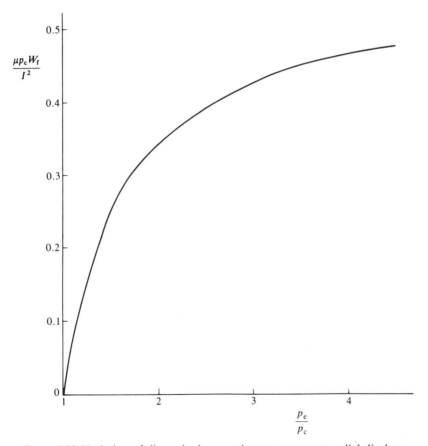

Figure 5.22 Variation of dimensionless maximum permanent radial displacement of a cylindrical shell, with $\omega^2 = 5$ and rigid reinforcing rings, which is subjected to a dynamic pressure pulse with a dimensionless effective pressure p_e/p_c. The transition from equation (5.186) to (5.187) occurs at $p_e/p_c = 1.57$.

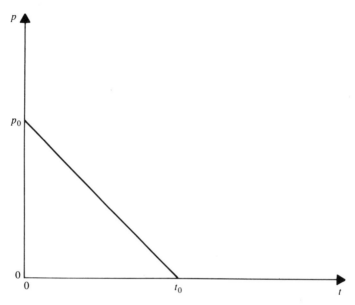

Figure 5.23 Pressure pulse with a linear decay.

and

$$p = 0, \qquad\qquad t \geqslant t_0, \qquad\qquad (5.188b)$$

which is illustrated in Figure 5.23. Clearly,† the total impulse (per unit area) is

$$I = p_0 t_0/2. \qquad\qquad (5.189)$$

Equation (4.188) gives the centroid of the pressure pulse $t_c = t_0/3$ and, therefore, the effective pressure is

$$p_e = 3p_0/4 \qquad\qquad (5.190)$$

according to equation (4.186).

Thus, equations (5.186) and (5.187), with $p_e = 3p_0/4$ from equation (5.190), predict, respectively,

$$W_f = 3I^2(4p_c/3p_0 - 1)/4\mu p_c \qquad\qquad (5.191)$$

and

$$W_f = I^2\{4p_c/3p_0 - (8 + \omega^2)/(6 + \omega^2)\}/2\mu p_c \qquad\qquad (5.192)$$

for the linearly decreasing pressure pulse in Figure 5.23 with $I = p_0 t_0/2$ according to equation (5.189). Alternatively, the theoretical predictions may be obtained directly from the curve in Figure 5.22.

Youngdahl[5.22] has presented the exact theoretical predictions for reinforced cylindrical shells subjected to a wide range of pulse shapes, as shown in Figure 5.24.

---

† Plastic deformation commences at $t = 0$, provided $p_0 \geqslant p_c$. Therefore, $t_y = 0$ in equation (4.187).

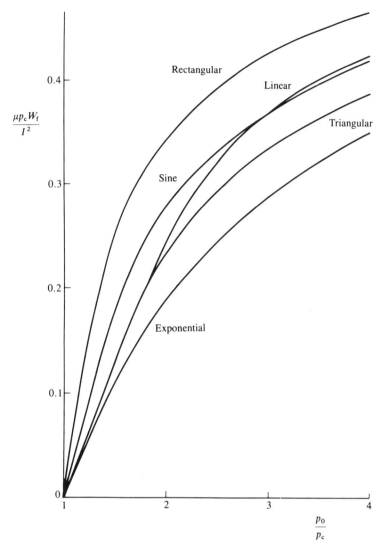

Figure 5.24 Variation with p_0/p_c of the dimensionless maximum permanent radial displacement for a reinforced cylindrical shell having $\omega^2 = 5$, and for various pressure pulse shapes.

The wide spread of the dimensionless maximum radial displacements for the different pulses in Figure 5.24 virtually disappears in Figure 5.25 when using the correlation parameter p_e/p_c. Thus, equations (5.186) and (5.187), or the curve in Figure 5.22, may be used to predict the maximum permanent radial displacement of a reinforced cylindrical shell subjected to a wide range of pressure pulses.

The correlation parameters, which are defined by equations (4.186) to (4.188) in § 4.10 and applied to a reinforced cylindrical shell above, are associated with an

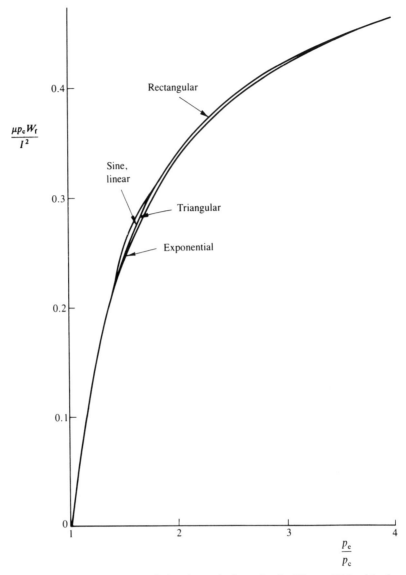

Figure 5.25 The variation of the theoretical results in Figure 5.24 with the dimensionless effective pressure p_e/p_c.

axisymmetric dynamic pressure pulse which is uniform in the axial sense (x) along a cylindrical shell. However, this situation is not always encountered in practice. Thus, Youngdahl[5.23] examined the influence of an axial variation in the dynamic pressure pulse and introduced an additional correlation parameter.

Now, consider an axisymmetric dynamic pressure pulse

$$p(x, t) = \Phi(x)\psi(t), \qquad 0 \leqslant |x| \leqslant L_1 \qquad (5.193a)$$

and

$$p(x, t) = 0, \qquad\qquad |x| \geqslant L_1, \qquad\qquad (5.193b)$$

which is distributed over a total axial length $2L_1$, and symmetric about a vertical plane at $x = 0$. $\Phi(x)$ and $\psi(t)$ are known[5.23] as the load shape and pulse shape, respectively. The parameter $\psi(t)$ is related to $P(t)$ in equation (4.187) of § 4.10 when $\Phi(x) = 1$.

Youngdahl[5.23] seeks to replace the actual axial distribution of the pressure pulse by a uniform distribution over a total length $2L_e$ as shown in Figure 5.26(a). The actual pressure-time characteristic is replaced by an equivalent rectangular pressure pulse with a magnitude ψ_e commencing at t_y, when plastic yielding starts, and ceasing at the time t_e later, as illustrated in Figure 5.26(b).

The total impulse of a dynamic pressure pulse (per unit circumferential length) is

$$I = \int_{t_y}^{t_f} \int_{-L_1}^{L_1} p(x, t) \, dx \, dt, \qquad\qquad (5.194a)$$

which may be written in the form

$$I = 2p_e L_e t_e, \qquad\qquad (5.194b)$$

where p_e, L_e and t_e are the effective pressure load, effective half length of the external pressure loading and effective pulse duration, respectively. The times t_y and t_f are associated with the initiation of plastic flow and the duration of motion, respectively. Youngdahl[5.23] defines the effective pulse duration as

$$t_e = 2 \int_{t_y}^{t_f} (t - t_y)\psi(t) \, dt \Big/ \int_{t_y}^{t_f} \psi(t) \, dt \qquad\qquad (5.195)$$

and

$$\psi_e = \int_{t_y}^{t_f} \psi(t) \, dt / t_e \qquad\qquad (5.196)$$

as the effective pulse shape. It is evident, when $\Phi(x) = 1$, that $t_e = 2t_c$, where t_c is defined by equation (4.188), and $\psi_e = p_e = P_e/(2\pi R 2 L_1)$. Youngdahl[5.23] also defines an effective load shape

$$\Phi_e = \left[\int_0^{\bar{x}} \Phi(x) \, dx \right]^2 \Big/ \left[2 \int_0^{\bar{x}} x\Phi(x) \, dx \right], \qquad\qquad (5.197)$$

where $\bar{x}$ is the initial plastic hinge location. It follows that the effective half-length of the loaded region is

$$L_e = \int_0^{L_1} \Phi(x) \, dx / \Phi_e, \qquad\qquad (5.198)$$

while the effective pressure load is

$$p_e = \Phi_e \psi_e \qquad\qquad (5.199)$$

according to equations (5.193a) and (5.194b).

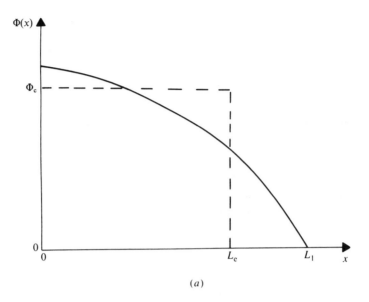

(a)

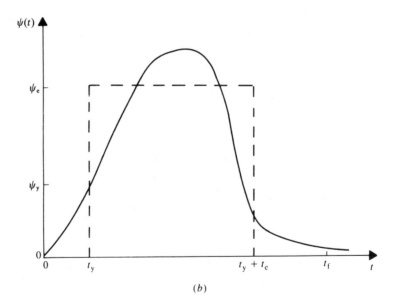

(b)

Figure 5.26 (a) Actual load shape with an equivalent rectangular-shaped distribution over an effective length L_e. (b) Actual pulse shape with an equivalent rectangular-shaped history with an effective time t_e.

Youngdahl[5.23] has used the theoretical procedure outlined above to study a long circular cylindrical shell which is subjected to an axisymmetric dynamic pressure pulse over a region with a total axial length $2L_1$. The theoretical predictions reveal that the dimensionless maximum permanent radial displacements are sensitive to the pulse and load shapes ($\psi(t)$ and $\Phi(x)$), while the correlation parameters allow the results for a wide range of shapes to collapse virtually onto a common curve.

The introduction of correlation parameters into the theoretical solution of a cylindrical shell which is subjected to a uniformly distributed rectangular pressure pulse enables the behaviour for more complex dynamic pressure distributions, with various pressure–time histories, to be obtained in an appealingly simple fashion. The validity of the general procedure has been demonstrated only for two particular cylindrical shell problems.[5.22, 5.23] It remains to be shown whether or not the method is also accurate for other cylindrical shells and different shell geometries subjected to dynamic loads which produce infinitesimal deflections.

More recently, Youngdahl and Krajcinovic[5.24] have obtained correlation parameters for an infinite plate which is subjected to an axisymmetric pressure pulse which may not be separated into a product of a function of position and a function of time, as assumed by equation (5.193a) for a cylindrical shell. Youngdahl[5.25] has devised a simplified theoretical procedure which has been used to examine a simply supported circular plate subjected to a pressure distribution which is an arbitrary function of radius and time.

Zhu et al.[5.26] have examined the correlation parameters of Youngdahl[5.22] and have introduced an alternative set in order to simplify calculations.

5.10 Final remarks

The transverse shear force (Q_x) is retained as a reaction in the equilibrium equations (5.1) and (5.2). Thus, it is not included in the yield condition (e.g., Figure 5.2), since the corresponding transverse shear strain is neglected. However, the influence of the transverse shear force on the plastic yielding of a simply supported cylindrical shell subjected to an impulsive velocity loading is examined in § 6.6.

The equilibrium equations (5.1) and (5.2) and the corresponding strains and change of curvature relations (5.3) to (5.5) were developed for the infinitesimal behaviour of cylindrical shells. In other words, the influence of geometry changes was ignored in this chapter. It is noted in § 7.9.1 that a significant strengthening occurs for pressure loadings when a cylindrical shell is restrained axially at the supports and the maximum radial displacement exceeds the shell thickness, approximately.

The material is assumed to be perfectly plastic in the various theoretical analyses presented in this chapter. It is well known that the plastic flow of many materials is sensitive to the magnitude of strain rate. This phenomenon is known as material strain rate sensitivity and is discussed in Chapter 8. Some work on shells is noted briefly in § 8.5.5. Further studies have been undertaken by Duffey and Krieg,[5.27] who investigated the influence of material strain hardening and a linear form of strain rate sensitivity on the dynamic behaviour of an elastic-plastic cylindrical shell. Incidentally, they noted, for a particular example, that the permanent deflections calculated for a rigid-plastic material were within 20 per cent of the corresponding elastic-plastic values when the ratio of plastic to elastic energy was greater than three. Perrone[5.28] presented a theoretical analysis for a tube loaded impulsively which avoids the important restriction of linearised strain rate sensitivity introduced in reference 5.27.

The theoretical analyses in this chapter focus on the stable behaviour of cylindrical and spherical shells with internal dynamic pressures. However, dynamic plastic buckling, or unstable behaviour, may occur for external dynamic pressures. This phenomenon is discussed in Chapter 10.

The theoretical solutions in this chapter were developed for cylindrical shells, subjected to axisymmetric pressure pulses, and spherical caps and spherical shells subjected to spherically symmetric pressure pulses. Non-symmetric behaviour leads to considerable complexity because inplane shear effects must be included in the yield condition and the equilibrium equations. An approximate theoretical rigid, perfectly plastic procedure was developed in reference 5.13 in order to estimate the stable dynamic response of arbitrarily shaped shells. Encouraging agreement was found with experimental results for impulsively loaded cylindrical panels.[5.14] This theoretical method has also been used to examine the dynamic plastic behaviour of shell intersections which consist of cylindrical nozzles, or cylindrical shells, intersection either spherical or cylindrical shells.[1.15, 5.29, 5.30]

The five publications listed under reference 5.31 contain a review of other theoretical and numerical work on the dynamic non-symmetric response of shells. Although theoretical rigid-plastic methods are attractive for predicting the dynamic response of axisymmetric shell problems, they are cumbersome for non-symmetric shells, and numerical schemes often become necessary. However, it should be remarked that numerical schemes for shells are complex and must be thoroughly checked before any results may be accepted for design purposes. Even then, they should be accepted only tentatively because there is a paucity of experimental data on the constitutive equations for materials under dynamic multi-axial stress states, and a shortage of reliable test data on many shell geometries subjected to dynamic loads which produce an inelastic response.

PROBLEMS

5.1 Obtain equations (5.27) and (5.28) for the energy dissipated in the long cylindrical shell which is examined in § 5.3. Prove that the total energy dissipated plastically equals the total work done by the dynamic pressure pulse which is shown in Figure 5.3.

5.2 Prove that the theoretical solution in § 5.4 for a long reinforced cylindrical shell is statically and kinematically admissible provided inequalities (5.42) and (5.43) are satisfied.

5.3 Integrate equation (5.65) and obtain equation (5.66).

Sketch the variation of the dimensionless plastic hinge speed $\tau\dot{\xi}/L$ against the dimensionless time t/τ for given values of η and ω, and several values of ξ/L, where ξ/L is obtained from equation (5.66).

5.4 Show that the second phase of motion in § 5.5.3 for a fully clamped short cylindrical shell is kinematically admissible. (See § 5.5.6.)

5.5 Show that energy is conserved during each of the first three phases of motion for the elastic, perfectly plastic spherical shell which is studied in § 5.6.4. (See § 5.6.4(f).

5.6 Repeat the theoretical analysis in § 5.6.4 for an elastic, perfectly plastic spherical shell with $t_1 \leqslant \tau$, where t_1, given by equation (5.105), is the time when the shell becomes plastic.

5.7 (a) Obtain the static collapse pressure for a simply supported, uniformly loaded, flat circular plate from equation (5.178) for a shallow shell. Use the corollaries of the limit theorems (§ 2.4.4) to bound the exact collapse pressure and compare the result with equation (2.18), which was obtained for the Tresca yield condition.

(Note that the static collapse pressure for a shallow shell, which is made from a rigid perfectly plastic material obeying the two-moment limited interaction yield condition in Figure 5.21(a), is obtained in reference 5.9. This expression reduces to equation (2.18) for a flat circular plate.)

(b) A shallow spherical shell has a rise which equals the shell thickness. Estimate the thickness of a flat circular plate which is required to support the same pressure over the same circular opening. Assume that both have simple supports. Compare the masses of the flat plate and the shallow spherical shell.

6

Influence of transverse shear and rotatory inertia

6.1 Introduction

It is well known that, normally, transverse shear forces do not influence markedly the static plastic behaviour of structural members such as beams, frames, plates and shells. However, transverse shear effects may become important in short beams, for example, and design codes are available to provide a designer with some guidance.[6.1]

The bending moment distribution in a simply supported beam, which is subjected to a uniformly distributed static pressure, is given by equation (1.26). Differentiation of this expression yields the transverse shear force $Q/M_0 = -2x/L^2$, or

$$Q_s = -2M_0/L \qquad (6.1)\dagger$$

at the supports ($x = L$). Equation (6.1) may be obtained directly from the overall vertical and moment equilibrium requirements for a simply supported rigid, perfectly plastic beam with a fully developed plastic hinge at the mid-span.

If a beam has a rectangular cross-section with a thickness H and is made from a ductile material with a yield stress σ_0, then $M_0 = \sigma_0 H^2/4$ and $Q_0 \cong \sigma_0 H/2$, where M_0 and Q_0 are the respective bending moment (per unit length) and transverse shear force (per unit length) necessary to develop independently a fully plastic cross-section. In this circumstance, equation (6.1) becomes

$$Q_s/Q_0 = -H/L \qquad (6.2)$$

and, therefore, $|Q_s/Q_0| \ll 1$ for a beam with $H \ll L$. Thus, the static plastic behaviour of the beam is governed primarily by bending effects with negligible transverse shear effects.

Now, let us examine a similar beam which is subjected to a uniformly distributed dynamic pressure pulse having the rectangular pressure–time history shown in

† Figure 1.1 shows the positive sense of Q.

Figure 3.5. Equation (3.32) gives the corresponding bending moment distribution in the outer region of the beam during the first phase of motion. This expression may be differentiated with respect to x, according to equation (3.1), to predict the transverse shear force

$$Q/M_0 = -\eta(x/L - \bar{\xi}_0)^2/(1 - \bar{\xi}_0)L, \qquad \xi_0 \leqslant x \leqslant L, \tag{6.3}$$

where $\bar{\xi}_0$ is the dimensionless position of the stationary plastic hinge shown in Figure 3.7(a), and is given by equations (3.30) and (3.31) (i.e., $(1 - \bar{\xi}_0)^2 = 3/\eta$). Thus, equation (6.3) with $x = L$ at the supports gives

$$Q_d/Q_0 = -(H/2L)(3\eta)^{1/2} \tag{6.4}$$

for a beam with a rectangular cross-section of thickness H having $M_0 = \sigma_0 H^2/4$ and $Q_0 \cong \sigma_0 H/2$.

The theoretical analysis leading to equation (6.4) is valid for dynamic pressure pulses with $\eta \geqslant 3$, and, therefore, predicts $|Q_d/Q_0| \geqslant 1$ when $\eta \geqslant (2L/H)^2/3$. In fact, it is evident that $|Q_d| \to \infty$ when $\eta \to \infty$ for the impulsive velocity loading case discussed in § 3.5.1.

The vertical equilibrium equation for the static loading of a simply supported beam with a rectangular cross-section predicts a small value for Q_s/Q_0 according to equation (6.2), while the conservation of linear momentum for the impulsive velocity case gives rise to an infinitely large transverse shear force at the supports. This demonstrates clearly that transverse shear forces are potentially more important for dynamic problems, particularly for large pressure pulses with a short duration. Indeed, the photograph in Figure 6.1 shows a beam which has failed due to the development of an excessive transverse shear force generated by a mass striking it near a support.

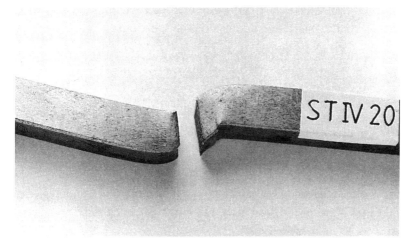

Figure 6.1 Failure of a beam due to an excessive transverse shear force caused by a dropped mass striking it near a support.[6.2]

Large transverse shear forces may also develop during the higher modal dynamic response of a beam.[6.3, 6.4] It is evident from equation (3.1) ($\partial M/\partial x = Q$) that $\partial M/\partial x$ and, therefore, the transverse shear force must increase for higher modes because the wavelength of the bending moment M decreases as the mode number increases. In other words, $M = fn_1(x/2L)$ for the first or primary mode response of a beam, while $M = fn_2\{2(x/2L)\}$ is a crude approximation for the second mode. In general, this line of reasoning gives $M = fn_n(nx/2L)$ for the nth mode, which gives rise to a transverse shear force which is of order n times larger than that present in the first mode.†

The dynamic plastic response of ideal fibre-reinforced, or strongly anisotropic, beams is also governed primarily by transverse shear effects.[6.5, 6.6]

The governing equations for beams which retain transverse shear and rotatory inertia effects are presented in the next section, together with yield criteria, which include the influence of a transverse shear force on the plastic flow of a material. An exact theoretical solution is developed in § 6.3 for a simply supported beam which is subjected to a uniformly distributed impulsive velocity. This is followed by § 6.4, which contains a theoretical solution for the dynamic plastic response of a long beam struck by a mass G travelling with an initial velocity V_0. The influence of transverse shear effects on the response of an impulsively loaded, simply supported circular plate and an impulsively loaded, simply supported cylindrical shell are examined in §§ 6.5 and 6.6, respectively. Finally, the chapter ends with some general remarks, in § 6.7, on the influence of yield criteria and transverse shear failure.

6.2 Governing equations for beams

The equilibrium equations for the beam element in Figure 3.3 are given by equations (3.1) and (3.2). The vertical equilibrium equation (3.2) remains valid whether or not the effects of rotatory inertia and transverse shear influence the dynamic response of a beam, i.e.,

$$\partial Q/\partial x = -p + m\ddot{w}, \qquad (6.5a)$$

where

$$(\dot{\ }) = \partial(\ )/\partial t. \qquad (6.5b)$$

However, rotatory inertia effects contribute to the moment equilibrium equation (3.1), which now becomes

$$Q = \partial M/\partial x - I_r\ddot{\psi}, \qquad (6.6)$$

where $I_r = mk^2$ is the second moment of inertia for the cross-section when rotating about the mid-plane of the beam (per unit length), k is the radius of gyration and

† It is evident from Table 1 in reference 6.4 that the largest values of Q/Q_0 are 0.17, 0.24 and 0.35 for the first, second and third modes, respectively. This gives ratios of $1:1.41:2.06$ rather than the predictions of $1:2:3$ from the crude analysis, which is presented to provide a physical basis for the phenomenon.

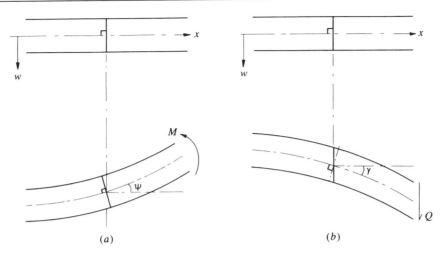

Figure 6.2 (a) Pure bending behaviour of a beam due to a bending moment M.
(b) Pure shearing behaviour of a beam due to a transverse shear force Q.

the total slope of the centre line of a beam is

$$\partial w/\partial x = \gamma - \psi, \tag{6.7}$$

as shown in Figure 6.2. The angle $\psi(x, t)$ is the usual rotation of a beam centre line due to bending alone, while $\gamma(x, t)$ is the shear angle of points along the beam centre line due to a transverse shear force alone. Thus, the curvature of the beam centre line associated with bending is

$$\kappa = \partial\psi/\partial x \tag{6.8}$$

and the transverse shear strain is γ.† It is evident that, in the absence of transverse shear strain, $\gamma = 0$, and equation (6.7) reduces to $\partial w/\partial x = -\psi$, which allows equation (6.8) to be written as equation (3.3).

Plastic flow of a beam without axial forces is controlled by a combination of the bending moment M and a transverse shear force Q. If $Q = 0$, then the fully plastic carrying capacity of a beam is reached when $M = M_0$ (and $\dot{\kappa} \geqslant 0$), which is the yield condition employed in Chapter 3 for beams loaded dynamically. On the other hand, if $M = 0$, then plastic yielding develops in a beam when $Q = Q_0$ (and $\dot{\gamma} \geqslant 0$), where Q_0 is the maximum transverse shear force for the cross-section of a beam which is made from a perfectly plastic material. Clearly, an interaction curve, or yield curve, which relates M and Q, is required for beams which are subjected to M and Q simultaneously.

It is assumed that a rigid, perfectly plastic material obeys the square yield

† It is shown in Appendix 4 that the equilibrium equations (6.5) and (6.6) are consistent with the curvature change (κ) and transverse shear strain (γ) relations according to the principle of virtual velocities.

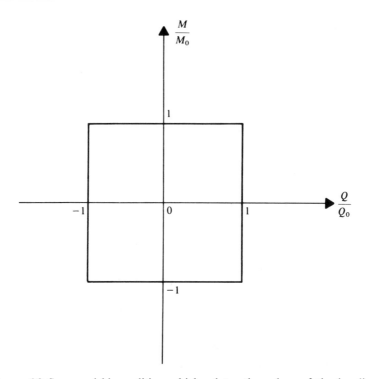

Figure 6.3 Square yield condition which relates the values of the bending moment (M) and transverse shear force (Q) required for the plastic flow of a beam cross-section.

criterion illustrated in Figure 6.3. Section 6.7.2 contains some further discussion on yield criteria.

6.3 Transverse shear effects in a simply supported beam loaded impulsively

6.3.1 Introduction

It was observed in § 6.1 that the transverse shear force in a simply supported beam, which is subjected to a dynamic pressure pulse with a rectangular shaped pressure–time history, exceeded the corresponding plastic carrying capacity of a solid cross-section when $\eta \geqslant (2L/H)^2/3$ and that $|Q_d| \to \infty$ for an initial impulsive velocity with $\eta \to \infty$. Thus, transverse shear effects should have the greatest significance for a simply supported beam which is subjected to an impulsive velocity and, therefore, the beam shown in Figure 6.4(a) is now examined in this section.

It is convenient to introduce a dimensionless parameter

$$v = Q_0 L/2M_0, \tag{6.9}$$

which is a ratio between the transverse shear strength and the bending moment

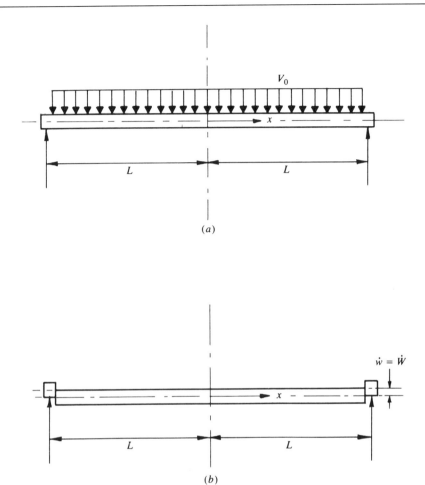

Figure 6.4 (a) Simply supported beam which is subjected to a uniformly distributed impulsive velocity of magnitude V_0. (b) Transverse shear slide at supports of an impulsively loaded beam with $v \leqslant 1$.

capacity of a beam cross-section. Small values of v ($v \ll 1$) imply that the transverse shear strength of a beam is relatively small, while a predominantly bending response would be associated with large values of v ($v \gg 1$), since the transverse shear strength is relatively large compared with the bending moment capacity. It might be anticipated that both bending and transverse shear effects would contribute to the behaviour of beams with intermediate values of v of order unity.

It transpires that three different theoretical solutions[6.7] are required when the beam, illustrated in Figure 6.4(a), is made from a material which flows plastically according to the square yield condition in Figure 6.3. These cases $0 \leqslant v \leqslant 1$, $1 \leqslant v \leqslant 1.5$ and $v \geqslant 1.5$ are now examined in the following sections.

6.3.2 Beams with $v \leqslant 1$

A bending hinge may develop in a perfectly plastic beam when $M = \pm M_0$, as shown in Figures 3.4, 3.7 and 3.14, for example. These hinges are idealised as infinitesimally small regions over which there is a finite angular change with $|\dot{\kappa}| \to \infty$. In the same spirit, a transverse shear hinge may develop when $Q = \pm Q_0$ with $|\dot{\gamma}| \to \infty$. This hinge, or slide, is an infinitesimally short plastic zone with discontinuities in the transverse displacement and velocity.

Now, the beam in Figure 6.4(a) is relatively weak in shear when $v \leqslant 1$. Thus, it is assumed that transverse shear hinges develop at both ends where there are discontinuities in transverse velocity at $t = 0$ due to the stationary supports. This transverse velocity field is illustrated in Figure 6.4(b) and clearly

$$\dot{w} = \dot{W}, \qquad 0 \leqslant x \leqslant L, \tag{6.10}$$

when focusing attention on only one-half of a beam owing to symmetry.

Equations (6.5a) and (6.10) give

$$\partial Q/\partial x = m\ddot{W} \tag{6.11}$$

since $p = 0$ and, therefore,

$$Q = m\ddot{W}x, \qquad 0 \leqslant x \leqslant L, \tag{6.12}$$

because $Q = 0$ at $x = 0$ for symmetry. However, the transverse shear slide at $x = L$ requires $Q = -Q_0$, so that equation (6.12) predicts

$$\ddot{W} = -Q_0/mL \tag{6.13}$$

and, therefore,

$$\dot{W} = -Q_0 t/mL + V_0 \tag{6.14}$$

for an initial impulsive velocity V_0 at $t = 0$, and

$$W = -Q_0 t^2/2mL + V_0 t \tag{6.15}$$

when $W = 0$ at $t = 0$.

Motion ceases at

$$T = mLV_0/Q_0 \tag{6.16}$$

according to equation (6.14) with $\dot{W} = 0$, and the associated uniform permanent transverse displacement is

$$w_f = mLV_0^2/2Q_0. \tag{6.17}$$

Equations (6.12) and (6.13) predict that

$$Q/Q_0 = -x/L, \qquad 0 \leqslant x \leqslant L, \tag{6.18}$$

which, together with equations (6.6) and (6.9), gives

$$M/M_0 = v\{1 - (x/L)^2\} \tag{6.19}$$

since $M = 0$ for simple supports at $x = L$. It is evident from equation (6.18) that $Q \leqslant |Q_0|$ throughout the beam, while equation (6.19) leads to a yield violation in M (i.e., $M \geqslant M_0$ at $x = 0$) when $v \geqslant 1$.

6.3.3 Beams with $1 \leqslant v \leqslant 1.5$

It is evident from equations (6.18) and (6.19) that the square yield condition in Figure 6.3 is penetrated at the mid-span ($x = 0$) when $v \geqslant 1$. This yield violation suggests that a stationary plastic bending hinge develops at $x = 0$ for $v \geqslant 1$ in addition to stationary transverse shear sliding at the supports ($x = L$), as shown in Figure 6.5(a), or

$$\dot{w} = \dot{W}_s + (\dot{W} - \dot{W}_s)(1 - x/L), \qquad 0 \leqslant x \leqslant L. \tag{6.20}$$

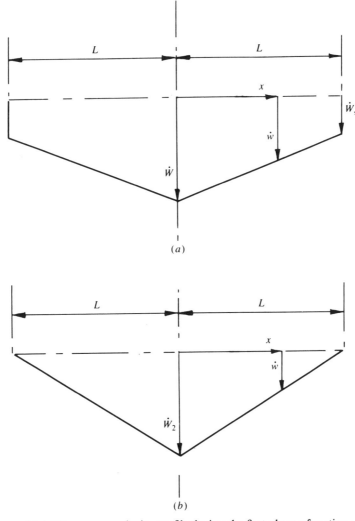

(a)

(b)

Figure 6.5 (a) Transverse velocity profile during the first phase of motion of an impulsively loaded beam with $1 \leqslant v \leqslant 1.5$. (b) Transverse velocity profile during the second phase of motion of an impulsively loaded beam with $1 \leqslant v \leqslant 1.5$.

Thus, equation (6.5a) becomes

$$\partial Q/\partial x = m\ddot{W}_s + m(\ddot{W} - \ddot{W}_s)(1 - x/L), \tag{6.21}$$

which, when integrating with respect to x and satisfying the symmetry condition $Q = 0$ at $x = 0$, leads to

$$Q = m\ddot{W}_s x + m(\ddot{W} - \ddot{W}_s)(x - x^2/2L), \qquad 0 \leqslant x \leqslant L. \tag{6.22}$$

Equations (6.6) and (6.22) predict the bending moment

$$M = m\ddot{W}_s x^2/2 + m(\ddot{W} - \ddot{W}_s)(x^2/2 \div x^3/6L) + M_0, \qquad 0 \leqslant x \leqslant L, \tag{6.23}$$

when $I_r = 0$ and $M = M_0$ at $x = 0$.

The transverse shear hinges at the supports demand that equation (6.22) yields $Q = -Q_0$ at $x = L$, or

$$\ddot{W}_s + \ddot{W} = -2Q_0/mL, \tag{6.24}$$

while the simply supported boundary condition $M = 0$ at $x = L$ gives

$$\ddot{W}_s + 2\ddot{W} = -6M_0/mL^2 \tag{6.25}$$

according to equation (6.23). Equations (6.24) and (6.25), therefore, predict that

$$\ddot{W} = 2M_0(2v - 3)/mL^2 \tag{6.26}$$

and

$$\ddot{W}_s = 2M_0(3 - 4v)/mL^2, \tag{6.27}$$

which, when integrated with respect to time, yield

$$\dot{W} = 2M_0(2v - 3)t/mL^2 + V_0, \tag{6.28a}$$

$$\dot{W}_s = 2M_0(3 - 4v)t/mL^2 + V_0, \tag{6.28b}$$

$$W = M_0(2v - 3)t^2/mL^2 + V_0 t \tag{6.29a}$$

and

$$W_s = M_0(3 - 4v)t^2/mL^2 + V_0 t, \tag{6.29b}$$

since $\dot{W} = \dot{W}_s = V_0$ and $W = W_s = 0$ at $t = 0$ for impulsively loaded beams.

It is evident from equation (6.28b) that transverse shear sliding ceases at the supports when

$$T_s = mL^2 V_0/2M_0(4v - 3), \tag{6.30}$$

at which time

$$\dot{W} = 6V_0(v - 1)/(4v - 3) \tag{6.31}$$

according to equation (6.28a). Thus, $\dot{W} \geqslant 0$ for $v \geqslant 1$, so that the beam has a kinetic energy at $t = T_s$ which must be dissipated during a further phase of motion.

It is assumed that during the subsequent phase of motion the plastic bending hinge remains stationary at the mid-span, while plastic shearing ceases at the supports. A suitable transverse velocity profile is illustrated in Figure 6.5(b), or

$$\dot{w} = \dot{W}_2(1 - x/L), \qquad 0 \leqslant x \leqslant L, \tag{6.32}$$

and, therefore, equation (6.5a) now becomes

$$\partial Q/\partial x = m\ddot{W}_2(1 - x/L), \tag{6.33}$$

which gives

$$Q = m\ddot{W}_2(x - x^2/2L), \qquad 0 \leqslant x \leqslant L, \tag{6.34}$$

when satisfying $Q = 0$ at $x = 0$, while equation (6.6) with $I_r = 0$ and $M = M_0$ at $x = 0$ yields

$$M = m\ddot{W}_2(x^2/2 - x^3/6L) + M_0, \qquad 0 \leqslant x \leqslant L. \tag{6.35}$$

The simply supported boundary condition demands that equation (6.35) satisfy $M = 0$ at $x = L$, or

$$\ddot{W}_2 = -3M_0/mL^2 \tag{6.36}$$

and, therefore,

$$\dot{W}_2 = -3M_0t/mL^2 + 3V_0/2 \tag{6.37}$$

when ensuring continuity with equation (6.31) at $t = T_s$. Motion finally ceases when $\dot{W}_2 = 0$, or

$$T = mL^2V_0/2M_0. \tag{6.38}$$

It may be shown that the permanent transverse displacement profile is

$$w_f = mL^2V_0^2\{1 + 6(v - 1)(1 - x/L)\}/\{4M_0(4v - 3)\}, \tag{6.39}$$

while the generalised stresses for $0 \leqslant t \leqslant T_s$ are

$$Q/Q_0 = \{2v - 3 - 3(v - 1)(x/L)\}(x/L)/v, \qquad 0 \leqslant x \leqslant L, \tag{6.40a}$$

and

$$M/M_0 = 1 + (2v - 3)(x/L)^2 - 2(v - 1)(x/L)^3, \qquad 0 \leqslant x \leqslant L, \tag{6.40b}$$

and, for $T_s \leqslant t \leqslant T$,

$$Q/Q_0 = -3(1 - x/2L)(x/L)/2v, \qquad 0 \leqslant x \leqslant L, \tag{6.41a}$$

and

$$M/M_0 = 1 - (3 - x/L)(x/L)^2/2, \qquad 0 \leqslant x \leqslant L. \tag{6.41b}$$

Now, equation (6.40b) gives $M = M_0$ and $\partial M/\partial x = Q = 0$ at $x = 0$. Moreover, $(L^2/M_0) \partial^2 M/\partial x^2 = 2(2v - 3)$ at $x = 0$, from which it is evident that $\partial^2 M/\partial x^2 \geqslant 0$ when $v \geqslant 1.5$. This would lead to a yield violation near the mid-span so that the foregoing theoretical analysis is valid for $1 \leqslant v \leqslant 1.5$.

6.3.4 Beams with $v \geqslant 1.5$

6.3.4(a) First phase of motion, $0 \leqslant t \leqslant T_s$

It was noted that the theoretical analysis in § 6.3.3 is valid for impulsively loaded beams having $1 \leqslant v \leqslant 1.5$, since a yield violation occurs near the mid-span for $v > 1.5$ when using the transverse velocity profile in Figure 6.5(a). The theoretical solution in § 3.5 ignores transverse shear effects and is, therefore, valid for impulsively loaded beams with $v \to \infty$ according to equation (6.9). A plastic bending hinge develops at the supports during the first phase of motion and travels inwards towards the mid-span, where it remains stationary throughout a final phase of motion while the remaining kinetic energy is dissipated plastically. These two cases for $v = 1.5$ and $v = \infty$, therefore, suggest the transverse velocity profile in Figure 6.6(a) for the first phase of motion of an impulsively loaded beam with an intermediate value of v. Thus,

$$\dot{w} = V_0, \qquad 0 \leqslant x \leqslant \xi_0, \tag{6.42a}$$

and

$$\dot{w} = \dot{W}_s + (V_0 - \dot{W}_s)(L - x)/(L - \xi_0), \qquad \xi_0 \leqslant x \leqslant L, \tag{6.42b}$$

where ξ_0 is the position of the stationary plastic bending hinge.

Equations (6.5a) and (6.42a) give $\partial Q/\partial x = 0$, or

$$Q = 0, \qquad 0 \leqslant x \leqslant \xi_0, \tag{6.43}$$

since $Q = 0$ at $x = 0$ for symmetry. Thus, equation (6.6), with $I_r = 0$, predicts that

$$M = M_0, \qquad 0 \leqslant x \leqslant \xi_0, \tag{6.44}$$

in order for a stationary plastic bending hinge to develop at $x = \xi_0$.

Now, it may be shown that equations (6.5a) and (6.42b), with continuity of Q at $x = \xi_0$ (i.e., $Q = 0$ at $x = \xi_0$), predict that

$$Q = m\dot{W}_s(x - \xi_0)[1 + \{(x + \xi_0)/2 - L\}/(L - \xi_0)], \qquad \xi_0 \leqslant x \leqslant L, \tag{6.45}$$

which, when a transverse shear slide develops at $x = L$ (i.e., $Q = -Q_0$ at $x = L$), gives

$$\ddot{W}_s = -2Q_0/\{m(L - \xi_0)\}. \tag{6.46}$$

Moreover, equation (6.6), with $I_r = 0$, and equation (6.42b) predict

$$M = m\ddot{W}_s(L - x)[\xi_0 - (L + x)/2 + \{(2L^2 + 2Lx - x^2)/6$$
$$- \xi_0(L - \xi_0/2)\}/(L - \xi_0)], \qquad \xi_0 \leqslant x \leqslant L, \tag{6.47}$$

when satisfying the simply supported boundary condition (i.e., $M = 0$ at $x = L$). Equation (6.47) gives

$$\ddot{W}_s = -6M_0/\{m(L - \xi_0)^2\}, \tag{6.48}$$

since M is continuous at the plastic bending hinge (i.e., $M = M_0$ at $x = \xi_0$). Thus, equations (6.46) and (6.48) predict that

$$1 - \xi_0/L = 3/2v \tag{6.49}$$

and

$$\ddot{W}_s = -8v^2 M_0/3mL^2. \tag{6.50}$$

Equation (6.50) may be integrated with respect to time to predict a transverse shear displacement

$$W_s = -4v^2 M_0 t^2/3mL^2 + V_0 t \tag{6.51}$$

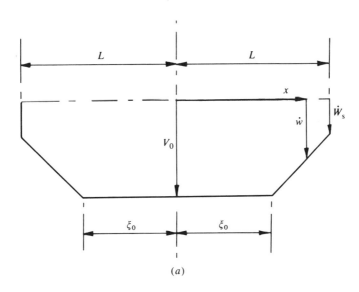

(a)

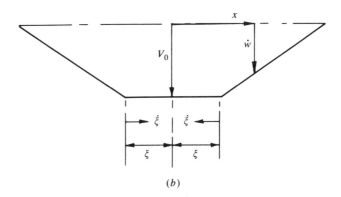

(b)

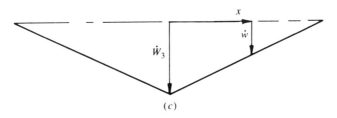

(c)

Figure 6.6 Transverse velocity profiles for beams having $v \geqslant 1.5$ during the (a) first phase of motion, (b) second phase of motion, (c) third phase of motion.

which ceases at

$$T_s = 3mL^2 V_0/8v^2 M_0 \tag{6.52}$$

when $\dot{W}_s = 0$.

6.3.4(b) Second phase of motion, $T_s \leqslant t \leqslant T_1$

Now, equation (6.42) with $\dot{W}_s = 0$ at $t = T_s$ indicates that the beam is still travelling and, therefore, has a finite kinetic energy which must be dissipated as plastic work when $t \geqslant T_s$. Thus, it is assumed that the transverse velocity profile is

$$\dot{w} = V_0, \qquad 0 \leqslant x \leqslant \xi, \tag{6.53a}$$

and

$$\dot{w} = V_0(L - x)/(L - \xi), \qquad .\xi \leqslant x \leqslant L, \tag{6.53b}$$

where ξ is the time-dependent position of a plastic bending hinge, as shown in Figure 6.6(b), and which travels from $\xi = \xi_0$ at $t = T_s$ towards the beam mid-span at $t = T_1$.

Equations (6.5a) and (6.6) together with equation (6.53a) give

$$Q = 0 \qquad \text{and} \qquad M = M_0, \qquad 0 \leqslant x \leqslant \xi, \tag{6.54a, b}$$

when satisfying the symmetry requirements and the conditions for a plastic bending hinge at $x = \xi$. Similarly, it may be shown that

$$Q = mV_0\dot{\xi}(x - \xi)\{L - (x + \xi)/2\}/(L - \xi)^2, \qquad \xi \leqslant x \leqslant L, \tag{6.55}$$

and

$$M = M_0 + mV_0\dot{\xi}(x - \xi)^2(L/2 - x/6 - \xi/3)/(L - \xi)^2, \qquad \xi \leqslant x \leqslant L, \tag{6.56}$$

when satisfying the continuity requirements at $x = \xi$.

The simply supported boundary condition $M = 0$ at $x = L$ and equation (6.56) leads to the plastic hinge velocity

$$\dot{\xi} = -3M_0/\{mV_0(L - \xi)\}, \tag{6.57}$$

which may be integrated with respect to time to give

$$t = T_s + mV_0\{(L - \xi)^2 - (L - \xi_0)^2\}/6M_0. \tag{6.58}$$

The two travelling plastic bending hinges reach the mid-span and coalesce at $t = T_1$ when $\xi = 0$, where

$$T_1 = mV_0L^2/6M_0 \tag{6.59}$$

according to equations (6.49), (6.52) and (6.58).

It is evident from equations (6.42a), (6.53a) and (6.59), when $t = T_1$, that the maximum transverse displacement occurs at the mid-span and is V_0T_1, or

$$W(T_1) = mL^2V_0^2/6M_0. \tag{6.60}$$

6.3.4(c) Third phase of motion, $T_1 \leqslant t \leqslant T$

The beam has a finite kinetic energy at the end of the second phase of motion since $\dot{w} = V_0$ at the mid-span when $t = T_1$. Thus, for $t \geqslant T_1$, it is assumed that

$$\dot{w} = \dot{W}_3(1 - x/L), \qquad 0 \leqslant x \leqslant L, \tag{6.61}$$

which is illustrated in Figure 6.6(c).

Equations (6.5a), (6.6) and (6.61) predict that

$$Q = m\ddot{W}_3(x - x^2/2L), \qquad 0 \leqslant x \leqslant L, \tag{6.62}$$

and

$$M = M_0 + m\ddot{W}_3(x^2/2 - x^3/6L), \qquad 0 \leqslant x \leqslant L, \tag{6.63}$$

when satisfying the requirements $Q = 0$ and $M = M_0$ at the central stationary plastic bending hinge. Equation (6.63), with the simply supported boundary condition $M = 0$ at $x = L$, predicts that

$$\ddot{W}_3 = -3M_0/mL^2. \tag{6.64}$$

Integrating equation (6.64) with respect to time and preserving velocity continuity at $t = T_1$ gives

$$\dot{W}_3 = V_0 - 3M_0(t - T_1)/mL^2, \tag{6.65}$$

so that motion ceases when

$$T = mV_0L^2/2M_0. \tag{6.66}$$

Finally, another integration of equation (6.65) predicts the mid-span displacement

$$W_3 = V_0t - 3M_0(t - T_1)^2/2mL^2, \tag{6.67}$$

which leads to a maximum final transverse displacement

$$W_f = mV_0^2L^2/3M_0. \tag{6.68}$$

It is left as an exercise for an interested reader to obtain the permanent deformed profile of the beam.†

6.3.5 Static admissibility

6.3.5(a) $v \leqslant 1$

Equations (6.18) and (6.19) for Q/Q_0 and M/M_0, respectively, show that the theoretical analysis in § 6.3.2 satisfies the yield criterion provided $v \leqslant 1$.

6.3.5(b) $1 \leqslant v \leqslant 1.5$

It is straightforward to show that equations (6.40) and (6.41) for the transverse shear force and bending moment do not penetrate the yield curve in Figure 6.3 when $v \leqslant 1.5$.

† See Problem 6.6.

6.3.5(c) $v \geqslant 1.5$

Equations (6.43) to (6.48) for the first phase of motion predict generalised stresses which do not penetrate the yield curve in Figure 6.3. It may be shown with the aid of equation (6.57) that equations (6.54) to (6.56) do not violate the yield condition throughout the second phase of motion. Finally, the generalised stresses in § 6.3.4(c) lie on or within the yield condition throughout the final phase of motion.

6.3.6 Conditions at bending hinges and shear slides

Some general expressions are developed initially then specialised later in order to examine the kinematic admissibility of the foregoing theoretical solution.

Continuity of the transverse displacement and of the angular deformation associated with the bending of a beam at a moving interface respectively requires

$$[w]_\xi = 0 \qquad \text{and} \qquad [\psi]_\xi = 0, \qquad (6.69a, b)\dagger$$

where $[X]_\xi$ means the difference in X on either side of an interface at ξ which travels with a velocity $\dot{\xi}$, as shown in Figure 3.9. Equations (6.69a) and (6.69b) imply that

$$[Dw/Dt]_\xi = [\partial w/\partial t]_\xi + \dot{\xi}[\partial w/\partial x]_\xi = 0 \qquad (6.70a)$$

and

$$[D\psi/Dt]_\xi = [\partial\psi/\partial t]_\xi + \dot{\xi}[\partial\psi/\partial x]_\xi = 0, \qquad (6.70b)$$

where $\partial\psi/\partial x$ is the curvature (κ) associated with bending as defined by equation (6.8). Thus

$$[\partial w/\partial t]_\xi + \dot{\xi}[\gamma]_\xi = 0 \qquad (6.71)$$

according to equations (6.7), (6.69b) and (6.70a).

It may be shown that, in the absence of transverse shear strains when $\gamma = 0$, equations (6.69a, b), (6.70a, b) and (6.71) are identical to equations (3.54), (3.56), (3.55), (3.59) and (3.57), respectively.

Now, conservation of momentum across a travelling discontinuity requires

$$[Q]_\xi = -m\dot{\xi}[\partial w/\partial t]_\xi \qquad (6.72a)$$

and

$$[M]_\xi = -I_r\dot{\xi}[\partial\psi/\partial t]_\xi. \qquad (6.72b)$$

Equations (6.71) and (6.72a) predict

$$[Q]_\xi = m\dot{\xi}^2[\gamma]_\xi, \qquad (6.73a)$$

while equation (6.72b) becomes

$$[M]_\xi = I_r\dot{\xi}^2[\partial\psi/\partial x]_\xi \qquad (6.73b)$$

when using equation (6.70b).

$\dagger$ ψ is defined in Figure 6.2(a).

Symonds[6.8] examined a square yield criterion which relates M and Q for a rigid, perfectly plastic material as illustrated in Figure 6.3 and observed that

$$[Q]_\xi = 0 \quad \text{and} \quad [M]_\xi = 0 \qquad (6.74a, b)$$

at a moving discontinuity. In this circumstance, equations (6.73a) and (6.73b) with $I_r \neq 0$ show that $[\gamma]_\xi = 0$ and $[\partial\psi/\partial x]_\xi = 0$ (i.e., $[\kappa] = 0$) unless $\dot\xi = 0$. In other words, a discontinuity in γ may develop only at a stationary interface, while a plastic bending hinge must also remain stationary in a perfectly plastic beam.

However, if rotatory inertia is neglected then $I_r = 0$ and equation (6.73b) shows that a plastic bending hinge may travel. Nevertheless, equations (6.73a) and (6.74a) still require $[\gamma]_\xi = 0$. Thus, equation (6.71) requires $[\partial w/\partial t]_\xi = 0$, which, following the interface at $x = \xi$, becomes $d[\partial w/\partial t]_\xi/dt = 0$, or

$$[\partial^2 w/\partial t^2]_\xi + \dot\xi[\partial^2 w/\partial x\,\partial t]_\xi = 0. \qquad (6.75)$$

Equation (6.70b) gives $[\partial\psi/\partial t]_\xi = -[\partial^2 w/\partial x\partial t]_\xi = -\dot\xi[\partial\psi/\partial x]_\xi$ so that equation (6.75) may be written

$$[\partial^2 w/\partial t^2]_\xi + \dot\xi^2[\kappa]_\xi = 0 \qquad (6.76)$$

when using equation (6.8). Equation (6.75) is identical to equation (3.58) for travelling plastic bending hinges in beams without transverse shear effects. It may also be shown that equations (6.75) and (6.76), with $\gamma = 0$, may be cast into the form of equation (3.59).

Now, returning to the theoretical analysis in §§ 6.3.2 to 6.3.4, it is evident that the bending hinges and transverse shear slides are stationary for all values of v and all phases of motion except for the plastic bending hinge which travels during the second phase of motion in § 6.3.4(b). In the case of the stationary bending hinges and shear slides, the conditions $[Q]_\xi = [M]_\xi = 0$ have been used in the analysis as well as $[w]_\xi = [\dot w]_\xi = 0$ at a plastic bending hinge.

The influence of rotational inertia has been disregarded in the theoretical analysis in §§ 6.3.2 to 6.3.4† so that the travelling plastic bending hinge in § 6.3.4(b) is admissible since $[\gamma]_\xi = [\dot w]_\xi = 0$, as required. Moreover, it is straightforward to show that equations (6.75) and (6.76) are satisfied.

6.3.7 Energy ratios

It is evident from § 6.3.2 that all the initial kinetic energy is dissipated as transverse shear sliding at the supports when $v \leqslant 1$. Thus,

$$D_s/K_e = 1, \quad v \leqslant 1, \qquad (6.77)$$

where K_e and D_s are the initial kinetic energy and the energy dissipated due to transverse shear effects, respectively.

---

† In fact, the theoretical analysis in § 6.3.2 for $v \leqslant 1$ is independent of I_r since $\psi = 0$ according to equation (6.10) and Figure 6.4(b).

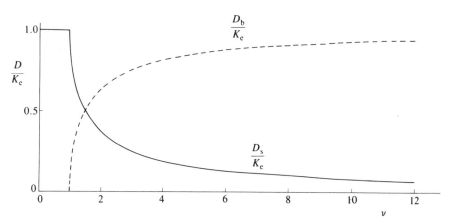

Figure 6.7 Variation of the dimensionless transverse shear (D_s/K_e) and bending (D_b/K_e) energies with the parameter v. ———: D_s/K_e given by equations (6.77, 6.78, 6.80). – – –: D_b/K_e given by equations (6.79, 6.81–6.83). K_e = initial kinetic energy.

The transverse shear energy dissipated during the first phase of motion in § 6.3.3 for $1 \leqslant v \leqslant 1.5$ is $2 \int_0^{T_s} Q_0 \dot{W}_s \, dt$, or

$$D_s/K_e = v/(4v - 3), \qquad 1 \leqslant v \leqslant 1.5, \tag{6.78}$$

where v is defined by equation (6.9). The energy absorbed in the central plastic bending hinge is $2 \int_0^T M_0(\dot{W} - \dot{W}_s) \, dt/L$, or $2M_0\{W_f - W_s(T_s)\}/L$, which gives

$$D_b/K_e = 3(v - 1)/(4v - 3), \qquad 1 \leqslant v \leqslant 1.5. \tag{6.79}$$

A transverse shear slide is only active at the supports during the first phase of motion in § 6.3.4(a) for $v \geqslant 1.5$. Thus, $D_s = 2Q_0 W_s$, or

$$D_s/K_e = 3/4v, \qquad v \geqslant 1.5. \tag{6.80}$$

The remaining kinetic energy is dissipated by stationary plastic bending hinges during the first (D_{b1}) and final (D_{b3}) phases of motion and a travelling bending hinge during the second phase (D_{b2}). It may be shown that

$$D_{b1}/K_e = 1/4v, \tag{6.81}$$

$$D_{b2}/K_e = (2v - 3)/3v \tag{6.82}†$$

and

$$D_{b3}/K_e = 1/3. \tag{6.83}$$

The proportions of the initial kinetic energy (K_e) which are absorbed by transverse shear sliding (D_s) and plastic bending (D_b) are shown in Figure 6.7.

† The energy dissipated in the two travelling plastic bending hinges in § 6.3.4(b) and Figure 6.6(b) may be written $D_{b2} = 2 \int\int M_0(\partial^2 \dot{w}/\partial x^2) \, dx \, dt = -2M_0 \int (\partial \dot{w}/\partial x) \, dt = 2M_0 \int V_0 \, dt/(L - \xi)$ when using equation (6.53b). However, $\dot{\xi} = d\xi/dt$ is given by equation (6.57) so that $D_{b2} = -2M_0 V_0 \int_{\xi_0}^0 mV_0 \, d\xi/3M_0$ which, with equation (6.49), leads to equation (6.82).

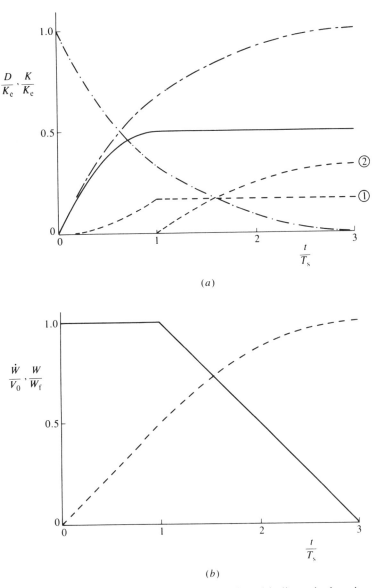

(a)

(b)

Figure 6.8 (a) Variation of dimensionless energies with dimensionless time for impulsively loaded simply supported beams with $v = 1.5$. ——: D_s/K_e. ——— ①: D_{b1}/K_e. — — — ②: D_{b2}/K_e. – – –: D_T/K_e. —·—: $K(t)/K_e$. D_s, D_{b1}, D_{b2}, D_T, $K(t)$ and K_e are transverse shear energy, energy dissipated due to bending during the first phase of motion, energy dissipated due to bending during the second phase of motion, total shear and bending energy, kinetic energy at time t and the initial kinetic energy, respectively. (Note that $T = 3T_s$ according to equations (6.30) and (6.38) with $v = 1.5$.) (b) Variation with dimensionless time of dimensionless transverse velocity and dimensionless transverse displacement at the mid-span of an impulsively loaded simply supported beam having $v = 1.5$. ——: $\dot{W}/V_0$. – – –: W/W_f.

Clearly, transverse shear effects dominate the energy absorbed in beams having small values of v, while the influence of bending is important for large values of v with less than 10 per cent of the initial kinetic energy being absorbed due to transverse shear effects when $v \geqslant 7.5$.

It is left as an exercise for an interested reader to obtain the temporal variations of the kinetic, shear and bending energies. The theoretical predictions for the particular case when $v = 1.5$ may be obtained from the equations in § 6.3.3 and are shown in Figure 6.8. It is evident that one-half of the initial kinetic energy is absorbed by transverse shear sliding at the supports within the first third of the response time and that one-third of the initial kinetic energy is absorbed during the last two-thirds of the response duration.

6.3.8 General comments

The theoretical analysis in §§ 6.3.2 to 6.3.4 for the impulsively loaded beam, which is illustrated in Figure 6.4(a), is both statically and kinematically admissible according to the discussion in §§ 6.3.5 and 6.3.6 and is, therefore, exact for the square yield curve shown in Figure 6.3. This analysis was first derived by Nonaka.[6.7] It may be shown that the present theoretical predictions agree with those in reference 6.7 and are a special case of reference 6.9.

In the particular case when $v \to \infty$, then the transverse shear strength is infinitely larger than the bending strength according to equation (6.9), so that bending effects govern the behaviour. This is the well-known bending-only solution which predicts $T_s = 0$ according to equation (6.52), while the response time and maximum permanent transverse displacement are respectively given by equations (6.66) and (6.68), which are identical to equations (3.63a, b), as expected.

The dimensionless maximum permanent transverse displacements according to equations (6.17), (6.39) and (6.68) for $v \leqslant 1$, $1 \leqslant v \leqslant 1.5$ and $v \geqslant 1.5$, respectively, are plotted in Figure 6.9. The bending-only solution, which is given by equation (3.63b), is independent of the parameter v, while the transverse shear effects are responsible for smaller permanent transverse displacements when $0.75 \leqslant v \leqslant 1.5$ and larger values when $v \leqslant 0.75$. Indeed, the maximum transverse displacement for beams having $v = 1$ is only 75 per cent of the theoretical predictions for the bending only case, while it may be many times larger for values of v smaller than 0.75.

Now, according to equation (6.9), $v = (2L/H)/2$ for a beam with a span $2L$ and a rectangular cross-section of depth H having $Q_0 \cong \sigma_0 H/2$ and $M_0 = \sigma_0 H^2/4$. Thus, small values of v are associated with short and stubby beams having rectangular cross-sections for which a beam theory might not remain valid. However, the parameter v for wide-flanged I-beams may be quite small even for relatively long span-to-depth ratios, as discussed in reference 6.10.

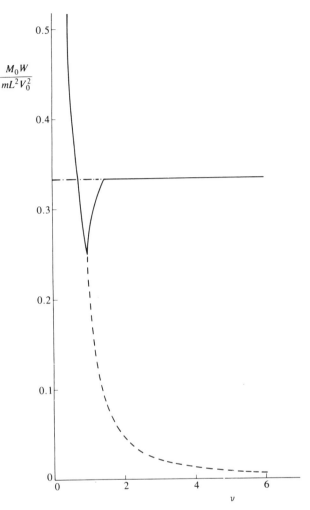

Figure 6.9 Variation of dimensionless maximum permanent transverse displacement with v for an impulsively loaded simply supported beam. $---$: $M_0 W_s / mL^2 V_0^2$ according to equations (6.17), (6.29b) and (6.51). ———: $M_0 W_f / mL^2 V_0^2$ according to equations (6.17), (6.39) and (6.68). $-\cdot-\cdot-$: theoretical analysis which neglects transverse shear effects (equation (3.63b)).

The response duration for $v \geqslant 1$, which is predicted by equations (6.38) and (6.66), is identical to the bending-only case (equation (3.63a)) and is, therefore, independent of the parameter v. However, for $v \leqslant 1$, the response time is $T = mL^2 V_0 / 2vM_0$ according to equation (6.16) and is, therefore, inversely proportional to the parameter v.

The influence of the rotatory inertia term in equation (6.6) has not been retained in the theoretical analysis in §§ 6.3.2 to 6.3.4. However, it transpires for beams with

$v \leqslant 1$, which are studied in § 6.3.2, that $\dot{\psi} = 0$ for the transverse velocity field described by equation (6.10) and illustrated in Figure 6.4(b). Thus, $\dot{\psi} = 0$ and the theoretical analysis in § 6.3.2 is, therefore, valid whether rotatory inertia effects are retained, or not.

The influence of rotatory inertia on the response of an impulsively loaded simply supported beam has been examined in reference 6.10. As noted above, the response is governed by transverse shear effects when $0 \leqslant v \leqslant 1$ and the theoretical

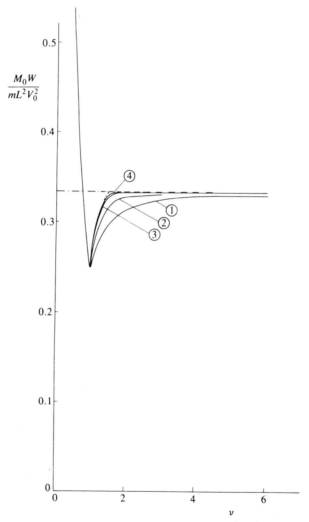

Figure 6.10 The influence of rotatory inertia on the dimensionless maximum permanent transverse displacement of an impulsively loaded simply supported beam. ----: $I_r = 0$ (see Figure 6.9). ①: solid rectangular cross-section ($I_r \neq 0$). ②-④: wide-flanged I-beams ($I_r \neq 0$).[6.10] ---·---: theoretical analysis which neglects transverse shear and rotatory inertia effects (equation (3.63b)).

predictions are independent of rotatory inertia. If $v \geqslant 1$, then the response has two phases. Transverse shear sliding occurs at the supports and bending deformations develop in a central plastic zone of finite and constant length during the first phase of motion. The second phase of motion commences when shear sliding stops at the supports, and the plastic zone gradually shrinks and degenerates to a plastic hinge with zero length when all the kinetic energy is dissipated and motion ceases.

Some theoretical predictions from reference 6.10 are shown in Figure 6.10. The curve labelled ① is for a beam with a rectangular cross-section, while the curves ② to ④ correspond to wide-flanged I-beams, the details of which are discussed further in reference 6.10. It is evident from Figure 6.10 that transverse shear effects may exercise a dominant effect on the beam response, while the influence of rotatory inertia is less important. Similar conclusions are also reached in reference 6.11 for a beam which flows plastically according to the Ilyushin–Shapiro yield condition rather than the square yield criterion in Figure 6.3.

6.4 Impact of a mass on a long beam

6.4.1 Introduction

The theoretical analysis in § 3.8 was developed for a fully clamped beam of length $2L$ which was struck at the mid-span by a mass travelling with an initial velocity V_0, as shown in Figure 3.15(a). The bending moment distribution throughout the central portion of the beam $0 \leqslant x \leqslant \xi$ during the first phase of motion $0 \leqslant t \leqslant t_1$ is given by equation (3.112). Now, a straightforward differentiation of this equation gives the transverse shear force ($Q = \partial M / \partial x$) which is infinitely large underneath the striking mass immediately after impact.† Transverse shear effects are, therefore, expected to be important for the impact loading of beams which are made from materials having a finite transverse shear strength.

It is evident from equation (3.98) that all the initial kinetic energy is expended during the first phase of motion for an infinitely long beam with $2L \rightarrow \infty$ and that the duration of motion is infinitely long according to equation (3.97). Thus, it is somewhat simpler to explore the importance of transverse shear effects in an infinitely long beam since an analysis for the second phase of motion in § 3.8.3 is not required.

The impact problem in Figure 6.11(a) is examined in the following sections for a beam made from a material which is governed by the square yield condition in Figure 6.3 relating the transverse shear force and bending moment required for plastic flow. It transpires that stationary plastic bending hinges and transverse shear slides are active during a first phase of motion. Transverse shear sliding ceases at the end of the first phase of motion and the bending hinges then travel outwards during a final phase.

† See Problem 6.8.

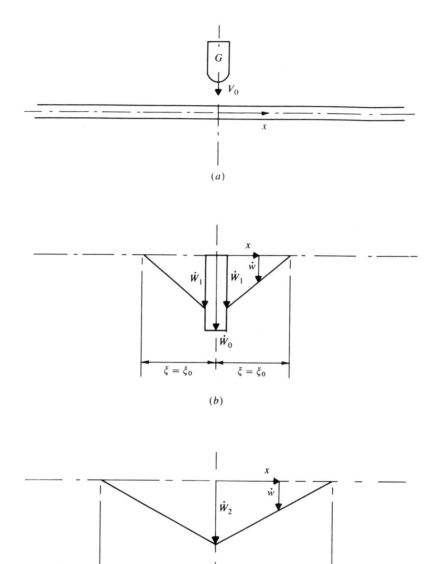

Figure 6.11 (a) Long beam struck with a mass G travelling with an initial velocity V_0. (b) Transverse velocity profile during the first phase of motion $0 \leqslant t \leqslant T_1$. (The width of the striking mass is infinitesimally small in the theoretical analysis.) (c) Transverse velocity profile during the final phase of motion $t \geqslant T_1$.

6.4.2 First phase of motion, $0 \leqslant t \leqslant T_1$

In view of the comments in § 6.4.1, it is likely that transverse shear slides will form on either side of the impact mass G, as indicated in Figure 6.11(b). Thus, the transverse velocity is $\dot{W}_0$ for the mass G and the beam material immediately underneath the striker, while the adjacent material travels with a transverse velocity $\dot{W}_1$, or

$$\dot{w} = \dot{W}_0, \qquad x = 0, \tag{6.84a}$$

$$\dot{w} = \dot{W}_1(1 - x/\xi), \qquad 0 \leqslant x \leqslant \xi, \tag{6.84b}$$

and

$$\dot{w} = 0, \qquad x \geqslant \xi, \tag{6.84c}$$

where ξ is the location of the plastic bending hinge.

It transpires that moment equilibrium of the portion $0 \leqslant x \leqslant \xi$ follows equations (3.88) to (3.90) in § 3.8.2, or

$$t = m\xi^2 \dot{W}_1/12M_0. \tag{6.85}$$

Transverse equilibrium for the portion of the beam between the two plastic bending hinges ($x = \pm \xi$) follows closely equations (3.84) to (3.86) except that the acceleration of the striking mass G is now $\ddot{W}_0$. Thus,

$$G\dot{W}_0 + m\xi\dot{W}_1 = GV_0. \tag{6.86}$$

The transverse equilibrium equation for the striking mass is

$$G\ddot{W}_0 = -2Q_0, \tag{6.87}$$

which may be integrated to give

$$G\dot{W}_0 - GV_0 = -2Q_0 t. \tag{6.88}$$

Now, combining equations (6.86) and (6.88) gives $2Q_0 t = m\xi\dot{W}_1$ which, using equation (6.85), predicts

$$\xi_0 = 6M_0/Q_0 \tag{6.89}\dagger$$

so that the plastic bending hinges remain stationary at $x = \pm\xi_0$ during the first phase of motion.

The transverse shear slides at the mid-span remain active provided $\dot{W}_0 \geqslant \dot{W}_1$, which, using equations (6.85), (6.88) and (6.89), is satisfied for $t \leqslant T_1$, where

$$T_1 = GV_0/\{2Q_0(1 + G/m\xi_0)\} \tag{6.90}$$

is the time when $\dot{W}_0 = \dot{W}_1$ and transverse shear sliding ceases.

A time integration of equation (6.88) predicts the displacement–time history of the striking mass

$$GW_0 = GV_0 t - Q_0 t^2, \tag{6.91a}$$

---

† Equation (6.89) predicts that $\xi_0 = 3H$ when taking $M_0 = \sigma_0 H^2/4$ and $Q_0 = \sigma_0 H/2$ for the particular case of a beam with a solid rectangular cross-section of depth H. The deformation is, therefore, highly localised during the first phase of motion.

which gives

$$W_0(T_1) = (GV_0^2/4Q_0)(1 + 2G/m\xi_0)/(1 + G/m\xi_0)^2 \qquad (6.91b)$$

at the end of the first phase of motion. Equations (6.85) and (6.89) predict that

$$W_1 = Q_0^2 t^2/6mM_0 \qquad (6.92a)$$

and

$$W_1(T_1) = (G^2 V_0^2/24mM_0)/(1 + G/m\xi_0)^2. \qquad (6.92b)$$

6.4.3 Second phase of motion $t \geqslant T_1$

It is evident from equations (6.85), (6.88), (6.89) and (6.90) that $\dot{W}_0 = \dot{W}_1 > 0$ at $t = T_1$ when transverse shear sliding ceases. Thus, the beam in Figure 6.11(a) has a finite kinetic energy at $t = T_1$ which remains to be absorbed plastically in plastic bending hinges during a second phase of motion. Thus, the response is assumed to be governed by the transverse velocity field

$$\dot{w} = \dot{W}_2(1 - x/\xi), \qquad 0 \leqslant x \leqslant \xi, \qquad (6.93a)$$

and

$$\dot{w} = 0, \qquad x \geqslant \xi, \qquad (6.93b)$$

where ξ is the time-dependent location of the travelling plastic hinge, as shown in Figure 6.11(c).

Equations (6.85) and (6.86) now become

$$t = m\xi^2 \dot{W}_2/12M_0 \qquad (6.94)$$

and

$$G\dot{W}_2 + m\xi\dot{W}_2 = GV_0, \qquad (6.95)$$

respectively. Thus,

$$\dot{W}_2 = V_0/(1 + m\xi/G) \qquad (6.96)$$

and

$$t = m\xi^2 V_0/\{12M_0(1 + m\xi/G)\}. \qquad (6.97)$$

It is evident from equation (6.96) that $\dot{W}_2 \to 0$ as $\xi \to \infty$ and, therefore, $t \to \infty$ according to equation (6.97).

A time differentiation of equation (6.97) gives the plastic bending hinge speed,

$$\dot{\xi} = (12M_0/mV_0\xi)(1 + m\xi/G)^2/(2 + m\xi/G), \qquad (6.98)$$

which reduces with time to the value

$$\dot{\xi} = 12M_0/GV_0, \qquad (6.99)$$

when $\xi \to \infty$. The transverse displacement at the mid-span of the beam is

$$W = W_0(T_1) + \int_{T_1}^{t} \dot{W}_2 \, dt \qquad (6.100)$$

which, using equations (6.91b), (6.96) and (6.98), becomes

$$W = (GV_0^2/4Q_0)(1 + 2G/m\xi_0)(1 + G/m\xi_0)^{-2}$$
$$+ (G^2V_0^2/12mM_0)[\log_e\{(1 + m\xi/G)/(1 + m\xi_0/G)\}$$
$$- (1 + m\xi_0/G)^{-2}/2 + (1 + m\xi/G)^{-2}/2]. \qquad (6.101)$$

If a beam has a sufficient transverse shear strength so that $\xi_0 \to 0$, then only bending deformations develop and equation (6.101) reduces to equation (3.95) with $x = 0$.

6.4.4 Discussion

It may be shown that the theoretical analysis in §§ 6.4.2 and 6.4.3 is both statically and kinematically admissible and is, therefore, exact for the beam illustrated in Figure 6.11(a) when it is made from a rigid, perfectly plastic material which obeys the square yield condition in Figure 6.3. This theoretical analysis was first published by Symonds.[6.8]

If a beam is made from a material that is relatively strong in shear compared with the bending strength, then equation (6.89) shows that $\xi_0 \to 0$ and, therefore, the duration of the first phase of motion vanishes (i.e., $T_1 \to 0$ according to equation (6.90)). In this circumstance, the behaviour is dominated by bending effects and is described by the second phase of motion in § 6.4.3. The theoretical solution, with $m\xi_0/G \to 0$, is then identical to that developed in § 3.8.2 (with $2L \to \infty$) which neglects transverse shear effects.

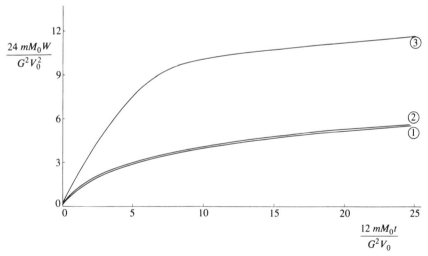

Figure 6.12 Variation of the dimensionless maximum transverse displacement with dimensionless time for an infinitely long beam impacted by a mass G travelling with an initial velocity V_0. ①: Equations (6.91a) and (6.101) with $m\xi_0/G = 0.1$. Also bending-only analysis (equation (3.95)). ②, ③: Equations (6.91a) and (6.101) with $m\xi_0/G = 1$ and 10, respectively.

The influence of transverse shear effects is, therefore, confined to the first phase of motion ($0 \leqslant t \leqslant T_1$), during which transverse shear sliding occurs underneath the impact mass. This leads to dimensionless maximum transverse deflections which, at the same dimensionless time, are larger than those produced when transverse shear effects are neglected, as shown in Figure 6.12 for various values of the parameter $m\xi_0/G$. Further comparisons are made in Figure 6.13 for the dimensionless position of the plastic bending hinge and the dimensionless transverse velocities.

Now, $\xi_0 = 3H$ for the particular case of a beam with a solid rectangular cross-section, as noted in the footnote to equation (6.89). Thus, $m\xi_0/G = 3mH/G$ and small values of $m\xi_0/G$, therefore, correspond to impact masses G which are large relative to the mass of the deformed region of a beam having a length ξ_0. It is evident from Figure 6.12 that the theoretical predictions of §§ 6.4.2 and 6.4.3 are similar to the bending only analysis in § 3.8.2 when $m\xi_0/G \leqslant 1$, approximately. This situation is associated with beams which are either struck by large impact masses, or are strong in shear relative to the bending strength. However, the dimensionless maximum transverse deflections in Figure 6.12 at a given dimensionless time are larger than the corresponding bending-only values when $m\xi_0/G = 10$. This value is associated with light-impact masses or beams with a relatively weak transverse shear strength.

The mechanism of energy absorption in impact-loaded beams changes quite significantly with increase in the dimensionless parameter $m\xi_0/G$. For small values of the ratio $m\xi_0/G$, the initial kinetic energy is absorbed largely through the plastic bending hinges, while it is absorbed through transverse shear sliding for large values of $m\xi_0/G$. It is evident from Figure 6.14 that about 9 per cent of the initial kinetic energy is expended in the two transverse shear slides which form underneath the impact mass when $m\xi_0/G = 0.1$ and $t \leqslant T_1$, while over 90 per cent is dissipated by transverse shearing when $m\xi_0/G = 10$ and $t \leqslant T_1$.

The influence of rotatory inertia on the dynamic plastic response of the particular impact problem in Figure 6.11(a) has been examined in reference 6.10. It transpires that, for convenience, the beam response may be divided into two phases of motion. The first phase of motion is similar to that in § 6.4.2 with stationary plastic hinges except that the rigid region, with $x \geqslant \xi$, is replaced by a plastic zone. Transverse shear sliding ceases at the end of the first phase of motion and the inner boundary of the plastic zone propagates throughout a second phase of motion. However, it is observed[6.10] that rotatory inertia effects do not exercise an important influence on the response, particularly when compared with the effect of retaining transverse shear effects in the yield condition. It is also noted in reference 6.10 that the effect of rotatory inertia on the dynamic plastic response of beams is sensitive to the kind of boundary conditions and type of dynamic loading. Oliveira[6.12] arrived at similar conclusions when examining the influence of

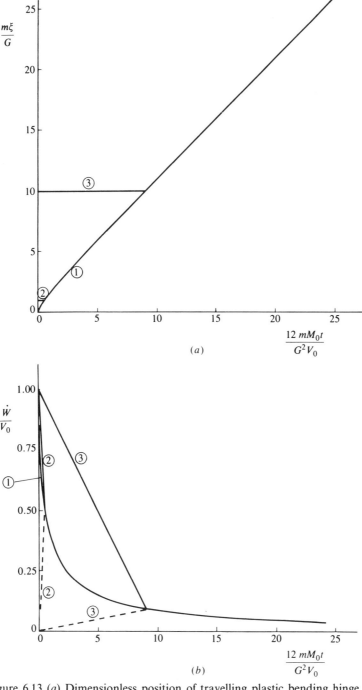

Figure 6.13 (a) Dimensionless position of travelling plastic bending hinge in Figure 6.11(c). ①: Equations (6.89) and (6.97) with $m\xi_0/G = 0.1$. Also bending-only analysis (equation (3.91) in § 3.82). ②, ③: Equations (6.89) and (6.97) with $m\xi_0/G = 1$ and 10, respectively. (b) Temporal variation of dimensionless transverse velocity. ——: transverse velocity of mass G ($\dot{W}_0/V_0$ $(0 \leqslant t \leqslant T_1)$, $\dot{W}_2/V_0$ $(t \geqslant T_1)$) (equations (6.88) and (6.96)). ---: $\dot{W}_1/V_0$ $(0 \leqslant t \leqslant T_1)$ (equation (6.85)). ①: $m\xi_0/G = 0.1$. Also, the curve —— is the bending-only analysis (equation (3.87) in 3.8.2). ②, ③: $m\xi_0/G = 1$ and 10, respectively.

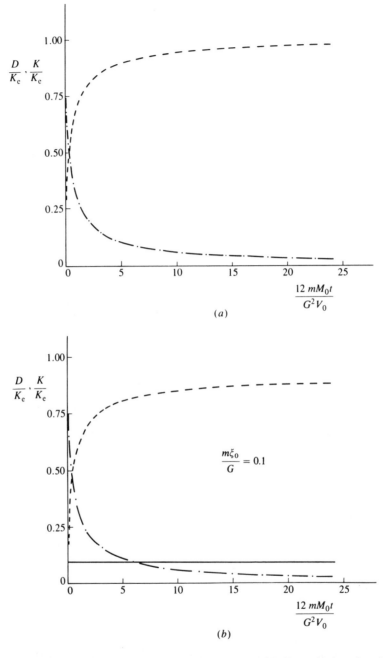

$$\frac{D}{K_e}, \frac{K}{K_e}$$

(a)

$$\frac{D}{K_e}, \frac{K}{K_e}$$

$$\frac{m\xi_0}{G} = 0.1$$

$$\frac{12\,mM_0 t}{G^2 V_0}$$

(b)

Figure 6.14 Variation of the dimensionless energies with dimensionless time for an infinitely long beam impacted by a mass G travelling with an initial velocity V_0. (a) Bending-only analysis (§ 3.8.2). (b), (c), (d) Analysis in § 6.4 with $m\xi_0/G = 0.1$, 1 and 10, respectively. (Note that $12mM_0 t_1/G^2 V_0 = 0.009\,09$, 0.5

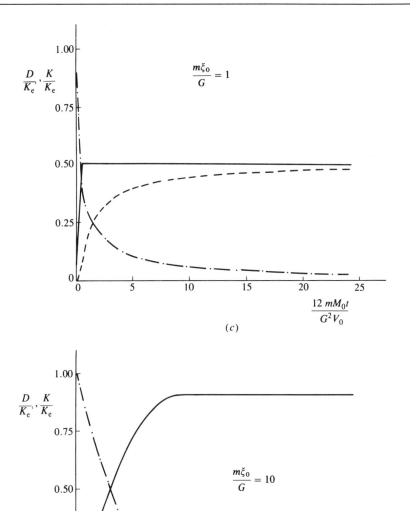

(c)

(d)

Caption for Fig. 6.14 (*cont.*)
and 9.0909, respectively.) ———: D_s/K_e. ———: D_b/K_e. —·—·—: K/K_e. D_s, D_b, K
and K_e are the transverse shear energy, energy dissipated due to bending,
kinetic energy at time t and the initial kinetic energy, respectively.

transverse shear and rotatory inertia on the dynamic plastic response of beams which were either perfectly clamped or simply supported and subjected to lateral projectile impact at the mid-span.

6.5 Transverse shear effects in a simply supported circular plate

6.5.1 Introduction

The dynamic plastic response of a simply supported circular plate subjected to a uniformly distributed pulse with a rectangular shaped pressure–time history was examined in §§ 4.4 and 4.5. The influence of the transverse shear force on the plastic flow of this particular problem is examined in this section.

The transverse shear force in a circular plate loaded axisymmetrically is given by equation (4.1), or

$$Q_r = (1/r)\, \partial(rM_r)/\partial r - M_\theta/r. \tag{6.102}$$

Now equations (4.36a), (4.49) and (4.24) predict that

$$M_\theta/M_0 = 1, \qquad 0 \leqslant r \leqslant R, \tag{6.103}$$

and

$$M_r/M_0 = 1 + (\eta - 2)(r/R)^2 - (\eta - 1)(r/R)^3, \qquad 0 \leqslant r \leqslant R, \tag{6.104}$$

where

$$\eta = p_0/p_c \tag{6.105}$$

is a pressure ratio which must satisfy $1 \leqslant \eta \leqslant 2$ for the theoretical analysis to remain statically admissible. Thus, equation (6.102) gives the transverse shear force

$$Q_r/Q_0 = \{3(\eta - 2) - 4(\eta - 1)r/R\}(r/R)/2v, \qquad 0 \leqslant r \leqslant R, \tag{6.106}$$

when introducing the parameter

$$v = Q_0 R/2M_0, \tag{6.107}$$

where Q_0 is the transverse shear force necessary for plastic flow of the plate cross-section (per unit length). The parameter v is a dimensionless ratio of the transverse shear strength Q_0 relative to the plastic bending strength M_0 of the plate cross-section.

Equation (6.106) shows that the greatest magnitude of the transverse shear force is developed around the outer boundary $(r = R)$ where

$$Q_r/Q_0 = -(\eta + 2)/2v, \tag{6.108}$$

or

$$Q_r/Q_0 = -2/v \tag{6.109}$$

when $\eta = 2$. Clearly, $|Q_r/Q_0| \geqslant 1$ when $v \leqslant 2$. Thus, it is anticipated that transverse shear effects might exercise an important influence on the dynamic plastic response of circular plates having $v \leqslant 2$.

It may be shown, with the aid of the equations in § 4.5.5 for a circular plate, which is subjected to a dynamic pressure pulse with $\eta \geqslant 2$, that equation (6.102) predicts a transverse shear force

$$Q_r/Q_0 = -R^2(r - \xi_0)^2(2r + \xi_0)/\{vr(R - \xi_0)^3(R + \xi_0)\} \qquad (6.110)$$

in the outer plastic zone $\xi_0 \leqslant r \leqslant R$. The radius ξ_0 is defined by equation (4.63), which gives $\xi_0 = 0$ when $\eta = 2$ and $\xi_0 \to R$ for impulsive pressures with $\eta \to \infty$. Equation (6.110) becomes

$$Q_r/Q_0 = -R(2R + \xi_0)/\{v(R^2 - \xi_0^2)\} \qquad (6.111)$$

at the plate supports ($r = R$) and reduces to equation (6.109) when $\eta = 2$, while

$$Q_r/Q_0 = -3R^2/\{v(R^2 - \xi_0^2)\} \qquad (6.112)$$

when $\eta \to \infty$ for an impulsive velocity loading. Thus, $|Q_r/Q_0| \to \infty$ regardless of the magnitude of the dimensionless parameter v which is defined by equation (6.107). In other words, the transverse shear force around the outer boundary of an impulsively loaded simply supported circular plate is infinitely large at the start of motion regardless of the magnitude of the transverse shear strength. The influence of transverse shear forces on the dynamic response of an impulsively loaded circular plate is, therefore, examined in the remainder of this section.

6.5.2 Circular plates with $v \leqslant 1.5$

The equilibrium equations (4.1) and (4.2) remain valid for the dynamic plastic behaviour of circular plates with transverse shear effects. These equations may be recast to give the transverse shear force according to equation (6.102) and

$$\partial^2(rM_r)/\partial r^2 - \partial M_\theta/\partial r + rp - \mu r \, \partial^2 w/\partial t^2 = 0, \qquad (6.113)$$

where μ is the mass per unit area.

It is assumed that plastic flow in an axisymmetrically loaded circular plate is controlled by the yield condition in Figure 6.15 which was used by Sawczuk and Duszek[6.13] to examine the influence of the transverse shear force on the static plastic behaviour of circular plates.

It was shown in § 6.5.1 that the transverse shear force is infinitely large around the boundary of a circular plate which has an infinite transverse shear strength and which is subjected to an impulsive velocity loading. Thus, a transverse shear slide is likely to develop around the supporting boundary of a circular plate with a finite transverse shear strength. Moreover, if the transverse shear strength of a circular plate is relatively weak compared with the associated plastic bending strength, then transverse shear effects will dominate bending effects and lead to the transverse velocity profile shown in Figure 6.16, or

$$\dot{w} = \dot{W}, \qquad 0 \leqslant r \leqslant R. \qquad (6.114)$$

The entire plate, therefore, remains rigid except around the supporting boundary.

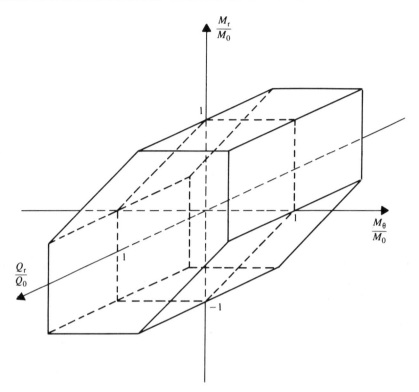

Figure 6.15 Yield surface for a circular plate.

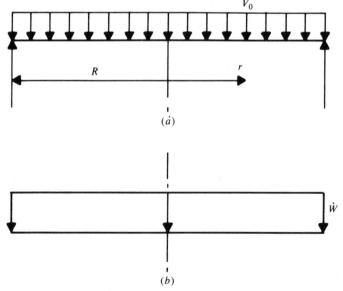

Figure 6.16 (a) Impulsively loaded simply supported circular plate. (b) Transverse velocity profile for circular plates with $v \leqslant 1.5$.

Now, taking the circumferential bending moment M_θ as constant,† then equations (6.113) and (6.114) give

$$\partial^2(rM_r)/\partial r^2 = \mu r \ddot{W} \tag{6.115}$$

or

$$M_r = \mu \ddot{W}(r^2 - R^2)/6 \tag{6.116}$$

since M_r is finite at $r = 0$ and $M_r = 0$ at $r = R$.

Equations (6.102) and (6.116) predict the transverse shear force

$$Q_r = \mu \ddot{W}(3r - R^2/r)/6 - M_\theta/r, \tag{6.117}$$

which gives

$$M_\theta = -\mu \ddot{W} R^2/6 \tag{6.118}$$

when satisfying the symmetry requirement $Q_r = 0$ at $r = 0$. Moreover, the condition for a transverse shear slide to form around the boundary demands $Q_r = -Q_0$ at $r = R$, which is satisfied by equations (6.117) and (6.118) when

$$\ddot{W} = -2Q_0/\mu R. \tag{6.119}$$

A straightforward integration of equation (6.119) with respect to time yields

$$W = -Q_0 t^2/\mu R + V_0 t \tag{6.120}$$

since $w = 0$ and $\dot{w} = V_0$ at $t = 0$, where V_0 is the magnitude of the uniformly distributed impulsive velocity. Transverse motion of the plate ceases when $\dot{W} = 0$ at

$$T = \mu V_0 R/2Q_0 \tag{6.121}$$

and the associated permanent transverse displacement is

$$w_f = \mu V_0^2 R/4Q_0, \qquad 0 \leqslant r \leqslant R. \tag{6.122}$$

Equations (6.116) to (6.119) together with equation (6.107) predict the generalised stresses

$$M_r/M_0 = 2v(1 - r^2/R^2)/3, \qquad 0 \leqslant r \leqslant R, \tag{6.123a}$$
$$Q_r/Q_0 = -r/R, \qquad 0 \leqslant r \leqslant R, \tag{6.123b}$$

and

$$M_\theta/M_0 = 2v/3, \qquad 0 \leqslant r \leqslant R. \tag{6.123c}$$

It is evident from equations (6.123) that $Q_r \leqslant Q_0$ for $0 \leqslant r \leqslant R$ and

$$v \leqslant 1.5 \tag{6.124}$$

ensures that $M_r \leqslant M_0$ and $M_\theta \leqslant M_0$ for $0 \leqslant r \leqslant R$. Thus, the generalised stresses in the foregoing theoretical solution lie on or within the yield condition in Figure 6.15 and are, therefore, statically admissible provided inequality (6.124) is satisfied. Moreover, the theoretical solution may be shown to be kinematically admissible and is, therefore, exact for the yield condition illustrated in Figure 6.15.

---

† It transpires that this assumption leads to an exact theoretical solution. However, if the theoretical analysis using equation (6.114) and taking M_θ as constant had not been statically admissible, then alternative assumptions would have been required.

6.5.3 Circular plates with $1.5 \leqslant v \leqslant 2$

It is evident from equations (6.123) that M_θ violates the yield condition in Figure 6.15 throughout a plate and M_r penetrates the yield surface in a central circular region when $v \geqslant 1.5$. Thus, the first stage of motion for the present case is governed by the transverse velocity profile sketched in Figure 6.17(b) which gives plastic bending throughout a plate with a stationary transverse shear slide at the supports. This phase of motion is completed when transverse shear sliding ceases at the supports and is followed by a final stage of motion, with the transverse velocity profile illustrated in Figure 6.17(c).

It is shown in reference 6.14 that motion finally ceases when

$$T = \mu V_0 R^2 / 6M_0, \tag{6.125}$$

with the associated permanent transverse displacement

$$w_f = \mu V_0^2 R^2 \{(3 - 2v)r/R + (4v - 5)/2\}/\{12M_0(v - 1)\}. \tag{6.126}$$

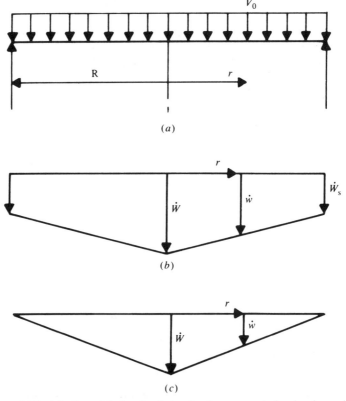

(a)

(b)

(c)

Figure 6.17 (a) Impulsively loaded simply supported circular plate. (b) Transverse velocity profile during the first phase of motion for circular plates with $1.5 \leqslant v \leqslant 2$.[6.14] (c) Transverse velocity profile during the final phase of motion for circular plates with $1.5 \leqslant v \leqslant 2$.[6.14]

6.5.4 Circular plates with $v \geqslant 2$

It is shown in reference 6.14 that a yield violation occurs at the plate centre in the theoretical analysis in § 6.5.3 when $v \geqslant 2$. This yield violation is avoided when a plate has a rigid central circular zone $0 \leqslant r \leqslant r_1$ as illustrated in Figure 6.18(b). The outer radius r_1 of this central zone remains stationary throughout the first phase of

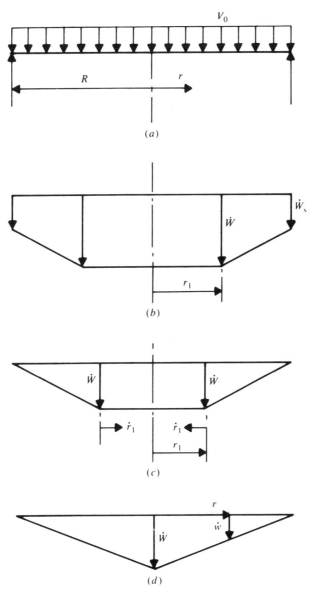

Figure 6.18 (a) Impulsively loaded simply supported circular plate. (b), (c), (d) Transverse velocity profiles during the first, second and third phases of motion, respectively, for circular plates with $v \geqslant 2$.[6.14]

motion until transverse shear sliding ceases around the outer boundary of the plate. The radius (r_1) of the central rigid zone decreases during a second phase of motion, as indicated in Figure 6.18(c). This phase of motion is completed when $r_1 = 0$ and is followed by a third and final phase of motion with the transverse velocity profile in Figure 6.18(d).

The theoretical analysis in reference 6.14 shows that the response duration is identical to equation (6.125), while the associated permanent transverse deflection is

$$w_f = \mu V_0^2 R^2 (3 + 2r/R + r^2/R^2)(1 - r/R)/24M_0, \qquad 0 \leqslant r \leqslant r_1, \quad (6.127a)$$

and

$$w_f = \mu V_0^2 R^2 \{(1 + 2r_1/R + 3r_1^2/R^2)(1 - r/R) + (1 - r_1^2/R^2)(2 - r_1/R - r/R)\}/24M_0,$$
$$r_1 \leqslant r \leqslant R, \quad (6.127b)$$

where

$$r_1/R = \{(1 - 8v + 4v^2)^{1/2} - 1\}/2v. \qquad (6.128)$$

This theoretical solution is exact[6.14] for the yield condition illustrated in Figure 6.15.

6.5.5 Comments

In the particular case when $v \to \infty$, where v is the dimensionless parameter which is defined by equation (6.107), then the transverse shear strength of a plate cross-section is infinitely larger than the corresponding plastic bending strength. In this circumstance, it is shown in reference 6.14 that the foregoing theoretical analysis reduces to the bending only solution of Wang.[6.15] For example, the response time according to equation (6.125), which is valid for $v \to \infty$, is the same as equation (4.92). Moreover, equation (6.128) predicts that $r_1 = R$ and, therefore, equation (6.127a), for the maximum permanent transverse deflection is identical to equation (4.93a).

The theoretical results in Figure 6.19 show the variation of the dimensionless maximum permanent transverse displacements of the plate centre (W_f) according to equations (6.122), (6.126) and (6.127) and at the plate supports (W_s) with the dimensionless strength parameter v. It is evident that the overall features are similar to those in Figure 6.9 for a simply supported beam which is loaded impulsively. The variation with v of the plastic bending energy and the energy which is absorbed in transverse shear deformations at the plate supports is shown, in Figure 6.20, as a proportion of the initial kinetic energy. It is interesting to observe that the maximum permanent transverse deflections are similar to the corresponding bending-only theoretical predictions in § 4.5.7 and of Wang,[6.15] when $v \geqslant 5$, approximately. However, the results in Figure 6.20 reveal that a

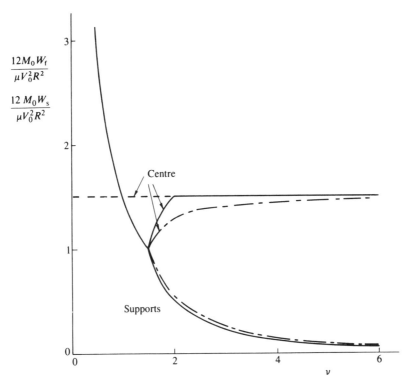

Figure 6.19 Dimensionless permanent transverse displacements at centre (W_f) and supports (W_s) of impulsively loaded simply supported circular plates.[6.14] — — —: bending-only case.[6.15] ———: transverse shear effects are retained in the yield condition. — ⋅ —: transverse shear and rotatory inertia effects are retained in the analysis for a solid homogeneous cross-section.

significant portion of the initial kinetic energy is dissipated through transverse shearing deformations at the supports for all values of v shown in the figure.

The theoretical solution in reference 6.13 for a simply supported circular plate subjected to a uniformly distributed static pressure indicates that transverse shear effects do not influence the static collapse behaviour for the yield surface in Figure 6.15 when $v \geqslant 1.5$. Thus, the present study demonstrates that transverse shear effects are more important for the dynamic case than for the corresponding static problem. This was also found in reference 6.8 for beams, and is discussed in reference 6.10.

It should be noted that $v = R/H$ for the particular case of a circular plate having a solid homogeneous cross-section with $Q_0 = \sigma_0 H/2$ and $M_0 = \sigma_0 H^2/4$.

On the other hand, if a circular plate is constructed with a sandwich cross-section, then an inner core of thickness h and a shear yield stress τ_0 supports a maximum transverse shear force $Q_0 = \tau_0 h$ (per unit length), while thin exterior

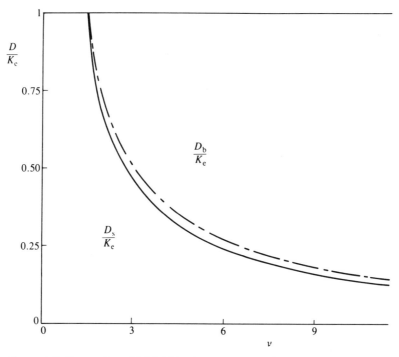

Figure 6.20 Proportion of initial kinetic energy (K_e) absorbed due to shearing (D_s) and bending (D_b) deformations (at $t = T$) in an impulsively loaded simply supported circular plate.[6.14] ———: without rotatory inertia. — · — · —: with rotatory inertia for a solid homogeneous cross-section.

sheets of thickness t can carry independently a maximum bending moment $M_0 = \sigma_0 t(h + t)$, where σ_0 is the corresponding tensile yield stress. In this circumstance, $v = Q_0 R/2M_0$ gives

$$v = \left\{ \frac{R}{H} \frac{\tau_0}{\sigma_0/2} \right\} \left\{ \frac{h/H}{1 - (h/H)^2} \right\} \tag{6.129}$$

when $H = h + 2t$. Thus, a sandwich plate with $2R/H = 15$, $\sigma_0/2\tau_0 = 8$, and $h/H = 0.735$ (e.g., a 0.5 in (12.7 mm) thick core with 0.1 in (2.54 mm) sheets gives $h/H = 0.714$) gives $v = 1.5$, for which transverse shear effects are important according to the results in Figure 6.19.

Kumar and Reddy[6.16] have examined the influence of transverse shear effects on the dynamic plastic response of a simply supported circular plate when subjected to a uniformly distributed pressure pulse with a rectangular pressure–time history.

It is evident that the transverse velocity field illustrated in Figure 6.16 and used to study a simply supported circular plate in § 6.5.2 does not produce any rotation of the plate elements. This theoretical analysis remains valid, therefore, when both transverse shear and rotatory inertia effects are retained in the basic equations for circular plates with $v \leqslant 1.5$.

The influence of rotatory inertia on the dynamic plastic response of simply supported circular plates with $v \geqslant 1.5$ is examined in reference 6.14. It transpires that the theoretical analysis for a uniform impulsive velocity loading has two distinct phases of motion. Transverse shear sliding occurs at the supports during the first phase, together with a stationary central circular plastic zone. The transverse shear sliding ceases at the end of the first phase and the second phase of motion commences with the boundary of the plastic zone propagating radially inwards until it reaches the centre, when motion ceases.

It is evident from Figure 6.19 that the inclusion of rotatory inertia in the governing equations and the retention of transverse shear as well as bending effects in the yield criterion leads to an increase in the permanent transverse shear sliding at the plate supports and a decrease in the maximum final transverse displacement which occurs at the plate centre. However, the inclusion of rotatory inertia effects gives rise to respective changes in these quantities of approximately 11.5 and 14.2 per cent at most. Thus, the simpler theoretical analysis in §§ 6.5.3 and 6.5.4, without rotatory inertia, would probably suffice for most practical purposes. If greater accuracy is required, then it is only necessary to include rotatory inertia effects for circular plates with $1.5 \leqslant v \leqslant 4$, approximately.

The response durations given by equation (6.121) for $v \leqslant 1.5$ and equation (6.125) for $v \geqslant 1.5$ are independent of rotatory inertia. In fact, equation (6.125) is independent of both v and rotatory inertia when $v \geqslant 1.5$.

6.6 Transverse shear effects in cylindrical shells

6.6.1 Introduction

The dynamic plastic response of some axisymmetric rigid, perfectly plastic cylindrical shells was examined in §§ 5.3 to 5.5. The plastic flow of the material in these shells was controlled by combinations of the circumferential membrane force (N_θ) and the longitudinal bending moment (M_x) which obey the square yield condition shown in Figure 5.2. The transverse shear force Q_x was taken as a reaction in these analyses and, therefore, did not influence the plastic flow of the material.

Equation (5.1) gives the transverse shear force

$$Q_x = \partial M_x / \partial x \qquad (6.130)$$

in a cylindrical shell.

Equations (5.35) and (5.36) are developed for a long cylindrical shell which is reinforced by equally spaced rigid rings at longitudinal intervals of $2L$, as shown in Figure 5.5, and is subjected to a dynamic pressure pulse with a rectangular shaped pressure–time history. These equations, together with equation (6.130), predict that

$$Q_x/Q_0 = \{8 + (\eta - 1)(2 + c^2)\}/4v \qquad (6.131)$$

at $x = L$, where

$$v = Q_0 L/2M_0, \qquad (6.132a)$$

$$c^2 = N_0 L^2/2M_0 R, \qquad (6.132b)\dagger$$

$$p_c = N_0/R + 4M_0/L^2 \qquad (6.132c)\ddagger$$

and

$$\eta = p_0/p_c. \qquad (6.132d)$$

Equations (6.132a), (6.132c) and (6.132d) are the transverse, or radial, shear force to plastic longitudinal bending moment strength ratio, uniform static collapse pressure and dynamic to static pressure ratio, respectively.

Equation (6.131) indicates that the magnitude of the transverse shear force at the supports may exceed the associated plastic collapse value (Q_0) either for sufficiently small values of the strength parameter v, or for sufficiently large pressure ratios η.§ Incidentally, equation (6.131) also illustrates that transverse shear effects are potentially more important for dynamic loadings since equation (6.131), with $\eta = 1$, gives $Q_x/Q_0 = 2/v$ for the corresponding statically loaded case. In fact, the theoretical analysis in § 5.5.2 for large dynamic pressure pulses predicts a transverse shear force

$$Q_x/Q_0 = 3/\{v(1 - \xi_0/L)\} \qquad (6.133)$$

at the supports, where ξ_0 is defined by equation (5.53). It may be shown that $\xi_0 \to L$ as $\eta \to \infty$ for an impulsive velocity in which circumstance $Q_x/Q_0 \to \infty$.

The foregoing comments reveal that transverse shear forces may exercise an important influence on the dynamic plastic response of cylindrical shells, particularly when subjected to large pressure pulses. In order to explore this effect further, the equilibrium equations (5.1) and (5.2) are recast in the form of equation (6.130) and

$$\partial^2 M_x/\partial x^2 + N_\theta/R - p - \mu\, \partial^2 w/\partial t^2 = 0, \qquad (6.134)$$

respectively. Plastic flow in a cylindrical shell is assumed to be governed by the simple cube-shaped yield surface in Figure 6.21.

6.6.2 Transverse shear effects in a simply supported cylindrical shell with $v \leqslant 1$

Consider a cylindrical shell which is simply supported across a span $2L$ and subjected to a uniformly distributed radially outwards impulsive velocity V_0, as

---

† If $N_0 = \sigma_0 H$ and $M_0 = \sigma_0 H^2/4$ for a cylindrical shell with a solid homogeneous cross-section, then $c^2 = 2L^2/RH$, which is the same as ω^2 defined by equation (5.31).

‡ This is the same as the static collapse pressure given by equation (2.66b).

§ The pressure ratio η cannot violate inequality (5.42).

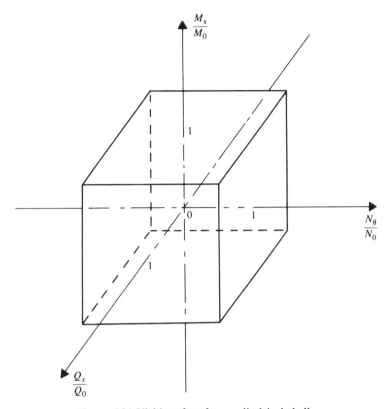

Figure 6.21 Yield surface for a cylindrical shell.

shown in Figure 6.22(a). In view of the observations in § 6.6.1, it is possible that transverse shear sliding would develop at the supports of a cylindrical shell. The shell, therefore, is assumed to deform with the radial velocity profile sketched in Figure 6.22(b), or

$$\dot{w} = -\dot{W}, \quad 0 \leqslant x \leqslant L. \tag{6.135}$$

It is evident from equations (5.4) and (5.5) that $\dot{\epsilon}_\theta \geqslant 0$ and $\dot{\kappa}_x = 0$ for $0 \leqslant x \leqslant L$. Thus, in order to satisfy the normality requirements of plasticity, the generalised stresses lie on the side

$$N_\theta = N_0 \tag{6.136}$$

of the yield surface in Figure 6.21. In this circumstance, equations (6.134) to (6.136) become

$$\partial^2 M_x / \partial x^2 = -(N_0/R + \mu \ddot{W})$$

or

$$\partial M_x / \partial x = -(N_0/R + \mu \ddot{W})x, \tag{6.137}$$

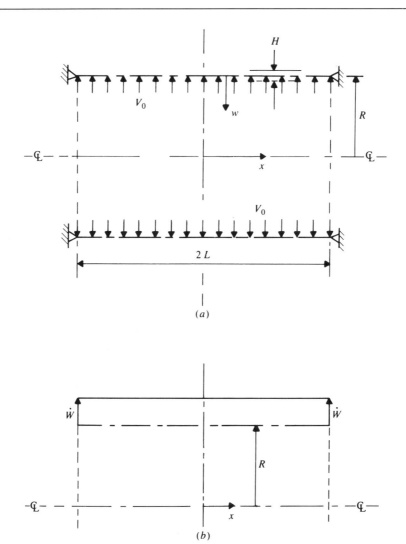

Figure 6.22 (a) Impulsively loaded simply supported cylindrical shell. (b) Axisymmetric radial velocity profile for an impulsively loaded simply supported cylindrical shell with $v \leqslant 1$.

since $Q_x = \partial M_x / \partial x = 0$ at $x = 0$ for symmetry and

$$M_x = (N_0/R + \mu \ddot{W})L^2(1 - x^2/L^2)/2, \qquad (6.138)$$

because $M_x = 0$ at $x = L$ for simple supports. Moreover, $Q_x = Q_0$ at $x = L$ in order for transverse shear sliding to develop at the supports. This is satisfied by equations (6.130) and (6.137), when

$$\ddot{W} = -2M_0(v + c^2)/\mu L^2, \qquad (6.139)$$

where v and c^2 are defined by equations (6.132a) and (6.132b), respectively.

Now,

$$\dot{W} = -2M_0(v + c^2)t/\mu L^2 + V_0 \qquad (6.140)$$

when integrating equation (6.139) and satisfying the initial condition $\dot{w} = -V_0$, or $\dot{W} = V_0$. Equation (6.140) shows that motion ceases when

$$T = \mu V_0 L^2/\{2M_0(v + c^2)\}. \qquad (6.141)$$

Another integration gives

$$W = -M_0(v + c^2)t^2/\mu L^2 + V_0 t \qquad (6.142)$$

because $W = 0$ at $t = 0$. Equations (6.141), (6.142) and (6.135) predict the permanent radial displacement

$$w_f = -\mu V_0^2 L^2/\{4M_0(v + c^2)\}. \qquad (6.143)$$

The generalised stresses associated with this theoretical solution are given by equations (6.136), (6.138) and (6.130), together with equation (6.139), or

$$N_\theta/N_0 = 1, \qquad 0 \leqslant x \leqslant L, \qquad (6.144a)$$

$$M_x/M_0 = -v(1 - x^2/L^2), \qquad 0 \leqslant x \leqslant L, \qquad (6.144b)$$

and

$$Q_x/Q_0 = x/L, \qquad 0 \leqslant x \leqslant L. \qquad (6.144c)$$

It is evident from equation (6.144b) that the longitudinal bending moment would penetrate the yield surface in Figure 6.21 at $x = 0$ when $v > 1$. Thus, the foregoing theoretical solution is restricted to cylindrical shells which are made from rigid, perfectly plastic materials with a plastic strength parameter $v \leqslant 1$.

The initial kinetic energy (K_e) is absorbed through transverse shear displacements at the supports (D_s) and in plastic circumferential membrane behaviour throughout the shell (D_m). It may be shown that

$$D_s/K_e = v/(v + c^2) \qquad (6.145)$$

and

$$D_m/K_e = c^2/(v + c^2). \qquad (6.146)$$

This behaviour should be contrasted with the beam and circular plate problems in §§ 6.3.2 and 6.5.2, in which all of the initial kinetic energy is absorbed by transverse shear sliding at the supports. For example, if $v = c^2 = 1$, then one-half of the initial kinetic energy is absorbed through circumferential membrane stretching, leaving only one-half to be absorbed by transverse shear sliding at the supports of a cylindrical shell. If $c^2 = 10$ and $v = 1$, then only 9 per cent of the initial kinetic energy would be absorbed in transverse shear sliding at the supports.

6.6.3 Cylindrical shell with $1 \leqslant v \leqslant 1.5$

The plastic yield violation at $x = 0$ in equation (6.144b), when $v \geqslant 1$, is removed in reference 6.17 by employing the transverse velocity field sketched in Figure 6.23.

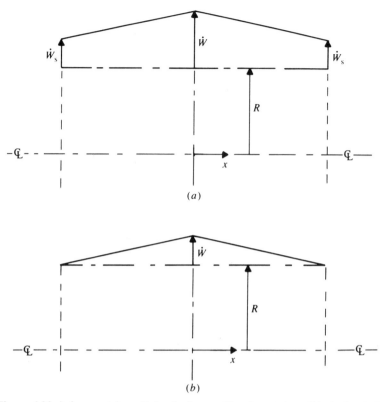

Figure 6.23 Axisymmetric radial velocity profiles for an impulsively loaded-simply supported cylindrical shell with $1 \leqslant v \leqslant 1.5$. (a) First phase of motion. (b) Second phase of motion.

Stationary transverse shear slides develop at the supports and a plastic bending hinge forms at the mid-span during the first phase of motion. Transverse shear sliding ceases but motion continues during a second phase of motion, with a stationary plastic bending hinge at $x = 0$, as indicated in Figure 6.23(b).

It is shown in reference 6.17 that the response duration is

$$T = \mu V_0 L^2 / \{2(1 + c^2)M_0\}, \tag{6.147}$$

with an associated maximum permanent radial displacement

$$w_f = -\mu V_0^2 L^2 (c^2 + 6v - 5) / \{4(1 + c^2)(4v + c^2 - 3)M_0\}. \tag{6.148}$$

This theoretical solution is statically admissible provided $c^2 \leqslant 3 + 2\sqrt{3}$ and $1 \leqslant v \leqslant 1.5$.

6.6.4 Cylindrical shell with $v \geqslant 1.5$

A theoretical solution is presented in reference 6.17 in order to avoid the yield violation which occurs in the analysis outlined in § 6.6.3, when $v \geqslant 1.5$. This is

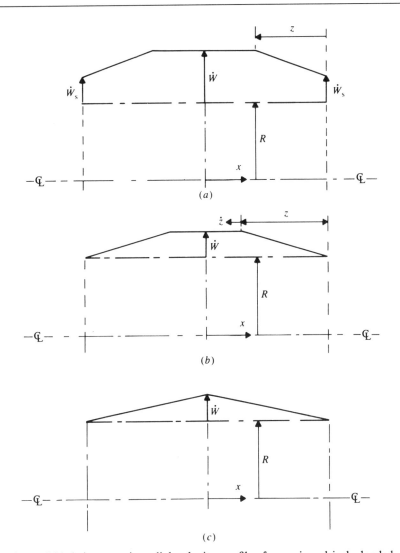

Figure 6.24 Axisymmetric radial velocity profiles for an impulsively loaded simply supported cylindrical shell with $v \geqslant 1.5$. (a) First phase of motion. (b) Second phase of motion. (c) Third phase of motion.

developed using the transverse velocity fields in Figure 6.24 for three distinct phases of motion. It transpires that the response duration is identical to equation (6.147), while the maximum permanent radial displacement is

$$w_f = -\mu V_0^2 L^2 (4 + c^2)/\{4(1 + c^2)(3 + c^2)M_0\}. \tag{6.149}$$

This theoretical solution is exact for the yield surface in Figure 6.21, provided

$$c^2 \leqslant 9, \tag{6.150}$$

as discussed in reference 6.17.

6.6.5 Comments

Despite the fact that transverse shear sliding occurs during the first phase of motion of the theoretical solution in § 6.6.4, as indicated by the transverse velocity profile in Figure 6.24, the response duration and maximum permanent transverse, or radial, displacement are independent of the strength parameter v according to equations (6.147) and (6.149), respectively. It may be shown that these expressions agree with

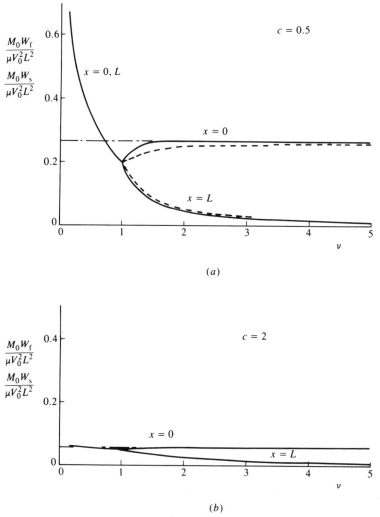

(a)

(b)

Figure 6.25 Variation of dimensionless permanent radial displacements at mid-span (W_f, $x = 0$) and supports (W_s, $x = L$) in an impulsively loaded simply supported cylindrical shell. −−−−: elementary solution.[6.18] ———: transverse shear effects are retained in the yield condition. −−−: transverse shear and rotatory inertia effects are retained in the analysis for a solid homogeneous cross-section. (a) $c = 0.5$, (b) $c = 2$.

those developed in reference 6.18 for a similar impulsively loaded simply supported cylindrical shell but made from a material with a infinite transverse shear strength (i.e., $v \rightarrow \infty$).

The theoretical results in Figure 6.25 show the variation of the dimensionless maximum permanent transverse displacements at the mid-span (W_f) according to equations (6.143), (6.148) and (6.149) and at the cylindrical shell supports (W_s), with the dimensionless strength parameter v. This figure indicates that the permanent transverse shearing displacements at the shell supports decrease with increase in v and vanish for $v \rightarrow \infty$.

If the shell cross-section is solid and homogeneous, then $M_0 = \sigma_0 H^2/4$, $N_0 = \sigma_0 H$ and $Q_0 = \sigma_0 H/2$, giving $v = L/H$ and $c^2 = 2vL/R$. Thus, $v = 1$ corresponds to a very short shell or ring with a total length $2L = 2H$ and $c = 1$ when $L = R/2$. On the other hand, transverse shear effects would be more important for a cylindrical shell having a sandwich† cross-section with an inner core of thickness h and a shear yield stress τ_0, and thin exterior sheets of thickness t and tensile yield stress σ_0. If the inner core alone supports transverse shear forces, then $Q_0 = \tau_0 h$, while $M_0 = \sigma_0 t(h + t)$ if the exterior sheets alone provide bending resistance. Thus, $v = Q_0 L/2M_0$, according to equation (6.132a), gives

$$v = \left(\frac{L}{H}\right)\left(\frac{\tau_0}{\sigma_0/2}\right)\left\{\frac{h/H}{1 - (h/H)^2}\right\}, \tag{6.151}$$

when $H = h + 2t$. A sandwich shell with $2L/H = 10$, $\sigma_0/2\tau_0 = 8$ and $h/H = 0.735$ (e.g., a 0.5 in (12.7 mm) thick core with 0.1 in (2.54 mm) sheets gives $h/H = 0.714$) gives $v = 1$, for which transverse shear effects are important according to the results in Figures 6.25 and 6.26. However, equation (6.132b), with $N_0 = 2\sigma_0 t$, gives

$$c^2 = \frac{2L^2}{RH(1 + h/H)} \tag{6.152}$$

which, with the above values, predicts $2L/R = 0.087$ when $c = 0.5$. The combination of these parameters for this particular case gives $R/H = 115$.

It is evident from equation (6.135) and Figure 6.22(b) that rotatory inertia effects would not influence the theoretical analysis in § 6.6.2 which is, therefore, valid for $0 \leqslant v \leqslant 1$ whether or not rotatory inertia effects are retained in the basic equations. A theoretical solution, which explores the importance of rotatory inertia effects in cylindrical shells having $v \geqslant 1$, is presented in reference 6.17. These theoretical predictions are compared in Figure 6.25 with the analyses in §§ 6.6.2 to 6.6.4.

It is evident from Figure 6.25 that the inclusion of rotatory inertia in the governing equations leads to an increase in the permanent transverse shear sliding at the shell supports and a decrease in the maximum final transverse displacement

---

† The sandwich cross-section is assumed to have the same mass per unit area (μ) as a shell with a solid and homogeneous cross-section. Possible crushing of the core is disregarded.

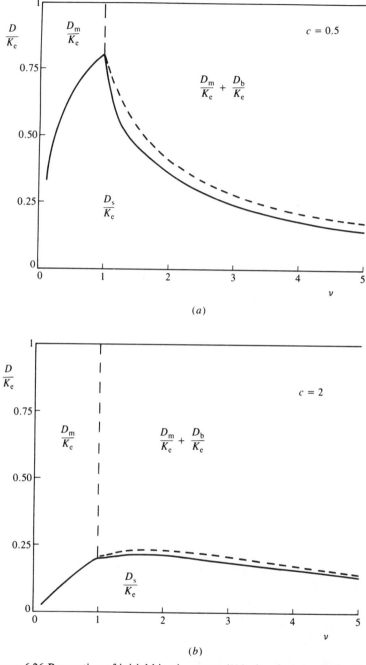

Figure 6.26 Proportion of initial kinetic energy (K_e) absorbed due to bending (D_b), membrane (D_m) and shearing (D_s) deformations (at $t = T$) in an impulsively loaded, simply supported cylindrical shell. ———: without rotatory inertia. – – –: with rotatory inertia for a solid homogeneous cross-section. (a) $c = 0.5$, (b) $c = 2$.

when $v \geqslant 1$. However, the changes brought about by the consideration of rotatory inertia are relatively small so that the simpler theoretical analysis in this section, which includes transverse shear effects without rotatory inertia, would probably suffice for most practical purposes. Transverse shear effects are even small for cylindrical shells with solid cross-sections and having practical dimensions, so that the theoretical predictions in reference 6.18, which retain only the influence of bending moments and membrane forces in the yield condition, might be adequate for most designs, except, possibly, for shells with sandwich cross-sections. Notwithstanding these comments, transverse shear effects have an important influence on the proportion of energy dissipated in various modes, as indicated in Figure 6.26.

The response durations which are predicted by equations (6.141) and (6.147) remain unchanged whether or not rotatory inertia effects are retained[6.17] in the theoretical analysis. Moreover, equation (6.147) is identical to the response duration predicted by a theoretical analysis in reference 6.18 for the dynamic infinitesimal response of an impulsively loaded simply supported cylindrical shell which does not retain the influence of transverse shear forces in the yield criterion.

6.7 Final remarks

6.7.1 Comparison of the response for impulsively loaded beams, circular plates and cylindrical shells

The dimensionless plastic strength parameters v have similar forms for impulsively loaded beams, circular plates and cylindrical shells, according to equations (6.9), (6.107) and (6.132a), respectively. However, different generalised forces govern the plastic flow of beams, circular plates and cylindrical shells, as indicated by the yield criteria in Figures 6.3, 6.15 and 6.21, respectively. In particular, the yield condition for a beam is two-dimensional, while the plastic flow in circular plates and cylindrical shells is controlled by three-dimensional yield criteria.

It is interesting to observe that the impulsively loaded beams, circular plates and cylindrical shells with simple supports and no rotatory inertia, which are examined in § 6.3, § 6.5 and § 6.6, have three types of response, depending on the magnitude of the plastic strength parameter v. The three inequalities $v \leqslant 1$, $1 \leqslant v \leqslant 1.5$ and $v \geqslant 1.5$ are the same for beams and cylindrical shells, while the shearing-only phase for a circular plate occurs when $v \leqslant 1.5$ and the two remaining inequalities are $1.5 \leqslant v \leqslant 2$ and $v \geqslant 2$.

The general characteristics of the maximum permanent transverse displacements at the mid-span and the transverse shear displacements at the supports are similar for impulsively loaded beams, circular plates and cylindrical shells, as shown in Figures 6.9, 6.19 and 6.25(a), respectively.

It is evident that rotatory inertia effects are relatively unimportant for all three problems so that the simpler theoretical analyses, without rotatory inertia, in this chapter, would probably suffice for most practical purposes. Moreover, transverse shear effects are not important for values of the dimensionless plastic strength parameter $v \geqslant 5$, approximately, in which circumstance the corresponding theoretical analyses, without transverse shear effects, would be adequate in design. Nevertheless, it is possible, for large values of v, that the actual shear sliding at the supports could become excessive for large impact velocities and lead to failure, as discussed in § 6.7.3.

Martin[6.19] obtained an upper bound theorem for the permanent displacement and a lower bound theorem for the response duration of a rigid, perfectly plastic continuum which undergoes infinitesimal displacements when subjected to an impulsive velocity. This important paper was followed by many other studies on bounding methods and, in particular, Morales and Nevill,[6.20] who derived the associated lower bound theorem for the permanent displacement.

It is evident from the theoretical analyses in §§ 6.3 to 6.6 that the influence of transverse shear forces is important during the early stages of motion when the displacements of a structure remain small. The theorems of Martin[6.19] and Morales and Nevill[6.20] were developed for continua which undergo infinitesimal displacements. Thus, they might provide more accurate estimates for problems with important transverse shear effects. The theorems were derived to bound the behaviour of an impulsively loaded continuum, but they remain valid for structural members provided both transverse shear and rotatory inertia effects are retained.

It was found in reference 6.21 that, when transverse shear effects are important, the simple bound theorems provide excellent estimates of the response durations and permanent transverse displacements of the impulsively loads beams, circular plates and cylindrical shells which were examined in §§ 6.3, 6.5 and 6.6, respectively.

6.7.2 Yield criteria

It appears that there is considerable uncertainty in the published literature on the precise role of transverse shear forces on the yielding of rigid, perfectly plastic beams loaded statically. Indeed, it has been demonstrated that interaction curves relating the bending moment (M) and the transverse shear force (Q) are not proper yield curves and, as further support of this view, interaction curves for I-beams have been constructed which are not convex.[6.22] Nevertheless, the finite transverse shear strength of beams loaded statically is catered for in practical design, as remarked in § 6.1, and several formulae are available which, in the case of an I-beam, assume that the maximum transverse shear force equals the carrying capacity of the web alone.

The role of transverse shear forces on the plastic yielding of beams was examined in reference 6.23 and some justification was given for using convex yield curves for I-beams within the setting of engineering or classical beam theory. A suitable compromise from an engineering viewpoint between the simple local (stress resultant) and more rigorous non-local (plane stress, plane strain) theories might be achieved for I-beams when using a local theory[6.24] with a maximum transverse shear force based only on the web area. It is evident from Figure 6.27 that Hodge's[6.24] revised theoretical results now provide an inscribing lower bound curve in the $M/M_0 - Q/Q_0$ plane which, because of its simplicity, might be acceptable for many theoretical studies on beams. Furthermore, the theoretical predictions of Heyman and Dutton[6.22] and of Ranshi, Chitkara and Johnson[6.25] and others are reasonably well approximated by a square yield curve, which has been used in this chapter. In fact, Hodge's[6.24] revised results and a square yield curve provide two simple methods for essentially bounding the actual yield curve for an I-beam.

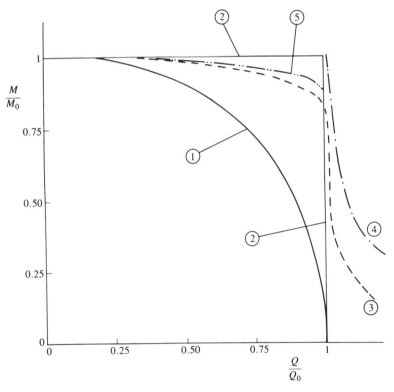

Figure 6.27 Yield curves for a 8WF40 wide-flanged I-beam (Q_0 is the transverse shear load calculated using area of web). (1): Hodge [6.24] with Q_0 calculated using web area. (2): square yield curve. (3): lower bound after Neal.[6.26] (4): upper bound after Neal.[6.26] (5): Heyman and Dutton[6.22]

Fortunately, it shown in Figure 6.28 that a number of local and non-local theories given similar curves in the $M/M_0 - Q/Q_0$ plane for beams with rectangular cross-section so, in this case, one may select whichever theory is the most convenient.

Ilyushin[6.27] developed a yield surface for thin shells with solid cross-sections using the von Mises yield criterion and the usual assumptions of thin shell theory. Shapiro[6.28] extended this theoretical work in order to cater for transverse shear effects. The Ilyushin–Shapiro[6.27–6.29] yield curve, for the plastic yielding of a beam which is subjected to a bending moment and a transverse shear force, is drawn in Figure 6.28. It may be demonstrated, when using standard relations between inverse hyperbolic and logarithmic functions, that Hodge's[6.24] yield curve for a beam with a rectangular cross-section is identical to the Ilyushin–Shapiro theoretical predictions.

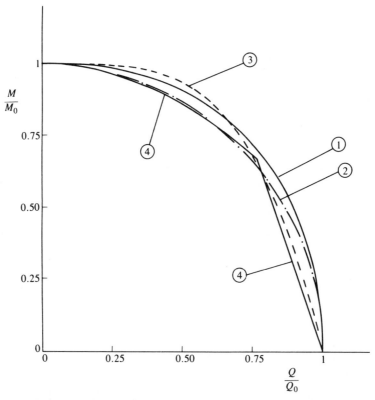

Figure 6.28 Yield curves for beams with rectangular cross-sections.[6.23] ①: Hodge,[6.24] Ilyushin–Shapiro.[6.27]–[6.29] ②: Robinson,[6.29] $Y'_1 = 1$, $Y'_3 = 1$. ③: Equations (27) of Neal[6.26] and Drucker.[6.30] ④: Equations (46) of Reckling.[6.31]

Robinson[6.29] has recently examined various approximations to the Ilyushin–Shapiro yield surface and found that the approximation $Y'_1 = 1$ or $Y'_3 = 1$ inscribes the Ilyushin–Shapiro yield curve and would provide an acceptable approximation for most practical problems, as indicated in Figure 6.28.

Numerical predictions are presented in reference 6.11 for the dynamic plastic response of a simply supported beam, which is made from a rigid, perfectly plastic material, governed by the Ilyushin–Shapiro yield curve[6.27-6.29] in Figure 6.28 and subjected to a uniform impulsive velocity V_0 as shown in Figure 6.4(a). The results, which include the effects of transverse shear and rotatory inertia, are virtually indistinguishable from the corresponding theoretical values presented in reference 6.10 for a square yield curve.

It may be shown that the generalised stress profiles, when neglecting rotatory inertia, which are given in § 6.3.5 for a square yield curve, lie within or on Robinson's circular yield curve,[6.23, 6.29] which, in turn, inscribes the Ilyushin–Shapiro yield curve. The theoretical results in § 6.3 are identical to the corresponding theoretical predictions using an Ilyushin–Shapiro yield curve.

Numerical predictions are also given in reference 6.11 for the response of a rigid, perfectly plastic beam hit by a mass travelling with an initial velocity V_0 as shown in Figure 6.11(a). The theoretical results, with transverse shear retained according to the Ilyushin–Shapiro yield condition but rotatory inertia neglected, are compared with the theoretical predictions for a square yield condition, presented in § 6.4, and the simple bending only solution. It transpires that transverse shear effects lead to a dramatic reduction in the angle underneath the striker and a significant increase in the maximum transverse displacement underneath the striker. Moreover, the angle is quite sensitive to the actual shape of the yield curve, while the maximum displacement is less sensitive.

The numerical results for the combined influence of transverse shear and rotatory inertia on the response of beams having various cross-sections are compared to the corresponding theoretical results presented in reference 6.11 for a square yield curve. Again the shape of the yield curve exercises an important effect on the angle, while the maximum transverse displacement is fairly insensitive. Thus, an analysis with a square yield curve is adequate when the maximum transverse displacements are of interest. However, the Ilyushin–Shapiro yield curve must be used when the rotation under the striker is required.

6.7.3 Transverse shear failure

Menkes and Opat[6.32] conducted an experimental investigation into the dynamic plastic response and failure of fully clamped metal beams which were subjected to uniformly distributed velocities over the entire span, as shown in Figure 6.4(a) for the simply supported case. Menkes and Opat observed that the beams responded in a ductile manner and acquired permanently deformed profiles when subjected to velocities less than a certain value.† However, when the impulsive velocities were equal to this critical value, then the beams failed owing to tearing of the beam material at the supports. As the impulsive velocities were further increased beyond this critical value, failure occurred, and the plastic deformation of the beams became more localised near the supports until another critical velocity, which was associated with a transverse shear failure at the supports, was reached. On the basis of these experimental tests, which are shown in Figure 6.29, Menkes and Opat[6.32] classified the three failure modes for fully clamped beams with rectangular cross-sections as follows:

Mode 1: large inelastic deformation of the entire beam;
Mode 2: tearing (tensile failure) of the beam material at the supports;
Mode 3: transverse shear failure of the beam material at the supports.

It should be noted that some of the beams exhibited failures which involved both the tearing and shearing modes when subjected to the impulsive velocities lying between the smallest velocities required for the Mode 2 and Mode 3 failures. However, the foregoing definitions of Modes 2 and 3 are used simply to define the smallest, or threshold, impulsive velocities which are necessary to cause tensile tearing and pure shear failures at the supports, respectively.

The remainder of this section focuses on the Mode 3 type of structural failure. A simplified theoretical procedure, which was developed in reference 6.33, is used to estimate the threshold velocities required for a Mode 3 transverse shear failure.

Now, the maximum transverse shear force occurs at the supports in the three impulsive loading problems which are examined in this chapter and underneath the striker for the beam impact problem in Figure 6.11(a). It is evident that transverse shear slides develop during the first phase of motion for all these problems and cease at the end of the phase. Thus, in order to avoid severance of the structure, it is necessary to ensure that

$$\int_0^T \dot{W}_s \, dt \leqslant kH \tag{6.153}$$

for the impulsive loading problems in §§ 6.3, 6.5 and 6.6, and

$$\int_0^{T_1} (\dot{W}_0 - \dot{W}_1) \, dt \leqslant kH \tag{6.154}$$

for the impact problem in § 6.4, where k is a constant $(0 < k \leqslant 1)$. Clearly, complete severance occurs in a structure with a thickness H, when $k = 1$, but transverse shear failure is likely to develop for smaller values of k.[6.34]

The elementary method which is developed in reference 6.33 for a beam may be

---

† The impulsive velocities in the tests were always larger than those necessary to produce a wholly elastic response.

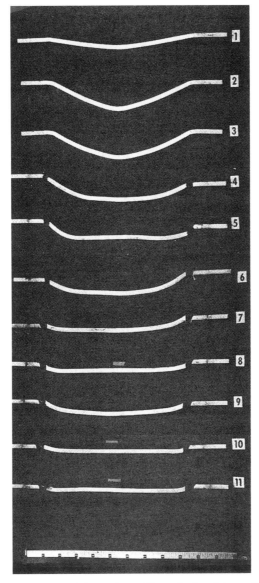

Figure 6.29 Permanent deformed profiles of aluminium 6061-T6 beams examined by Menkes and Opat[6.32], illustrating the transition from a Mode 1 to a Mode 3 response with increasing impulsive velocity.

used to obtain the conditions for transverse shear failure of any structural member. It is now illustrated for the impulsively loaded simply supported beam which is shown in Figure 6.4(a) and examined in § 6.3. Substituting equations (6.51) and (6.52) for a beam with $v \geqslant 1.5$ into equation (6.153) gives

$$3mL^2 V_0^2/(16v^2 M_0) \leqslant kH$$

or

$$V_0 \leqslant V_0^s, \tag{6.155}$$

where

$$V_0^s = (16v^2 M_0 kH/3mL^2)^{1/2}. \tag{6.156}$$

In other words, a transverse shear failure occurs at the supports of the beam in Figure 6.4(a), with $v \geqslant 1.5$, when the magnitude of the uniformly distributed impulsive velocity equals the threshold value V_0^s given by equation (6.156).

It may be shown that equations (6.153) and (6.155), together with the theoretical results in §§ 6.3.2 and 6.3.3, predict the threshold impulsive velocities

$$V_0^s = (4v M_0 kH/mL^2)^{1/2} \tag{6.157}$$

and

$$V_0^s = \{4(4v - 3)M_0 kH/mL^2\}^{1/2} \tag{6.158}$$

for simply supported beams having $v \leqslant 1$ and $1 \leqslant v \leqslant 1.5$, respectively.

In the particular case of a beam with a solid rectangular cross-section, then $Q_0 \cong \sigma_0 H/2$ and $M_0 = \sigma_0 H^2/4$ for a beam of unit width and, therefore, equations (6.157), (6.158) and (6.156) predict that

$$V_0^s = (k\sigma_0/\rho)^{1/2}(H/L)^{1/2},$$
$$V_0^s = (k\sigma_0/\rho)^{1/2}(4L/H - 3)^{1/2}(H/L)$$

and

$$V_0^s = (k\sigma_0/\rho)^{1/2}2/\sqrt{3}$$

for $v \leqslant 1$, $1 \leqslant v \leqslant 1.5$ and $v \geqslant 1.5$, respectively. However, $v = L/H$ for a beam with a rectangular cross-section according to equation (6.9). Thus, the threshold velocity from equation (6.156) for beams with $v > 1.5$ is of practical importance, while equations (6.157) and (6.158) refer to very short stubby beams. Nevertheless, equations (6.157) and (6.158) are important for beams with sandwich cross-sections.

Now, consider a beam which is constructed with a sandwich cross-section. An inner core of thickness h and a shear yield stress τ_0 could support a maximum plastic transverse shear force $Q_0 = \tau_0 h$ (per unit width). Thin exterior sheets of thickness t could support independently a maximum plastic bending moment $M_0 = \sigma_0 t(h + t)$ (per unit width), where σ_0 is the corresponding tensile yield stress. In this circumstance, equation (6.9) becomes

$$v = \left(\frac{\tau_0}{\sigma_0/2}\right)\left\{\frac{h/H}{1 - (h/H)^2}\right\}\left(\frac{L}{H}\right), \tag{6.159}$$

when $H = h + 2t$. Thus, a sandwich beam with $\sigma_0/2\tau_0 = 8$ and $h/H = 0.714$† gives

$$v = 0.182L/H \tag{6.160}$$

---

† For example, a 0.5 in (12.7 mm) thick core with 0.1 in (2.54 mm) sheets gives $h/H = 0.714$.

and, therefore, $v = 1$, for example, when the span to depth ratio ($2L/H$) is 11. It may be shown that equations (6.156) to (6.160) predict

$$V_0^s(\rho/\sigma_0 k)^{1/2} = 0.127/v^{1/2},$$

$$V_0^s(\rho/\sigma_0 k)^{1/2} = 0.127(4v - 3)^{1/2}/v$$

and

$$V_0^s(\rho/\sigma_0 k)^{1/2} = 0.147$$

for $v \leqslant 1$, $1 \leqslant v \leqslant 1.5$ and $v \geqslant 1.5$, respectively, as indicated in Figure 6.30. Initial impulsive velocities which lie below the threshold line in this figure would not lead to a transverse shear failure according to the simplified approach taken in this section.

There is a paucity of experimental work on the dynamic plastic failure of structures.[6.34, 6.35] However, the theoretical predictions using the above simplified procedure for fully clamped beams with $k = 1$[6.33] gave an encouraging agreement with the experimental results obtained by Menkes and Opat[6.32] on impulsively loaded aluminium 6061-T6 beams.

The same general procedure may be used to estimate the threshold velocities for the beam impact, circular plate and cylindrical shell problems which are examined in §§ 6.4 to 6.6, respectively. It was found in reference 6.36 that the same general approach and $k = \frac{1}{2}$, together with the theoretical analysis in § 6.5.2 for a circular plate having $v \leqslant 1.5$, gave good agreement with Recht and Ipson's[6.37] predictions

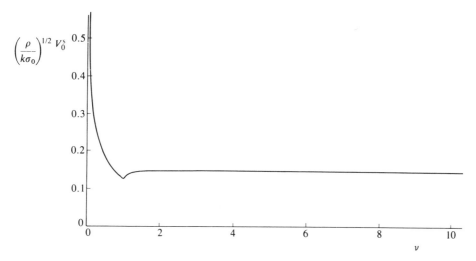

Figure 6.30 Variation of dimensionless threshold impulsive velocities for a simply supported beam with a sandwich cross-section having $\sigma_0/2\tau_0 = 8$ and $h/H = 0.714$.

for the perforation of a plate struck by a blunt cylindrical projectile. However, fair agreement was observed[6.36] between the simplified method, with $k = \frac{3}{4}$, and the threshold perforation velocities on aluminium-2024 T3 plates, obtained experimentally by Liss, Goldsmith and Kelly.[6.38]

PROBLEMS

6.1 Determine the transverse shear force at the supports of the simply supported beam in Figure 1.7. The beam is subjected to a uniformly distributed static pressure which causes plastic collapse. Assume that the beam is made from a rigid, perfectly plastic material and that the beam cross-section is solid and square. (See equation (6.2) and § 1.5.)

6.2 Determine the transverse shear force at the supports of the simply supported beam in Figure 3.4. The beam is subjected to a uniformly distributed pressure with a rectangular shaped pressure–time history with $p_0 \leqslant 3p_c$. (See § 3.3.)

6.3 Repeat Problem 6.2 when $p_0 \geqslant 3p_c$. (See equation (6.4) and § 3.4.)

6.4 Use the principle of virtual velocities to show that the equilibrium equations (6.5a) and (6.6) for a beam are consistent with the shear strain and curvature relations (6.7) and (6.8). (See Appendix 4.)

6.5 Obtain the permanent transverse displacement profile for the impulsively loaded simply supported beam which is examined in § 6.3.3. (See equation (6.39) for $1 \leqslant v \leqslant 1.5$.)

6.6 Repeat Problem 6.5 for beams with $v \geqslant 1.5$ which are examined in § 6.3.4.

6.7 Obtain the temporal variations of the kinetic energy, transverse shear energy dissipation and bending energy dissipation for the impulsively loaded beams in § 6.3 with $v \leqslant 1$, $1 \leqslant v \leqslant 1.5$ and $v \geqslant 1.5$.

6.8 Show that the initial transverse shear force is infinitely large underneath the striking mass in Figure 3.15(a).

6.9 Show that the theoretical analysis in §§ 6.4.2 and 6.4.3 for the beam in Figure 6.11(a) is both statically and kinematically admissible.

6.10 Show that the transverse shear force around the boundary of the impulsively loaded simply supported circular plate in § 4.5.7 is infinitely large when motion commences. (See equation (6.112).)

6.11 Show that the initial transverse shear force is infinitely large at the supports of the cylindrical shell in § 5.5.2 when the loading is idealised as impulsive. (See equation (6.133).)

6.12 Obtain equations (6.145) and (6.146) for the energy dissipated through transverse shear displacements and circumferential membrane forces in a simply supported cylindrical shell having $v \leqslant 1$.

6.13 Use equation (6.153) to obtain the threshold velocity for the impulsively loaded simply supported circular plate, with $v \leqslant 1.5$, which is examined in § 6.5.2.

6.14 Repeat Problem 6.13 for the cylindrical shell with $v \leqslant 1$ which is examined in § 6.6.2.

7

Influence of finite displacements

7.1 Introduction

The theoretical solutions for the static and dynamic plastic behaviour of beams, plates and shells were developed in the previous chapters for infinitesimal displacements. In other words, the equilibrium equations, which govern the response of these structures, were obtained using the original undeformed configuration. For example, the equilibrium equations (1.1) and (1.2) were derived for the beam element shown in Figure 1.1 by ignoring any deformations which would develop under the action of the external loads. It turns out that theoretical analyses, which incorporate this simplification, are often capable of predicting static plastic collapse loads which agree with the corresponding experimental results, as indicated in Figure 1.14 for the transverse loading of metal beams without any axial restraints at the supports.

It may be recalled from § 1.9 that the concentrated load which may be supported by an axially restrained rigid, perfectly plastic beam exceeds the associated static plastic collapse load for finite transverse displacements. In fact, it is evident from Figure 1.15 that the external load is about twice the associated static plastic collapse value when the maximum permanent transverse displacement equals the beam thickness. However, an element of the original beam cross-section is deformed severely and displaced considerably from its initial position, as sketched in Figure 7.1. This change of geometry is ignored in the theoretical methods which are presented in the previous chapters. It is evident that the equilibrium equations (1.1) and (1.2) no longer control the behaviour of a deformed beam element, and different equations are required. It transpires that the most important effect of this change is the development of the membrane, or in-plane, force, N, which is shown in Figure 7.1.

If the supports of a transversely loaded beam are restrained axially, then the centre line of the beam must be longer in the deformed configuration. This

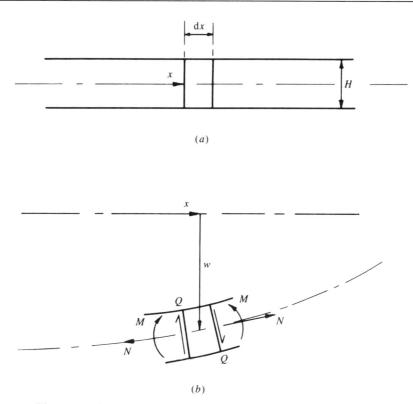

(a)

(b)

Figure 7.1 Element of a beam. (a) Initial shape. (b) After deformation.

stretching gives rise to an axial membrane strain and an associated membrane force N which is indicated in Figure 7.1.

The equilibrium of a beam undergoing small displacements is governed by a bending moment M and a transverse shear force Q as shown in Figure 1.1. However, as the transverse displacement increases, the importance of M and Q diminish and the membrane force N develops until it dominates the behaviour at a sufficiently large transverse displacement. This is known as a string response. In this circumstance, the bending moment M and transverse shear force Q do not play an important role and a perfectly plastic beam flows plastically at the fully plastic membrane force N_0 for the cross-section.

The phenomenon of geometry changes, which is described above for a beam, may also occur during the static response of circular and rectangular plates, as shown in Figures 2.23 and 2.24, respectively. It also plays a role in the behaviour of cylindrical shells which are loaded radially and restrained axially at the supports.

The static plastic collapse, or limit, theorems of plasticity are valid only for infinitesimal displacements, as already noted in § 2.12.1. No general theorems are available for the finite-displacement case of interest in this chapter, although many attempts have been made and several proofs do exist for special cases. Thus, the theoretical methods which have been developed for the finite-displacement behaviour of structures are not rigorous and are, therefore, approximate. However, good agreement has been obtained with the corresponding experimental results in some cases.

An energy ratio E_r was defined in Chapter 3 (see equation (3.126)) in order to characterise the energy which is associated with the external dynamic loads which act on beams. This energy ratio must be sufficiently large to ensure that a theoretical analysis using a rigid, perfectly plastic material gives reasonable predictions. The requirement for a large energy ratio would give rise to finite transverse displacements and the introduction of membrane forces in axially restrained beams. Thus, it is clear that the strengthening influence of finite-displacements, or geometry changes, may also play an important role during the response of structures loaded dynamically.

The next section examines the finite-displacement behaviour of a perfectly plastic beam which is subjected to a centrally located concentrated static load, as shown in Figure 7.2. Two further examples of statically loaded circular and rectangular plates are studied in §§ 7.3 and 7.4, respectively. An approximate theoretical procedure is used in § 7.5 to examine the dynamic plastic response of several rectangular plates and comparisons are made with the corresponding experimental results. This theoretical procedure is also used in § 7.6 to examine the behaviour of beams. The dynamic plastic response of circular plates and a circular membrane is examined in §§ 7.7 and 7.8, respectively, and the chapter closes with some final remarks and by introducing the phenomena of pseudo-shakedown and tensile tearing failure.

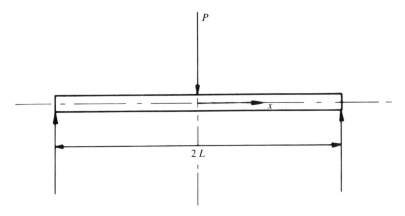

Figure 7.2 Simply supported beam with a central concentrated load P.

7.2 Static plastic behaviour of a beam subjected to a concentrated load

7.2.1 Introduction

It was remarked in § 7.1 that the membrane force N in Figure 7.1 develops for finite transverse displacements of the beam in Figure 7.2 when restrained axially at the supports. The bending moment M dominates the response for small transverse displacements, but decreases in importance as the transverse displacement increases. Thus, a yield condition is required for the combinations of M and N which cause yielding of a perfectly plastic beam cross-section. It is assumed that the transverse shear force Q does not influence plastic yielding.

A procedure, which is somewhat similar to that in § 2.9 and Figure 2.15, is employed in the next section to obtain a yield condition. The equilibrium equations are derived in § 7.2.3 and used with the yield condition in § 7.2.4 to assess the importance of finite-deflections, or geometry changes, on the load–displacement behaviour of the axially restrained beam in Figure 7.2.

7.2.2 Yield condition

In order to obtain a yield condition for the problem in Figure 7.2 with finite-deflections, it is necessary to find the combinations of the bending moment M and membrane force N which cause the cross-section of a perfectly plastic beam to become fully plastic.

If a perfectly plastic beam has a rectangular cross-section of breadth B and thickness H as shown in Figure 7.3(a), then a pure bending moment causes plastic collapse when

$$M_0 = \sigma_0 BH^2/4, \tag{7.1}$$

as shown in Figure 7.3(b) and as noted already in § 1.2 and equation (1.5). In the absence of any bending, an axial membrane force

$$N_0 = \sigma_0 BH \tag{7.2}$$

may be supported, as indicated in Figure 7.3(c).

It might be anticipated that the combined influence of a bending moment M and a membrane force N for plastic collapse would give rise to the stress distribution across the beam thickness in Figure 7.3(d). The neutral axis is located at a distance ηH from the bottom of the beam and, therefore, no longer coincides with the centroidal axis for a rectangular cross-section. For the purpose of analysis, this stress distribution may be separated into the two parts indicated in Figures 7.3(e) and 7.3(f).

The stress distribution in Figure 7.3(e) has no net axial force and is a pure bending moment about the centroidal axis

$$M = \sigma_0 B(H - \eta H)\eta H$$

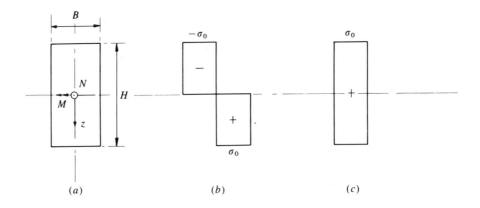

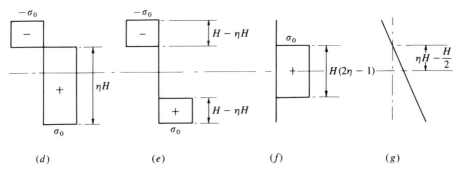

Figure 7.3 Combinations of axial force (N) and bending moment (M) on a rigid, perfectly plastic beam with a rectangular cross-section. (a) Cross-section of beam. (b) Stress distribution associated with a pure bending moment (M_0). (c) Stress distribution associated with an axial membrane force (N_0). (d) Stress distribution associated with a combined bending moment (M) and membrane force (N) where the stress distributions due to (M) and (N) alone are shown in (e) and (f), respectively. (g) Strain distribution across depth of beam (H) due to the stress distribution in (d).

or

$$M/M_0 = 4\eta(1 - \eta) \tag{7.3}$$

when using equation (7.1). The stress distribution in Figure 7.3(f) gives an axial force which acts on the centroidal axis

$$N = \sigma_0 B\{H - 2(H - \eta H)\}$$

or

$$N/N_0 = 2\eta - 1 \tag{7.4}$$

when using equation (7.2). Equation (7.4) may be rearranged in the form

$$\eta = (1 + N/N_0)/2. \tag{7.5}$$

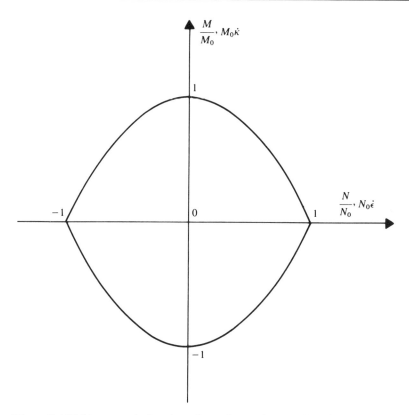

Figure 7.4 Yield curve relating the values of M and N required for plastic flow in a perfectly plastic beam with a rectangular cross-section.

Thus, eliminating η from equation (7.3), when using equation (7.5), gives

$$M/M_0 + (N/N_0)^2 = 1, \tag{7.6}$$

which is shown in Figure 7.4 and is the required yield condition relating M and N for the plastic flow of a rectangular-shaped beam cross-section. The foregoing derivation is associated with the upper right-hand quadrant in Figure 7.4, but the yield curve in the remaining quadrants may be constructed in a similar manner.

It is assumed usually that the cross-section of a beam remains plane throughout deformation as indicated in Figure 7.3(g). Thus, if $\dot{\epsilon}$ is the axial or membrane strain rate at the centroidal axis of the cross-section and $\dot{\kappa}$ is the associated curvature rate, then

$$\dot{\epsilon} = \dot{\kappa}(\eta H - H/2), \tag{7.7}\dagger$$

† Elementary beam bending theory gives $\sigma_x/z = E/R$, where E is Young's modulus and R is the radius of curvature produced by a pure bending moment M. Thus, $\sigma_x/E = z/R$, or $\epsilon_x = z\kappa$, where κ is the change of curvature. This expression may be obtained directly using geometrical arguments associated with initially plane cross-sections remaining plane during the deformation.

or

$$\dot{\epsilon}/\dot{\kappa} = (2\eta - 1)H/2, \tag{7.8}$$

which, using equation (7.4), becomes

$$\dot{\epsilon}/\dot{\kappa} = (H/2)(N/N_0), \tag{7.9}$$

or

$$\dot{\epsilon}/\dot{\kappa} = 2NM_0/N_0^2. \tag{7.10}$$

Now, if the yield condition according to equation (7.6) is differentiated with respect to N, then

$$(\mathrm{d}M/\mathrm{d}N)/M_0 + 2N/N_0^2 = 0,$$

which gives the slope

$$\mathrm{d}M/\mathrm{d}N = -2NM_0/N_0^2. \tag{7.11}$$

The bending moment M and the membrane force N are generalised stresses, while $\dot{\kappa}$ and $\dot{\epsilon}$ are the corresponding generalised strain rates, according to § 2.2. Thus, equation (7.10) is locally normal to a tangent of the yield curve (equation (7.11)) since, as shown in Figure 7.4, the generalised strain rates $\dot{\kappa}$ and $\dot{\epsilon}$ are parallel to the M and N axes, respectively. This observation may have been anticipated from the normality relation which was introduced in § 2.3.4. The associated energy dissipation rate is

$$\dot{D} = M\dot{\kappa} + N\dot{\epsilon}. \tag{7.12}†$$

It should be observed that the slope of the dimensionless yield curve in Figure 7.4 is $\mathrm{d}(M/M_0)/\mathrm{d}(N/N_0) = -2N/N_0$, while the normality relation which is given by equation (7.10) may be written $N_0\dot{\epsilon}/M_0\dot{\kappa} = 2N/N_0$. In other words, $M_0\dot{\kappa}$ and $N_0\dot{\epsilon}$ are the generalised strain rates along the M/M_0 and N/N_0 axes of Figure 7.4, respectively, which ensures that the energy dissipation rate

$$\dot{D} = (M/M_0)(M_0\dot{\kappa}) + (N/N_0)(N_0\dot{\epsilon})$$

agrees with equation (7.12).

7.2.3 Equilibrium equations

Now, consider equilibrium of the beam element in Figure 7.5 which is subject to transverse loads which produce finite transverse displacements. Axial, or longitudinal, equilibrium requires

$$\mathrm{d}N/\mathrm{d}x = 0, \tag{7.13}$$

while moment equilibrium demands

$$\mathrm{d}M/\mathrm{d}x = Q \tag{7.14}$$

when the deformations remain moderate. Transverse equilibrium of the element in

---

† See equation (2.3).

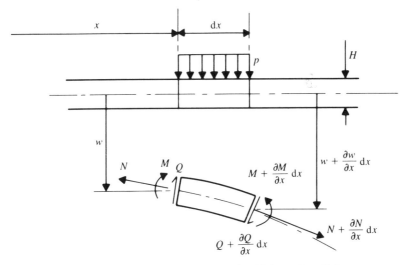

Figure 7.5 Element of a beam subjected to loads which produce finite transverse displacements.

Figure 7.5 leads to the equation

$$dQ/dx + d(N \, dw/dx)/dx + p = 0. \tag{7.15}†$$

Equation (7.13) predicts that the membrane force N is constant throughout the span of a beam. The membrane force N is zero in a beam without any axial forces and undergoing infinitesimal displacements, as indicated in Figure 1.1. In this circumstance, the non-linear term $d(N \, dw/dx)/dx = 0$ so that equation (7.15) reduces to equation (1.2), while equation (7.14) is identical to equation (1.1).

7.2.4 Simply supported beam subjected to a concentrated load

7.2.4(a) $0 \leqslant W/H \leqslant \frac{1}{2}$

It was demonstrated in § 1.7 that the beam in Figure 7.2 collapses with the transverse velocity profile shown in Figure 1.7(b). Thus, it is assumed that the shape of the transverse displacement profile for finite-displacements has the same form, or

$$w = W(1 - x/L) \tag{7.16}$$

as shown in Figure 7.6.

Now, the change in length of the beam centre line in Figure 7.6 is

$$2\{(L^2 + W^2)^{1/2} - L\}$$

---

† The membrane force N has a vertical component $N \sin(dw/dx) \cong N \, dw/dx$ at x on the left-hand side of the beam element in Figure 7.5. Thus, the vertical force at $x + dx$ on the right-hand side of the element is $N \, dw/dx + (d/dx) \, (N \, dw/dx)dx$. The difference between these two vertical forces is $(d/dx) \, (N \, dw/dx)dx$.

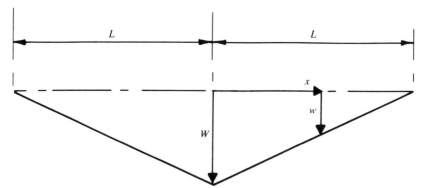

Figure 7.6 Transverse displacement profile for the beam in Figure 7.2 when subjected to a concentrated load at the mid-span.

which gives rise to a membrane strain

$$\epsilon = 2L\{(1 + W^2/L^2)^{1/2} - 1\}/l \qquad (7.17)$$

when the extension occurs in a single central hinge of length l. Equation (7.17) with $(W/L)^2 \ll 1$ may be written with the aid of the binomial series in the form

$$\epsilon \cong (W/L)^2(L/l),$$

which predicts a membrane strain rate

$$\dot{\epsilon} = 2W\dot{W}/Ll. \qquad (7.18)$$

The total angular change across the central hinge in Figure 7.6 is $2W/L$, approximately, which gives rise to a mean curvature rate

$$\dot{\kappa} = 2\dot{W}/Ll \qquad (7.19)$$

across the hinge of length l.

Equations (7.18) and (7.19) predict that

$$\dot{\epsilon}/\dot{\kappa} = W. \qquad (7.20)$$

A comparison between equations (7.10) and (7.20) reveals that

$$2NM_0/N_0^2 = W$$

or $$N/N_0 = 2W/H \qquad (7.21)\dagger$$

when using equations (7.1) and (7.2).

Now, the governing equation for the beam in Figure 7.2 is obtained by substituting equation (7.14) into equation (7.15), or

$$d^2M/dx^2 + (dN/dx)(dw/dx) + N\,d^2w/dx^2 = 0 \qquad (7.22)$$

since $p = 0$. However, equations (7.13) and (7.16) show that the two non-linear terms in equation (7.22) are zero and, therefore,

$$d^2M/dx^2 = 0 \qquad (7.23)$$

† Alternatively, equation (7.21) may be obtained by identifying a rotation $\theta = W/L$ with M and an extension Δ with N, where $\Delta/\theta = W = -dM/dN$, instead of introducing a central hinge l in equations (7.17) and (7.19).

or

$$M = Ax + B, \tag{7.24}$$

where A and B are constants of integration.

The beam in Figure 7.2 is symmetric about the mid-span so that only the portion $0 \leqslant x \leqslant L$ is examined in the following theoretical analysis. It is simply supported at both ends, or $M = 0$ at $x = L$, and equation (7.24) becomes

$$M = A(x - L). \tag{7.25}$$

In addition, the vertical concentrated load at the mid-span must be supported by the vertical components of N and Q at $x = 0$ as indicated in Figure 7.7, or

$$P/2 = -Q \cos(dw/dx) - N \sin(dw/dx),$$

which becomes

$$P/2 \cong -Q - N \, dw/dx, \tag{7.26}$$

provided the finite-displacements remain moderate and where all the quantities are evaluated at $x = 0$. Equation (7.26) may be rewritten

$$P/2 \cong -dM/dx - N \, dw/dx \tag{7.27}$$

with the aid of equation (7.14).

Equation (7.27) together with equations (7.16) and (7.25) at $x = 0$ gives

$$P/2 = -A + NW/L$$

or

$$A = NW/L - P/2$$

which allows equation (7.25) to be written

$$M = (NW/L - P/2)(x - L). \tag{7.28}$$

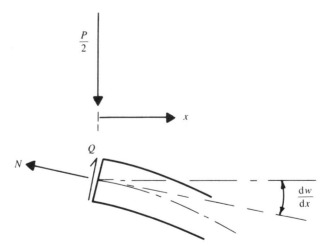

Figure 7.7 Forces at the mid-span of the beam in Figure 7.2.

Equation (7.28) with N eliminated using equation (7.21) predicts that

$$M = (P/2 - 2N_0 W^2/HL)L \tag{7.29}$$

at $x = 0$. Now, substituting both equations (7.21) and (7.29) into the yield condition (7.6) gives

$$(P/2 - 2N_0 W^2/HL)L/M_0 + (2W/H)^2 - 1 = 0,$$

which may be recast in the form

$$P/P_c = 1 + 4W^2/H^2 \tag{7.30a}$$

when using the static plastic collapse load

$$P_c = 2M_0/L, \tag{7.30b}$$

which is defined by equation (1.36).

Equation (7.21) shows that the membrane force N is zero when $W = 0$ and equation (7.30a) gives $P = P_c$ as expected. If equation (7.21) is substituted into the yield condition (7.6), then it is evident that $M = M_0$ at $W = 0$. Thus, the plastic flow of a perfectly plastic beam with a rectangular cross-section and simply supported at both ends is at point A on the yield curve in Figure 7.8 when it first

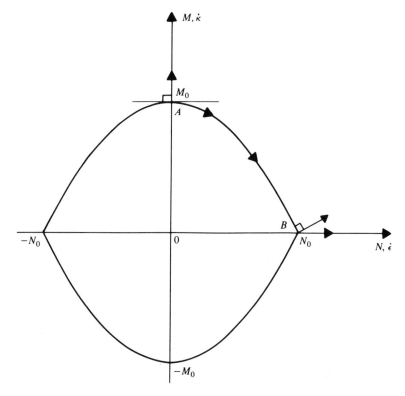

Figure 7.8 Yield curve for a rigid, perfectly plastic beam with a rectangular cross-section.

collapses. The fully plastic membrane state $N = N_0$ is reached when $W = H/2$ according to equation (7.21) and, therefore, equation (7.6) gives $M = 0$. It is evident that, as the transverse displacement increases, the plastic hinge moves from A around the yield curve in Figure 7.8 and reaches the point B, when the maximum transverse displacement equals one-half the beam thickness.

Equation (7.20) with $W = 0$ shows that $\dot{\epsilon}/\dot{\kappa} = 0$ which is satisfied when $\dot{\kappa} \to \infty$ and agrees with the normality requirement at A in Figure 7.8. Moreover, equation (7.20) with $W = H/2$ gives $\dot{\epsilon}/\dot{\kappa} = H/2$, which satisfies the normality requirement at B in Figure 7.8.

It is necessary to examine whether or not the generalised stress field is statically admissible. Now, equation (7.21) shows that $N/N_0 \leqslant 1$ provided $W \leqslant H/2$. Moreover, N is constant across the beam span according to equation (7.13). Thus, the membrane force N is statically admissible for $0 \leqslant W \leqslant H/2$. If equations (7.21) and (7.30a) are substituted into equation (7.28), then

$$M/M_0 = (1 - 4W^2/H^2)(1 - x/L), \qquad 0 \leqslant W/H \leqslant \tfrac{1}{2}, \tag{7.31}$$

which is statically admissible.

Equation (7.30a) predicts the exact static load-carrying capacity of the beam in Figure 7.2 with $0 \leqslant W/H \leqslant \tfrac{1}{2}$ since the theoretical solution is both statically and kinematically admissible.

7.2.4(b) $W/H \geqslant \tfrac{1}{2}$

It is evident from the discussion in § 7.2.4(a) that equation (7.30a) is valid only for $0 \leqslant W \leqslant H/2$ and it becomes necessary, therefore, to extend the theoretical solution for transverse displacements with $W \geqslant H/2$. In this case, the generalised stresses remain at the point B in Figure 7.8, or

$$N = N_0 \qquad \text{and} \qquad M = 0, \tag{7.32a, b}$$

which satisfy the equilibrium equations (7.13) to (7.15) when using equation (7.16). Clearly, the simply supported boundary condition at $x = L$ and the yield condition (7.6) at $x = 0$ are both satisfied by equations (7.32).

Vertical equilibrium of the beam underneath the concentrated load is governed by equation (7.27) at $x = 0$, which, when using equations (7.16) and (7.32a, b), gives

$$P = 2N_0 W/L \tag{7.33}$$

or

$$P/P_c = 4W/H, \qquad W/H \geqslant \tfrac{1}{2}, \tag{7.34}$$

where P_c is defined by equation (7.30b).

It is observed that equation (7.20) still controls the behaviour of the beam so that the generalised stresses remain at the point B of the yield curve in Figure 7.8 and the generalised strain rate vector rotates towards the abscissa (N-axis) as the transverse displacement W increases. The generalised strain rate vector always remains within the fan at B in Figure 7.8 and, therefore, obeys the normality requirement of plasticity which is discussed in § 2.3.4 and illustrated in Figure 2.3.

Equation (7.34) is the exact static load-carrying capacity for the beam in Figure 7.2 when $W/H \geqslant \frac{1}{2}$ since the theoretical solution is both statically and kinematically admissible.

The dimensionless static load–transverse displacement characteristics according to equations (7.30a) and (7.34) are plotted in Figure 7.9 and compared with the corresponding static plastic collapse load P_c which is given by equation (7.30b).

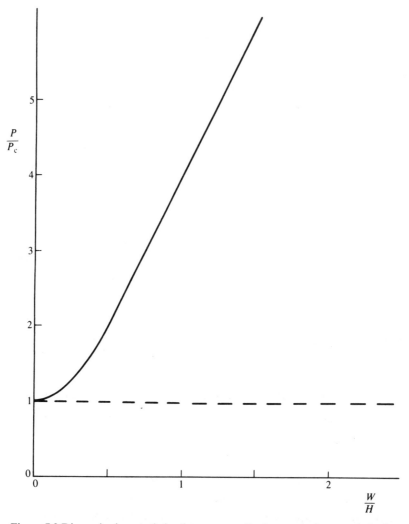

Figure 7.9 Dimensionless static load-transverse displacement characteristics for the beam in Figure 7.2 according to equations (7.30a) and (7.34) (——). The line – – – represents the static plastic collapse load for infinitesimal displacements according to equation (7.30b).

The beam enters the membrane or string state when the maximum transverse displacement equals one-half of the beam thickness and the associated external concentrated load is double the corresponding static plastic collapse value.

It is straightforward to show that the total external energy absorbed by the beam, $E_T = \int_0^W P \, dW$, may be expressed in the dimensionless form

$$E_T/P_c H = W/H + 4(W/H)^3/3, \qquad 0 \leqslant W/H \leqslant \tfrac{1}{2}, \qquad (7.35a)$$

and

$$E_T/P_c H = \tfrac{1}{6} + 2(W/H)^2, \qquad W/H \geqslant \tfrac{1}{2}. \qquad (7.35b)$$

This energy is absorbed owing to bending rotations about the central plastic hinge in Figure 7.6 and membrane stretching, which leads to equation (7.17). The energy absorbed due to bending is $D_b = \int_0^W M2 \, dW/L$, which may be written

$$D_b/P_c H = W/H - 4(W/H)^3/3, \qquad 0 \leqslant W/H \leqslant \tfrac{1}{2}, \qquad (7.36a)$$

and

$$D_b/P_c H = \tfrac{1}{3}, \qquad W/H \geqslant \tfrac{1}{2}. \qquad (7.36b)$$

Similarly, the dimensionless energy absorbed by stretching deformations is

$$D_m/P_c H = 8(W/H)^3/3, \qquad 0 \leqslant W/H \leqslant \tfrac{1}{2}, \qquad (7.37a)$$

and

$$D_m/P_c H = 2(W/H)^2 - \tfrac{1}{6}, \qquad W/H \geqslant \tfrac{1}{2}. \qquad (7.37b)$$

If the bending deformations were disregarded and the beam was assumed to respond as a membrane, or string, throughout deformation, then $D_b = 0$ and

$$E_T/P_c H = D_T/P_c H = E_m/P_c H = 2(W/H)^2, \qquad W/H \geqslant 0. \qquad (7.38)$$

Equations (7.35) to (7.38) are compared in Figure 7.10. It is evident that the contributions to the energy absorbed through bending and membrane behaviour are equal when the maximum transverse displacement equals one-half the beam thickness.

7.2.5 Fully clamped beam subjected to a concentrated load

A theoretical procedure, which is similar to that in § 7.2.4, may be used to examine the problem in Figure 7.2 but with fully clamped supports.[7.1] In this case, it may be shown that

$$P/\bar{P}_c = 1 + W^2/H^2, \qquad 0 \leqslant W/H \leqslant 1, \qquad (7.39a)$$

and

$$P/\bar{P}_c = 2W/H, \qquad W/H \geqslant 1, \qquad (7.39b)$$

where

$$\bar{P}_c = 4M_0/L \qquad (7.39c)$$

is the corresponding static plastic collapse load according to equation (1.37). The theoretical predictions of equations (7.39) are compared in Figure 1.15 with the experimental results recorded by Haythornthwaite[7.1] on steel beams.

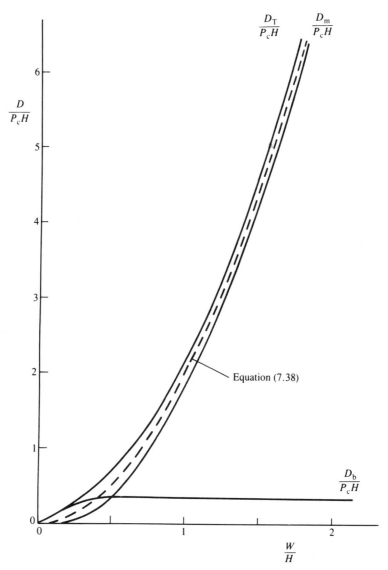

Figure 7.10 Variation of dimensionless energies with transverse displacement according to equations (7.35) to (7.38).

It should be noted that the static plastic collapse load for infinitesimal deflections, according to equation (7.39c), is double equation (7.30b) for the simply supported case. Thus, equations (7.34) and (7.39b) predict identical transverse load–displacement behaviour in the string or membrane phase for simply supported and fully clamped beams.

7.3 Static plastic behaviour of circular plates

7.3.1 Introduction

The static plastic collapse behaviour of several rigid, perfectly plastic circular plates were studied in § 2.6. The theoretical solutions were developed for plates which undergo infinitesimal displacements and with plastic flow controlled by radial and circumferential bending moments, as indicated in Figures 2.5 and 2.6(*b*).

Onat and Haythornthwaite[7.2] examined the carrying capacities of circular plates which were subjected to loads which produce finite transverse displacements. An approximate theoretical method, which incorporates the influence of geometry changes, was developed and compared with experimental test results on mild steel plates.

Onat and Haythornthwaite[7.2] assume that the transverse displacement field, which is associated with the static collapse load, governs the behaviour of the same problem when the external loads produce finite transverse displacements. The rate of internal energy dissipation is then calculated with the aid of the normality rule of plasticity which is discussed in § 2.3.4. This is equated to the external work rate to obtain the load-carrying capacity of a plate. The equilibrium equations have not been satisfied and, therefore, this theoretical approach is approximate.

7.3.2 Partially loaded circular plate with simple supports

The theoretical analysis in § 2.6.5 predicts that the circular plate problem in Figure 7.11(*a*) collapses with the conical transverse velocity profile in Figure 7.11(*b*). Onat and Haythornthwaite[7.2] use this transverse velocity profile and assume that the plate is freely supported around the outer boundary. It turns out that the circumferential membrane strain rate and circumferential curvature rate are the only non-zero generalised strain rates for this problem. The internal energy dissipation according to the maximum shear stress yield condition and the associated flow rule, is then equated to the external work rate to give

$$P/P_c = 1 + 4(W/H)^2/3, \qquad 0 \leqslant W/H \leqslant \tfrac{1}{2}, \tag{7.40a}$$

and

$$P/P_c = 2W/H + (H/W)/6, \qquad W/H \geqslant \tfrac{1}{2}, \tag{7.40b}$$

where P_c is the corresponding static plastic collapse load which is defined by equation (2.20a).

Equations (7.40) are discussed in § 2.12.1 and compared, in Figure 2.23, with the corresponding experimental results of Onat and Haythornthwaite[7.2] which were recorded on mild steel plates.

Onat and Haythornthwaite[7.2] have also studied the behaviour of a circular plate with fully clamped supports and compared the theoretical predictions with experimental results on steel plates.

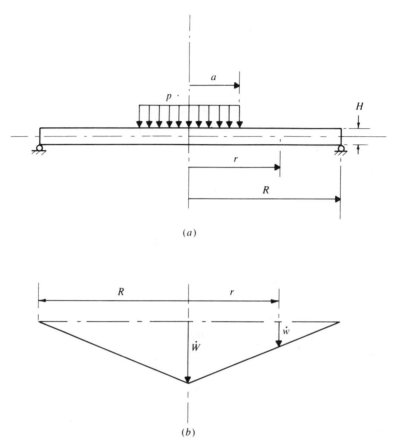

(a)

(b)

Figure 7.11 (a) Simply supported circular plate subjected to a central circular
distribution of pressure. (b) Transverse velocity profile.

7.4 Static plastic behaviour of rectangular plates

7.4.1 Introduction

It is difficult to obtain exact theoretical solutions for the static plastic collapse of
rectangular plates as already noted in § 2.8. Consideration of the influence of
finite-deflections, or geometry changes, would introduce membrane forces and
cause further complexity. However, the topic is one of practical importance and,
therefore, several approximate methods have been developed in the literature.

The next section contains a method which was developed in references 7.3 to 7.5
for arbitrarily shaped flat plates subjected to transverse, or lateral, loads which
produce finite displacements. This particular procedure assumes that the shape of
the static plastic collapse profile for infinitesimal displacements remains unchanged
during finite-displacements which is a simplification first introduced by Onat and
Haythornthwaite[7.2], as discussed in § 7.3.1.

The approximate method is used in § 7.4.3 to examine a beam which is subjected to a concentrated load. A uniformly loaded rectangular plate with simple or fully clamped supports is studied in § 7.4.4 and § 7.4.5, respectively, while § 7.4.6 contains a brief discussion.

7.4.2 Approximate method of analysis

It was shown in references 7.3 to 7.5 that

$$\int_A p\dot{w}\,\mathrm{d}A = \sum_{m=1}^{r} \int_{l_m} (M + Nw)\dot{\theta}_m\,\mathrm{d}l_m \tag{7.41}$$

for an initially flat rigid, perfectly plastic plate which deforms into a number of rigid regions separated by r straight-line plastic hinges each of length l_m. The area A on the left-hand side of equation (7.41) is the total area of the plate which is subjected to the external pressure p, $\dot{\theta}_m$ is the relative angular velocity across a straight line hinge, w is the transverse displacement along a line hinge, and N and M are the membrane force and bending moment which act on a plane which passes through a hinge and is transverse to the mid-surface of a plate.

It is straightforward to show that equation (7.41) is simply a statement of energy conservation.[†] The left-hand side of equation (7.41) is the external work rate, while the term $M\dot{\theta}_m$ on the right-hand side is related to the internal energy dissipation due to bending at the plastic hinge lines. If w and M are defined as shown in Figure 7.5, then the associated value of $\dot{\theta}_m$ is positive, as indicated in Figure 7.12(a). The term $Nw\dot{\theta}_m$ on the right-hand side of equation (7.41) is related to the internal energy dissipation of a plate arising from the membrane (in-plane) forces at the plastic hinge lines. This may be demonstrated with simple geometrical arguments for moderate transverse displacements and zero in-plane displacements. For example, the rigid region in Figure 7.12(b) rotates about a simple support with $w = 0$ and produces an axial extension rate $w\dot{\theta}_m$ in the plastic hinge at the other end.[‡] The associated energy dissipation rate is, therefore, $N(w\dot{\theta}_m)$.

The integrand on the right-hand side of equation (7.41)

$$\dot{D} = (M + Nw)\dot{\theta}_m \tag{7.42}$$

may be interpreted as the internal energy dissipation rate per unit length of a straight-line hinge. The explicit form of the dissipation function $(\dot{D})$ depends on the type of supports around the boundary of a plate and on the yield condition for the material.

---

† Equation (7.41) has also been derived by Taya and Mura[7.6] with the aid of a variational method. Equation (7.41) gives an upper bound to the exact collapse pressure for a plate which undergoes infinitesimal displacements according to the upper bound theorem of plasticity in § 2.4.3.

‡ The extension δ at the hinge in Figure 7.12(b) is given by

$$(L^2 + w^2)^{1/2} - L = L\{(1 + w^2/L^2)^{1/2} - 1\} \cong L\{1 + w^2/2L^2 + \cdots - 1\} = w^2/2L.$$

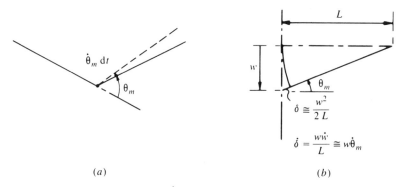

Figure 7.12 (*a*) Positive value of $\dot{\theta}_m$ at a plastic hinge. (*b*) Axial extension rate at a plastic hinge.

If the maximum normal stress yield criterion in Figure 7.4 is selected, then

$$M = M_0\{1 - (N/N_0)^2\} \tag{7.43}$$

according to equation (7.6), where, according to equation (7.21),

$$N/N_0 = 2w/H \tag{7.44}$$

when a plastic hinge forms at the mid-span of a simply supported section. Thus, substituting equations (7.43) and (7.44) into equation (7.42) gives

$$\dot{D} = M_0(1 + 4w^2/H^2)\dot{\theta}_m \tag{7.45}$$

for a solid rectangular cross-section with $M_0 = N_0H/4$ from equations (7.1) and (7.2). Equation (7.44) is valid provided $0 \leqslant w \leqslant H/2$ in order to ensure that $0 \leqslant N \leqslant N_0$. Thus, $N = N_0$ and $M = 0$ when $w \geqslant H/2$ and equation (7.42), therefore, predicts

$$\dot{D} = 4M_0 w\dot{\theta}_m/H. \tag{7.46}$$

The case of fully clamped supports may be examined in a similar manner, except now the axial extension calculated in § 7.2.4(a) is shared equally by the plastic hinges at both ends of a rigid region. Thus, following the procedure in § 7.2.4(a) leads to

$$N/N_0 = w/H, \tag{7.47}$$

which, together with equation (7.43), allows equation (7.42) to be written

$$\dot{D} = M_0(1 + 3w^2/H^2)\dot{\theta}_m \tag{7.48}$$

for an interior hinge. It is evident from equation (7.47) that $w \leqslant H$ to avoid a yield violation. Thus, equation (7.46) is also valid for clamped supports when $w/H \geqslant 1$.

7.4.3 Illustration of approximate method for the concentrated loading of beams

Now consider a simply supported beam which is subjected to a concentrated load at the mid-span as shown in Figure 7.2. The static plastic collapse profile is shown in Figure 7.6 and is assumed to remain valid for a beam undergoing geometry changes when using the approximate theoretical procedure in § 7.4.2. In this particular case, the variables in equation (7.41) take on the values $r = 1$, $\theta_1 = 2\dot{\theta}$, $w = W$, $\dot{W} = L\dot{\theta}$ and $l_1 = B$, so that equations (7.41) and (7.45) give

$$PL\dot{\theta} = M_0(1 + 4W^2/H^2)2\dot{\theta},$$

which reduces to equations (7.30) when $W/H \leqslant \frac{1}{2}$, while for $W/H \geqslant \frac{1}{2}$, equation (7.34) is recovered from equations (7.41) and (7.46).

If the beam in Figure 7.2 has fully clamped ends, then the collapse mechanism in Figure 7.6 is also associated with the exact static collapse load. In this case, $r = 3$ in equation (7.41) and

$$\dot{D}_1 = \dot{D}_3 = M\dot{\theta} \qquad\qquad (7.49a, b)\dagger$$

for the hinges at the two supports with $w = 0$, which may be recast in the form

$$\dot{D}_1 = \dot{D}_3 = M_0(1 - W^2/H^2)\dot{\theta} \qquad\qquad (7.49c, d)$$

with the aid of equations (7.43) and (7.47). The energy dissipation function for the hinge at the mid-span is

$$\dot{D}_2 = M_0(1 + 3W^2/H^2)2\dot{\theta} \qquad\qquad (7.50)$$

according to equation (7.48). Finally, substituting equations (7.49c, d) and (7.50) into equation (7.41) gives

$$PL\dot{\theta} = 2M_0(1 - W^2/H^2)\dot{\theta} + 2M_0(1 + 3W^2/H^2)\dot{\theta},$$

which reduces to equation (7.39a) and is valid for $W/H \leqslant 1$ in order to maintain $N \leqslant N_0$ in equation (7.47).

Now, $M = 0$ and $N = N_0$ when $W \geqslant H$ for fully clamped supports. Thus, equation (7.42) predicts $\dot{D}_1 = \dot{D}_3 = 0$ and $\dot{D}_2 = N_0W2\dot{\theta}$, which when substituted into equation (7.41) gives

$$PL\dot{\theta} = N_0W2\dot{\theta},$$

which reduces to equation (7.39b).

It is evident from the above two calculations that the approximate method in § 7.4.2 predicts the exact static load-carrying capacities which were developed in § 7.2.4 and § 7.2.5 for simply supported and fully clamped beams subjected to concentrated loads.

7.4.4 Simply supported rectangular plate subjected to a uniform pressure

7.4.4(a) Introduction

Bounds on the static plastic collapse pressure were derived in §§ 2.8.2 to 2.8.4 for the simply supported rectangular plate which is illustrated in Figure 2.9. It was

---

† The signs of M and θ are negative at the supports according to the definitions in Figures 7.5 and 7.12, respectively.

observed in Figure 2.12 that the lower and upper bounds are close for a rectangular plate made from a rigid, perfectly plastic material which obeys the Johansen yield condition shown in Figure 2.4. However, a consideration of geometry changes would introduce the membrane forces, N_x, N_y and $N_{xy} = N_{yx}$ and, therefore, require the satisfaction of a yield condition which involves all of the variables in equation (2.1) except the two transverse shear forces. The theoretical procedure which is introduced in § 7.4.2 in order to examine the influence of finite displacements, or geometry changes, avoids this complexity. This method assumes that the collapse mechanism for the present problem with finite-deflections is the same as that used in § 2.8.3 to obtain an upper bound to the exact static collapse pressure.

7.4.4(b) $W/H \leqslant \frac{1}{2}$

Equation (7.41) and the static plastic collapse mechanism, which is illustrated in Figure 2.10, is used to examine the influence of finite transverse displacements in this section. It is clear that the external work rate on the left-hand side of equation (7.41) is identical to equation (2.39). Moreover, the plate is simply supported around the edges so that the internal energy dissipation on the right-hand side of equation (7.41) is controlled by equation (7.45) which is valid for $0 \leqslant W \leqslant H/2$. Thus, using the notation in § 2.8.3 and Figure 2.10, equation (7.41) becomes

$$2B^2 p(L/B - \tan\phi/3)\dot{W} = M_0(1 + 4W^2/H^2)2\dot{\theta}_2 2(L - B\tan\phi)$$

$$+ 4M_0 \int_0^{B\tan\phi} (1 + 4w^2/H^2)\dot{\theta}_3 \, dx'/\sin\phi, \qquad (7.51a)$$

where

$$w = W(B\tan\phi - x')/B\tan\phi, \qquad 0 \leqslant x' \leqslant B\tan\phi, \qquad (7.51b)$$

$$\dot{\theta}_2 = \dot{W}/B \qquad (7.51c)$$

and

$$\dot{\theta}_3 = \dot{W}/B\sin\phi. \qquad (7.51d)$$

Thus,

$$p/p_c = 1 + 4\{\xi_0 + (3 - 2\xi_0)^2\}(W/H)^2/3(3 - \xi_0), \qquad (7.52)$$

where

$$\xi_0 = \beta\tan\phi, \qquad (7.53a)$$

$$\beta = B/L \qquad (7.53b)$$

and

$$p_c = 6M_0/[B^2\{(3 + \beta^2)^{1/2} - \beta\}^2] \qquad (7.53c)$$

is the static plastic collapse pressure according to equation (2.43).

7.4.4(c) $W/H \geqslant \frac{1}{2}$

Now, when $W/H \geqslant \frac{1}{2}$, $N = N_0$ and $M = 0$ according to equation (7.44), and the internal energy dissipation function (7.45) must be replaced by equation (7.46). A

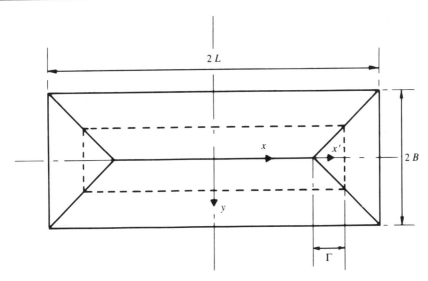

Figure 7.13 Plan view of the plastic hinge line pattern on a uniformly loaded rectangular plate with the boundary (– – –) between the inner and outer regions at $x' = \Gamma$ where $w = H/2$.

rectangular plate, therefore, contains two major regions with a common boundary at $x' = \Gamma$, where $w = H/2$, as illustrated in Figure 7.13. Any plastic hinges which lie within the central region with $w/H \geqslant \frac{1}{2}$ are controlled by $N = N_0$, $M = 0$ and equation (7.46), while those lying within the outer region with $w/H \leqslant \frac{1}{2}$ are governed by equation (7.45). The external work rate on the left-hand side of equation (7.41) is again given by equation (2.39). Thus,

$$2B^2 p(L/B - \tan \phi/3)\dot{W} = N_0 W(2\dot{\theta}_2)2(L - B \tan \phi)$$

$$+ 4 \int_0^\Gamma N_0 W\{(B \tan \phi - x')/B \tan \phi\}\dot{\theta}_3 \, dx'/\sin \phi$$

$$+ 4 \int_\Gamma^{B \tan \phi} M_0[1 + 4\{W(B \tan \phi - x')/HB \tan \phi\}^2]\dot{\theta}_3 \, dx'/\sin \phi \quad (7.54)$$

or

$$p/p_c = 4(W/H)\{1 + \xi_0(\xi_0 - 2)(1 - H^2/12W^2)/(3 - \xi_0)\}. \quad (7.55)$$

Equations (7.52) and (7.55) predict the same pressure when $W/H = \frac{1}{2}$ as expected.

7.4.5 Fully clamped rectangular plate subjected to a uniform pressure

7.4.5(a) Introduction

The static plastic collapse pressure was obtained in § 2.8.5 for a rectangular plate with clamped supports. The general method outlined in the previous section for the simply supported case is used in this section to examine the influence of finite-deflections on the behaviour of rectangular plates with fully clamped boundaries.

7.4.5(b) $W/H \leqslant 1$

The dissipation function is given by equation (7.48) for an interior hinge in a fully clamped region. Thus, using equation (2.39) for the external work rate and substituting in equation (7.41), together with equations (7.43), (7.47) and (7.48), leads to

$$2B^2 p(L/B - \tan \phi/3)\dot{W} = 4 \int_0^B M_0(1 - w^2/H^2)\dot{\theta}_1 \, dy$$

$$+ 4 \int_0^{B \tan \phi} M_0(1 - w^2/H^2)\dot{\theta}_2 \, dx' + M_0(1 - W^2/H^2)\dot{\theta}_2 4(L - B \tan \phi)$$

$$+ M_0(1 + 3W^2/H^2)2\dot{\theta}_2 2(L - B \tan \phi)$$

$$+ 4 \int_0^{B \tan \phi} M_0(1 + 3w^2/H^2)\dot{\theta}_3 \, dx'/\sin \phi, \tag{7.56a}$$

which, using equations (7.51),

$$w = W(B - y)/B, \qquad 0 \leqslant y \leqslant B, \tag{7.56b}$$

and

$$\dot{\theta}_1 = \dot{W}/B \tan \phi \tag{7.56c}$$

gives

$$p/p_c = 1 + \{\xi_0 + (3 - 2\xi_0)^2\}(W/H)^2/3(3 - \xi_0), \tag{7.57}$$

where ξ_0 is defined by equation (7.53a) and

$$p_c = 12M_0/[B^2\{(3 + \beta^2)^{1/2} - \beta\}^2] \tag{7.58}$$

is the static plastic collapse pressure according to equation (2.51).

7.4.5(c) $W/H \geqslant 1$

It is necessary to divide a plate into two regions, as shown in Figure 7.13. However, in this case, $w = H$ on the boundary at $x' = \Gamma$ between the two regions.

It is left as an exercise to a reader to show that equation (7.41) predicts

$$p/p_c = (2W/H)\{1 + \xi_0(2 - \xi_0)(H^2/3W^2 - 1)/(3 - \xi_0)\}. \tag{7.59}$$

Equations (7.57) and (7.59) predict the same pressure when $W/H = 1$.

7.4.6 Discussion

The theoretical predictions of equations (7.52), (7.55), (7.57) and (7.59) are presented in Figure 7.14 for the uniform pressure loading of simply supported and fully clamped rectangular plates with an aspect ratio of $\beta = 0.5$. It is evident that the load-carrying capacity for moderate transverse displacements is significantly larger than the corresponding static collapse value, which is predicted in § 2.8 for

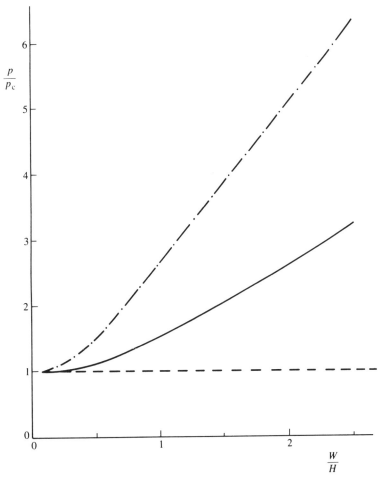

Figure 7.14 Dimensionless pressure–transverse displacement characteristics for rectangular plates with $\beta = 0.5$. — — —: static collapse behaviour for infinitesimal displacements. —·—·—: equations (7.52) and (7.55) for simple supports. ———: equations (7.57) and (7.59) for fully clamped case.

infinitesimal displacements.† The variation of the dimensionless pressure carrying capacity with aspect ratio β is illustrated in Figure 7.15 for fully clamped plates with various levels of maximum permanent transverse displacement, or damage (iso-damage curves).

The theoretical predictions of equations (7.57) and (7.59) are compared in Figure 2.24 with some experimental results on fully clamped mild steel plates having aspect ratios of $\beta = 0.5$ and $\beta = 1$ (square plate).

---

† It is evident from equations (7.53c) and (7.58) that the static plastic collapse pressure for a fully clamped plate is twice as large as that for a rectangular plate with simply supported boundaries.

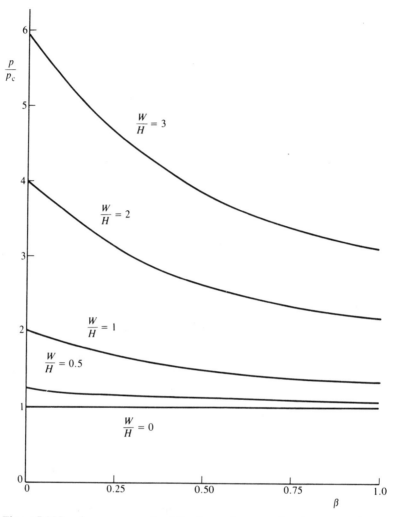

Figure 7.15 Iso-damage curves for fully clamped rectangular plates according to equations (7.57) and (7.59).

Now, when $\beta \to 0$, then $L \gg B$ and the theoretical predictions for rectangular plates reduce to the corresponding results for uniformly loaded beams, as illustrated in Figure 2.13. In this circumstance, equation (2.42) gives $\tan \phi = \sqrt{3}$ and $\zeta_0 \to 0$ according to equation (7.53a). Thus, equations (7.52) and (7.55) predict that

$$p/p_c = 1 + 4(W/H)^2, \qquad 0 \leqslant W/H \leqslant \tfrac{1}{2}, \tag{7.60a}$$

and

$$p/p_c = 4W/H, \qquad W/H \geqslant \tfrac{1}{2}, \tag{7.60b}$$

for a uniformly loaded beam with simple supports,† while equations (7.57) and (7.59) reduce to

$$p/p_c = 1 + (W/H)^2, \qquad 0 \leqslant W/H \leqslant 1, \qquad (7.61a)$$

and

$$p/p_c = 2W/H, \qquad\qquad W/H \geqslant 1, \qquad (7.61b)$$

for the fully clamped case.‡ It should be noted that equations (7.53c) and (7.58) with $\beta = 0$ predict that $p_c = 2M_0/B^2$ and $p_c = 4M_0/B^2$ for the simply supported and fully clamped cases, respectively.

7.5 Dynamic plastic behaviour of rectangular plates

7.5.1 Introduction

It is evident from §§ 7.2 to 7.4 that geometry changes play an important role in the behaviour of several structural members and may lead to a significant increase in the static load-carrying capacity for moderate transverse displacements. It is anticipated that the influence of finite-displacements, or geometry changes, may also play an important role for some structures loaded dynamically.

The approximate theoretical method, developed in references 7.3 to 7.5 for arbitrarily shaped plates, was specialised in § 7.4.2 for the static behaviour of plates which deform into a number of rigid regions separated by straight line plastic hinges. A version of this theoretical method, which retains dynamic effects, is introduced in the next section and is used in § 7.5.3 to study a simply supported rectangular plate loaded dynamically. The theoretical procedure is repeated for the same problem in § 7.5.4 but for a simplified yield condition, while a fully clamped rectangular plate is examined in § 7.5.5. The particular case of an impulsive velocity loading is examined in § 7.5.6 and comparisons are made with experimental results on metal plates in § 7.5.7.

7.5.2 Approximate method of analysis

An approximate theoretical procedure was developed in reference 7.4 and may be written in the simplified form

$$\int_A (p - \mu\ddot{w})\dot{w} \, dA = \sum_{m=1}^{r} \int_{l_m} (M + Nw)\dot{\theta}_m dl_m \qquad (7.62)\S$$

for an initially flat, rigid, perfectly plastic plate which deforms with a number of rigid regions separated by r stationary straight line plastic hinges each of length l_m.

---

† Equations (7.60) provide excellent agreement with the exact numerical solution of Gürkök and Hopkins,[7.7] as shown in Figure 1 of reference 7.8.

‡ Equations (7.61) are compared with the experimental results of Young[7.9] in Figure 5 of reference 7.5.

§ μ is the mass per unit surface area.

Equation (7.62) reduces to equation (7.41) for static loads. Moreover, the various dissipation relations and other quantities which were developed in § 7.4.2 remain valid for the dynamic case.

7.5.3 Dynamic response of a simply supported rectangular plate subjected to a uniform pressure pulse

7.5.3(a) Introduction

The static plastic behaviour of a simply supported rectangular plate subjected to a uniformly distributed pressure, which produces finite transverse displacements, or geometry changes, was examined in § 7.4.4 using equation (7.62) for static loads (i.e., equation (7.41)). Equation (7.62) is employed in the next section to study the response of this problem when subjected to a dynamic pressure pulse with a rectangular pressure–time history, as shown in Figure 7.16, or

$$p = p_0, \qquad 0 \leqslant t \leqslant \tau, \tag{7.63a}$$

and

$$p = 0, \qquad t \geqslant \tau. \tag{7.63b}$$

7.5.3(b) $W/H \leqslant \frac{1}{2}$

It is assumed that the dynamic pressure pulse does not produce a maximum transverse displacement larger than one-half of the plate thickness. Thus, it is necessary to consider two phases of motion $0 \leqslant t \leqslant \tau$ and $\tau \leqslant t \leqslant T$.

7.5.3(b)(i) First phase of motion, $0 \leqslant t \leqslant \tau$

It is assumed that the shape of the transverse velocity field, which is associated with the dynamic pressure pulse in Figure 7.16, is the same as the one used in § 7.4.4 for a static pressure (see Figure 2.10). Thus, all the terms in equation (7.62), with the exception of the inertia term ($\mu \ddot{w}$), are the same as those in § 7.4.4.

Now, consider the inertia term in equation (7.62), which, using equations (2.33) and (2.34), may be written

$$\int_A \mu \ddot{w} \dot{w} \, dA = 4\mu \int_0^{B \tan \phi} \int_0^{x' \cot \phi} \ddot{W}_1 \dot{W}_1 (B \tan \phi - x')^2 \, dy \, dx' / B^2 \tan^2 \phi$$

$$+ 4\mu \int_0^B \int_0^{y \tan \phi} \ddot{W}_1 \dot{W}_1 (B - y)^2 \, dx' \, dy / B^2$$

$$+ 4\mu \int_0^B \int_0^{L - B \tan \phi} \ddot{W}_1 \dot{W}_1 (B - y)^2 \, dx \, dy / B^2 \tag{7.64}$$

and integrated and simplified to give

$$\int_A \mu \ddot{w} \dot{w} \, dA = 4\mu \ddot{W}_1 \dot{W}_1 (LB/3 - B^2 \tan \phi / 6), \tag{7.65}$$

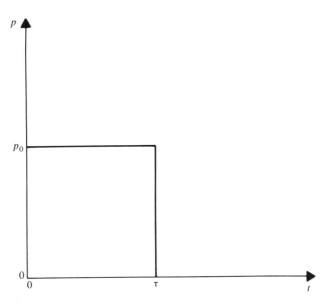

Figure 7.16 Rectangular pressure pulse.

where $\dot{W}_1$ is the peak value of the transverse velocity profile during the first phase of motion which has the same form as Figure 2.10. Thus, substituting equations (7.51a)† and (7.65) into equation (7.62) yields

$$a_1 \ddot{W}_1 + a_2 W_1^2 = \eta - 1, \qquad (7.66)‡$$

where

$$a_1 = \mu B^2 (2 - \beta \tan \phi)/\{6M_0(1 + \beta \cot \phi)\}, \qquad (7.67a)$$

$$a_2 = 4\{1 + 2(1 - \beta \tan \phi)/(1 + \beta \cot \phi)\}/3H^2 \qquad (7.67b)$$

and

$$\eta = p_0/p_c \qquad (7.67c)$$

is a dynamic pressure ratio, where p_c is the static plastic collapse pressure which is given by equation (7.53c).

Equation (7.66) with the initial conditions $W_1 = \dot{W}_1 = 0$ at $t = 0$ is a non-linear differential equation which may be solved using the method of successive approximations. Now, when writing equation (7.66) in the form

$$\ddot{W}_1 + h_1 W_1^2 = d_1, \qquad (7.68a)$$

† With p, W and $\dot{W}$ replaced by p_0, W_1 and $\dot{W}_1$, respectively.
‡ Equation (7.66) without the inertia term ($\ddot{W}_1$) reduces to equation (7.52) for the static load-carrying capacity when finite deflection effects are retained in the basic equations. Moreover, if the non-linear term (W_1^2) is disregarded, then equation (7.66) with $\beta = 1$ reduces to the governing equation for the dynamic response of a square plate with $p_c \leqslant p_0 \leqslant 2p_c$, which is given by equation (4.128). Finally, if both the inertial and the non-linear terms in equation (7.66) are neglected, then the static plastic collapse pressure according to equation (7.53c), or equation (2.43), is recovered.

where

$$h_1 = a_2/a_1 \tag{7.68b}$$

and

$$d_1 = (\eta - 1)/a_1, \tag{7.68c}$$

the first approximation is obtained by disregarding the non-linear term, or

$$\ddot{W}_1^{(1)} = d_1, \tag{7.69a}$$

which gives

$$W_1^{(1)} = d_1 t^2/2 \tag{7.69b}$$

since

$$\dot{W}_1^{(1)} = W_1^{(1)} = 0 \qquad \text{when} \quad t = 0.$$

The second approximation $W_1^{(2)}$ is obtained from

$$\ddot{W}_1^{(2)} + h_1 (W_1^{(1)})^2 = d_1 \tag{7.70a}$$

or

$$W_1^{(2)} = d_1 t^2 (1 - h_1 d_1 t^4/60)/2 \tag{7.70b}$$

because $\dot{W}_1^{(2)} = W_1^{(2)} = 0$ when $t = 0$.

Clearly, this procedure may be repeated† in order to obtain higher order terms, but it is not taken any further here. Thus, at the end of the first phase of motion when the pressure pulse is released

$$W_1 = \bar{W}_1 = d_1 \tau^2 (1 - h_1 d_1 \tau^4/60)/2 \tag{7.71a}$$

and

$$\dot{W}_1 = \dot{\bar{W}}_1 = d_1 \tau (1 - h_1 d_1 \tau^4/20). \tag{7.71b}$$

7.5.3(b)(ii) Second phase of motion, $\tau \leqslant t \leqslant T$

The transverse pressure $p = 0$ according to equation (7.63b) and, therefore, equation (7.62) gives equation (7.66) with $\eta = 0$, or

$$a_1 \ddot{W}_2 + a_2 W_2^2 = -1, \tag{7.72}‡$$

which may be written in the form

$$\ddot{W}_2 + h_1 W_2^2 = d_2, \tag{7.73a}$$

where

$$d_2 = -1/a_1 \tag{7.73b}$$

and W_2 is the transverse displacement during the second phase of motion. Again

---

† The nth approximation $W_1^{(n)}$ is obtained from

$$\ddot{W}_1^{(n)} + h_1 (W_1^{(n-1)})^2 = d_1$$

with $\dot{W}_1^{(n)} = W_1^{(n)} = 0$ at $t = 0$.

‡ Equation (7.72) with $\beta = 1$ and without the non-linear term (W_2^2) reduces to the governing equation (4.136) for the dynamic response of a square plate with $p_c \leqslant p_0 \leqslant 2p_c$.

the method of successive approximations may be used to solve equation (7.73a) and satisfy the initial conditions at $t = \tau$, which are given by equations (7.71).

It may be shown that motion ceases at $t = T$ when $\dot{W}_2 = 0$, or

$$\dot{W}_1 + (d_2 - h_1\bar{W}_1^2)\tau\rho - h_1\dot{\bar{W}}_1\bar{W}_1\tau^2\rho^2 - (h_1 d_2\bar{W}_1 + h_1\dot{\bar{W}}_1^2)\tau^3\rho^3/3$$
$$- h_1 d_2\dot{\bar{W}}_1\tau^4\rho^4/4 - h_1 d_2^2\tau^5\rho^5/20 = 0, \qquad (7.74a)\dagger$$

where

$$\rho = (T - \tau)/\tau \qquad (7.74b)$$

is the dimensionless duration of the second phase of motion. The associated dimensionless maximum permanent transverse displacement is

$$W_f/H = \bar{W}_1/H + \dot{\bar{W}}_1\tau\rho/H + (d_2 - h_1\bar{W}_1^2)\tau^2\rho^2/2H$$
$$- h_1\bar{W}_1\dot{\bar{W}}_1\tau^3\rho^3/3H - (h_1 d_2\bar{W}_1 + h_1\dot{\bar{W}}_1^2)\tau^4\rho^4/12H$$
$$- h_1 d_2\dot{\bar{W}}_1\tau^5\rho^5/20H - h_1 d_2^2\tau^6\rho^6/120H. \qquad (7.75)$$

Equations (7.74a) and (7.75) may be rewritten in the form

$$\alpha_3 + (2\alpha_4 - \alpha_2^2\alpha_5)\rho - \alpha_2\alpha_3\alpha_5\rho^2 - \alpha_5(2\alpha_2\alpha_4 + \alpha_3^2)\rho^3/3$$
$$- \alpha_3\alpha_4\alpha_5\rho^4/2 - \alpha_4^2\alpha_5\rho^5/5 = 0 \qquad (7.76a)$$

and

$$W_f/H = \alpha_2 + \alpha_3\rho + (\alpha_4 - \alpha_2^2\alpha_5/2)\rho^2 - \alpha_2\alpha_3\alpha_5\rho^3/3$$
$$- \alpha_5(\alpha_2\alpha_4 + \alpha_3^2/2)\rho^4/6 - \alpha_3\alpha_4\alpha_5\rho^5/10 - \alpha_4^2\alpha_5\rho^6/30, \qquad (7.76b)$$

respectively, where

$$\alpha_1 = (\eta - 1)\{\xi_0 + (3 - 2\xi_0)^2\}(3 - \xi_0)I'^4/\{45\eta^4(2 - \xi_0)^2\}, \qquad (7.76c)$$

$$\alpha_2 = (1 - \alpha_1)(\eta - 1)(3 - \xi_0)I'^2/2\eta^2(2 - \xi_0), \qquad (7.76d)$$

$$\alpha_3 = (1 - 3\alpha_1)(\eta - 1)(3 - \xi_0)I'^2/\eta^2(2 - \xi_0), \qquad (7.76e)$$

$$\alpha_4 = -(3 - \xi_0)I'^2/2\eta^2(2 - \xi_0), \qquad (7.76f)$$

$$\alpha_5 = 4\{\xi_0 + (3 - 2\xi_0)^2\}I'^2/3\eta^2(2 - \xi_0) \qquad (7.76g)$$

and

$$I' = p_0\tau/(\mu H p_c)^{1/2}. \qquad (7.76h)$$

7.5.3(c) $W/H \geqslant \frac{1}{2}$

The theoretical analysis of § 7.5.3(b) is valid provided the maximum transverse displacement, which is predicted by equation (7.76b), does not exceed one-half of the plate thickness. This restriction arises because the energy dissipation relation given by equation (7.45) for simple supports is valid only for $0 \leqslant W \leqslant H/2$ in order to maintain $0 \leqslant N \leqslant N_0$. The dynamic response now consists of three phases of motion which are discussed in the next three sections.

---

† A straightforward iterative procedure may be used to obtain the actual value of ρ from equation (7.74a) for a given problem.

7.5.3(c)(i) First phase of motion, $0 \leqslant t \leqslant \tau$

It is assumed that $\overline{W}_1/H \leqslant \frac{1}{2}$ throughout the first phase of motion while the pressure pulse is active.† In this circumstance, the theoretical analysis remains identical to that presented in § 7.5.3(b)(i).

7.5.3(c)(ii) Second phase of motion, $\tau \leqslant t \leqslant t_2$

The theoretical analysis in § 7.5.3(b)(ii) governs the response until the time t_2 when $W_2 = H/2$. The corresponding transverse velocity $(\dot{W}_2)$ at t_2 is found from the theoretical solution of equation (7.73a) using the method of successive approximations.

7.5.3(c)(iii) Third phase of motion, $t_2 \leqslant t \leqslant T$

Now, the regions of the rectangular plate having $W/H \geqslant \frac{1}{2}$ are governed by the dissipation relation (7.46), as noted already in § 7.4.4(c) for the corresponding static finite-deflection case. It may be shown that the governing equation is

$$C_1 \ddot{W}_3 + C_2 W_3 + C_3/W_3 = 0, \qquad (7.77a)$$

where

$$C_1 = 2\mu(2 - \xi_0)/p_c(3 - \xi_0), \qquad (7.77b)$$
$$C_2 = 4\{1 + (1 - \xi_0)(3 - 2\xi_0)/(3 - \xi_0)\}/H, \qquad (7.77c)$$
$$C_3 = 2H\xi_0(2 - \xi_0)/3(3 - \xi_0) \qquad (7.77d)$$

and W_3 is the maximum transverse displacement during the third phase of motion. Equation (7.77a) has the general solution‡

$$(\dot{W}_3)^2 = -C_2 W_3^2/C_1 - 2(C_3/C_1) \log_e W_3 + C, \qquad (7.78)$$

where the constant of integration C is obtained from the continuity requirements at $t = t_2$.

Now, equation (7.78) with $\dot{W}_3 = 0$ when motion ceases gives the transcendental equation

$$\alpha_{10}(1/4 - W_f^2/H^2) - 2\alpha_{11} \log(2W_f/H) + (\dot{W}_2\tau/H)^2 = 0 \qquad (7.79a)$$

for the dimensionless maximum permanent transverse displacement, where

$$\alpha_{10} = 4\{1 + (\xi_0 - 1)(\xi_0 - 2)\}I'^2/\eta^2(2 - \xi_0), \qquad (7.79b)$$
$$\alpha_{11} = \xi_0 I'^2/3\eta^2 \qquad (7.79c)$$

and $\dot{W}_2$ is the maximum transverse velocity at $t = t_2$.

† $\overline{W}_1$ is defined by equation (7.71a). The procedure in § 7.5.3(c) may be extended to examine the case for $\overline{W}_1/H \geqslant 1/2$.

‡ If $p = \dot{W}_3$ and $\ddot{W}_3 = (dp/dW_3)(dW_3/dt) = p \, dp/dW_3$ are substituted into equation (7.77a), then $p \, dp = -(C_2 W_3/C_1 + C_3/C_1 W_3) \, dW_3$ and a straightforward integration leads to equation (7.78).

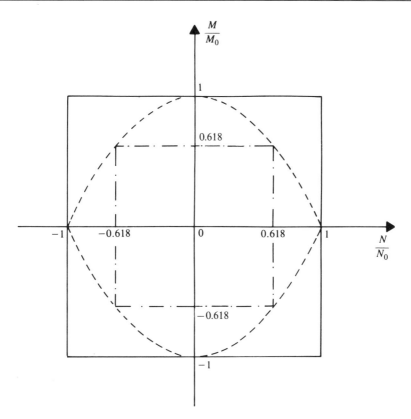

Figure 7.17 Exact and approximate yield curves for a beam with a rectangular cross-section. $---$: exact yield curve (maximum normal stress yield criterion). ——: circumscribing square yield curve. $-\cdot-\cdot-$: inscribing square yield curve.

7.5.4 Simplified analysis for the uniform dynamic pressure loading of a simply supported rectangular plate

7.5.4(a) Introduction

The theoretical analysis, which was presented in § 7.5.3, is simplified in this section by using the square yield condition in Figure 7.17. The exact maximum normal stress yield condition is given by equation (7.6) and is bounded by the square yield surface and another which is 0.618 times as large.† It was mentioned in § 2.4.4 that upper and lower bounds on the exact collapse pressure are provided by yield criteria which circumscribe and inscribe the exact yield surface, respectively.

Now, $\dot{\epsilon} \geqslant 0$ and $\dot{\kappa} \geqslant 0$ at the plastic hinge lines in the collapse mechanism, which is sketched in Figure 2.10. Thus, the normality relation suggests the region

$$N = N_0 \quad \text{and} \quad M = M_0 \qquad (7.80a, b)$$

of the yield condition in Figure 7.17.

† Hodge[7.10] discusses methods for constructing bounds on yield criteria.

7.5.4(b) First phase of motion, $0 \leqslant t \leqslant \tau$

The right-hand side of equation (7.62), together with equations (7.80a, b), may be written

$$\sum_{m=1}^{r} \int_{l_m} (M_0 + N_0 w)\dot{\theta}_m \mathrm{d}l_m = (M_0 + N_0 W_1)2(L - B \tan \phi)2\dot{\theta}_2$$
$$+ 4M_0\dot{\theta}_3 B/\cos \phi + 4N_0\dot{\theta}_3$$
$$\times \int_0^{B \tan \phi} W_1\{(B \tan \phi - x')/B \tan \phi\} \, \mathrm{d}x'/\sin \phi, \quad (7.81)$$

or

$$\sum_{m=1}^{r} \int_{l_m} (M_0 + N_0 w)\dot{\theta}_m \mathrm{d}l_m = 4M_0\dot{W}_1(L/B + \cot \phi)$$
$$+ 8M_0 W_1 \dot{W}_1(2L/B + \cot \phi - \tan \phi)/H. \quad (7.82)\dagger$$

Now, substituting both equation (2.39) for the external work rate and the inertial term according to equation (7.65) into the governing equation (7.62), together with equation (7.82), gives

$$2B^2 \dot{W}_1 p_0(L/B - \tan \phi/3) - 4\mu \ddot{W}_1 \dot{W}_1(LB/3 - B^2 \tan \phi/6)$$
$$= 4M_0 \dot{W}_1(L/B + \cot \phi) + 8M_0 \dot{W}_1 W_1(2L/B + \cot \phi - \tan \phi)/H,$$

or

$$a_1 \ddot{W}_1 + a_2 W_1 = \eta - 1, \quad (7.83a)\ddagger$$

where

$$a_1 = \mu B^2(2 - \beta \tan \phi)/\{6M_0(1 + \beta \cot \phi)\}, \quad (7.83b)$$

$$a_2 = 2\{1 + (1 - \beta \tan \phi)/(1 + \beta \cot \phi)\}/H \quad (7.83c)$$

and η is the pressure ratio which is defined by equation (7.67c). Equation (7.83a) is a linear ordinary differential equation and predicts the transverse displacement

$$W_1 = (1 - \eta)(\cos a_3 t - 1)/a_1 a_3^2 \quad (7.84a)\S$$

when satisfying the initial conditions $W_1 = \dot{W}_1 = 0$ at $t = 0$ and where

$$a_3^2 = a_2/a_1 = 12M_0(2 + \beta \cot \phi - \beta \tan \phi)/\{\mu H B^2(2 - \beta \tan \phi)\}. \quad (7.84b)$$

This phase of motion is completed at $t = \tau$ when the pressure pulse is removed as shown in Figure 7.16.

---

† If finite deflection effects are neglected then the second term on the right-hand side is zero and equation (7.82) reduces to equation (2.38).

‡ Equations (7.66) and (7.83a) are identical when the finite-deflection terms are neglected.

§ This equation gives $\dot{W}_1 = 0$ when $a_3 t = \pi$. Thus, the analysis is valid only for $a_3 t \leqslant \pi$ when $a_3 \tau \leqslant \pi$. For pressure pulses with $a_3 \tau \geqslant \pi$, the plate remains rigid for $t \geqslant \pi/a_3$ with a maximum permanent transverse displacement W_1, which is twice the corresponding static value.

7.5.4(c) Second phase of motion, $\tau \leqslant t \leqslant T$

The governing equation of motion is the same as equation (7.83a), except $\eta = 0$ according to equation (7.63b). Thus, it may be shown that

$$W_2 = \{\eta \sin a_3\tau \sin a_3 t + (\eta - 1)(\cos a_3\tau - 1) \cos a_3 t + \cos a_3\tau \cos a_3 t - 1\}/a_2$$

(7.85)

when ensuring that the transverse displacement and velocity are continuous between the two phases of motion at $t = \tau$.

Finally, motion ceases at $t = T$ when $\dot{W}_2 = 0$, where

$$\tan a_3 T = \eta \sin a_3\tau/(1 - \eta + \eta \cos a_3\tau)$$

(7.86a)

and the associated maximum permanent transverse displacement is

$$W_f/H = (3 - \xi_0)[\{1 + 2\eta(\eta - 1)(1 - \cos a_3\tau)\}^{1/2} - 1]/[4\{1 + (\xi_0 - 2)(\xi_0 - 1)\}].$$

(7.86b)

7.5.5 Dynamic response of a fully clamped rectangular plate subjected to a uniform pressure pulse

7.5.5(a) Introduction

Theoretical solutions, which use the approximate methods in §§ 7.5.3 and 7.5.4, have been published for a rectangular plate[7.4], which is fully clamped around the boundaries and subjected to a uniform dynamic pressure pulse. However, only the simplified theoretical method, which uses the square yield condition in Figure 7.17, is presented in this section for the pressure pulse described by equations (7.63) and shown in Figure 7.16. The theoretical development follows closely § 7.5.4 with the transverse velocity profile which is illustrated in Figure 2.10, except now the energy dissipation around the fully clamped boundaries is included on the right-hand side of equation (7.62).

7.5.5(b) First phase of motion, $0 \leqslant t \leqslant \tau$

In addition to the internal energy dissipation terms in equations (7.81) and (7.82) for a simply supported rectangular plate, it is now necessary to consider the energy dissipated in the plastic hinges at the four boundaries, or $4M_0\dot{\theta}_2(L + B \cot \phi)$, which allows equation (7.62) to be recast in the form

$$\ddot{W}_1 + a_3^2 W_1 = (\eta - 1)/b_1,$$

(7.87a)†

where

$$b_1 = a_1/2,$$

(7.87b)

and η is defined by equation (7.67c) when p_c is the static plastic collapse pressure for a fully clamped rectangular plate, which is defined by equation (7.58).

---

† If both the inertia and finite-deflection terms are neglected, then the static plastic collapse pressure given by equation (2.51) is recovered.

Now, integrating the ordinary linear differential equation (7.87a) and satisfying the initial conditions $W_1 = \dot{W}_1 = 0$ at $t = 0$ gives the transverse displacement

$$W_1 = (1 - \eta)(\cos a_3 t - 1)/b_2, \tag{7.88a}†$$

where

$$b_2 = a_2/2. \tag{7.88b}$$

7.5.5(c) Second phase of motion, $\tau \leqslant t \leqslant T$

The dynamic pressure pulse is removed at $t = \tau$ so that $p = 0$ when $t \geqslant \tau$ according to equation (7.63b) and, therefore, equation (7.87a) with $\eta = 0$ now governs the response. Thus, solving the differential equation and ensuring continuity of the transverse displacement and transverse velocity between the two phases of motion at $t = \tau$ gives

$$W_2 = \{(1 - \eta + \eta \cos a_3 \tau) \cos a_3 t + \eta \sin a_3 \tau \sin a_3 t - 1\}/b_2. \tag{7.89}$$

It is straightforward to show that all motion ceases at the time $t = T$ when $\dot{W}_2 = 0$, or

$$\tan a_3 T = \eta \sin a_3 \tau/(1 - \eta + \eta \cos a_3 \tau). \tag{7.90}$$

The associated maximum permanent transverse displacement is

$$W_f/H = (3 - \xi_0)[\{1 + 2\eta(\eta - 1)(1 - \cos a_3 \tau)\}^{1/2} - 1]$$
$$\div [2\{1 + (\xi_0 - 1)(\xi_0 - 2)\}], \tag{7.91a}$$

where

$$a_e \tau = (2)^{1/2} I'\{1 - \xi_0 + 1/(2 - \xi_0)\}^{1/2}/\eta \tag{7.91b}$$

and I' is defined by equation (7.76h).

7.5.6 Impulsive loading

Impulsive loading has been examined already in several chapters and is the extreme case of a rectangular pressure pulse with a large magnitude $\eta \to \infty$ and a short duration $\tau \to 0$ with a finite impulse $I = p_0 \tau$ (per unit area). Thus, when conserving linear momentum, the procedure in § 4.5.7 gives

$$p_0 \tau = \mu V_0, \tag{7.92}$$

where V_0 is the magnitude of an initial impulsive velocity distributed uniformly over the entire plate.

Now, if $\eta \gg 1$ and $\tau \to 0$ for an impulsive loading, then equation (7.86b) predicts the maximum permanent transverse displacement for a simply supported rectangular plate

$$W_f/H = (3 - \xi_0)\{(1 + 2\eta^2 a_3^2 \tau^2/2)^{1/2} - 1\}/[4\{1 + (\xi_0 - 2)(\xi_0 - 1)\}] \tag{7.93}$$

since $1 - \cos a_3 \tau \cong a_3^2 \tau^2/2$. Thus, using equations (7.53a), (7.53c), (7.76h), (7.84b)

† See footnote § on p. 308.

and (7.92), equation (7.93) may be recast in the form

$$W_f/H = (3 - \xi_0)[\{1 + 2\lambda\xi_0^2(1 - \xi_0 + 1/(2 - \xi_0))/3\}^{1/2} - 1]$$
$$\div [4\{1 + (\xi_0 - 2)(\xi_0 - 1)\}] \tag{7.94a}$$

where

$$\lambda = \mu V_0^2 L^2/M_0 H \tag{7.94b}$$

is a dimensionless initial kinetic energy.

Similarly, equation (7.91a) for a fully clamped rectangular plate, which is subjected to a uniform impulsive velocity, is

$$W_f/H = (3 - \xi_0)[\{1 + \lambda\xi_0^2(1 - \xi_0 + 1/(2 - \xi_0))/6\}^{1/2} - 1]$$
$$\div [2\{1 + (\xi_0 - 1)(\xi_0 - 2)\}] \tag{7.95}$$

7.5.7 Comparison with experimental results

Experimental results have been recorded on impulsively loaded fully clamped rectangular ductile metal plates and reported in references 7.11 and 7.12. It is evident from Figure 7.18 that the theoretical predictions, which are based on the

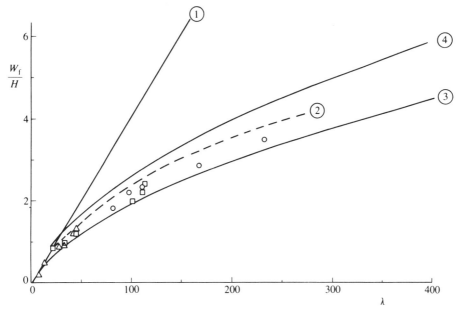

Figure 7.18 Maximum permanent transverse displacements of fully clamped rectangular plates with $\beta = 0.593$ and subjected to a uniformly distributed impulsive velocity. $\triangle$, $\square$, $\bigcirc$: experimental results on aluminium 6061-T6 plates.[7.11] (1): theoretical analysis using equation (7.62) for infinitesimal displacements (i.e., Nw term is neglected). (2): finite-deflection solution using equation (7.62) and the maximum normal stress yield criterion.[7.4] (3): equation (7.95) for a circumscribing yield criterion. (4): equation (7.95) with σ_0 replaced by $0.618\sigma_0$ for an inscribing yield criterion.

maximum normal stress yield criterion,† give reasonable agreement with the corresponding experimental results. The straight line in Figure 7.18 corresponds to the theory which disregards the influence of geometry changes (i.e., the term containing Nw in equation (7.62) is neglected). Clearly, the influence of finite transverse displacements are important for maximum permanent transverse displacements larger than the plate thickness, approximately. The simple infinitesimal displacement theory is, therefore, suitable only for small permanent displacements. For displacements larger than the plate thickness, it overpredicts the maximum transverse displacement and underpredicts the energy which may be absorbed plastically by a plate for a given permanent transverse displacement.

The simpler theoretical prediction in § 7.5.5 was obtained with the aid of the square yield condition in Figure 7.17, which circumscribes the maximum normal stress yield condition. Equation (7.95) predicts smaller permanent displacements in Figure 7.18 for a given dimensionless kinetic energy consistent with a circumscribing yield surface which gives a stronger plate. A square yield condition with all dimensions multiplied by 0.618 (i.e., the plastic flow stress is $0.618\sigma_0$) inscribes[7.10] the maximum normal stress yield condition and it is evident from Figure 7.18 that the corresponding theoretical predictions are larger. Thus, the simple theoretical method in § 7.5.5 provides upper and lower bounds on the maximum permanent transverse displacements which were recorded on rectangular metal plates and predicted by a more exact theory.

7.6 Dynamic plastic behaviour of beams

7.6.1 Introduction

The two approximate theoretical procedures, which are outlined in § 7.5, may be used to examine the dynamic behaviour of beams and initially flat plates having any shape. However, the theoretical predictions for beams may be obtained as a special case from the theoretical predictions for rectangular plates, as already demonstrated for the static behaviour in § 7.4.6. The theoretical predictions for the dynamic behaviour of uniformly loaded beams are obtained in a similar way from the results in § 7.5 for the square yield condition. Theoretical predictions for the maximum normal stress yield condition are presented in reference 7.4.

7.6.2 Dynamic response of a simply supported beam subjected to a uniformly distributed pressure pulse

The theoretical analysis for a rectangular plate predicts the behaviour of a beam‡ with a span $2B$ when $\beta \to 0$, $\tan \phi = \sqrt{3}$ and $\xi_0 \to 0$ as discussed in § 7.4.6. Thus,

---

† The theoretical predictions for the fully clamped case are presented in reference 7.4 and are obtained in a manner which is similar to that in § 7.5.3 for a rectangular plate with simple supports.

‡ See also Figure 2.13.

equation (7.86a) for the response duration T reduces to

$$\tan \gamma T = \eta \sin \gamma \tau / (1 - \eta + \eta \cos \gamma \tau), \qquad (7.96a)$$

where

$$\gamma = (12M_0/\mu H B^2)^{1/2}, \qquad (7.96b)\dagger$$

while equation (7.86b) predicts the maximum permanent transverse displacement

$$W_{\mathrm{f}}/H = [\{1 + 2\eta(\eta - 1)(1 - \cos \gamma \tau)\}^{1/2} - 1]/4. \qquad (7.97)$$

Now, $\lambda \xi_0^2 = (\mu V_0^2 L^2 / M_0 H)(\beta \tan \phi)^2 = 3\mu V_0^2 B^2 / M_0 H$ for a beam with a span $2B$. Thus, $\lambda \xi_0^2 = 3\lambda$ when the beam span $2B$ is relabelled $2L$, where λ is defined by equation (7.94b). Equation (7.94a) for a uniformly distributed impulsive velocity may, therefore, be written in the form

$$W_{\mathrm{f}}/H = \{(1 + 3\lambda)^{1/2} - 1\}/4. \qquad (7.98)\dagger$$

7.6.3 Dynamic response of a fully clamped beam subjected to a uniformly distributed pressure pulse

The general procedure in § 7.6.2 may be used to obtain the dynamic response of fully clamped beams from the theoretical predictions for rectangular plates in § 7.5.5. Thus, equations (7.90) and (7.91a) predict

$$\tan \gamma T = \eta \sin \gamma \tau / (1 - \eta + \eta \cos \gamma \tau) \qquad (7.99)\ddagger$$

and

$$W_{\mathrm{f}}/H = [\{1 + 2\eta(\eta - 1)(1 - \cos \gamma \tau)\}^{1/2} - 1]/2 \qquad (7.100)$$

respectively. Equation (7.95) for impulsive loading reduces to

$$W_{\mathrm{f}}/H = \{(1 + 3\lambda/4)^{1/2} - 1\}/2. \qquad (7.101)$$

7.6.4 Comparison with experimental results

The experimental results, which were reported in references 7.11 and 7.13 for impulsively loaded fully clamped and axially restrained beams, are compared in Figure 7.19 with various theoretical predictions. It is evident that the discussion in § 7.5.7 for rectangular plates also applies to the results in Figure 7.19 for beams. In particular, the theoretical predictions using circumscribing and inscribing square yield criteria do bound the more exact theoretical solution and experimental results.

---

† Note that $m = \mu B$, where m is the mass per unit length of a beam and M_0 is the plastic bending moment per unit breadth of a beam.

‡ It is noted from the comments in § 7.5.5(b) that η is the ratio of the magnitude of the dynamic pressure pulse (p_0) and the static plastic collapse pressure for a fully clamped beam.

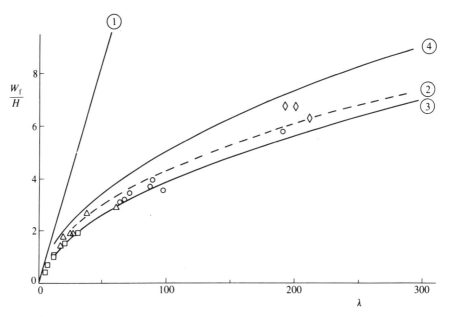

Figure 7.19 Maximum permanent transverse displacements of fully clamped beams subjected to a uniformly distributed impulsive velocity. △, □, ○, ◇: experimental results on aluminium 6061-T6 beams.[7.13] ①: theoretical analysis according to equation (3.82) for infinitesimal displacements. ②: theoretical analysis with finite-displacements according to equation (7.62) for the maximum normal stress yield criterion.[7.4] ③: equation (7.101) for a circumscribing yield criterion. ④: equation (7.101) with σ_0 replaced by $0.618\sigma_0$ for an inscribing yield criterion.

7.7 Dynamic plastic behaviour of circular plates

7.7.1 Introduction

It was remarked in § 7.5.1 that the theoretical method which was developed in reference 7.4 could be used to study the dynamic response of arbitrarily shaped rigid, perfectly plastic plates. The form of this theoretical procedure in § 7.5.2 is suitable for plates which deform into rigid regions separated by stationary straight line plastic hinges. This version was used to examine the dynamic plastic response of rectangular plates and beams in § 7.5 and § 7.6, respectively, and found to give good agreement with the corresponding experimental results in Figures 7.18 and 7.19.

However, the zones between the plastic hinge lines in some plates may deform plastically rather than remaining rigid. For example, a uniformly distributed dynamic pressure pulse produces a conical shaped transverse velocity profile with plastic deformations throughout a simply supported circular plate, as shown in Figure 4.8 and § 4.4. Thus, the more general formulation in reference 7.4 is required for such problems, or

$$\int_A (p - \mu\ddot{w})\dot{w} \, dA = \sum_{m=1}^{r} \int_{C_m} (M_{ij} + N_{ij}w)\dot{w},_{i}n_j \, dC_m + \int_A (M_{ij} + N_{ij}w)\dot{\kappa}_{ij} \, dA.$$

$$(7.102)\dagger$$

The integral on the left-hand side is the work rate due to the external dynamic pressure and inertia force, while the first integral on the right-hand side accounts for the energy dissipated at any plastic hinges. The last integral on the right-hand side is the energy dissipated in any continuous deformation fields.

The behaviour of circular plates which undergo an axisymmetric response is examined in the remainder of this chapter. In this case, equation (7.102) may be transformed from rectangular coordinates to polar coordinates to give

$$\int_A (p - \mu\ddot{w})\dot{w} \, dA = \sum_{m=1}^{r} \int_{C_m} (M_r + N_r w)(\partial\dot{w}/\partial r) \, dC_m$$

$$+ \int_A \{(M_r + N_r w)\dot{\kappa}_r + (M_\theta + N_\theta w)\dot{\kappa}_\theta\} \, dA$$

$$(7.103)\ddagger$$

when restricted to axisymmetric behaviour and where $\partial\dot{w}/\partial r$ is the angular velocity across an axisymmetric hinge, N_r and N_θ are the radial and circumferential membrane forces, respectively, and the remaining quantities are defined in § 2.5 and Figure 2.5.

It is difficult to solve many problems with the aid of equation (7.103), despite the simplifications which have been incorporated. To simplify the presentation in this chapter, it is assumed that the plastic flow is controlled by the simple square yield condition in Figure 7.20.

7.7.2 Dynamic response of a simply supported circular plate subjected to a pressure pulse

7.7.2(a) Introduction

Now, consider the response of the simply supported circular plate in Figure 4.8 when subjected to the dynamic pressure pulse with the rectangular pressure-time history in Figure 7.16, or

$$p = p_0, \quad 0 \leqslant t \leqslant \tau, \tag{7.104a}$$

† See equation (8) in reference 7.4. The summation convention is used on the right-hand side of equation (7.102) and i, j have the values 1, 2 which correspond to the coordinates x_1 and x_2 which lie in the mid-plane of a plate.

‡ It is straightforward to transform equation (7.102) into the form of equation (7.103) but rather tedious. This operation is not necessary for an understanding of the theoretical work in this chapter. Equation (7.103) could be obtained by analogy from equation (7.62) and recognising that it is a statement of energy conservation. Equation (7.103) reduces to equation (7.62) when all the plastic hinges are straight and there are no continuous deformation fields.

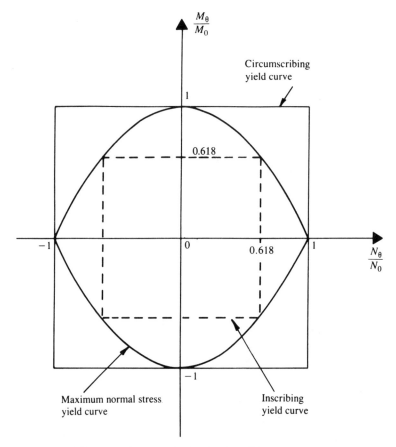

Figure 7.20 Yield curves for a perfectly plastic solid rectangular cross-section.

and
$$p = 0, \qquad \tau \leqslant t \leqslant T. \tag{7.104b}$$

The infinitesimal displacement behaviour of this particular plate was examined in §§ 4.4 and 4.5, while the finite displacement response is studied in this section using equation (7.103).

It is assumed that the transverse velocity of the plate is the same as the static collapse profile in Figure 2.7(b), or

$$\dot{w} = \dot{W}(1 - r/R), \qquad 0 \leqslant r \leqslant R, \tag{7.105}$$

which gives $\dot{\kappa}_r = 0$ and $\dot{\kappa}_\theta = \dot{W}/rR$ according to equations (4.3) and (4.4). The transverse displacement profile has the same form as equation (7.105).

7.7.2(b) First phase of motion, $0 \leqslant t \leqslant \tau$

Equation (7.103), together with equations (7.104a) and (7.105), gives

$$\int_0^R \{p_0 - \mu \ddot{W}(1 - r/R)\} \dot{W}(1 - r/R)2\pi r \, dr =$$

$$\int_0^R \{M_\theta + N_\theta W(1 - r/R)\}(\dot{W}/rR)2\pi r \, dr, \quad (7.106)$$

which, using the square yield condition in Figure 7.20 with $M_\theta = M_0$ and $N_\theta = N_0$, becomes

$$\ddot{W} + a^2 W = b, \quad (7.107)\dagger$$

where

$$a^2 = 24M_0/\mu R^2 H, \quad (7.108a)$$

$$b = 12(\eta - 1)M_0/\mu R^2, \quad (7.108b)$$

$$\eta = p_0/p_c \quad (7.108c)$$

and

$$p_c = 6M_0/R^2 \quad (7.108d)$$

is the static collapse pressure according to equation (4.33).

It is straightforward to show that equation (7.107) has the solution

$$W = A \cos at + B \sin at + b/a^2, \quad (7.109)$$

or

$$W = b(1 - \cos at)/a^2 \quad (7.110)\ddagger$$

when satisfying the initial conditions $W = \dot{W} = 0$ at $t = 0$.

7.7.2(c) Second phase of motion, $\tau \leqslant t \leqslant T$

Equation (7.106) still governs the behaviour, except now $p_0 = 0$ according to equation (7.104b). Thus, equation (7.107) is again obtained except b is replaced by

$$c = -12M_0/\mu R^2 \quad (7.111)$$

and therefore

$$W = C \cos at + D \sin at + c/a^2. \quad (7.112)$$

The displacement and velocity from equations (7.110) and (7.112) must be continuous at $t = \tau$, which leads to

$$W/H = \{(1 - \eta + \eta \cos a\tau) \cos at + \eta \sin a\tau \sin at - 1\}/2. \quad (7.113)$$

Motion ceases at $t = T$ when $\dot{W} = 0$, where

$$\tan aT = \eta \sin a\tau/(1 - \eta + \eta \cos a\tau) \quad (7.114)$$

---

† If finite-displacement effects are disregarded (i.e., $a^2 = 0$), then equation (7.107) reduces to equation (4.39) for infinitesimal displacements.

‡ This equation is valid for $at \leqslant \pi$ because $\dot{W} = 0$ when $at = \pi$.

and equation (7.113) gives a maximum permanent transverse displacement

$$W_f/H = [\{1 + 2\eta(\eta - 1)(1 - \cos a\tau)\}^{1/2} - 1]/2. \qquad (7.115)$$

7.7.2(d) Impulsive loading

It was noted in § 7.5.6 that equation (7.92) is satisfied for impulsive loading with $\eta \gg 1$ and $\tau \to 0$. Now, $1 - \cos a\tau \cong (a\tau)^2/2$ when $\tau \to 0$ so that equation (7.115) predicts

$$W_f/H = \{(1 + a^2\eta^2\tau^2)^{1/2} - 1\}/2, \qquad (7.116)$$

which, using equations (7.92) and (7.108a, c, d), can be written in the form

$$W_f/H = \{(1 + 2\lambda/3)^{1/2} - 1\}/2, \qquad (7.117)\dagger$$

where

$$\lambda = \mu V_0^2 R^2/M_0 H. \qquad (7.118)$$

7.7.2(e) Comparison with experimental results

Florence[7.15] has conducted experimental tests on simply supported aluminium 6061-T6 circular plates subjected to uniformly distributed impulsive velocities which produce inelastic behaviour and permanent transverse displacements. These experimental results are compared with the theoretical predictions of equation (7.117) in Figure 7.21. This approximate theoretical analysis was simplified with the aid of the square yield condition in Figure 7.20 which circumscribes the exact yield condition. Another yield condition, which is 0.618 times as large, would inscribe the exact one and, in this case, equation (7.117) becomes

$$W_f/H = [\{1 + 2\lambda/(3 \times 0.618)\}^{1/2} - 1]/2, \qquad (7.119)$$

which is also shown in Figure 7.21.

It is evident that equations (7.117) and (7.119) provide reasonable estimates of the experimental results of Florence.[7.15] However, the more complete theoretical analysis which is developed in reference 7.16 provides better bounds of the experimental values, although equations (7.117) and (7.119) would be adequate for preliminary design purposes. The theoretical predictions of equation (4.93a) for infinitesimal displacements are also presented in Figure 7.21, but overestimate significantly the maximum permanent transverse displacements for large dimensionless impulses.

7.8 Dynamic plastic behaviour of a circular membrane

The result in Figure 7.21 show that the influence of finite transverse displacements, or geometry changes, dominates the response for large permanent deflections. In other words, membrane forces play a major role in the response, and the

† This theoretical result was first published by Guedes Soares[7.14] who used the theoretical procedure developed in reference 7.4 and reported in § 7.5.2.

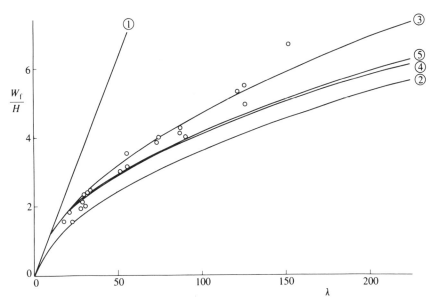

Figure 7.21 Maximum permanent transverse displacements of simply supported circular plates subjected to a uniformly distributed impulsive velocity.
○: experimental results on aluminium 6061-T6 circular plates.[7.15] ①:
theoretical analysis according to equation (4.93a) for infinitesimal displacements. ②: theoretical analysis with finite-displacements according to equation
(7.117) for a circumscribing square yield criterion. ③: theoretical analysis with
finite-displacements according to equation (7.119) for an inscribing square yield
criterion. ④: equation (7.129), membrane solution. ⑤: equation (7.130),
membrane solution according to Symonds and Wierzbicki.[7.18]

importance of bending moments is restricted to circular plates with small
permanent displacements. This phenomenon is also apparent in Figures 7.18 and
7.19 for impulsively loaded rectangular plates and beams, respectively. In fact, it
was shown in § 7.5.3(c) that bending moments do not contribute to the energy
dissipation in any zones of simply supported rectangular plates with transverse
displacements larger than one-half of the plate thickness.

These observations suggest, therefore, that, for sufficiently large transverse
displacements of a circular plate, the influence of the bending moments may be
neglected entirely and the response controlled by membrane forces alone. In this
circumstance, equation (7.103) becomes

$$\int_A (p - \mu \ddot{w}) \dot{w} \, dA = \sum_{m=1}^{r} \int_{C_m} N_r w (\partial \dot{w}/\partial r) \, dC_m + \int_A (N_r w \dot{\kappa}_r + N_\theta w \dot{\kappa}_\theta) \, dA$$

$$(7.120)$$

for the axisymmetric behaviour of a rigid, perfectly plastic circular membrane. If a membrane is supported† around the boundary and $N_r = N_\theta = N_0$,‡ then

$$\int_0^R (p - \mu\ddot{w})\dot{w}2\pi r \, dr = -N_0 \int_0^R w\{\partial^2\dot{w}/\partial r^2 + (1/r)\partial\dot{w}/\partial r\}2\pi r \, dr \quad (7.121)$$

when using equations (4.3) and (4.4).

Now, if equation (7.105) is selected for the transverse velocity profile of a circular membrane, together with a similar form for the transverse displacement profile, then equation (7.121) may be recast in the form

$$\ddot{W} + a^2 W = d, \qquad 0 \leqslant t \leqslant \tau, \qquad (7.122a)$$

when using equation (7.104a) and where

$$d = 12\eta M_0/\mu R^2 \qquad (7.122b)$$

and a^2 and η are defined by equations (7.108a) and (7.108c), respectively. Equation (7.122a) has the solution

$$W = \eta H(1 - \cos at)/2, \qquad 0 \leqslant t \leqslant \tau, \qquad (7.123)*$$

when satisfying the initial conditions $w = \dot{w} = 0$ at $t = 0$.

Equation (7.122a) with $d = 0$ governs the response during the second phase of motion since $p = 0$. Thus, it may be shown that

$$W/H = \eta\{(\cos a\tau - 1)\cos at + \sin a\tau \sin at\}/2 \qquad (7.124)$$

when ensuring continuity of the transverse displacement and velocity between the two phases of motion at $t = \tau$.

Motion ceases finally when $\dot{W} = 0$ which occurs at a time T given by

$$\tan aT = \sin a\tau/(\cos a\tau - 1). \qquad (7.125)$$

The permanent transverse displacement according to equations (7.124) and (7.125) is

$$W_f/H = \eta\{(1 - \cos a\tau)/2\}^{1/2}. \qquad (7.126)$$

Equation (7.126) reduces to the form

$$W_f/H = (\lambda/6)^{1/2} \qquad (7.127)§$$

when using the procedure in § 7.7.2(d) for impulsive loading and where λ is defined by equation (7.118), or

$$\lambda = 4\mu V_0^2 R^2/N_0 H^2. \qquad (7.128)$$

† The first term on the right-hand side of equation (7.120) is zero for both simply supported and clamped membranes with $w = 0$ around the boundary.

‡ The in-plane (radial) equilibrium equation for the axisymmetric response of a circular membrane is $N_\theta - \partial(rN_r)/\partial r = 0$ when the radial inertia is neglected.[7.17] However, $\dot{\epsilon}_r \geqslant 0$, which requires $N_r = N_0$ according to the normality requirement for the Tresca yield condition. Thus, the in-plane equilibrium equation is satisfied when $N_\theta = N_0$.

* See footnote ‡ on page 317.

§ Equation (7.117) for a simply supported circular plate reduces to equation (7.127) for large impulsive loadings with $\lambda \gg 1$.

Equations (7.127) and (7.128) give

$$W_f/H = \{2\rho V_0^2 R^2/3\sigma_0 H^2\}^{1/2},$$ (7.129)

where ρ is the density of the material.

Symonds and Wierzbicki[7.18] have examined impulsively loaded circular plates and obtained a more exact theoretical membrane mode solution, which may be written in the form

$$W_f/H = 1.0186\{2\rho V_0^2 R^2/3\sigma_0 H^2\}^{1/2}.$$ (7.130)†

Clearly, equation (7.130) is only 1.86 per cent larger than the simple estimate given by equation (7.129). However, the theoretical predictions of equations (7.129) and (7.130) lie below most of the experimental results of Florence[7.15] in Figure 7.21. Nevertheless, Symonds and Wierzbicki[7.18] found good agreement with experimental results on fully clamped steel and titanium circular plates when the influence of material strain rate sensitivity‡ was considered.

7.9 Final remarks

7.9.1 General comments

Theoretical investigations into the influence of finite displacements, or geometry changes, on the static and dynamic behaviour of rigid, perfectly plastic structures have been hampered by the lack of simple theorems. Thus, the response of various structural members has been explored on an *ad hoc* basis, though there are several exact§ theoretical solutions for some special cases of simple structures loaded statically[7.1, 7.19] and dynamically.[7.20] However, it is shown in §§ 7.4 to 7.8 that approximate theoretical procedures, which use a kinematic method, may give reasonable agreement with experimental results on beams and plates which are subjected to static or dynamic loads. This theoretical method has also been used to explore some practical problems such as the bow[7.21] and slamming[7.22] damage of plating on ships and marine vehicles. Again, reasonable agreement was found between the theoretical predictions and the corresponding experimental results.

Kaliszky[7.23–7.25] has also developed an approximate theoretical procedure in order to examine the influence of material elasticity, strain hardening and strain rate effects, as well as geometry changes, on the behaviour of dynamically loaded structures and continua which are assumed to respond with a time-independent velocity profile.

The influence of in-plane displacements has not been retained in the basic equations which are used in this chapter. However, it was shown in reference 7.26

---

† See equations (12) and (13a) in reference 7.18.

‡ The phenomenon of material strain rate sensitivity is examined in Chapter 8.

§ If a theoretical solution is kinematically admissible and statically admissible for an exact or approximate yield curve, then it is considered to be exact.

that very small in-plane displacements at the supports of beams, which suffer finite transverse displacements, may have a profound influence on the static transverse load-carrying capacity. For example, a beam of length $2L$ with any depth H and a maximum transverse displacement W of $2L/100$ requires an in-plane displacement at each support of the magnitude $W/100$ in order to achieve a freely supported boundary condition. On the other hand, when $H = 2L/100$, the load-carrying capacity would be four times larger with zero in-plane displacements at the supports of a rotationally free beam. It is, therefore, imperative for designers to know precisely the characteristics of the axial boundary restraints. However, it is shown in reference (7.26) that this phenomenon is less important for large transverse displacements since all beams with some axial restraint eventually reach the membrane, or string, state. A similar sensitivity to in-plane displacements is likely to occur in the dynamic case, although this phenomenon has not been studied for dynamic loadings.

The simply supported annular plate, which was subjected to a dynamic transverse pressure loading and examined in reference 7.20, is assumed to have a zero radial strain. This requires the consideration of in-plane radial displacements when the influence of finite-deflections is retained in the basic equations.†

The influence of finite-displacements, or the phenomenon of geometry changes, is also important for the static and dynamic plastic behaviour of certain shells. For example, an axially restrained cylindrical shell, which is made from a rigid, perfectly plastic material, has been examined in references 7.27 and 7.28 when subjected to static and dynamic radial pressures, respectively. A significant strengthening influence of finite displacements is observed for both problems. Thus, the approximate theoretical procedure, which is discussed in §§ 7.4 to 7.8 for arbitrarily shaped plates, has been extended in reference 7.29 in order to examine the influence of finite-displacements on the dynamic plastic response of rigid, perfectly plastic arbitrarily shaped shells. Reid[7.30] has used different methods of analysis to examine the static and dynamic crushing of cylindrical shells.

7.9.2 Pseudo-shakedown

It was observed in reference 7.22 that a phenomenon known as pseudo-shakedown could occur in a rigid, perfectly plastic rectangular plate which strengthens with finite displacements when subjected to repeated dynamic loads having a triangular-shaped pressure–time history. However, this phenomenon may develop for any structure which suffers from changes in geometry when subjected to repeated dynamic loads having any load–time history which produces plastic deformations.

---

† The radial strain in a circular plate with finite displacements is $\epsilon_r = u' + w'^2/2$, where $(\ )' = \partial(\ )/\partial r$ and u is the radial displacement. Thus, $\epsilon_r = 0$ gives $u' = -w'^2/2$.

For example, if a simply supported rigid, perfectly plastic beam is subjected to a rectangular pressure pulse with a magnitude p_m and a short duration τ, then the dimensionless maximum permanent transverse displacement according to equation (7.97) is

$$W_1/H = [\{1 + 2\eta_m(\eta_m - 1)(1 - \cos \gamma\tau)\}^{1/2} - 1]/4, \qquad (7.131a)$$

where

$$\eta_m = p_m/p_c, \qquad (7.131b)$$

$$\gamma = (12M_0/\mu HL^2)^{1/2} \qquad (7.131c)\dagger$$

and

$$p_c = 2M_0/L^2 \qquad (7.131d)$$

are obtained respectively from equation (7.67c) with p_0 replaced by p_m, equation (7.96b) and equation (7.53c) with $\beta = 0$ for a beam with a span $2L$. The dimensionless transverse displacement of a simply supported beam, which is subjected to a uniformly distributed static pressure, is

$$W_s/H = (\eta_m - 1)/4, \qquad (7.132)$$

according to equation (7.83a) with $\ddot{W}_1 = 0$ and $\beta = 0$.

It is evident, from the results in Figure 7.22(a), that the maximum permanent transverse displacement from equation (7.131a) for a simply supported beam under the dynamic pressure p_m with small values of $\gamma\tau$, is smaller than the corresponding static value from equation (7.132). Now, under the action of the above dynamic load, a beam is deformed permanently with the dimensionless maximum permanent transverse displacement W_1/H given by equation (7.131a). Thus, the static pressure p_1, which produces a permanent transverse displacement W_1 of a beam, is obtained from

$$W_1/H = (\eta_1 - 1)/4, \qquad (7.133a)$$

where

$$\eta_1 = p_1/p_c \qquad (7.133b)$$

by analogy with equations (7.132) and (7.131b), respectively. It is evident when $\gamma\tau$ is sufficiently small that $W_1/H < W_s/H$ according to equations (7.131a) and (7.132), or Figure 7.22(a). Thus, $1 \leqslant \eta_1 \leqslant \eta_m$, or $p_c \leqslant p_1 \leqslant p_m$ and the static plastic collapse pressure for the permanently deformed beam after a single dynamic pressure pulse is now p_1 instead of p_c.

If the same dynamic pressure pulse (with a peak value p_m and a duration τ) were repeated, then, the rigid, perfectly plastic beam might reach a final state with a maximum permanent transverse deflection W_2, which is still smaller than W_s, as shown in Figure 7.22(b). It is evident that this process continues with incremental

---

† See footnote on equation (7.96b).

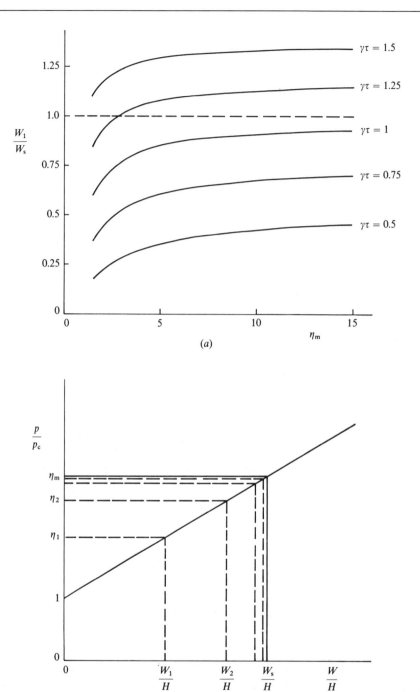

Figure 7.22(a) Ratio of equations (7.131a) and (7.132) for several values of $\gamma\tau$. (b) Pseudo-shakedown for a simply supported beam subjected to a dynamic pressure pulse having a dimensionless peak value η_m. ——: $W/H = (\eta - 1)/4$ for static pressure loading.

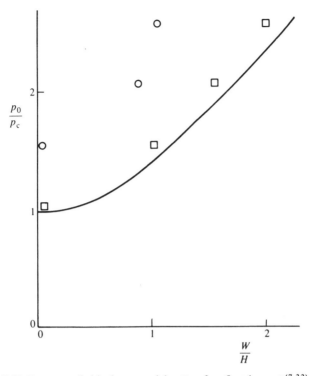

Figure 7.23 Damage of ship bow models. ◯: after first impact.[7.32] ▢: after repeated impact[7.32] (saturated, or pseudo-shakedown condition). ——: equations (7.57) and (7.59) with $\beta = 0.705$.

contributions to the permanent set, which tends towards W_s after an infinite number of repetitions, as shown in Figure 7.22(b), provided a fatigue failure does not occur. A rigid, perfectly plastic beam would then have reached a pseudo-shakedown state.†

It is evident for a given dimensionless pressure ratio η_m in Figure 7.22(a) that there exists a dimensionless pulse duration $\gamma\tau^*$ such that the permanent deflections according to equation (7.131a) for $\gamma\tau > \gamma\tau^*$ exceed the transverse deflection which would be predicted by equation (7.132) for the same pressure p_m supported statically. In this case, a rigid, perfectly plastic beam remains rigid for all further repetitions of the same dynamic pressure pulse, and pseudo-shakedown does not occur.

The phenomenon of pseudo-shakedown has been observed for the repeated wave impact of ship bows,[7.21] and Figure 7.23 shows the results of some tests conducted on steel panels by Yuhara.[7.32] The open circles give the dimensionless permanent transverse displacement after the first impact, while the open squares

† It should be noted that the classical shakedown theorem describes a different phenomenon and is discussed in reference 7.31.

represent the dimensionless permanent transverse displacement after a sufficient number of identical repeated impacts with the same impact pressure for which no further growth was observed. The static rigid, perfectly plastic theoretical predictions (i.e., W_s) in Figure 7.23 are given by equations (7.57) and (7.59) with $\beta = 0.705$ for the fully clamped rectangular plate specimens of Yuhara.[7.32]

It is evident from the foregoing discussion that two kinds of response are possible for plastic structures which suffer changes in geometry when subjected to repeated identical loads having any pressure–time history. For a given structure, pulse duration times τ^* may be associated with the peak values of a dynamic pressure pulse, which marks the transition between the following two cases:

(a) $\tau < \tau^*$. In this case, the transverse displacements acquired during the first loading–unloading cycle are smaller than those predicted by an analysis for the same maximum pressure applied statically. The transverse displacements increase during subsequent loading–unloading cycles and tend towards the associated permanent transverse displacements predicted by a static analysis. This phenomenon is defined as pseudo-shakedown and is illustrated in Figures 7.22(b) and 7.23 for a simply supported beam and a fully clamped rectangular plate, respectively.

(b) $\tau \geqslant \tau^*$. In this case,† the maximum permanent transverse displacement during the first loading cycle may be equal to or larger than that given by a quasi-static analysis, as illustrated in Figure 7.22(a) for a simply supported beam. Thus, the final deformed shape is reached during the initial loading cycle and no further plastic deformations occurs for subsequent repetitions of the same loading–unloading cycle provided reversed plastic yielding‡ does not occur.

7.9.3 Tensile tearing failure

The static and dynamic ductile behaviour of several structures has been examined in this chapter when the influence of finite displacements, or changes in geometry, play an important role. In some practical applications, failure, or damage, is associated with excessive transverse displacements of a structure. In this circumstance, the iso-damage curves in Figure 7.15 for a fully clamped plate loaded statically, for example, give the magnitude of the maximum permissible applied loading, or allow a plate to be designed to avoid exceeding a critical value of the transverse displacement.

The theoretical predictions are developed in this chapter for structures which are made from a perfectly plastic material with unlimited ductility. In practice, a

† See footnotes on equations (7.84a), (7.88a), (7.110) and (7.123) for relatively long duration pressure pulses.
‡ The phenomenon of reversed plastic yielding is discussed in reference 7.31.

structure may tear and fail due to excessive local strains as illustrated for beams in Figure 6.29.[7.33]

The transverse shear failure of structures (Mode 3) has been examined already in § 6.7.3. Another failure mode due to excessive tensile tearing, which was identified as a Mode 2 failure in the last chapter, is now examined in this section.

The theoretical procedure leading to equation (7.101) was reasonably successful in predicting the dynamic plastic response of fully clamped beams which exhibited a Mode 1 behaviour as shown in Figure 7.19. Thus, this particular method is modified in order to predict the onset of a Mode 2 behaviour.

Equation (7.101) was obtained using the theoretical procedure in § 7.5.2 and the kinematically admissible transverse displacement profile illustrated in Figure 7.24. Plastic hinges which involve both axial strain (ϵ) and curvature (κ) form at the locations a, b and c. The plastic hinges at the ends of the rigid members ab and bc are assumed to have a length l and the total hinge length at the centre of the beam is assumed to be $2l$. Thus, the maximum total strain (ϵ_m) at the supports of a fully clamped beam is

$$\epsilon_m = \epsilon + H\kappa/2, \qquad (7.134a)$$

where

$$\epsilon \cong W^2/4Ll \qquad (7.134b)\dagger$$

and

$$\kappa \cong W/Ll \qquad (7.134c)\ddagger$$

when the angle θ in Figure 7.24 remains small and using a procedure similar to that in § 7.2.4(a).

The actual values of l are not required when seeking the transverse displacement-time history in reference 7.4 (and therefore equation (7.101) here) because the normality requirements of plasticity involve the ratios $\dot{\epsilon}/\dot{\kappa}$.§ However, in order to evaluate the maximum strain (ϵ_m) from equation (7.134a) it is clear that the value of l must be known. It was shown in reference 7.4 that the plastic hinges at a, b and c in Figure 7.24 were located near the positions $M = \pm M_0$, $N = 0$ on the yield surface shown in Figure 7.4 when the displacements were very small. As the transverse displacements increased, then the plasticity conditions of the hinges at a, b and c moved around the exact yield curve from the locations $M = \pm M_0$, $N = 0$ to the position $M = 0$, $N = N_0$, as discussed in § 7.2.4(a) and shown in Figure 7.8 for

---

† Equation (7.17) becomes $\epsilon = 2L\{(1 + W^2/L^2)^{1/2} - 1\}/4l$ for a fully clamped beam, where $4l$ is the total length of the three plastic hinges which must accommodate the axial extension. Expansion of this expression with the aid of the binomial series leads to equation (7.134b) which is the axial strain at all three plastic hinges.

‡ $\kappa = \theta/l$ at both supports and $\kappa = 2\theta/2l$ at the mid-span. Thus, equation (7.134c) follows when $\theta \cong W/L$.

§ See Appendix 2 for some comments on the path-dependence of an inelastic material.

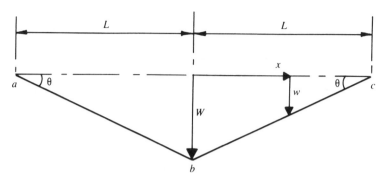

Figure 7.24 Transverse displacement profile for a fully clamped beam subjected to a uniformly distributed impulsive velocity.

a simply supported beam. It turns out for the fully clamped case that the membrane condition ($M = 0$, $N = N_0$) is reached when $W/H = 1$.

The plastic hinges at a and c when $W/H = 0$ and $M = \pm M_0$, $N = 0$, are assumed by Nonaka[7.34] to have a length $l = H$, as indicated in Figure 7.25(a). As W/H increases, then the membrane force (N) increases, as discussed previously, and axial stretching (ϵ) of a beam occurs until the hinges at a and c take the form which was assumed by Nonaka[7.34] and sketched in Figure 7.25(b). Thus, the hinge length l is a function of time. However, since l varies over a relatively small range, the following discussion is simplified by using the average value of the two extremes, i.e.,

$$l = 3H/2 \qquad (7.135)$$

for $0 \leqslant W/H \leqslant 1$, provided $W_f/H \geqslant 1$. Clearly, if $W_f/H < 1$, then a smaller value of l should be used which could be obtained in a similar manner as the average of

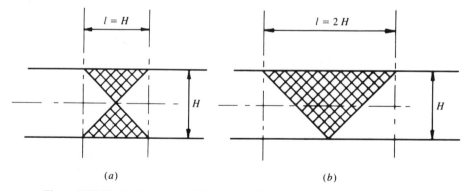

Figure 7.25 Plastic hinges in a fully clamped beam with thickness H according to Nonaka.[7.34] (a) $W/H = 0$. (b) $W/H = 1$ at the initiation of the membrane phase.

$l = H$ and the value of l corresponding to the terminal displacement W_f. It is evident from equations (7.134) and (7.135) that the maximum strain at the beam supports is

$$\epsilon_{m1} = H/2L \qquad (7.136)$$

when $W/H = 1$.

A fully clamped beam commences to behave as a string once the membrane state $N = N_0$ has been reached at $W/H = 1$, as discussed in references 7.4 and 7.35. Moreover, the axial membrane force N in a beam is constant according to the equation of motion† when surface tractions and axial inertia forces are absent. Thus, it appears reasonable to assume for $1 \leqslant W/H \leqslant W_f/H$ that axial stretching occurs over the entire span $2L$. In this circumstance,

$$l = L/2 \qquad (7.137)$$

which, together with equation (7.134b), gives

$$\epsilon_m = W^2/2L^2 - H^2/2L^2 \qquad (7.138)$$

for the strain accumulated between $W/H = 1$ and W/H. Now, $W = W_f$ when the permanent deformed profile is reached, so that equation (7.138) becomes

$$\epsilon_{m2} = \{(W_f/H)^2 - 1\}(H^2/2L^2). \qquad (7.139)$$

The total tensile strain at the supports when $W = W_f$ is, therefore, the sum of equations (7.136) and (7.139), or

$$\epsilon_m = 2\{(W_f/H)^2 + L/H - 1\}(H/2L)^2, \qquad (7.140)$$

where

$$W_f/H = \{(3\lambda)^{1/2}/2 - 1\}/2 \qquad (7.141)‡$$

from equation (7.101) when $3\lambda/4 \gg 1$. Substituting equation (7.141) into equation (7.140) gives

$$\lambda^{1/2} = 2[1 + (2)^{1/2}\{2 + \epsilon_m(2L/H)^2 - 2L/H\}^{1/2}]/3^{1/2}. \qquad (7.142)‡$$

If ϵ_m is the tensile strain at which a material fractures in a uniaxial tensile test, then equation (7.142) predicts the value of λ required for the onset of a Mode 2 response in a beam with a given $2L/H$. Equations (7.94b) and (7.142) can be rearranged to give the initial impulsive velocity

$$V_0 = 2[1 + (2)^{1/2}\{2 + \epsilon_m(2L/H)^2 - 2L/H\}^{1/2}](H/2L)(\sigma_0/\rho)^{1/2}/3^{1/2} \qquad (7.143)$$

and the initial impulse per unit area

$$I = \rho H V_0. \qquad (7.144)$$

Thus, the threshold impulse for the Mode 2 response of a beam with given values of $2L/H$ and material properties does not depend on L but only on H as observed by Menkes and Opat[7.33] during their experimental tests.

---

† See equation (A.65) in Appendix 4.
‡ See footnote to equation (7.96b).

Table 7.1

Comparison of the theoretical threshold impulses for a Mode 2 response from equations (7.143) and (7.144) with the corresponding experimental results of Menkes and Opat.[7.33]

$\dfrac{H}{mm}$	$\dfrac{2L}{mm}$	$\dfrac{I \text{ (equations (7.143) and (7.144))}}{(ktaps)^a}$	$\dfrac{I \text{ (equations (7.143) and (7.144) with } 0.618\sigma_0)}{(ktaps)^a}$	$\dfrac{I \text{ (experiments}^{(7.33)})}{(ktaps)^a}$
4.75	203.2	27.1	21.3	26
4.75	101.6	26.4	20.8	26
6.35	203.2	35.9	28.2	32
6.35	101.6	34.7	27.3	32
9.53	203.2	52.9	41.6	45
9.53	101.6	50.6	39.7	45

^{a} 1 tap = 1 dyne s/cm^2 = 0.1 Pa s, 69.5 ktaps $\cong$ 1 lbf s/in^2.

Equations (7.142) to (7.144) were derived using a square yield curve which circumscribes the exact or maximum normal stress yield curve, as shown in Figure 7.17. As pointed out earlier in § 7.5.4(a), another square yield curve which is 0.618 times as large would inscribe completely the maximum normal stress yield curve. Equations (7.142) to (7.144) remain unchanged for this case except that σ_0 should be replaced by $0.618\sigma_0$. It appears reasonable to assume that the theoretical predictions according to the exact yield curve would lie between the predictions of the circumscribing and inscribing square yield curves, as indicated in Figure 7.19 for Mode 1 responses. Unfortunately, no theorems have been proved to show that this is true for dynamic plasticity, as discussed in § 5.7.3(d).

Equations (7.142) to (7.144) with $\epsilon_m = 0.17$ were used in reference 7.36 to predict the onset of Mode 2 behaviour for the aluminium 6061-T6 beams which were examined by Menkes and Opat.[7.33] It is evident from Table 7.1 that the experimental results of Menkes and Opat for the onset of Mode 2 type failure are bounded by the predictions of the circumscribing and inscribing square yield curves. It should be pointed out, however, that Menkes and Opat admit that the selection of the individual tearing and shearing mode thresholds is highly subjective. Nevertheless, the theoretical predictions of equations (7.142) to (7.144) are simple calculations and permit rapid estimates of the Mode 2 threshold which should be useful for a designer.

Equation (7.143) is plotted in Figure 7.26 for several values of the tensile failure strain ϵ_m. The curves in this figure were obtained for a square yield curve which circumscribes the exact yield curve as shown in Figure 7.17. However, these curves

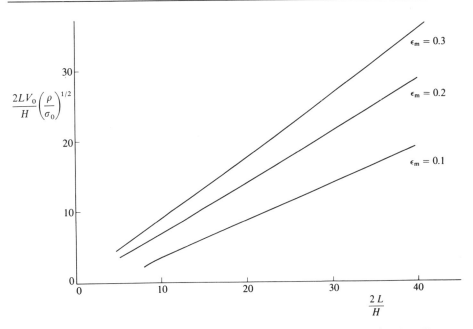

Figure 7.26 Dimensionless threshold velocities according to equation (7.143) for fully clamped beams with various tensile failure strains (ϵ_m).

may also be used for an inscribing yield curve by simply replacing σ_0 in the ordinate by $0.618\sigma_0$. The theoretical method does not lie on a firm foundation despite the fact that the general theoretical procedure predicts reasonable agreement with the experimental results of Menkes and Opat[7.33], as shown in Table 7.1. Further theoretical studies and experimental tests are required in order to gain confidence with the method. A summary of the literature on the tensile tearing of beams subjected to impulsive velocities and impact loads is presented in reference 7.37, while Duffey[7.38] explores the dynamic rupture of shells, Atkins[7.39] examines the interaction of plastic flow and fracture in sheet tearing and the splitting of metal tubes, and Yu et al.[7.40] study the tearing of thin metal sheets.

PROBLEMS

7.1 Prove that equation (7.13) is the axial equilibrium equation for a beam subjected to static transverse loads which produce finite transverse displacements.

7.2 Use the principle of virtual velocities in Appendix 3 to obtain the strain and change of curvature relations which are consistent with the equilibrium equations (7.13) to (7.15). (See also Appendix 4.)

7.3 Show that equation (7.20), for a simply supported beam subjected to a concentrated load, satisfies the normality requirement associated with the yield condition in Figure 7.8, when $0 \leqslant W \leqslant H/2$.

7.4 Show that the total external work done by a static concentrated load on a simply supported beam, which produces finite displacements, is given by equations (7.35a, b). Obtain equations (7.36) and (7.37) for the bending and stretching energy absorbed in a simply supported beam, respectively.

7.5 Repeat the theoretical procedure in § 7.2.4 to obtain the static plastic collapse behaviour of a beam with fully clamped supports and subjected to a concentrated load at the mid-span. (See equations (7.39).)

7.6 Derive equations (7.40) for the finite displacement behaviour of a simply supported circular plate which is subjected to a uniform static pressure within a central circular zone.

7.7 Use physical arguments to show that equation (7.41) is an energy balance.

7.8 Show that equations (7.52) and (7.55) predict the same external pressure when $W/H = \frac{1}{2}$. Use these equations to predict the finite-deflection behaviour of a simply supported beam subjected to a uniform static pressure. (See Figure 2.13.)

7.9 Use equation (7.41) to obtain the static finite-deflection characteristics of a fully clamped rectangular plate subjected to a uniformly distributed pressure. (See equations (7.57) and (7.59).)

7.10 Show that equation (7.78) is the solution of the differential equation (7.77a).

7.11 Show that equation (7.92) is valid for a circular plate.

7.12 (a) Show that equation (7.117), for a simply supported circular plate with large impulsive loadings ($\lambda \gg 1$), reduces to equation (7.127) for an impulsively loaded circular membrane. (b) Repeat § 7.7.2 for the fully clamped case. (c) repeat (a) using the result in (b).

7.13 Prove that the radial strain in an axisymmetric circular plate problem is $\epsilon_r = u' + w'^2/2$, where $(\ )' = \partial(\ )/\partial r$ and u is the radial displacement.

8

Strain rate sensitive behaviour of materials

8.1 Introduction

The theoretical analyses developed in the previous chapters have examined the influence of inertia on the response of various elementary structures subjected to dynamic loads which cause plastic behaviour. The yield criteria, which govern the plastic flow in these structural problems, were assumed to be independent of the rate of strain ($\dot{\epsilon}$). However, the plastic flow of some materials is sensitive to strain rate, which is known as material strain rate sensitivity, or viscoplasticity.[8.1] This phenomenon is illustrated in Figure 8.1 for mild steel specimens which were tested at various uniaxial compressive strain rates.[8.2]

It is evident from Figure 8.1 that the plastic behaviour of mild steel is highly sensitive to strain rate. The strain rates in Figure 8.1 are realistic and are encountered in practical engineering problems. For example, consider a vertical 1 m long mild steel bar ($L = 1$ m) which is struck at one end with a large mass dropped from a height of 5 m and having an axial velocity of 10 m/s on impact.† An average axial strain rate of $\dot{\epsilon} = \epsilon/t = (\delta/L)/t = V/L = 10 \text{ s}^{-1}$, approximately, is generated in the bar when neglecting any stress wave effects. The plastic flow stress corresponding to a strain rate of 10 s^{-1} is approximately double the static flow stress according to the experimental results in Figure 8.1 for strains up to about 0.02.

The theoretical predictions in § 7.6.3 for the permanent transverse displacements of an impulsively loaded fully clamped beam agree reasonably well with the experimental results on aluminium 6061-T6 beams reported in Figure 7.19. However, the same theoretical predictions are larger than the corresponding maximum permanent transverse displacements shown in Figure 8.2, which were

---

† Impact velocity $V = \sqrt{2g \times 5} \cong 10$ m/s. Average axial strain (ϵ) in bar $\cong \delta/L$, where δ is the axial deformation at the struck end. Axial velocity at struck end (V) $\cong \delta/t$, where t is time.

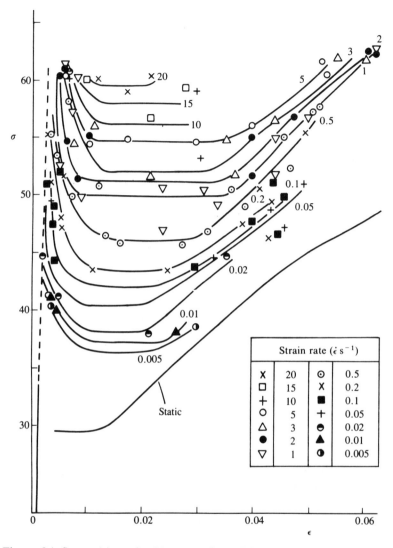

Figure 8.1 Stress (σ)–strain (ϵ) curves for mild steel at various uniaxial compressive strain rates according to Marsh and Campbell.[8.2] 1 unit of ordinate is 10^3 lbf/in^2 or 6.895 MN/m^2.

obtained from a series of tests conducted on mild steel beams.[8.3] It is evident, though, that the dimensionless term $\lambda = \rho V_0^2 L^2 / \sigma_0 H^2$ in the abscissa contains the strain-rate-independent, or static, yield stress (σ_0) in the denominator. Thus, any increase in the plastic flow stress of the test specimens due to strain rate effects, as shown in Figure 8.1 for mild steel, means that the value of λ for the experimental test points should be smaller, thereby giving better agreement with the theoretical predictions.

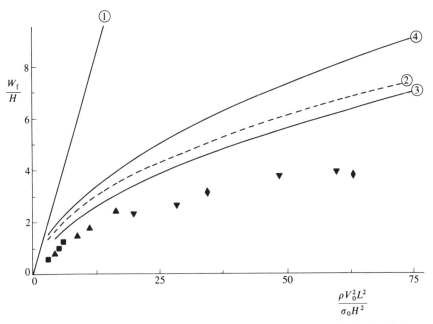

Figure 8.2 Maximum permanent transverse displacements (W_f) of a fully clamped beam subjected to an impulsive velocity of magnitude V_0 distributed uniformly across the entire span $2L$. ①: infinitesimal analysis (bending only) (equation (3.82)). ②, ③, ④: theoretical analyses which include the influence of finite deflections (see § 7.6.4 and Figure 7.19). ■, ▲, ♦, ▼: experimental results on mild steel beams.[8.3]

It is evident from Figure 8.2 that the influence of material strain rate sensitivity manifests itself as a strengthening effect in a structure.† This might suggest that it is a beneficial phenomenon since it provides an additional safety factor.‡ However, Perrone[8.5] remarked that energy absorption systems for enhancing the structural crashworthiness of vehicles, for example, could impart unacceptable forces on the human body which might otherwise have been acceptable for an identical material (and structure) but with strain-rate-independent material properties.

Material strain rate sensitivity is a material effect and is independent of the structural geometry.. It is impossible to review here the vast amount of literature which is available on the strain-rate-sensitive behaviour of materials. However, several excellent surveys on various aspects of the subject have been published.[8.1, 8.6-8.11] This chapter focuses, primarily, on the behaviour of metals which undergo moderate strain rates.

† Sometimes a structural mode change occurs which causes larger and not smaller associated permanent deformations, as discussed by Bodner and Symonds.[8.4]

‡ In fact, the associated fracture strain might decrease with increase in strain rate, as discussed later in § 8.2.3 and shown in Figure 8.6.

An introduction to the strain rate characteristics of some metals subjected to various loads (compression, tension, shear, bending, etc.) is presented in § 8.2, while § 8.3 contains some elementary constitutive equations which are particularly valuable for dynamic structural problems. Several useful approximations emerge from the analyses of two idealised problems examined in § 8.4. The influence of material strain rate sensitivity on the behaviour of several simple structural geometries is discussed in § 8.5.

8.2 Material characteristics

8.2.1 Introduction

It is the purpose of this section to introduce some of the strain rate characteristics of materials which play an important role in the dynamic plastic behaviour of structures. This is achieved by examining the behaviour of materials under various simple dynamic loads. No attempt is made to describe the underlying materials science. Nor is any attention given to the considerable practical difficulties encountered in high strain rate tests (e.g., stress wave effects and inertia of the test rigs and recording equipment[8.9, 8.12]).

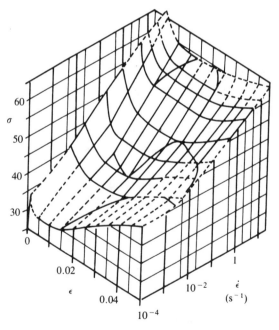

Figure 8.3 Isometric projection of stress (σ), strain (ϵ), strain rate ($\dot{\epsilon}$) surface for mild steel subjected to dynamic uniaxial compression.[8.2] 1 unit of ordinate is 10^3 lbf/in^2 or 6.895 MN/m^2.

8.2.2 Compression

The uniaxial compressive results of Marsh and Campbell[8.2] shown in Figure 8.1 are assembled in Figure 8.3 together with additional test values on mild steel. This figure shows clearly a significant increase in the upper yield stress with increase of strain rate. The stress at a strain of 0.05 also increases, but less markedly. Material strain hardening, therefore, decreases with increase of strain rate for this type of mild steel with strains up to at least 0.05.

The compressive strain rate behaviour of titanium and aluminium 6061-T6 according to Maiden and Green[8.13] is shown in Figure 8.4. The test results for the titanium in Figure 8.4(a) do not indicate the presence of an upper yield stress and are less sensitive to strain rate than the mild steel in Figures 8.1 and 8.3.

It is evident that the behaviour of the aluminium 6061-T6 in Figure 8.4(b) is essentially strain-rate-insensitive, which possibly accounts for the good agreement between rigid, perfectly plastic theoretical methods and the corresponding experimental results for beams and other structures (e.g., Figure 7.19). Nevertheless, the experimental compressive test results on a work-hardened aluminium by Hauser[8.14] presented in Figure 8.5 do show a dependence on strain rate, which, however, is noticeably less for a given change of strain rate than the mild steel in Figures 8.1 and 8.3.

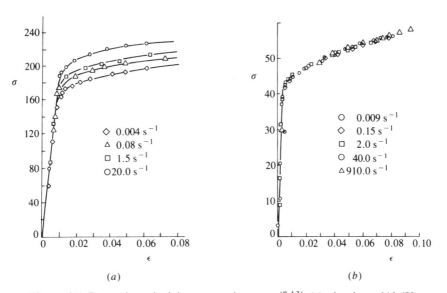

Figure 8.4 Dynamic uniaxial compression tests.[8.13] (a) titanium 6Al-4V. (b) aluminium 6061-T6, 1 unit of ordinate is 10^3 lbf/in^2 or 6.895 MN/m^2.

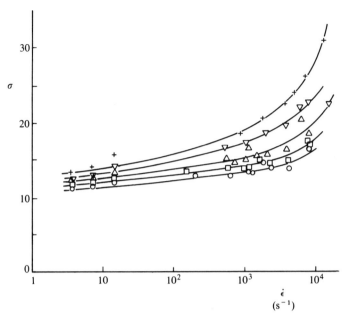

Figure 8.5 Dynamic uniaxial compressive tests on work-hardened alumi-nium[8.14] at various constant strains ($\circ$: $\epsilon = 0.01$; $\square$: $\epsilon = 0.02$; $\triangle$: $\epsilon = 0.04$; $\triangledown$: $\epsilon = 0.08$; $+$: $\epsilon = 0.16$). 1 unit of ordinate is 10^3 lbf/in^2 or 6.895 MN/m^2.

8.2.3 Tension

Manjoine[8.15] reported in 1944 on some tensile tests which he conducted on a low-carbon steel using a high-speed tension machine. His experimental results indicated that the lower yield stress and the ultimate tensile stress increased with increase in strain rate, the increase being more significant for the lower yield stress.

Campbell and Cooper[8.16] have examined the dynamic tensile behaviour of low-carbon mild steel specimens up to fracture, as shown in Figure 8.6. A summary of their results is presented in Figure 8.7. The upper and lower yield stresses increase with increase in strain rate, as observed by Manjoine.[8.15] However, the ultimate tensile stress also increases, but more slowly. Thus, the reduction in the importance of material strain hardening with increase in strain rate observed in Figure 8.3 and discussed in § 8.2.2 for dynamic compression is also found in mild steel with large tensile strains and large strain rates. Indeed, it appears, apart from the upper yield stress, that this material behaves as a perfectly plastic material with little or no strain hardening at high strain rates.

It is interesting to note from the results of Campbell and Cooper[8.16] shown in Figure 8.6 that the fracture strain decreases with increase in strain rate. In other words, the material becomes more brittle at higher strain rates.

Many authors have conducted dynamic tensile tests since the early experiments

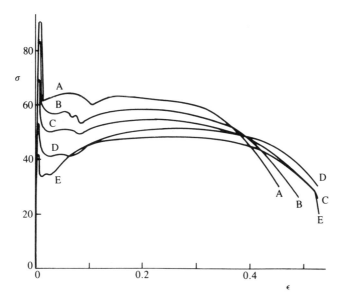

Figure 8.6 Dynamic uniaxial tensile tests on mild steel[8.16] at various mean plastic strain rates. A: $\dot{\epsilon} = 106$ s^{-1}; B: $\dot{\epsilon} = 55$ s^{-1}, C: $\dot{\epsilon} = 2$ s^{-1}; D: $\dot{\epsilon} = 0.22$ s^{-1}; E: $\dot{\epsilon} = 0.001$ s^{-1}. 1 unit of ordinate is 10^3 lbf/in^2 or 6.895 MN/m^2.

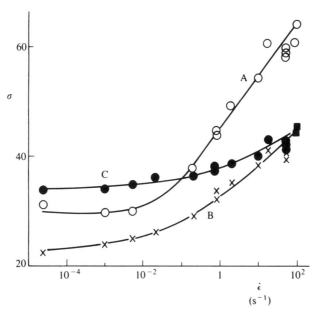

Figure 8.7 Variation of strength with strain rate for the dynamic uniaxial tensile behaviour of mild steel.[8.16] A: upper yield stress; B: lower yield stress; C: ultimate tensile stress. 1 unit of ordinate is 1 kg/mm^2 or 9.807 MN/m^2.

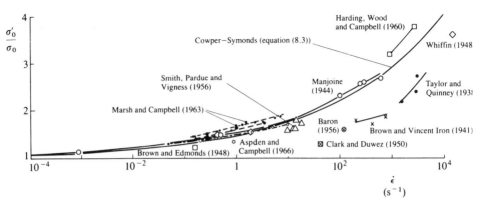

Figure 8.8 Variation of dynamic uniaxial lower yield stress of mild steel with strain-rate.[8.17] All test results were obtained in dynamic tension except the dynamic compression tests of Marsh and Campbell (1963), Aspden and Campbell (1966) and Whiffin (1948). Equation (8.3) is the Cowper–Symonds empirical relation which is introduced in § 8.3.2 ($D = 40.4 \text{ s}^{-1}$ and $q = 5$).

of Manjoine.[8.15] Symonds[8.17] has gathered together the dynamic lower yield or flow stresses for mild steel which have been recorded over a thirty-year period in a number of laboratories. These results are presented in Figure 8.8 and reveal a trend of increasing flow stress with increase of strain rate over a wide range of strain rates. The data have considerable scatter which is, undoubtedly, related to the range of different mild steels having different grain sizes and heat treatments, and the variety of testing machines and data recording equipment.

The dynamic tensile strain-rate-sensitive behaviour of several metals is presented in Figures 8.9 and 8.10, which are taken from Nicholas.[8.18]

8.2.4 Shear

Some dynamic shear tests have been reported using double-notch shear test specimens made from mild steel and α-uranium, as discussed by Harding and Huddart.[8.19] However, most experimental results on the dynamic shear behaviour of materials have been generated using thin-walled tubular specimens subjected to dynamic torsional loads.

Klepaczko[8.20] studied the strain rate behaviour of technically pure iron (with 0.05 per cent C, approximately) and obtained the behaviour shown in Figure 8.11. It is clear that the pure shear behaviour of this material is strongly strain-rate-sensitive.

Nicholas and Campbell[8.21] studied a high-strength aluminium alloy (BS HE 15 WP aluminium alloy which is equivalent to US 2014-T6) and found no significant variation in flow stress for increases of shear strain rate up to 10^3 s^{-1}. However, the

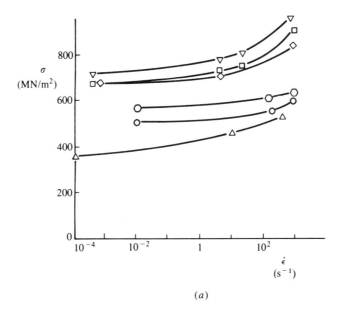

(a)

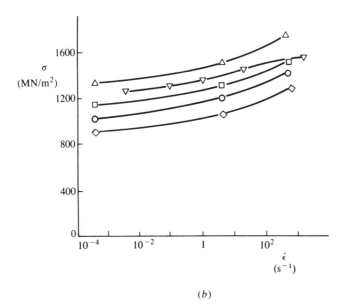

(b)

Figure 8.9 Variation of dynamic uniaxial tensile stress with strain rate.[8.18]
(a) Stainless steels with $\epsilon = 0.10$ (type of stainless steel: ▽ (410), □ (304), ◇ (321),
○ (347), ○ (304)). △: dynamic uniaxial compressive tests on 304 stainless steel
at $\epsilon = 0.04$. (b) Titanium alloys with $\epsilon = 0.04$ (type of titanium alloy: △ (6–6–2),
□ (7–4), ○ (6–4), ◇ (8–1–1)). ▽: dynamic uniaxial compressive tests on
Ti-6Al-4V.

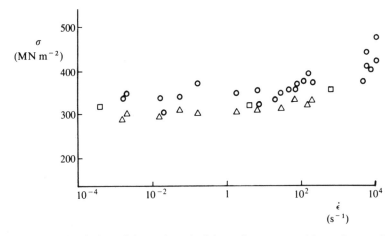

Figure 8.10 Variation of dynamic uniaxial tensile stresses with strain rate for aluminium 6061-T6.[8.18] ($\triangle$: yield stress; $\bigcirc$: ultimate stress; $\square$: $\epsilon = 0.04$).

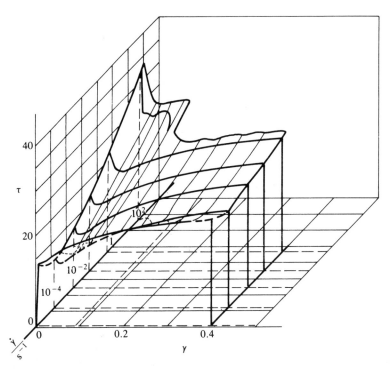

Figure 8.11 Isometric projection of shear stress (τ), shear strain (γ), shear strain rate ($\dot{\gamma}$) surface for iron subjected to dynamic pure shear.[8.20] 1 unit of ordinate is 1 kg/mm^2 or 9.807 MN/m^2.

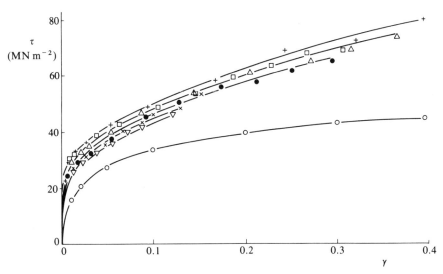

Figure 8.12 Dynamic shear stress (τ)–shear strain (γ) curves for commercially pure aluminium at various strain rates.[8.22] The nominal shear strain rates are given by the following symbols: $+: \dot{\gamma} = 2800 \text{ s}^{-1}$; $\triangle: \dot{\gamma} = 2200 \text{ s}^{-1}$; $\square: \dot{\gamma} = 1600 \text{ s}^{-1}$, $\bullet: \dot{\gamma} = 1450 \text{ s}^{-1}$; $\times: \dot{\gamma} = 800 \text{ s}^{-1}$; $\nabla: \dot{\gamma} = 600 \text{ s}^{-1}$; $\bigcirc: \dot{\gamma} = 0.002 \text{ s}^{-1}$.

experimental results of Tsao and Campbell[8.22] in Figure 8.12 do show some strain rate sensitivity for aluminium.

Duffy[8.23] has found a reasonable measure of agreement between independent dynamic shear and dynamic uniaxial tests on both lead and aluminium 1100-0 specimens. These comparisons were achieved by converting the dynamic shear test results (τ, γ) to the equivalent uniaxial case (σ, ϵ) using $\sigma = \sqrt{3}\tau$ and $\epsilon = \gamma/\sqrt{3}$ according to the von Mises criterion and material incompressibility.†

8.2.5 Biaxial

Gerard and Papirno[8.24] conducted tests on thin circular membranes subjected to lateral or transverse impact loads which produced strain rates of order 1 s^{-1}. The diaphragms were sufficiently thin so that the stress field was biaxial and uniform

---

† The von Mises equivalent stress

$$\sigma_e = [(\sigma_x - \sigma_y)^2 + (\sigma_y - \sigma_z)^2 + (\sigma_z - \sigma_x)^2 + 6(\tau_{xy}^2 + \tau_{yz}^2 + \tau_{zx}^2)]^{1/2}/\sqrt{2}$$

gives $\sigma_e = \sigma_x$ for uniaxial behaviour and $\sigma_e = \sqrt{3}\tau_{xy}$ for pure shear. The equivalent strain $\epsilon_e = \sqrt{2}[(\epsilon_x - \epsilon_y)^2 + (\epsilon_y - \epsilon_z)^2 + (\epsilon_z - \epsilon_x)^2 + 6(\epsilon_{xy}^2 + \epsilon_{yz}^2 + \epsilon_{zx}^2)]^{1/2}/3$ and the incompressibility condition $\epsilon_x + \epsilon_y + \epsilon_z = 0$ give $\epsilon_e = \epsilon_x$ for uniaxial behaviour and $\epsilon_e = 2\epsilon_{xy}/\sqrt{3}$ for pure shear. The engineering shear strain $\gamma_{xy} = 2\epsilon_{xy}$ so that $\epsilon_e = \gamma_{xy}/\sqrt{3}$ for pure shear.

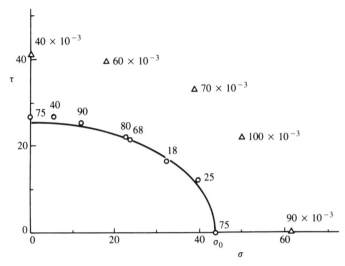

Figure 8.13 Comparison of static ($\circ$) and dynamic ($\triangle$) upper yield stresses for 1018 mild steel subjected to combined shear (τ) and tensile (σ) stresses.[8.25] ——: von Mises yield condition, $(\sigma/\sigma_0)^2 + 3\,(\tau/\sigma_0)^2 = 1$. Numbers represent time to yield in seconds. 1 unit on both axes is 10^3 lbf/in^2 or 6.895 MN/m^2.

throughout the thickness. No enhancement of the stress–strain characteristics of aluminium alloy 1100-0 membranes due to an increase in strain rate was found. However, similar dynamic tests on mild steel membranes gave rise to an enhancement of the associated stress–strain curve: the upper yield stress was increased above the corresponding static value by 80 per cent, approximately.

Gerard and Papirno[8.24] observed that unexpected failures of the mild steel membranes occurred occasionally in the dynamic tests.

Lindholm and Yeakley[8.25] have examined the dynamic behaviour of a 1018 mild steel tubing subjected to combined loads which gave rise to a biaxial stress field. The experimental results in Figure 8.13 compare the upper yield stresses in combined shear and tension for the static and dynamic loading cases. The static results are seen to obey a von Mises yield criterion, while the upper yield stresses associated with dynamic loadings are of the order of 50 per cent larger. No strain rate values were presented by Lindholm and Yeakley.[8.25]

Ng et al.[8.26] have also examined the dynamic behaviour of thin-walled tubular specimens. They found important strain rate effects for tubes made from aluminium 6061-T6. The results in Figure 8.14 contrast the static and dynamic yield stresses for biaxial stress states in a cylindrical tube with various ratios between the axial and circumferential stresses. The dynamic curve was evaluated for a strain rate of 40 s^{-1} and the authors claim that the enhancement is similar to that found by Hoge[8.27] in dynamic uniaxial tests on aluminium 6061-T6.

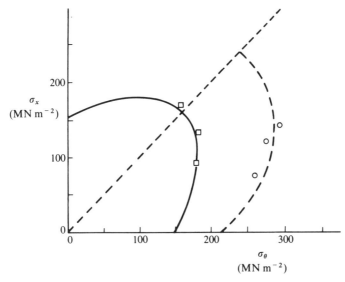

Figure 8.14 Comparison of static and dynamic yield stresses for aluminium 6061-T6 tubes subjected to biaxial tensile stresses.[8.26] ○: dynamic yield point. □: static yield point.

Lewis and Goldsmith[8.28] made some interesting observations on the dynamic biaxial behaviour of bone using a split Hopkinson bar. They found that plastic deformation without fracture can be achieved under combined dynamic torsion and compression, whereas it seldom occurs under either of these individual loadings.

8.2.6 Bending

It was the aim of the experiments reported in §§ 8.2.2 to 8.2.5 to produce a·uniform stress state† in the test specimens to aid interpretation of the results. However, many practical structural members are subjected to dynamic bending moments which do not cause a uniform state of stress across the thickness. The dynamic bending behaviour of a beam with a given cross-section is considered to give a fundamental constitutive equation which relates the bending moment to the corresponding curvature or rotation. In fact, most structural designs use generalised stresses (bending and twisting moments, membrane and shear forces) rather than stresses and generalised strains (curvatures, membrane strains, twisting strains) instead of strains.‡ It is often unnecessary in this circumstance to consider the actual stress and strain fields, and information on the generalised quantities is adequate, as already found for various structural members in the previous chapters.

† This is not achieved for large strains when local necking and fracture occurs.
‡ Generalised stresses and generalised strains are defined in § 2.2.

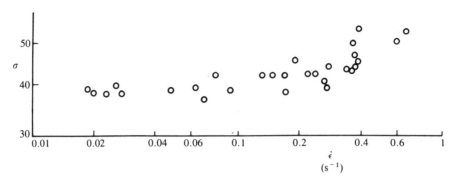

Figure 8.15 Lower yield stresses from dynamic bending tests on mild steel beams.[8.29] 1 unit on ordinate is 10^3 lbf/in^2 or 6.895 MN/m^2.

Rawlings[8.29] examined the dynamic behaviour of mild steel beams with rectangular cross-sections which were subjected to a uniform bending moment. The maximum strain rate produced by the experimental arrangement was only about 1 s^{-1}. Nevertheless, the experimental yield stresses in Figure 8.15 do show a marked increase with increase in strain rate.

Aspden and Campbell[8.30] constructed an apparatus in which beams were bent with maximum strain rates up to 20 s^{-1}, approximately. The static and dynamic bending moment-bending angle curves for various bending angle rates are shown in Figure 8.16(a), while the corresponding static and dynamic compression tests on the same steel are presented in Figure 8.16(b) for comparison. These results are discussed further in § 8.3.3.

More recently, Davies and Magee[8.31] have conducted dynamic bending tests on various steel and aluminium alloy sheet metal beams. The strain rates were not measured directly, but could be estimated from the machine cross-head velocities. A dynamic factor was defined as the flow stress at a machine cross-head speed of 5000 in/min. (5.0 mph, 8 kph, approximately) compared with the flow stress in a static test conducted at 0.1 in/min. Davies and Magee[8.31] recorded dynamic

Figure 8.16 Comparison of dynamic bending moment (M)–rotation (θ) curves and dynamic uniaxial compression σ–ϵ curves for mild steel.[8.30]

(a) Dynamic bending moment–rotation curves for various values of rotation rate ($\dot\theta$). — — —: theoretical predictions of equation (8.22) using $D = 40.4$ s^{-1} and $q = 5$ for ①: $\dot\theta = 4°$/ms ($\dot\epsilon \cong 12$ s^{-1}); ②: $\dot\theta = 0.05°$/ms ($\dot\epsilon \cong 0.15$ s^{-1}); ③: static. Numbers on curves are values of $\dot\theta$ (°/ms). 1 unit of ordinate is 1 lbf in or 0.113 Nm.

(b) Dynamic uniaxial compressive stress–strain curves for various strain rates ($\dot\epsilon$). Strain rate magnitudes ($\dot\epsilon$ s^{-1}) are: △ (20), ● (10), ▽ (5), × (2), ○ (1), + (0.5), ▲ (0.2), □ (0.1), ■ (0.05), ▼ (0.02). 1 unit of ordinate is 10^3 lbf/in^2 or 6.895 MN/m^2.

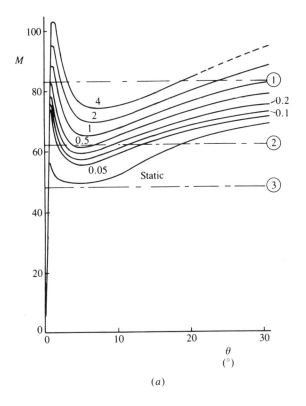

(a)

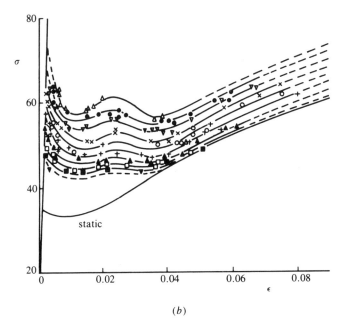

(b)

factors of 1.43, 1.12, 1.13 and 1.07 for carbon steel 1006, high strength low alloy YST 50 steel, stainless steel 302 and aluminium 6061-T6, respectively.

8.3 Constitutive equations

8.3.1 Introduction

Many different constitutive equations for the strain-rate-sensitive behaviour of materials have been proposed in the literature.[8.6–8.11] Careful experimental work is required in order to generate the various coefficients in these constitutive equations. Many authors have cast light on the characteristics of constitutive equations which is indispensable for guiding experimental test programmes. However, it is evident from the brief review of the experimental literature in § 8.2 that there is still considerable uncertainty and lack of reliable data even for some common materials. For example, some authors have observed that aluminium 6061-T6 is strain-rate-sensitive,[8.26, 8.27] while others have not.[8.13, 8.24, 8.32] In addition, insufficient data are available for the behaviour of materials under dynamic biaxial loadings and for the influence of generalised stresses (i.e., bending moments, membrane forces, etc.). Thus, only those constitutive relations which place relatively small demands on experimental test programmes, yet give reasonable agreement with the available experimental data, are extracted from the vast literature and discussed briefly in this chapter.

8.3.2 Cowper–Symonds constitutive equation

Cowper and Symonds[8.33] suggested the constitutive equation

$$\dot{\epsilon} = D\left(\frac{\sigma_0'}{\sigma_0} - 1\right)^q, \qquad \sigma_0' \geq \sigma_0, \tag{8.1}$$

where σ_0' is the dynamic flow stress at a uniaxial plastic strain rate $\dot{\epsilon}$, σ_0 is the associated static flow stress and D and q are constants for a particular material. Now, equation (8.1) may be written as

$$\log_e \dot{\epsilon} = q \log_e \left(\frac{\sigma_0'}{\sigma_0} - 1\right) + \log_e D \tag{8.2}$$

which is the equation of a straight line with $\log_e (\sigma_0'/\sigma_0 - 1)$ plotted against $\log_e \dot{\epsilon}$. The parameter q is the slope of this straight line, while the intercept on the ordinate is $\log_e D$.

Equation (8.1) may be recast in the form

$$\frac{\sigma_0'}{\sigma_0} = 1 + \left(\frac{\dot{\epsilon}}{D}\right)^{1/q}, \tag{8.3}$$

which, with $D = 40.4 \, \mathrm{s}^{-1}$ and $q = 5$, gives reasonable agreement with the experimental data for mild steel assembled by Symonds[8.17] and shown in Figure 8.8. There is considerable scatter of the experimental data in Figure 8.8, as

Table 8.1 *Coefficients of equation (8.3) for various materials*

Material	D (s^{-1})	q	Reference
Mild steel	40.4	5	Cowper and Symonds[8.33]
Aluminium alloy	6500	4	Bodner and Symonds[8.4]
α-Titanium (Ti 50A)	120	9	Symonds and Chon[8.34]
Stainless steel 304	100	10	Forrestal and Sagartz[8.35]

remarked in § 8.2.3. However, from an engineering viewpoint, equations (8.1) to (8.3) do provide a reasonable estimate of the strain rate sensitive uniaxial behaviour of mild steel.

The coefficients D and q in the Cowper–Symonds equation have also been determined for the materials listed in Table 8.1.

It is interesting to note from equation (8.3), that $\sigma_0' = 2\sigma_0$ when $\dot{\epsilon} = D$, regardless of the value of q, as indicated in Figure 8.17. Thus, the dynamic flow stress of mild steel is doubled at a strain rate of 40.4 s^{-1}, while a strain rate of 6500 s^{-1} is required to double the dynamic flow stress of an aluminium alloy. This large strain rate for an aluminium alloy contributes to the difficulty of detecting material strain-rate-sensitive effects and is partly responsible for the lack of agreement between experimental tests which was noted in § 8.3.1. In fact, it was observed[8.32]

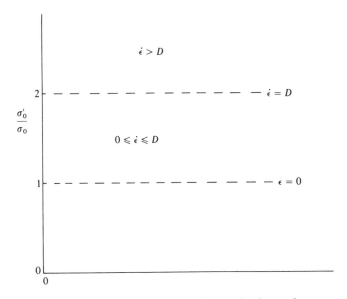

Figure 8.17 Cowper–Symonds rigid, perfectly plastic strain-rate-sensitive relation according to equation (8.3) with any values of D and q.

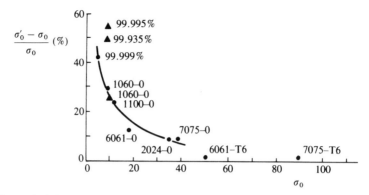

Figure 8.18 Percentage increase in the plastic flow stress at $\epsilon = 0.06$ between $\dot\epsilon = 10^{-3}\,\text{s}^{-1}$ and $\dot\epsilon = 10^3\,\text{s}^{-1}$ for various aluminium alloys.[8.7, 8.12] σ_0 is the static plastic flow stress at $\dot\epsilon = 10^{-3}\,\text{s}^{-1}$ and σ_0' is the dynamic flow stress at $\dot\epsilon = 10^3\,\text{s}^{-1}$. 1 unit on abscissa is $10^3\,\text{lbf/in}^2$ or $6.895\,\text{MN/m}^2$.

that equation (8.3), with $D = 1\,288\,000\,\text{s}^{-1}$ and $q = 4$, passes through the average of the widely scattered experimental data on aluminium 6061-T6. However, it is evident from Figure 8.18, which shows the strain rate sensitivity of various aluminium alloys, that it is vital to know the exact material specification.

The Cowper–Symonds constitutive equation (8.3) may be expressed in the form

$$\frac{\sigma_e'}{\sigma_0} = 1 + \left(\frac{\dot\epsilon_e}{D}\right)^{1/q} \tag{8.4}$$

where

$$\sigma_e' = [(\sigma_x' - \sigma_y')^2 + (\sigma_y' - \sigma_z')^2 + (\sigma_z' - \sigma_x')^2 + 6(\tau_{xy}'^2 + \tau_{yz}'^2 + \tau_{zx}'^2)]^{1/2}/\sqrt{2}$$

$$\text{(8.5)}\dagger$$

is the equivalent, or effective, dynamic flow stress and

$$\dot\epsilon_e = \sqrt{2}\,[(\dot\epsilon_x - \dot\epsilon_y)^2 + (\dot\epsilon_y - \dot\epsilon_z)^2 + (\dot\epsilon_z - \dot\epsilon_x)^2 + 6(\dot\epsilon_{xy}^2 + \dot\epsilon_{yz}^2 + \dot\epsilon_{zx}^2)]^{1/2}/3$$

$$\text{(8.6)}\dagger$$

is the associated equivalent, or effective, strain rate. The constants D and q are obtained from dynamic uniaxial, or pure shear, tests on the material and σ_0 is the corresponding static uniaxial flow stress.

Equation (8.5) reduces to $\sigma_e' = \sigma_x'$ in the uniaxial case, when all the stress components are zero except σ_x'. If the material in a dynamic uniaxial test with $\dot\epsilon_x > 0$ obeys the incompressibility relation ($\dot\epsilon_x + \dot\epsilon_y + \dot\epsilon_z = 0$), then $\dot\epsilon_y = \dot\epsilon_z = -\dot\epsilon_x/2$.

† The equivalent or effective stress and strain rate may also be written with the aid of the summation convention[8.36] in the tensorial forms $\sigma_e' = (3S_{ij}'S_{ij}'/2)^{1/2}$ and $\dot\epsilon_e = (2\dot\epsilon_{ij}\dot\epsilon_{ij}/3)^{1/2}$, respectively, where S_{ij}' is the dynamic deviatoric flow stress tensor, i and j have the range 1 to 3 when, 1, 2 and 3 are identified with the cartesian coordinates, x, y and z, respectively. Note that $\sigma_e' = (3J_2)^{1/2}$, where the second invariant of the deviatoric stress tensor $J_2 = S_{ji}'S_{ij}'/2$.

In this circumstance, equation (8.6) reduces to $\dot{\epsilon}_e = \dot{\epsilon}_x$ and, therefore, equation (8.3) is recovered from equation (8.4).

$\tau'_{xy} \neq 0$ and $\dot{\epsilon}_{xy} \neq 0$ (and $\dot{\epsilon}_{yx} \neq 0$) describe a state of pure shear in the x–y plane when the remaining stress and strain rate components are zero. In this case, equations (8.5) and (8.6) predict

$$\sigma'_e = \sqrt{3}\tau'_{xy} \quad \text{and} \quad \dot{\epsilon}_e = 2\dot{\epsilon}_{xy}/\sqrt{3}, \qquad (8.7\text{a, b})$$

respectively.

Equation (8.4) may be used to estimate the strain-rate-sensitive behaviour of a material in any dynamic uniaxial, biaxial or triaxial stress state. Of course, insufficient experimental data exist to demonstrate whether equation (8.4) is valid, except for uniaxial tension and compression. It is, nevertheless, a useful relation which employs assumptions which are similar to those customarily engaged to predict the multi-dimensional behaviour of materials from uniaxial static plastic and uniaxial creep properties. However, equation (8.4) does not provide a relation between the individual stress and strain rate components, except for some pathological cases (e.g., uniaxial and plane shear behaviour given by equations (8.3) and (8.7), respectively).

The Prandtl–Reuss constitutive equations for a plastic material[8.37] assume that an increment in plastic strain (e.g., $d\epsilon_x$) is directly proportional to the corresponding instantaneous deviatoric stress, which is the total stress minus the hydrostatic component (e.g., $S'_x = \sigma'_x - (\sigma'_x + \sigma'_y + \sigma'_z)/3$). Thus,

$$d\epsilon_x = d\lambda S'_x, \qquad (8.8)$$

where $d\lambda$ is a factor of proportionality which is related to the characteristics of a material and varies with stress and strain. In fact, it may be shown[8.37] that $d\lambda = \frac{3}{2}d\epsilon_e/\sigma'_e$, where σ'_e is defined by equation (8.5) and $d\epsilon_e$ has a form similar to equation (8.6).† Thus,

$$d\epsilon_x = \frac{3d\epsilon_e S'_x}{2\sigma'_e}, \qquad (8.9)$$

or

$$\frac{d\epsilon_x}{dt} = \frac{3}{2}\frac{d\epsilon_e}{dt}\frac{S'_x}{\sigma'_e}, \qquad (8.10)$$

which may be written

$$\dot{\epsilon}_x = \frac{3\dot{\epsilon}_e S'_x}{2\sigma'_e}. \qquad (8.11)$$

---

† Now, in addition to $d\epsilon_x = d\lambda S'_x$ given by equation (8.8), in the general case, $d\epsilon_y = d\lambda S'_y$, $d\epsilon_z = d\lambda S'_z$, $d\epsilon_{xy} = d\lambda \tau'_{xy}$ (since $S'_{xy} = \tau'_{xy}$), etc. Thus, if $d\epsilon_e$ is formed by substituting these expressions into equation (8.6), then it turns out that $d\epsilon_e = 2d\lambda\sigma'_e/3$, which gives $d\lambda = 3d\epsilon_e/2\sigma'_e$.

Note that for uniaxial tension $d\epsilon_x = d\lambda S'_x$, $d\epsilon_y = d\lambda S'_y$ and $d\epsilon_z = d\lambda S'_z$, where $S'_y = S'_z = -\sigma'_x/3$ and $S'_x = 2\sigma'_x/3$. Thus, $d\epsilon_y = d\epsilon_z = -d\epsilon_x/2$ and, therefore, $d\epsilon_x + d\epsilon_y + d\epsilon_z = 0$, as required for an incompressible material.

However, equation (8.4) predicts that

$$\dot{\epsilon}_e = D \left(\frac{\sigma'_e}{\sigma_0} - 1 \right)^q,$$

(8.12)

which, when substituted into equation (8.11), gives

$$\dot{\epsilon}_x = \frac{3D}{2\sigma'_e} \left(\frac{\sigma'_e}{\sigma_0} - 1 \right)^q S'_x$$

(8.13)

for a strain-rate-sensitive material. Similar expressions for $\dot{\epsilon}_y$ and $\dot{\epsilon}_z$ might be obtained by analogy with equation (8.13). In addition, the Prandtl–Reuss equations[8.37] give

$$d\epsilon_{xy} = d\lambda \tau'_{xy},$$

(8.14)

which becomes

$$\dot{\epsilon}_{xy} = \frac{3D}{2\sigma'_e} \left(\frac{\sigma'_e}{\sigma_0} - 1 \right)^q \tau'_{xy}$$

(8.15)

together with similar expressions for $\dot{\epsilon}_{xz}$ and $\dot{\epsilon}_{yz}$.†

Equations (8.1) to (8.4) retain neither material-strain-hardening nor elastic effects. Equation (8.1) could be refined to give

$$\dot{\epsilon} = D \left(\frac{\sigma'_0 - \sigma(\epsilon)}{\sigma_0} \right)^q, \qquad \sigma'_0 \geqslant \sigma(\epsilon),$$

(8.16)

where $\sigma(\epsilon)$ is the static uniaxial stress–strain curve which includes material strain hardening. This equation relates the plastic strain rate $\dot{\epsilon}$ to the dynamic overstress $\sigma'_0 - \sigma(\epsilon)$ which was first suggested by Malvern.[8.38] Another possible uniaxial relation[8.39] is

$$\dot{\epsilon} = D \left(\frac{\sigma'_0}{\sigma(\epsilon)} - 1 \right)^q.$$

(8.17)

It should be noted that material strain hardening is not expected to influence the data in Figure 8.8 for the dynamic lower yield stress of mild steel.

To cater for material elasticity, the linear elastic strain rate $\dot{\epsilon} = \dot{\sigma}/E$ is added to equations (8.1), (8.16) and (8.17). This gives the total strain rate when assuming that there is no coupling between elastic and plastic effects.

8.3.3 Bending

Bending effects are important in many structural problems and the strain rate sensitivity of steel in pure bending was discussed in § 8.2.6. A constitutive equation for the strain-rate-sensitive behaviour of a material subjected to a pure bending moment is developed in this section using the procedure outlined in references 8.29, 8.30 and 8.39.

---

† Equations (8.13) and (8.15), together with the other four strain rate components, can be written in the tensorial form $\dot{\epsilon}_{ij} = 3D(\sigma'_e/\sigma_0 - 1)^q S'_{ij}/2\sigma'_e$. This is a particular case of a more general strain-rate-sensitive constitutive equation discussed by Perzyna.[8.6]

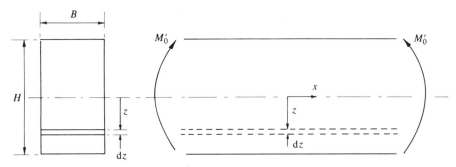

Figure 8.19 Pure bending of a beam with a rectangular cross-section $B \times H$ and made from a strain-rate-sensitive material.

Consider the beam in Figure 8.19 which has a rectangular cross-section with breadth B and depth H. The beam is made from a strain-rate-sensitive material and is subjected to a pure bending moment M'_0. The external dynamic bending moment

$$M'_0 = 2 \int_0^{H/2} \sigma'_x z B \, dz \tag{8.18}$$

when assuming that the material has the same behaviour in tension and compression. If the material obeys the Cowper–Symonds constitutive equation (8.3), then

$$\sigma'_x = \sigma_0 \{1 + (\dot{\epsilon}_x/D)^{1/q}\}, \tag{8.19}$$

where $\dot{\epsilon}_x$ is the axial, or longitudinal, strain rate at a distance z from the neutral axis of the beam. However, simple beam-bending theory, with plane cross-sections remaining plane, predicts that $\epsilon_x = z\kappa$, where κ is the change of curvature of the neutral axis. Thus, $\dot{\epsilon}_x = z\dot{\kappa}$, so that equations (8.18) and (8.19) yield the dynamic bending moment

$$M'_0 = 2B \int_0^{H/2} \sigma_0 \left\{1 + \left(\frac{z\dot{\kappa}}{D}\right)^{1/q}\right\} z \, dz,$$

or

$$M'_0 = 2B\sigma_0 \left[\frac{z^2}{2} + \frac{z^{2+1/q}}{2+1/q}\left(\frac{\dot{\kappa}}{D}\right)^{1/q}\right]_0^{H/2},$$

which becomes

$$\frac{M'_0}{M_0} = 1 + \frac{2q}{2q+1}\left(\frac{H\dot{\kappa}}{2D}\right)^{1/q}, \tag{8.20}†$$

† $H\dot{\kappa}/2$ is the maximum strain rate in the beam which occurs at the top and bottom surfaces of the beam in Figure 8.19. If the cross-section of a rectangular beam is idealised as a sandwich cross-section with only thin upper and lower flanges contributing to the bending moment, then it may be shown that $M'_0/M_0 = 1 + (H\dot{\kappa}/2D)^{1/q}$, where M_0 is now the static bending moment capacity of the two thin flanges. The form of this expression is similar to that of equation (8.20) and is identical when $q \gg 1$.

where

$$M_0 = \frac{\sigma_0 B H^2}{4} \tag{8.21}$$

is the static fully plastic bending moment.

Equation (8.20) predicts the behaviour of a beam with a rectangular cross-section when subjected to a dynamic pure bending moment. The influence of material strain rate sensitivity is assumed to obey the Cowper–Symonds constitutive equation (8.3) with the constants σ_0, D and q evaluated from a uniaxial tensile or compressive test. Thus, if $D = 40.4 \text{ s}^{-1}$ and $q = 5$ from Table 8.1 for mild steel, then equation (8.20) reduces to

$$\frac{M_0'}{M_0} = 1 + \frac{10}{11} \left(\frac{H\dot{\kappa}}{80.8} \right)^{0.2}, \tag{8.22}$$

where $\dot{\kappa} = \dot{\theta}/L$ when θ is the angular change across a beam of length L subjected to a pure bending moment M_0'.

The simple predictions of equation (8.22) are compared in Figure 8.16(a) with the experimental results of Aspden and Campbell.[8.30] Equation (8.22) does not include the influence of material strain hardening and, therefore, predicts a horizontal line in Figure 8.16(a). Also, it does not predict the upper yield stress because the coefficients D and q were obtained from data associated with the dynamic lower yield, or flow, stress. Notwithstanding these comments, equation (8.22) does give a reasonable engineering estimate of the strain-rate-sensitive moment–curvature characteristics of mild steel for moderate curvatures.

Aspden and Campbell[8.30] used the actual dynamic uniaxial compressive test data for the same material, which is shown in Figure 8.16(b), and integrated equation (8.18) numerically. This procedure gave somewhat better agreement with the actual moment–curvature characteristics than shown in Figure 8.16(a).†

8.3.4 Influence of strain hardening and strain rate sensitivity

In order to simplify theoretical analyses, Perrone[8.41] suggests that material-strain-hardening and strain-rate-sensitive effects could be decoupled and the corresponding constitutive equations expressed in the product form

$$\frac{\sigma_0'}{\sigma_0} = f(\dot{\epsilon})g(\epsilon), \tag{8.23}$$

where $f(\dot{\epsilon})$ could be the right-hand side of equation (8.3) in the uniaxial case and $g(\epsilon)$ caters for strain hardening. This relationship, which is plotted in Figure 8.20

---

† Gillis and Kelly[8.40] developed a procedure for extracting the dynamic uniaxial strain rate characteristics from dynamic bending tests. This method predicted surprisingly good agreement with the actual uniaxial data.

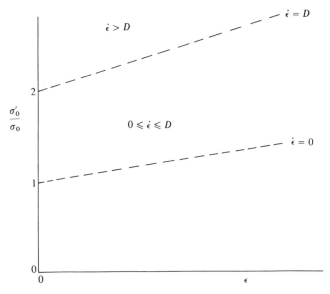

Figure 8.20 Equation (8.23) with linear strain hardening ($g(\epsilon)$) and with material strain rate sensitivity ($f(\dot{\epsilon})$) according to equation (8.3).

when $g(\epsilon)$ is taken as linear, neglects any coupling between strain hardening and strain rate effects. However, some coupling is evident in the experimental test data in Figure 8.7, since strain rate effects decrease with increase in strain.

8.4 Theoretical solutions of idealised models

8.4.1 Introduction

Some of the strain-rate-sensitive characteristics of materials are presented in § 8.2, while constitutive formulations are discussed in § 8.3. Several simple problems will be examined in this section in order to show the influence of strain rate sensitivity on the structural response and to introduce and justify further simplifications.

8.4.2 Wire holding an attached mass

8.4.2(a) Introduction

A mass M is attached to the end of a massless wire having an initial length L and subjected to an initial vertical velocity V_0 as shown in Figure 8.21. Wave propagation and material elasticity effects are ignored and the axial strain is taken to remain small.[8.42]

Vertical equilibrium of the mass M requires

$$M\frac{\mathrm{d}v}{\mathrm{d}t} = -A\sigma'_x,$$

(8.24)

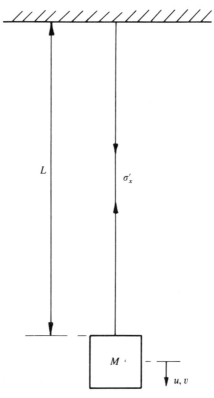

Figure 8.21 Mass M suspended by a wire of initial length L and subjected to a vertically downwards velocity V_0 at $t = 0$.

where v is the vertical velocity of the mass at any time t after impact and σ'_x is the dynamic axial stress in a wire having a cross-sectional area A.

8.4.2(b) Rigid, perfectly plastic case

A theoretical solution is sought when the wire in Figure 8.21 is made from a rigid, perfectly plastic material with a yield stress σ_0. Thus, $\sigma'_x = \sigma_0$ and equation (8.24) becomes

$$\frac{dv}{dt} = -\frac{A\sigma_0}{M}, \tag{8.25}$$

which predicts

$$v = V_0 - \frac{A\sigma_0 t}{M} \tag{8.26}$$

since $v = V_0$ at $t = 0$. Motion ceases at $t = t_f$ when $v = 0$, or

$$t_f = \frac{MV_0}{A\sigma_0}, \tag{8.27}$$

which is the duration of response.

Now, $v = du/dt$, so that integrating equation (8.26) and using the initial condition $u = 0$ when $t = 0$ gives

$$u = V_0 t - \frac{A\sigma_0 t^2}{2M}.$$ (8.28)

Equations (8.27) and (8.28) predict the final, or permanent, vertical displacement of the mass

$$u_f = \frac{MV_0^2}{2A\sigma_0}.$$ (8.29)

The plastic energy absorbed in the wire at time t is

$$D_a = \int_0^t \sigma_x' \dot{\epsilon}_x AL \, dt,$$ (8.30)

which, using $\sigma_x' = \sigma_0$ and $\dot{\epsilon}_x = v/L$ with v from equation (8.26), gives

$$D_a = A\sigma_0 \left(V_0 t - \frac{A\sigma_0 t^2}{2M} \right).$$ (8.31)

Equation (8.31) predicts that the initial kinetic energy $MV_0^2/2$ is absorbed by the wire when $t = t_f$, as expected.

8.4.2(c) Strain rate sensitive case: exact solution

A theoretical solution is now presented when the wire in Figure 8.21 is made from a rigid strain-rate-sensitive material which is governed by equation (8.3).

Equation (8.24) may be recast in the form

$$\frac{dv}{dt} = -\frac{A\sigma_0}{M} \left\{ 1 + \left(\frac{v}{DL} \right)^{1/q} \right\},$$ (8.32)

when using equation (8.3) with $\dot{\epsilon}_x = v/L$. Separating variables and integrating gives

$$\int_0^t dt = -\frac{M}{A\sigma_0} \int_{V_0}^v \frac{dv}{1 + (v/DL)^{1/q}},$$ (8.33)

or

$$t = \frac{M}{A\sigma_0} \int_v^{V_0} \frac{dv}{1 + (v/DL)^{1/q}},$$ (8.34)

which can be written

$$t = \frac{qMDL}{A\sigma_0} \int_{1 + (v/DL)^{1/q}}^{1 + (V_0/DL)^{1/q}} \frac{(h-1)^{q-1} \, dh}{h},$$ (8.35)

where

$$h = 1 + \left(\frac{v}{DL} \right)^{1/q}.$$ (8.36)

Equation (8.35) can be integrated using standard expressions† to give

$$
t = \frac{qMDL}{A\sigma_0} \left[\sum_{n=1}^{q-2} \left\{ \frac{(-1)^n}{(q-n)} (\lambda^{q-n} - \Lambda^{q-n}) \right\} \right.
$$

$$
\left. - (-1)^{q-1} \{ \Lambda - \lambda - \log_e(1 + \Lambda) + \log_e(1 + \lambda) \} \right], \tag{8.37}
$$

where

$$
\lambda = \left(\frac{v}{DL} \right)^{1/q} \tag{8.38a}
$$

and

$$
\Lambda = \left(\frac{V_0}{DL} \right)^{1/q}. \tag{8.38b}
$$

The duration of response $t = t_f$ is given by equation (8.37) with $\lambda = 0$.

Now, equation (8.32) may also be written

$$
\frac{d(v^2)}{du} = -\frac{2A\sigma_0}{M} \left\{ 1 + \left(\frac{v}{DL} \right)^{1/q} \right\} \tag{8.39}
$$

when recognising that $dv/dt = (dv/du)(du/dt) = (dv/du)v = d(v^2)/2du$. Separating variables and integrating gives

$$
\int_0^u du = -\frac{M}{2A\sigma_0} \int_{V_0^2}^{v^2} \frac{d(v^2)}{1 + (v^2/D^2L^2)^{1/2q}}, \tag{8.40}
$$

since the initial value of the vertical displacement is $u = 0$ at $t = 0$ when $v = V_0$. Equation (8.40) may be recast into the form

$$
u = \frac{M}{2A\sigma_0} \int_{v^2}^{V_0^2} \frac{d(v^2)}{1 + (v^2/D^2L^2)^{1/2q}}, \tag{8.41}
$$

or

$$
u = \frac{qMD^2L^2}{A\sigma_0} \int_{1+(v/DL)^{1/q}}^{1+(V_0/DL)^{1/q}} \frac{(h-1)^{2q-1}}{h} dh, \tag{8.42}
$$

where

$$
h = 1 + \left(\frac{v}{DL} \right)^{1/q}. \tag{8.43}
$$

The integrand in equation (8.42) has a form similar to that in equation (8.35). Thus, the vertical displacement of the mass is

$$
u = \frac{qMD^2L^2}{A\sigma_0} \left[\sum_{n=1}^{2q-2} \left\{ \frac{(-1)^n}{(2q-n)} (\lambda^{2q-n} - \Lambda^{2q-n}) \right\} + \Lambda - \lambda \right.
$$

$$
\left. - \log_e(1 + \Lambda) + \log_e(1 + \lambda) \right]. \tag{8.44}
$$

† For example, a repeated application of Case 142 in Selby.[8.43]

The final, or permanent, vertical displacement of the mass $u = u_f$ at $t = t_f$ is obtained from equation (8.44), with $v = 0$ (i.e., $\lambda = 0$).

It is simplest to substitute values of λ into the right-hand sides of equations (8.37) and (8.44) in order to calculate t and u, respectively, and hence the displacement–time (u–t) and velocity–time (v–t) histories. The dimensionless predictions of these two equations are presented in Figure 8.22 for the particular value of $q = 5$ for mild steel taken from Table 8.1. The parameter $\Lambda^q = V_0/DL$, which is defined by equation (8.38b), is the dimensionless initial strain rate in the wire, while $\lambda^q = v/DL$ from equation (8.38a) is the dimensionless strain rate at any time t. It is evident from Figure 8.22(c) that an increase in Λ for a given value of q leads to smaller permanent vertical displacements of the mass (u_f), as expected. A comparison of the theoretical predictions with equations (8.26), (8.28) and (8.29) for a rigid, perfectly plastic material clearly demonstrates the important influence of material strain rate sensitivity on the response of the mass.

8.4.2(d) Perrone's approximation

The velocity–time history (v–t) according to equation (8.37) is plotted in a dimensionless form in Figure 8.22(b). The quantity v/DL is the dimensionless strain rate which may be substituted into equation (8.3) in order to predict the associated dynamic axial flow stress shown in Figure 8.23(a). It is evident that the dynamic flow stress in the wire does not decrease significantly below the initial value throughout the greater part of the response. Furthermore, it is evident from Figure 8.23(b) that the dynamic flow stress departs appreciably from its initial value only after 90 per cent, or so, of the initial kinetic energy of the mass has been absorbed by plastic deformation in the wire.

These observations led Perrone[8.42] to speculate that an adequate estimate of the structural response could be obtained with a time-independent dynamic flow stress having a magnitude which is calculated from equation (8.3) for the initial strain rate. In this circumstance, the theoretical analysis in § 8.4.2(c) is simplified considerably since it reduces to the rigid, perfectly plastic solution in § 8.4.2(b), except that the constant flow stress (σ_0) is now evaluated from equation (8.3), using the initial value of the strain rate.

The initial dynamic axial flow stress in the wire suspending the mass in Figure 8.21 is

$$\sigma_0' = \sigma_0(1 + \Lambda) \tag{8.45}$$

according to equations (8.3) and (8.38b). Thus, σ_0 in equations (8.26) to (8.29) is replaced by σ_0' to give for the rigid strain-rate-sensitive case

$$v = V_0 - \frac{A\sigma_0(1 + \Lambda)t}{M}, \tag{8.46}$$

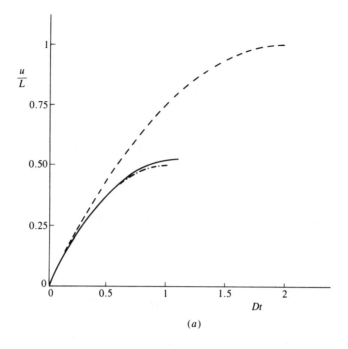

(a)

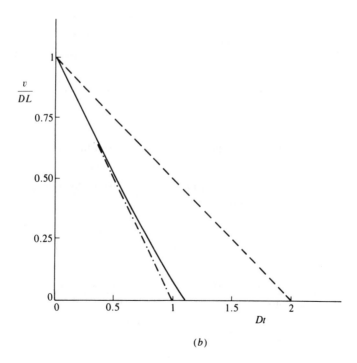

(b)

$$t_f = \frac{MV_0}{A\sigma_0(1 + \Lambda)}, \tag{8.47}$$

$$u = V_0 t - \frac{A\sigma_0(1 + \Lambda)t^2}{2M} \tag{8.48}$$

and

$$u_f = \frac{MV_0^2}{2A\sigma_0(1 + \Lambda)}, \tag{8.49}$$

respectively.

Notwithstanding the simplicity of this theoretical procedure, equations (8.46) to (8.49) do give reasonable agreement with the exact results from § 8.4.2(c) in Figure 8.22 and are a considerable improvement on the corresponding rigid, perfectly plastic analysis from § 8.4.2(b).

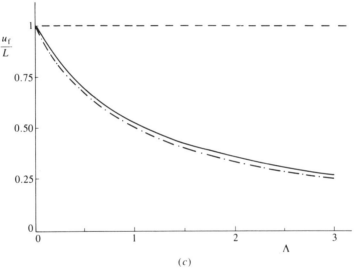

(c)

Figure 8.22 Comparison of theoretical predictions for a wire holding an attached mass as shown in Figure 8.21 with $q = 5$ and $\dfrac{2\bar{\sigma}_0 AL}{MV_0^2} = 1$. (a) Dimensionless displacement (u)–time (t) history with $\Lambda = \left(\dfrac{V_0}{LD}\right)^{1/q} = 1$. ——: exact solution (equation (8.44)). Also Symonds' viscous approximation (equation (8.59)). ———: rigid, perfectly plastic solution (equation (8.28)). —·—·—: Perrone's approximation (equation (8.48)). (b) Dimensionless velocity (v)–time (t) history with $\Lambda = 1$. ——: exact solution (equation (8.37)). Also Symonds' viscous approximation (equation (8.55)). ———: rigid, perfectly plastic solution (equation (8.26)). —·—·—: Perrone's approximation (equation (8.46)). (c) Dimensionless permanent displacement (u_f) versus $\Lambda = \left(\dfrac{V_0}{LD}\right)^{1/q}$. ——, ——— and —·—·— are defined in (a).

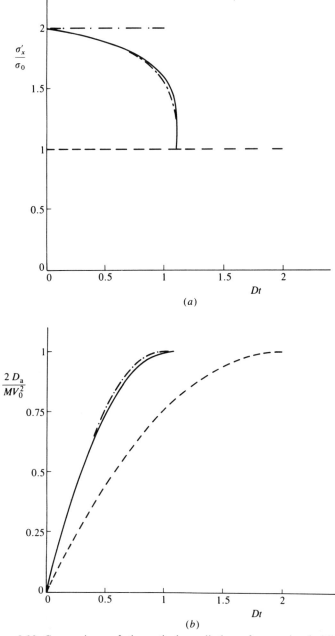

Figure 8.23 Comparison of theoretical predictions for a wire holding an attached mass as shown in Figure 8.21 with $q = 5$ and $\dfrac{2\sigma_0 AL}{MV_0^2} = 1$. (a) Dimensionless axial stress (σ'_x)–time (t) history with $\Lambda = 1$. ———: exact solution (§ 8.4.2(c)). ———: rigid, perfectly plastic solution (§ 8.4.2(b)). —·—·—: Perrone's approximation (equation (8.45)). ————: Symonds' viscous approximation (equations (8.50, 8.52)). (b) Dimensionless energy absorption (D_a)–time (t) history. ———: exact solution (§ 8.4.2(c)). Also Symonds' viscous approximation (§ 8.4.2(e)). ———: rigid, perfectly solution (equation (8.31)). —·—·—: Perrone's approximation (§ 8.4.2(d)).

8.4.2(e) Symonds' viscous approximation

It is evident from § 8.4.2(c) that material strain rate sensitivity complicates exact theoretical analyses even for the particularly simple problem shown in Figure 8.21. This arises from the non-linear character of equation (8.3) which Symonds[8.44] has suggested could be replaced by the simpler homogeneous viscous relation

$$\frac{\sigma_0'}{\sigma_0} = \mu \left(\frac{\dot{\epsilon}}{D} \right)^{1/vq}, \tag{8.50}$$

where D and q are defined in § 8.3.2.

Symonds[8.44] obtained the constants μ and v from the requirements that the initial stresses according to equations (8.3) and (8.50) should be equal, or

$$1 + \left(\frac{V_0}{DL} \right)^{1/q} = \mu \left(\frac{V_0}{DL} \right)^{1/vq}. \tag{8.51a}$$

In addition, Symonds[8.44] assumed that the initial slopes of the dimensionless forms of equations (8.3) and (8.50) were also equal, or

$$\frac{1}{q} \left(\frac{V_0}{DL} \right)^{1/q-1} = \frac{\mu}{vq} \left(\frac{V_0}{DL} \right)^{1/vq-1}. \tag{8.51b}$$

Equations (8.51a, b) respectively predict

$$v = \frac{1 + \Lambda}{\Lambda} \tag{8.52a}$$

and

$$\mu = \frac{1 + \Lambda}{\Lambda^{1/v}} \tag{8.52b}$$

when Λ is defined by equation (8.38b).

Now, with the aid of equation (8.50), the equilibrium equation (8.24) for the mass M in Figure 8.21 becomes

$$M \frac{dv}{dt} = -\mu A \sigma_0 \left(\frac{v}{DL} \right)^{1/vq}, \tag{8.53}$$

which may be rewritten

$$\int_0^t dt = \frac{M(DL)^{1/vq}}{\mu A \sigma_0} \int_v^{V_0} \frac{dv}{v^{1/vq}}, \tag{8.54}$$

or

$$t = \frac{M(DL)^{1/vq}}{\mu A \sigma_0 (1 - 1/vq)} (V_0^{1-1/vq} - v^{1-1/vq}). \tag{8.55}$$

Equation (8.55) may be rearranged in the form

$$\frac{du}{dt} = V_0 \left\{ 1 - \frac{\mu A \sigma_0 (vq - 1)t}{vqM(DL)^{1/vq}V_0^{1-1/vq}} \right\}^{vq/(vq-1)}, \tag{8.56}$$

or

$$u = \frac{-vqMD^2L^2\Lambda^{2q-1/v}}{\mu A\sigma_0(vq-1)} \int_1^k k^{vq/(vq-1)}\,dk, \tag{8.57}$$

where

$$k = 1 - \frac{\mu A\sigma_0(vq-1)t}{vqMDL\Lambda^{q-1/v}} \tag{8.58}$$

and Λ is defined by equation (8.38b). Finally, integrating equation (8.57) gives the vertical displacement of the mass M

$$u = \frac{vqMD^2L^2\Lambda^{2q-1/v}}{\mu A\sigma_0(2vq-1)}\left[1 - \left\{1 - \frac{\mu A\sigma_0(vq-1)t}{vqMDL\Lambda^{q-1/v}}\right\}^{\frac{2vq-1}{vq-1}}\right]. \tag{8.59}$$

The response duration

$$t_f = \frac{vqMDL\Lambda^{q-1/v}}{\mu A\sigma_0(vq-1)} \tag{8.60}$$

is obtained from equation (8.55) when $v = 0$, while the associated permanent displacement of the mass M

$$u_f = \frac{vqMD^2L^2\Lambda^{2q-1/v}}{\mu A\sigma_0(2vq-1)} \tag{8.61}$$

is calculated from equation (8.59) with $t = t_f$.

The theoretical predictions of this section are compared in Figure 8.22 with the exact solution from § 8.4.2(c). In fact, the predictions of Symonds' viscous approximation are virtually indistinguishable from the exact solution for the particular case shown in Figure 8.21.

8.4.3 Influence of large deflections

8.4.3(a) Introduction

The calculations in § 8.4.2 have disregarded the influence of finite-deflections, or geometry changes, which are important for certain structural problems as remarked in Chapter 7. The experimental and theoretical results for an impulsively loaded beam in Figure 8.24 indicate that finite-deflections are at least as important as material strain rate sensitivity, particularly for permanent transverse displacements larger than the structural thickness, approximately.

The approximate methods discussed in § 8.4.2(d) and § 8.4.2(e) do not account for mode changes which might occur in structures undergoing geometry changes.† A laterally loaded beam, for example, initially responds in a flexural manner which

---

† Mode changes may also occur in structures with infinitesimal deflections. For example, travelling hinges cause mode changes during the dynamic response of the cantilever beam which is examined in § 3.8.

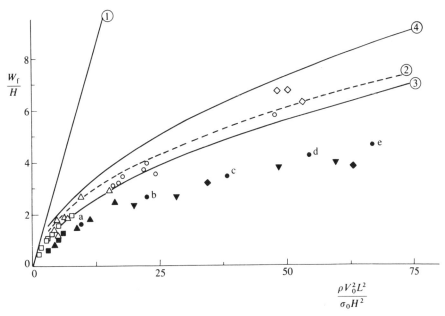

Figure 8.24 Maximum permanent transverse displacement (W_f) of a fully clamped beam subjected to an impulsive velocity of magnitude V_0 distributed uniformly across the entire span $2L$. ①: infinitesimal analysis (bending only) (equation (3.82). ②, ③, ④: theoretical analyses which include the influence of finite-deflections (see § 7.6.4 and Figure 7.19). □, △, ○, ◇: experimental results on aluminium 6061-T6 beams (see § 7.6.4). ■, ▲, ▼, ◆: experimental results on mild steel beams.[8.3] ● a–e: numerical finite-difference results of Witmer et al. (1963).[8.3]

is dominated by bending moments and transverse shear forces. However, a membrane force develops and grows as the transverse displacements increase, while the bending moments and transverse shear forces play a less important role. Clearly, Perrone's approximation in § 8.4.2(d) gives a dynamic flow stress σ_0' (or plastic bending moment M_0') which is evaluated from the initial strain rate (or initial curvature rate) and, therefore, is dominated by flexural effects. It appears that this flow stress has little relevance to the later response which is dominated by membrane forces. The constants μ and ν in equation (8.50) for Symonds'[8.44] viscous approximation in § 8.4.2(e) are also evaluated from the initial conditions which again might not be relevant for those problems dominated by membrane effects in the later stages.

Perrone and Bhadra[8.45] have examined the simple problem illustrated in Figure 8.25 in order to gain some insight into the behaviour of this class of structural members. The rigid mass M is secured across a span with two horizontal massless wires each of length L and made from a rigid, strain-rate-sensitive

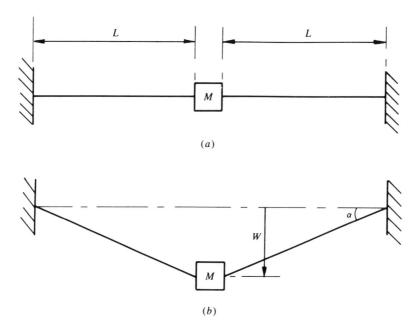

Figure 8.25 Mass M supported by two wires each of length L. (a) Initial position at $t = 0$ when the rigid mass is loaded impulsively with a transverse velocity V_0. (b) Deformed position at time t when the mass has reached a transverse displacement W.

material. Perrone and Bhadra[8.45] examined the influence of membrane forces and geometry changes in the wires when the mass is subjected to an impulsive velocity V_0 but ignored the effects of bending moments and transverse shear forces. Thus, the equation of motion for the mass M is

$$M \frac{\mathrm{d}^2 W}{\mathrm{d}t^2} = -2\sigma'_x A \sin \alpha, \tag{8.62}$$

where σ'_x is the dynamic flow stress in the wires which have a current cross-sectional area A.

8.4.3(b) Exact solution

The engineering strain in both wires in Figure 8.25 is

$$\epsilon = \frac{\sqrt{L^2 + W^2} - L}{L},$$

or

$$\epsilon \cong \frac{W^2}{2L^2}, \tag{8.63}$$

which gives a strain rate

$$\dot{\epsilon} = \frac{W\dot{W}}{L^2}, \tag{8.64}$$

where $(\dot{\ }) = d(\ )/dt$. If the initial cross-sectional area of the wires is A_0, then conservation of volume for both wires during plastic flow[8.37] requires

$$A_0 L = A\sqrt{L^2 + W^2},$$

or

$$A = \frac{A_0 L}{\sqrt{L^2 + W^2}}. \tag{8.65}$$

Equation (8.62), with equations (8.3), (8.64) and (8.65), becomes

$$M\frac{d(v^2)}{dW} = -\frac{4A_0 L \sigma_0 W}{L^2 + W^2}\left\{1 + \left(\frac{Wv}{DL^2}\right)^{1/q}\right\} \tag{8.66}\dagger$$

since

$$\sin \alpha = \frac{W}{\sqrt{L^2 + W^2}} \tag{8.67}$$

and

$$\frac{d^2 W}{dt^2} = \frac{dv}{dt} = \frac{1}{2}\frac{d}{dW}(v^2), \tag{8.68}$$

where $v = dW/dt$.

Perrone and Bhadra[8.45] integrated equation (8.66) numerically with the initial conditions $W = 0$ and $v = V_0$ at $t = 0$ and the final condition $v = 0$ at $t = t_f$ when motion ceases. The dimensionless final displacement (W_f/L) of the mass is plotted in Figure 8.26 with respect to the dimensionless initial kinetic energy $(MV_0^2/4\sigma_0 A_0 L)$ of the mass M.

8.4.3(c) Approximate solution of Perrone and Bhadra

It was found in § 8.4.3(b) that a numerical solution of equation (8.66) was required despite the simplicity of the problem illustrated in Figure 8.25. Practical structural problems are more complicated which would lead to even greater theoretical difficulties. This situation provides a strong motivation for seeking simplified theoretical analyses which are still valuable despite the fact that complex numerical solutions are now commonplace. Simple methods of analysis offer insight and the predictions are sometimes adequate because of various uncertainties in the external dynamic loading characteristics and structural details.

Various numerical results are presented in reference 8.45 but Figure 8.27, taken from this reference, is of considerable interest. This figure reveals that the dynamic flow stresses for both mild steel and aluminium 6061-T6, with the strain rate

---

† Strictly speaking, σ_0 in equation (8.66) is a true stress, though there is a negligible difference between the engineering and true stresses for small strains.

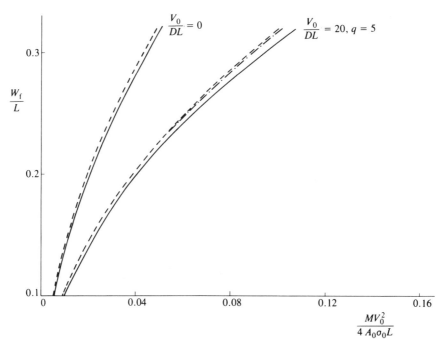

Figure 8.26 Dimensionless permanent transverse displacement versus dimensionless kinetic energy for the idealised structure in Figure 8.25. ———: exact solution, taken from Figure 12 of Perrone and Bhadra.[8.45] – – –: equations (8.71) and (8.81). —·—·—: equation (8.75) (same as – – – when $V_0/DL = 0$).

characteristics according to equation (8.3) and Table 8.1, remain virtually constant throughout the entire response except for relatively short initial and final periods. It appears reasonable, therefore, to use a constant correction factor for the dynamic flow stress throughout the response, as found in § 8.4.2(d) for the model in Figure 8.21. However, it is evident from Figure 8.27 that the dynamic flow stress corresponding to the initial strain rate is not appropriate in this case.

Perrone and Bhadra[8.45] also examined the variation of the strain rate in both wires, with the velocity of the mass M as shown in Figure 8.28. The maximum strain rate for both the mild steel and the aluminium 6061-T6 wires occurs when the velocity of the mass is $V_0/\sqrt{2}$, approximately. In other words, the maximum membrane strain rate occurs when about one-half of the initial kinetic energy of the mass M has been dissipated plastically in both wires. Unfortunately, the dynamic flow stress cannot be calculated easily from equation (8.3) because the strain rate depends on the transverse displacement W according to equation (8.64) (i.e., $\dot{\epsilon} = V_0 W/\sqrt{2}L^2$). However, Perrone and Bhadra[8.45] assumed that the 'average' strain rate throughout the response is equal to one-half of the maximum value

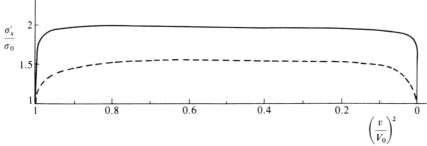

Figure 8.27 Variation of dimensionless stress (σ'_x/σ_0) with dimensionless kinetic energy $(v/V_0)^2$ for the model in Figure 8.25 with wires made from mild steel (———) or aluminium 6061-T6 (———) which obey equation (8.3) with the coefficients in Table 8.1.[8.45]

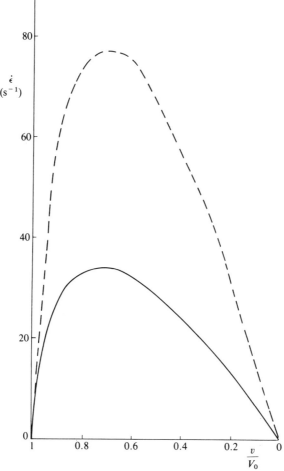

Figure 8.28 Variation of strain rate ($\dot{\epsilon}$) with dimensionless velocity (v/V_0) for the model in Figure 8.25 with wires made from mild steel (———) or aluminium 6061-T6 (———) described by equation (8.3) with the coefficients in Table 8.1.[8.45]

(i.e., $V_0 W/2\sqrt{2}L^2$), so that equations (8.3, 8.62, 8.64, 8.65, 8.67 and 8.68) become

$$M\frac{d(v^2)}{dW} = -4A_0 L\sigma_0 \left\{ 1 + \left(\frac{WV_0}{2\sqrt{2DL^2}} \right)^{1/q} \right\} \frac{W}{L^2 + W^2}, \qquad (8.69)$$

which may be integrated to give

$$v^2 - V_0^2 = -\frac{4A_0\sigma_0}{ML} \left[\frac{W^2}{2} - \frac{W^4}{4L^2} + \frac{W^6}{6L^4} + \left(\frac{V_0}{2\sqrt{2DL^2}} \right)^{1/q} \right.$$

$$\left. \times W^{2+1/q} \left\{ \frac{1}{2+1/q} - \frac{W^2}{L^2(4+1/q)} + \frac{W^4}{L^4(6+1/q)} + \cdots \right\} \right] \qquad (8.70)$$

when writing $(L^2 + W^2)^{-1}$ as $L^{-2}(1 - W^2/L^2 + W^4/L^4 + \cdots)$, approximately. The final, or permanent, transverse displacement of the mass M (W_f) is given by equation (8.70) with $v = 0$, or

$$\frac{2}{3}\left(\frac{W_f}{L} \right)^6 - \left(\frac{W_f}{L} \right)^4 + 2\left(\frac{W_f}{L} \right)^2 + 4q\left(\frac{V_0}{2\sqrt{2DL}} \right)^{1/q} \left(\frac{W_f}{L} \right)^{2+1/q}$$

$$\times \left\{ \frac{1}{2q+1} - \frac{1}{4q+1}\left(\frac{W_f}{L} \right)^2 + \frac{1}{6q+1}\left(\frac{W_f}{L} \right)^4 \right\} - \frac{MV_0^2}{A_0\sigma_0 L} = 0. \quad (8.71)$$

It is evident from Figure 8.26 that equation (8.71)† gives reasonable agreement with the exact values predicted in § 8.4.3(b). Equation (8.71) gives slightly different predictions from those presented by Perrone and Bhadra[8.45] who used an energy method instead of solving the governing differential equation (8.69).

8.4.3(d) Further simplification

It was noted in § 8.4.3(c) that the maximum strain rate in the wires of the idealised model in Figure 8.25 occurred when the transverse velocity of the mass M had reduced to $V_0/\sqrt{2}$. This velocity is reached when one-half of the initial kinetic energy has been absorbed as plastic work in the two wires. However, it is evident from Figure 8.29 that the maximum strain rate also occurs when the transverse displacement of the mass is about two-thirds of the final or permanent displacement (W_f).[8.46] Thus, it is possible to write the mean strain rate according to equation (8.64) as

$$\dot{\epsilon} \cong \left(\frac{2W_f}{3} \right)\left(\frac{V_0}{\sqrt{2}} \right)\left(\frac{1}{2L^2} \right) \qquad (8.72)$$

when taking the mean strain rate equal to one-half of the maximum value. In this circumstance, equation (8.69) becomes

$$\frac{d(v^2)}{dW} = -\frac{4A_0 L\sigma_0}{M} \left\{ 1 + \left(\frac{V_0 W_f}{3\sqrt{2DL^2}} \right)^{1/q} \right\} \frac{W}{L^2 + W^2}. \qquad (8.73)$$

† It is simplest to construct the curve in Figure 8.26 for particular values of V_0/DL and q by calculating $MV_0^2/A_0\sigma_0 L$ for a given value of W_f/L.

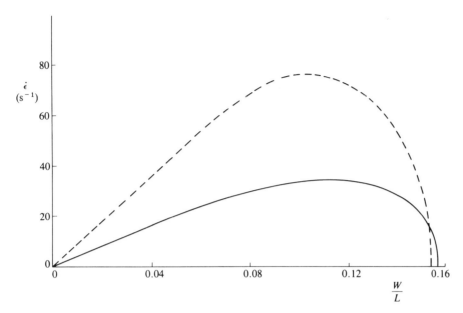

Figure 8.29 Variation of strain rate ($\dot{\epsilon}$) with dimensionless displacement (W/L) for the model in Figure 8.25 with wires made from mild steel (——) or aluminium 6061-T6 (– – –) which are idealised by equation (8.3) with the coefficients from Table 8.1.[8.45]

Now, integrating equation (8.73) yields

$$v^2 - V_0^2 = -\frac{4A_0 L\sigma_0}{M}\left\{1 + \left(\frac{W_f V_0}{3\sqrt{2}DL^2}\right)^{1/q}\right\}\left(\frac{W^2}{2L^2} - \frac{W^4}{4L^4} + \frac{W^6}{6L^6} + \cdots\right), \qquad (8.74)$$

or

$$\frac{MV_0^2}{2A_0 L\sigma_0} = \left\{1 + \left(\frac{V_0}{3\sqrt{2}DL}\right)^{1/q}\left(\frac{W_f}{L}\right)^{1/q}\right\}\left(\frac{W_f}{L}\right)^2\left\{1 - \frac{1}{2}\left(\frac{W_f}{L}\right)^2 + \frac{1}{3}\left(\frac{W_f}{L}\right)^4\right\},$$

$$(8.75)$$

since $W = W_f$ when $v = 0$.

This is a slightly simpler theoretical analysis than that in § 8.4.3(c) but gives equally acceptable predictions, as shown in Figure 8.26.

8.4.3(e) Extension of Symonds' viscous approximation

It was observed in § 8.4.2(e) that Symonds' viscous approximation gave theoretical predictions which were almost indistinguishable from the exact theoretical solution for the particular problem shown in Figure 8.21. It is, therefore, worthwhile to enquire whether or not the simplicity of this method can be combined with Perrone's and Bhadra's[8.45] observations in § 8.4.3(c) to examine the finite-displacement problem illustrated in Figure 8.25.

Equation (8.50) is taken to govern the response with the constants μ and v now found from the strain-rate-sensitive characteristics at $v = V_0/\sqrt{2}$ and $W = 2W_f/3$, as observed in §§ 8.4.3(c) and 8.4.3(d), respectively. Thus, following § 8.4.2(e), equations (8.3) and (8.50) with equation (8.64) give

$$1 + \left(\frac{2W_f V_0}{3\sqrt{2}DL^2}\right)^{1/q} = \mu\left(\frac{2W_f V_0}{3\sqrt{2}DL^2}\right)^{1/vq} \tag{8.76a}$$

and

$$\frac{1}{q}\left(\frac{2W_f V_0}{3\sqrt{2}DL^2}\right)^{1/q-1} = \frac{\mu}{vq}\left(\frac{2W_f V_0}{3\sqrt{2}DL^2}\right)^{1/vq-1}, \tag{8.76b}$$

or

$$v = \frac{1 + \left(\dfrac{\sqrt{2}V_0}{3DL}\right)^{1/q}\left(\dfrac{W_f}{L}\right)^{1/q}}{\left(\dfrac{\sqrt{2}V_0}{3DL}\right)^{1/q}\left(\dfrac{W_f}{L}\right)^{1/q}} \tag{8.77a}$$

and

$$\mu = \frac{1 + \left(\dfrac{\sqrt{2}V_0}{3DL}\right)^{1/q}\left(\dfrac{W_f}{L}\right)^{1/q}}{\left(\dfrac{\sqrt{2}V_0}{3DL}\right)^{1/vq}\left(\dfrac{W_f}{L}\right)^{1/vq}}. \tag{8.77b}$$

Equations (8.77) are the same as equations (8.52), with Λ replaced by

$$\left(\frac{\sqrt{2}V_0}{3DL}\right)^{1/q}\left(\frac{W_f}{L}\right)^{1/q}.$$

Now, equation (8.62) may be written with the aid of equations (8.50), (8.64), (8.65), (8.67) and (8.68) in the form

$$M\frac{d(v^2)}{dW} = -4A_0 L\sigma_0\mu\left(\frac{Wv}{DL^2}\right)^{1/vq}\frac{W}{L^2 + W^2} \tag{8.78}$$

which, with $v = V_0/\sqrt{2}$ and taking the mean strain rate as one-half the maximum value according to Perrone's and Bhadra's[8.45] observations in § 8.4.3(c), gives

$$v^2 - V_0^2 = -\frac{4A_0 L\sigma_0\mu}{M}\left(\frac{V_0}{2\sqrt{2}DL^2}\right)^{1/vq}\int_0^t \frac{W^{1+1/vq}}{L^2 + W^2}\,dW \tag{8.79}$$

or

$$v^2 - V_0^2 = -\frac{4A_0 L\sigma_0\mu}{M}\left(\frac{V_0}{2\sqrt{2}DL^2}\right)^{1/vq}\left\{\frac{W^{2+1/vq}}{L^2(2+1/vq)}\right.$$
$$\left. -\frac{W^{4+1/vq}}{L^4(4+1/vq)} + \frac{W^{6+1/vq}}{L^6(6+1/vq)}\right\} \tag{8.80}$$

when $(L^2 + W^2)^{-1} \cong (1 - W^2/L^2 + W^4/L^4)/L^2$. Thus,

$$V_0^2 = \frac{4vqA_0L\sigma_0}{M}\left(\frac{3}{4}\right)^{1/vq}\left\{1 + \left(\frac{\sqrt{2}V_0}{3DL}\right)^{1/q}\left(\frac{W_{\mathrm{f}}}{L}\right)^{1/q}\right\}$$

$$\times \left\{\frac{1}{(2vq + 1)}\left(\frac{W_{\mathrm{f}}}{L}\right)^2 - \frac{1}{(4vq + 1)}\left(\frac{W_{\mathrm{f}}}{L}\right)^4 + \frac{1}{(6vq + 1)}\left(\frac{W_{\mathrm{f}}}{L}\right)^6\right\} \quad (8.81)$$

since $W = W_{\mathrm{f}}$ when $v = 0$ and using equation (8.77b).

A straightforward iterative solution of equations (8.77a) and (8.81) gives the theoretical predictions presented in Figure 8.26.

8.5 Theoretical behaviour of strain-rate-sensitive structures

8.5.1 Introduction

Some simple theoretical procedures for estimating the influence of material strain rate sensitivity on the dynamic plastic response of structures have been suggested in § 8.4 after a careful analysis of the two idealised problems in Figures 8.21 and 8.25, respectively. This section contains several applications of these approximations.

8.5.2 Fully clamped beam loaded impulsively

Equation (7.101) predicts

$$\frac{W_{\mathrm{f}}}{H} = \frac{1}{2}\left\{\left(1 + 3\frac{\rho V_0^2 L^2}{\sigma_0 H^2}\right)^{1/2} - 1\right\} \quad (8.82)$$

for the maximum permanent transverse displacement (W_{f}) of a beam which is fully clamped at both ends and impulsively loaded, with a uniform velocity V_0 across the entire span ($2L$). This expression was developed using an approximate procedure which retains the influence of finite-deflections, or geometry changes, as well as bending moments, and gives reasonable agreement with the corresponding experimental results on strain-rate-insensitive aluminium 6061-T6 beams as shown in Figure 8.24 and Figure 7.19.

Now, it is shown in § 8.4.3(c) and § 8.4.3(d) that the influence of material strain rate sensitivity could be estimated using Perrone's and Bhadra's[8.45, 8.46] observations. It was found that the stress associated with a strain rate evaluated at a transverse deflection $2W_{\mathrm{f}}/3$ and a transverse velocity $V_0/\sqrt{2}$, when one-half of the initial kinetic energy was absorbed, gave good agreement with more exact theoretical solutions. In this circumstance, the static uniaxial yield stress σ_0 in equation (8.82) would be replaced by σ_0' from equation (8.3) with $\dot{\epsilon} = V_0 W_{\mathrm{f}}/3\sqrt{2}L^2$ according to equation (8.72). Thus, equation (8.82) becomes

$$\frac{W_{\mathrm{f}}}{H} = \frac{1}{2}\left\{\left(1 + \frac{3\rho V_0^2 L^2}{n\sigma_0 H^2}\right)^{1/2} - 1\right\}, \quad (8.83a)$$

where

$$n = \frac{\sigma_0'}{\sigma_0} = 1 + \left(\frac{V_0 W_f}{3\sqrt{2DL^2}}\right)^{1/q}. \tag{8.83b}$$

Perrone and Bhadra's[8.45, 8.46] method and, therefore, equation (8.83a) retain only the strain rate enhancement of the membrane force which, nevertheless, dominates the response for transverse displacements larger than the beam thickness, approximately, as indicated in Figure 7.18 and in Figure 8.24 for the strain-rate-insensitive case.

A straightforward iterative solution of equations (8.83a, b) for mild steel beams having two different thicknesses is compared in Figure 8.30 with the strain-rate-insensitive predictions of equation (8.82).

The influence of material strain rate sensitivity is a highly non-linear effect for mild steel with $q = 5$, which accounts for the success of the Perrone and Bhadra[8.45] approximation discussed in § 8.4.3(c) and used above. For example, a doubling of the strain rate $\dot{\epsilon}$ in equation (8.3) from $\dot{\epsilon} = D$ to $\dot{\epsilon} = 2D$ causes an

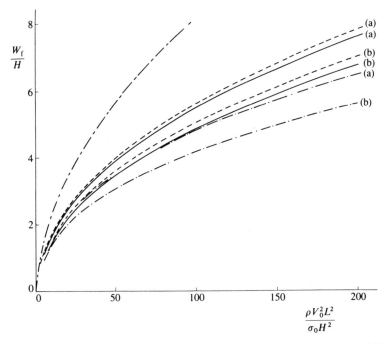

Figure 8.30 Maximum permanent transverse displacement (W_f) of a fully clamped beam loaded with a uniformly distributed impulsive velocity of magnitude V_0 over the entire span $2L$. ($q = 5$, $D = 40.4$ s^{-1}, $\rho = 7829$ kg/m^3, $\sigma_0 = 210.3$ MN/m^2, $L = 63.666$ mm, (a): $H = 2.342$ mm (0.0922 in) and (b): $H = 6.015$ mm (0.2368 in).) ———: equation (8.82). – – –: equations (8.83a, b). ——: equations (8.83a) and (8.85). – · – · –: equations (8.83a) and (8.86).

increase of 7.4 per cent in the dynamic flow stress when $q = 5$.† Thus, in order to avoid iteration, it is worth considering an approximation for n in equation (8.83a).

Equation (8.82) predicts

$$W_f \cong \frac{V_0 L}{2}\left(\frac{3\rho}{\sigma_0}\right)^{1/2}$$ (8.84)

when $W_f \gg H$. Substituting equation (8.84) into (8.83b) gives

$$n = 1 + \left\{\frac{V_0^2}{6\sqrt{2}DL}\left(\frac{3\rho}{\sigma_0}\right)^{1/2}\right\}^{1/q}.$$ (8.85)‡

Equations (8.83a) and (8.85) are simpler to use than equations (8.83a, b) yet predict similar results, as shown in Figure 8.30.

This particular beam problem was also examined in reference 8.3 wherein it was found, using an entirely different approach, that

$$n = 1 + \left(\frac{\rho V_0^3}{6\sigma_0 DH}\right)^{1/q}.$$ (8.86)

The theoretical predictions of equations (8.83), (8.85) and (8.86) for a strain-rate-sensitive material are compared in Figure 8.30 for two beam thicknesses. The yield criterion used in the derivation of equation (8.82) and, therefore, in equations (8.83), (8.85) and (8.86) circumscribes the exact yield curve. However, another yield criterion which is 0.618 times as large would inscribe the exact yield curve, as discussed in § 7.5.4(a) and shown in Figure 7.17. It is evident from Figure 8.31 that the upper and lower bounds of equations (8.83a) and (8.86) associated with inscribing and circumscribing yield criteria, respectively, bound almost all the corresponding experimental values for beams loaded impulsively.[8.3]

8.5.3 Circular membrane loaded impulsively

Equation (7.129) predicts the maximum permanent transverse displacement (W_f)

$$\frac{W_f}{H} = \left(\frac{2\rho V_0^2 R^2}{3\sigma_0 H^2}\right)^{1/2}$$ (8.87)

for a rigid, perfectly plastic membrane of radius R and thickness H which is clamped around the entire boundary and subjected to a uniformly distributed impulsive velocity V_0. This approximate theoretical procedure gives reasonable agreement with the more exact analysis of Symonds and Wierzbicki,[8.47] as shown in Figure 7.21.

Now, the influence of material strain rate sensitivity is explored using the method developed by Perrone and Bhadra[8.45, 8.46] which was employed in § 8.5.2 for a

---

† Equation (8.3) predicts $\sigma_{02}'/\sigma_{01}' = (1 + 2^{1/5})/(1 + 1) = 1.074$.

‡ A further improvement in accuracy is possible by replacing σ_0 by $n\sigma_0$. An iteration then yields the corresponding value of n.

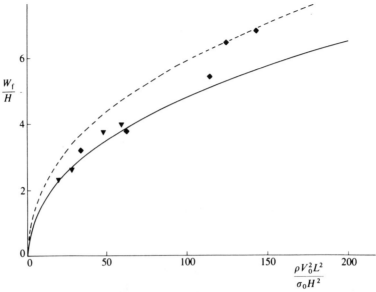

Figure 8.31 Maximum permanent transverse displacement (W_f) of a fully clamped beam loaded with a uniformly distributed impulsive velocity of magnitude V_0 over the entire span $2L$. ($q = 5$, $D = 40.4 \, \text{s}^{-1}$, $\rho = 7829 \, \text{kg/m}^3$, $\sigma_0 = 210.3 \, \text{MN/m}^2$, $L = 63.666$ mm and $H = 2.342$ mm (0.0922 in)). ——: lower bound on W_f; equations (8.83a) and (8.86). – – –: upper bound on W_f; equations (8.83a) and (8.86) with σ_0 replaced by $0.618\sigma_0$. ◆ ($H = 2.342$ mm (0.0922 in)), ▼ ($H = 2.786$ mm (0.1097 in)): experimental results for mild steel beams.[8.3]

fully clamped beam. Thus, equation (8.87) remains valid but with the flow stress σ_0 replaced by $n\sigma_0$, where n is given by equation (8.83b), and with R substituted for L. This procedure gives

$$\frac{W_f}{H} = \left(\frac{2\rho V_0^2 R^2}{3n\sigma_0 H^2}\right)^{1/2},$$ (8.88a)

where

$$n = 1 + \left\{\frac{V_0 W_f}{3\sqrt{2}DR^2}\right\}^{1/q}.$$ (8.88b)†

A simple iterative procedure is required to solve equations (8.88a, b).

If equation (8.87) is substituted into equation (8.88b) to eliminate W_f, then

$$n \cong 1 + \left\{\frac{V_0^2}{3DR}\left(\frac{\rho}{3\sigma_0}\right)^{1/2}\right\}^{1/q}$$ (8.89)

and an iterative method of solution is avoided.

It is evident from Figure 8.32 that the theoretical predictions according to

† $\epsilon_r = (\partial w/\partial r)^2/2$ for a circular plate with $u = 0$. Thus, $\dot{\epsilon}_r = (\partial w/\partial r)(\partial \dot{w}/\partial r)$, which, with $w = W(1 - r/R)$, becomes $\dot{\epsilon}_r = W\dot{W}/R^2$. The maximum strain rate occurs when $W = 2W_f/3$ and $\dot{W} = V_0/\sqrt{2}$ according to Perrone and Bhadra.[8.45, 8.46] If the average strain rate is one-half the maximum value, then $\dot{\epsilon} = (2W_f/3)(V_0/\sqrt{2})/2R^2$.

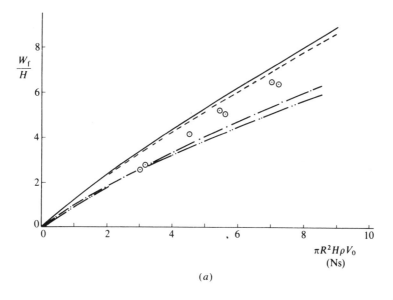

(a)

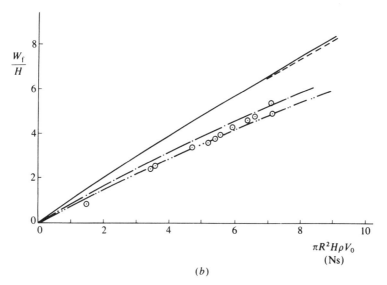

(b)

Figure 8.32 Maximum permanent transverse displacement (W_f) of a circular membrane with an impulsive velocity (V_0) distributed uniformly over the entire surface area. ———: equations (8.88a, b). – – –: equations (8.88a, 89). ⊙: experimental results[8.48]. —·—: theoretical predictions for average strain rate[8.47]. —··—: theoretical predictions of equation (11) in reference (8.46). (a) Steel plates: $R = 31.8$ mm, $H = 1.93$ mm, $\sigma_0 = 223$ MN/m^2, $\rho = 7850$ kg/m^3, $D = 40$ s^{-1}, $q = 5$. (b) Titanium plates: $R = 31.8$ mm, $H = 2.34$ mm, $\sigma_0 = 251$ MN/m^2, $\rho = 4520$ kg/m^3, $D = 120$ s^{-1}, $q = 9$.

equation (8.88a) are not markedly different for n calculated according to either equation (8.88b) or equation (8.89). However, equations (8.88a) and (8.89) are simpler to use and are, therefore, more suitable for design purposes. It is also evident from Figure 8.32 that the theoretical predictions lie above the experimental results obtained by Bodner and Symonds[8.48] and the theoretical predictions reported in references 8.46 and 8.47, though the present predictions are possibly acceptable for the steel plates.

8.5.4 Fully clamped rectangular plate loaded impulsively

Equation (7.95) predicts the maximum permanent transverse displacement (W_f)

$$\frac{W_f}{H} = \frac{(3 - \xi_0)\{(1 + \Gamma)^{1/2} - 1\}}{2\{1 + (\xi_0 - 1)(\xi_0 - 2)\}} \tag{8.90}$$

for a rectangular plate of length $2L$ and width $2B$ which is fully clamped around the entire boundary and subjected to a uniformly distributed impulsive velocity V_0, where

$$\Gamma = \frac{2\rho V_0^2 L^2 \beta^2}{3\sigma_0 H^2} (3 - 2\xi_0)\left(1 - \xi_0 + \frac{1}{2 - \xi_0}\right), \tag{8.91a}$$

$$\xi_0 = \beta\{(3 + \beta^2)^{1/2} - \beta\} \tag{8.91b}$$

and

$$\beta = \frac{B}{L}. \tag{8.91c}$$

This approximate theoretical procedure gives reasonable agreement with the experimental results recorded on the strain-rate-insensitive rectangular plates which are reported in Figure 7.18.

Now, the influence of material strain rate sensitivity is explored using the procedure of Perrone and Bhadra[8.45, 8.46] which was employed in § 8.5.2 for a fully clamped beam. Thus, equation (8.90) remains valid, but with σ_0 replaced by $n\sigma_0$ in Γ, where an estimate for n is given by equation (8.83b). This procedure gives

$$\frac{W_f}{H} = \frac{(3 - \xi_0)\{(1 + \Gamma/n)^{1/2} - 1\}}{2\{1 + (\xi_0 - 1)(\xi_0 - 2)\}}, \tag{8.92a}$$

where

$$n = 1 + \left(\frac{V_0 W_f}{3\sqrt{2}DB^2}\right)^{1/q}. \tag{8.92b}†$$

Equations (8.92a, b) may be simplified further when substituting

$$\frac{W_f}{H} \simeq \frac{(3 - \xi_0)\Gamma^{1/2}}{2\{1 + (\xi_0 - 1)(\xi_0 - 2)\}} \tag{8.93}$$

† B^2 is used in the denominator of equation (8.92b) instead of L^2 in equation (8.83b) because $0 \leqslant \beta \leqslant 1$ and a rectangular plate becomes a beam with a span $2B$ when $\beta \to 0$.

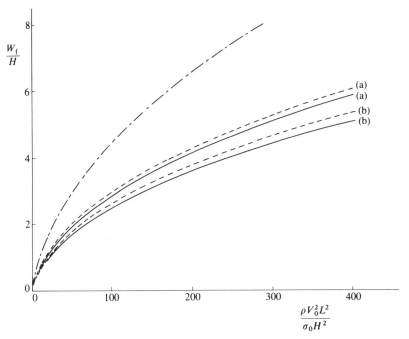

Figure 8.33 Maximum permanent transverse displacement (W_f) of a fully clamped rectangular plate subjected to an impulsive velocity of magnitude V_0 distributed uniformly over the entire plate with length $2L$ and breadth $2B$. ($\beta = 0.593$, $q = 5$, $D = 40.4\ \text{s}^{-1}$, $\rho = 7723\ \text{kg/m}^3$, $L = 64.29$ mm, (a): $H = 1.626$ mm (0.064 in) and $\sigma_0 = 247\ \text{MN/m}^2$ and (b): $H = 4.394$ mm (0.173 in) and $\sigma_0 = 253\ \text{MN/m}^2$.) ———: equation (8.90). —·—·—: equations (8.92a, b). ———: equations (8.92a) and (8.94).

from equation (8.90) when $W_f/H \gg 1$† in equation (8.83b) for n. In this case, W_f/H is predicted by equation (8.92a), with

$$n = 1 + \left[\frac{V_0 H (3 - \xi_0) \Gamma^{1/2}}{6\sqrt{2}DB^2\{1 + (\xi_0 - 1)(\xi_0 - 2)\}} \right]^{1/q} \qquad (8.94)‡$$

according to equations (8.92b) and (8.93).

It is evident from Figure 8.33 that the theoretical predictions according to equation (8.92a) are not markedly different for n calculated according to either equation (8.92b) or equation (8.94). Equations (8.92a) and (8.94) are simpler to use and are, therefore, more suitable for design purposes. However, it is evident from Figure 8.34 that the upper and lower bounds which are associated with inscribing

---

† The coefficient $\dfrac{3 - \xi_0}{2\{1 + (\xi_0 - 1)(\xi_0 - 2)\}}$ is $\frac{1}{2}$ and 1 for the extreme cases of a long narrow plate with β

= 0 (i.e., $\xi_0 = 0$) and a square plate with $\beta = 1$ (i.e., $\xi_0 = 1$), respectively.

‡ A further improvement in accuracy may be achieved by replacing σ_0 in Γ by $n\sigma_0$.

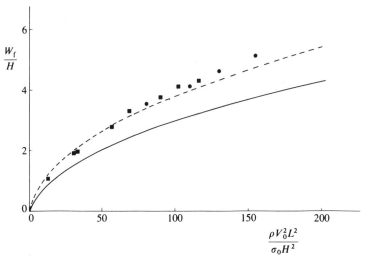

Figure 8.34 Maximum permanent transverse displacement (W_f) of a fully clamped rectangular plate subjected to an impulsive velocity of magnitude V_0 distributed uniformly over the entire plate surface of length $2L$ and breadth $2B$. ($\beta = 0.593$, $q = 5$, $D = 40.4\,\text{s}^{-1}$, $\rho = 7723\,\text{kg/m}^3$, $L = 64.29$ mm, $\sigma_0 = 247\,\text{MN/m}^2$ and $H = 1.626$ mm (0.064 in).) ———: lower bound on W_f; equations (8.92a, b). ---: upper bound on W_f; equations (8.92a, b) with σ_0 replaced by $0.618\sigma_0$. ●: $H = 1.626$ mm (0.064 in), $\sigma_0 = 247\,\text{MN/m}^2$ and ■: $H = 2.489$ mm (0.098 in), $\sigma_0 = 233\,\text{MN/m}^2$ (experimental results for mild steel rectangular plates[8.49]).

and circumscribing yield criteria, respectively, do not bound the experimental results recorded on mild steel rectangular plates,[8.49] except for $W_f/H < 2$, approximately. Nevertheless, the upper bound predictions appear acceptable for design purposes until further understanding of the strain-rate-sensitive behaviour of rectangular plates is achieved.

Equations (8.90) and (8.92) to (8.94), with $\beta \to 0$, reduce to the corresponding equations for a fully clamped beam in § 8.5.2 with a span $2B$.

8.5.5 Impulsive loading of shells

The simplified methods introduced in § 8.4 may also be used to reduce and present experimental data as illustrated in this section for shells.

The approximate procedure on the influence of material strain rate sensitivity in § 8.4.2(d) has been used in reference 8.32 to examine some experimental test data on impulsively loaded, fully clamped hemispherical shells and cylindrical shell panels. The influence of finite displacements, or geometry changes, is not important for moderate deformations of these two types of shell when subjected to uniform internal impulsive velocities.

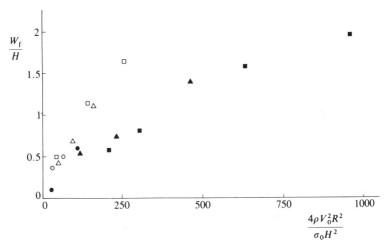

Figure 8.35 Maximum permanent radial displacement (W_f) of fully clamped hemispherical shells (mean radius R, wall thickness H) subjected to a uniformly distributed impulsive radial velocity V_0. $\circ$, $\square$, $\triangle$: experimental results for aluminium 6061-T6 hemispherical shells.[8.50] $\bullet$, $\blacksquare$, $\blacktriangle$: experimental results for mild steel hemispherical shells.[8.50]

The dimensionless maximum permanent radial displacements for the hemispherical shells are plotted in Figure 8.35 against the dimensionless initial kinetic energy, where V_0 is the initial impulsive velocity and σ_0 is the static uniaxial yield stress. This figure contains two distinct sets of data for the aluminium 6061-T6 (essentially strain-rate-insensitive) and mild steel (strain-rate-sensitive) specimens. However, if the experimental data in Figure 8.35 is replotted in Figure 8.36, using the dynamic flow stress σ_0' according to equation (8.4) instead of σ_0 in the abscissa,† then it is evident that these experimental results essentially collapse on to a common line apart from some experimental scatter. Similar results were also found for the cylindrical shell panels in reference 8.32.

The results in Figure 8.36 provide some experimental evidence for Perrone's suggestions in § 8.4.2(d). They also show that the dynamic behaviour of these shells, when made from a strain-rate-sensitive material, may be estimated using strain-rate-independent analyses with the static yield stress replaced by the initial dynamic flow stress from equation (8.4).

---

† The fully clamped hemispherical shells had a mean radius R, wall thickness H and were subjected to a uniformly distributed radially outwards impulsive velocity V_0. Thus, the initial in-plane strain-rates were $\dot{\epsilon}_\phi = \dot{\epsilon}_\theta = V_0/R$, while, for volume conservation (see § 8.3.2), the initial radial strain rate was $\dot{\epsilon}_r = -2V_0/R$. It is straightforward to show, when using equations (8.6) and (8.5), that $\dot{\epsilon}_e = 2V_0/R$ and $\sigma_e = \sigma_0'$, respectively, where σ_0' is the initial dynamic yield stress. Equation (8.4), therefore, becomes $\sigma_0'/\sigma_0 = 1 + (2V_0/(RD))^{1/q}$, where $D = 40.4 \text{ s}^{-1}$ and $q = 5$ for hot-rolled mild steel from Table 8.1, and $D = 1\,288\,000 \text{ s}^{-1}$ and $q = 4$ for aluminium 6061-T6 from § 8.3.2 and reference 8.32.

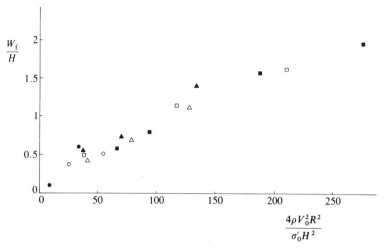

Figure 8.36 Maximum permanent radial displacements (W_f) of fully clamped hemispherical shells subjected to an impulsive radial velocity V_0. ○, □, △: experimental results for aluminium 6061-T6 hemispherical shells with σ_0' evaluated from equation (8.4) with $D = 1\,288\,000$ s^{-1} and $q = 4$.[8.32, 8.50] ●, ■, ▲: experimental results for mild steel hemispherical shells with σ_0' evaluated from equation (8.4) with $D = 40.4$ s^{-1} and $q = 5$ (see Table 8.1).[8.32, 8.50]

8.5.6 Further work

Perrone[8.42, 8.51] has studied the dynamic response of impulsively loaded rings and annular plates made from strain rate sensitive materials. He found that the theoretical behaviour, using the approximate method introduced in § 8.4.2(d) to account for material strain rate sensitivity, gave satisfactory agreement with the exact response.

Symonds[8.44] has utilised the procedure in § 8.4.2(e) to study the behaviour of a strain-rate-sensitive cantilever beam with a tip mass subjected to an impact load, as shown in Figure 3.18(a).

The introduction of the approximations discussed in § 8.4 into the above analyses made them considerably simpler than exact analyses, with little sacrifice in accuracy. These approximations, therefore, appear to have value for design purposes.

8.6 Final remarks

As already remarked in § 8.1, it is the object of this chapter to introduce the phenomenon of material strain rate sensitivity. A considerable body of knowledge has been generated and published on the form of the constitutive equations and on the properties of various materials, particularly for uniaxial dynamic loads. However, this chapter has been written from the viewpoint of an engineer interested in the dynamic response of structures with moderate values of strain rates. Thus, considerable emphasis has been placed on the simple constitutive equation of Cowper and Symonds[8.33] (equations (8.1, 8.3, 8.4)), which has been used extensively in engineering practice.

The coefficients for the Cowper–Symonds constitutive equation presented in Table 8.1 were obtained from uniaxial tests on specimens with small strains. However, it was mentioned in § 8.2.2 and § 8.2.3 and shown in Figures 8.3 and 8.7 that the strain-rate-sensitive characteristics of some materials is related to the strain magnitude. Indeed, several combinations of D and q have been suggested[8.52–8.54] for the strain rate properties of mild steel with large strains. Nevertheless, additional experimental studies are required to generate further data.[8.55]

The phenomenon of yield delay time has not been examined in this chapter because Symonds[8.17] and others have pointed out that the yield delay time and the enhanced upper yield stress disappear in impact tests when a specimen has a surface flaw, or a scratch. Various kinds of stress raisers are invariably present in real structures so that the yield delay times and upper yield stresses are rarely encountered. Nevertheless, it is interesting to note that Rabotnov and Suvorova[8.56] analysed this effect in various structures and observed that the theoretical prediction for the maximum permanent transverse displacement of a simply supported circular plate is slightly smaller than the theoretical solution in § 4.5.

Most of the materials test data presented in this chapter and in the literature were obtained during dynamic tests conducted at a constant strain rate. Clearly, the strain rate varies throughout the dynamic response of an actual structure, as shown in Figure 8.28 for an idealised model, so that several authors have explored the influence of strain rate history effects.[8.8, 8.10] Other groups have explored the influence of temperature, grain size and radiation on the strain-rate-sensitive behaviour of materials. The influence of material strain rate sensitivity on fracture initiation has been explored recently,[8.57, 8.58] while the variation of fracture strain with strain rate and inertia has been studied in references 8.59 to 8.63.

PROBLEMS

8.1 Show that the constitutive equation (8.4) reduces to equation (8.3) for the strain-rate-sensitive behaviour of an incompressible material in uniaxial tension.

8.2 Show that $d\lambda$ in the Prandtl–Reuss constitutive equation (8.8) may be expressed in the form $d\lambda = 3d\epsilon_e/2\sigma'_e$. (See the footnote on page 351.) Use this result, together with the Cowper–Symonds constitutive equation (8.4), to obtain equation (8.13) for $\dot{\epsilon}_x$.

8.3 Calculate the membrane strain rates in the complete thin-walled spherical shell, subjected to an internal pressure pulse with a rectangular-shaped pressure time history, which is discussed in § 5.6.5. The material is rigid, perfectly plastic and incompressible. Sketch the variation of the membrane strain rates and equivalent strain rate (write equation (8.6) in spherical coordinates) with time.

Sketch the variation of the equivalent stress according to equation (8.4) with time. Contrast the shape for the linear case with $q = 1$ and the highly nonlinear case having $q \gg 1$, say $q = 10$.

8.4 Now, it may be shown that $\epsilon_\theta \cong 0$ and $\epsilon_r \cong (\partial w / \partial r)^2 / 2$ for a thin circular membrane (e.g., equations (29) and (30) in reference 5.7 and the footnote to equation (8.88b)). Calculate the equivalent strain rate when the circular membrane, which is examined in § 7.8, is subjected to an impulsive velocity distributed uniformly throughout the entire membrane. Use equation (8.4) to give the corresponding variation with time of the equivalent stress when the membrane is made from a strain-rate-sensitive material which obeys the Cowper–Symonds law.

8.5 Show that equation (8.20), with the coefficient $2q/(q + 1)$ replaced by unity, gives the strain-rate-sensitive behaviour of a sandwich cross-section.

8.6 Calculate $\dot{\kappa}_\theta$ for the annular and circular plates studied in § 4.3 and § 4.4, respectively. Assume that equation (8.20) may be used to estimate the associated strain-rate-sensitive bending moment.

8.7 Evaluate the axial strain-rate history for the wire in § 8.4.2(b) when it is made from a rigid, perfectly plastic material. Use equation (8.3) to estimate the corresponding uniaxial stress history when the wire is made from a rate-sensitive material, and compare it with the constant value (equation (8.45)) which is proposed by Perrone in § 8.4.2(d).

8.8 Obtain v and μ according to equations (8.52a, b) for the Symonds' viscous approximation of the problem in Figure 8.21.

8.9 Obtain v and μ according to equations (8.77a, b) for the extension of the Symonds' viscous approximation for the finite-displacement problem in Figure 8.25.

8.10 A fully clamped hemispherical shell of mean radius, R, and wall thickness, H, is subjected to an internal uniformly distributed impulsive velocity, V_0. Calculate the initial membrane strain rates if the shell is thin (i.e., neglect curvature changes) and made from a rigid plastic material which is incompressible. Estimate the initial equivalent flow stress from equation (8.4) when the material is strain rate sensitive.

8.11 Repeat Problem 8.10 for a cylindrical shell roof having a mean radius, R, thickness, H, when subjected to a uniformly distributed impulsive velocity, V_0.

9

Dynamic progressive buckling

9.1 Introduction

The structural members examined in the previous chapters responded in a stable manner when subjected to dynamic loads. However, in practice, dynamic loads may cause an unstable response for which the methods of the previous chapters are

Figure 9.1 Static and dynamic axial crushing of a thin-walled mild steel cylindrical shell with a mean radius (R) of 27.98 mm, mean wall thickness (H) of 1.2 mm and an initial axial length (L) of 178 mm. The top three wrinkles developed as a result of static loading, while the remaining wrinkles were produced when the tube was struck by a mass (M) of 70 kg travelling at an impact velocity of 6.51 m/s.

unsuitable. This chapter and the next examine the unstable response of structures. In particular, the dynamic progressive buckling phenomenon[9.1], which is illustrated in Figure 9.1 for a circular tube, is studied in this chapter.

A thin-walled cylindrical shell, or tube, when subjected to a static axial load, as shown in Figure 9.2, may have force–axial displacement characteristics similar to those in Figure 9.3(a). It is evident that the tube exhibits an unstable behaviour after reaching the first peak load at point A. Most structural designs are based on a load equal to this peak load divided by a safety factor. The magnitude of this safety factor is selected by taking into account the slope AB of the load-deflection behaviour (post-buckling characteristics). However, thin-walled circular tubes are used in many practical situations to absorb impact energy. Indeed, Pugsley[9.2] examined the axial impact of thin-walled circular and square tubes in order to study the structural crashworthiness of railway coaches. In this circumstance, the

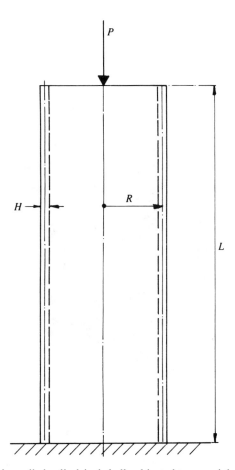

Figure 9.2 A thin-walled cylindrical shell subjected to an axial crushing force P.

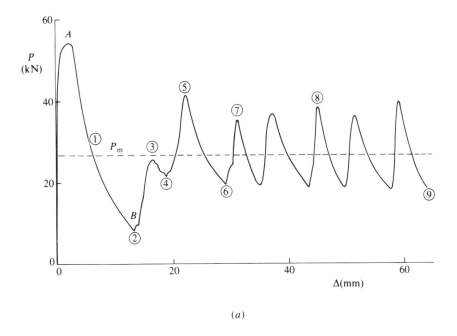

(a)

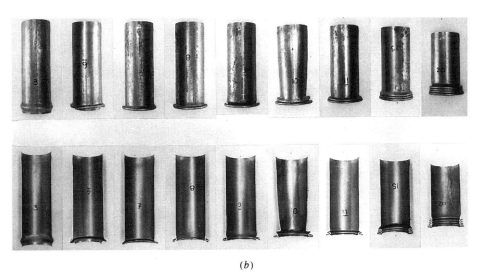

(b)

Figure 9.3 Static axial crushing behaviour of a thin-walled mild steel circular tube with a mean radius (R) of 27.98 mm, mean wall thickness (H) of 1.2 mm and an initial axial length (L) of 178 mm. (a) Axial force versus axial crushing distance. (b) Photographic record of the development of wrinkles during axial crushing. The photographs (from left to right) refer to the numbers ① to ⑨ in Figure 9.3(a). The upper row gives the outside views, while the lower row shows the specimens cut open across a diameter.

total axial displacement of a tube exceeds considerably the displacement associated with the load at B in Figure 9.3(a). Thus, an entirely different approach is required from that employed normally to examine the plastic buckling of structures.

It is evident from Figure 9.3(a) that the load–displacement behaviour exhibits a repeated pattern. In fact, each pair of peaks in Figure 9.3(a) is associated with the development of a wrinkle or buckle in Figure 9.3(b). Usually, these wrinkles, or buckles, develop sequentially from one end of a tube so that the phenomenon is known as progressive buckling. The most efficient use of the tube material occurs when as much as possible is crushed, as indicated in Figure 9.4 for a thin-walled tube with a square cross-section. For convenience, designers often ignore the fluctuations in the load-displacement characteristics and use a mean value (P_m), as indicated in Figure 9.3(a). Incidentally, an ideal energy-absorbing device is defined, for some purposes, as one which has a constant resistance and, therefore, offers a constant deceleration throughout the entire stroke.[9.4]

The axial impact of circular tubes with low velocities (up to tens of metres per second for metal tubes) is taken[9.1] as quasi-static, and the influence of inertia forces is, therefore, ignored. This is a reasonable simplification when the striking mass (M) is much larger than the mass of a tube (m). The axial inertia force of a striking mass is $M\ddot{u}$, where $\ddot{u}$ is the axial deceleration during the impact event. If the axial velocity–time history is continuous at the interface between a striking mass and the end of a tube, then the axial inertia force in the tube is of order $m\ddot{u}$, which is negligible compared with $M\ddot{u}$ when $m \ll M$.

The response of the impact problem described above is controlled by the phenomenon associated with static progressive buckling. Nevertheless, it is called dynamic progressive buckling in this chapter because material strain rate effects are important for a strain-rate-sensitive material.[9.1] If the influence of inertia effects of the tube are important in a practical problem, with larger axial impact velocities, then the phenomenon is known as dynamic plastic buckling, which is examined in the next chapter.

9.2 Static axial crushing of a circular tube

9.2.1 Introduction

A thin-walled circular tube of mean radius R and thickness H, when subjected to an axial force, as shown in Figure 9.2, may develop either axisymmetric buckles similar to those in Figure 9.1, or a non-axisymmetric (diamond) pattern, which is indicated in Figure 9.5. Various theoretical methods predict that thicker tubes with $R/H < 40$–45, approximately, deform axisymmetrically, while the thinner tubes, with larger values of R/H, buckle into a non-axisymmetric mode.[9.1] However, some tubes may switch, during a test, from an axisymmetric deformation mode into

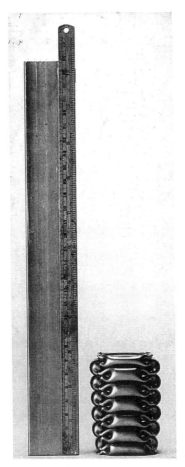

Figure 9.4 A thin-walled mild steel tube with a square cross-section before and after static axial crushing.[9.3]

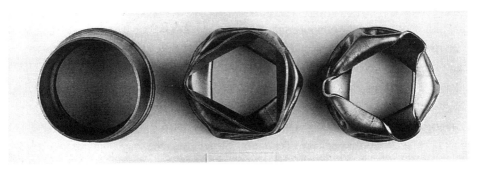

Figure 9.5 Axially crushed circular tube test specimens.[9.1] Axisymmetric, or concertina, deformation mode on left and non-axisymmetric, or diamond, deformation modes at centre and right.

Figure 9.6 Final deformation modes of circular cylindrical shells which were crushed dynamically with a drop weight. The first complete buckle (at the impacted end of each tube) is axisymmetric in each case. The subsequent buckles remain axisymmetric in the left-hand specimen, while the two non-axisymmetric buckles in the centre specimen have two corners and the non-axisymmetric buckles in the right-hand specimen have three corners.

a diamond pattern. This was observed by Mallock[9.5] as long ago as 1908 and more recently by Mamalis and Johnson[9.6] and is shown in Figure 9.6, which is taken from reference 9.7.†

Alexander[9.8] and Pugsley and Macaulay[9.9] presented static theoretical analyses for the axisymmetric and non-axisymmetric behaviour of circular tubes, respectively. The axisymmetric solution of Alexander[9.8] is outlined in the next section.

9.2.2 Axisymmetric crushing

9.2.2(a) Introduction

An approximate theoretical analysis for the axially loaded thin-walled circular tube in Figure 9.2 was obtained by Alexander.[9.8] He assumed that the tube was made from a rigid, perfectly plastic material and used the simplified axisymmetric

---

† In fact, all the circular tubes, which were loaded statically in references 9.1 and 9.7 and which exhibited a diamond pattern, commenced deformation with an axisymmetric mode of deformation.

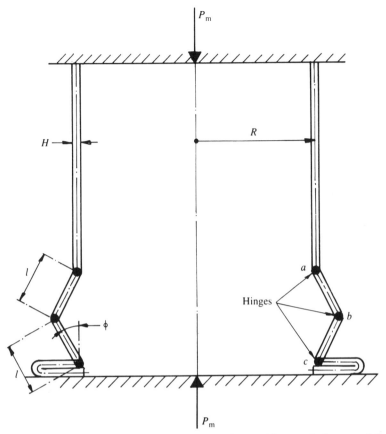

Figure 9.7 Idealised axisymmetric, or concentric, crushing mode for an axially compressed cylindrical shell.

deformation pattern with plastic hinges which is illustrated in Figure 9.7.† This collapse pattern is an idealisation of the actual behaviour, since it is evident from Figure 9.3(*b*) that the profiles of the wrinkles are curved rather than straight. Nevertheless, the theoretical predictions have some value for design purposes and have been extended in recent years to cater for various effects, giving improved agreement with experimental results. Moreover, this analysis illustrates the general approach which is employed for some dynamic progressive buckling problems and, therefore, has value as an introduction to the topic.

It is evident from Figure 9.7 that the work $P_m \times 2l$, which is expended by a constant or mean external force P_m to develop and flatten completely one axisymmetric wrinkle,† or buckle, is equal to the internal energy dissipated due to plastic deformation in the tube. The internal energy dissipation rate is derived in the next section.

† The dimension l in Figure 9.7 is taken as constant, regardless of the number of wrinkles.

9.2.2(b) Internal energy dissipation

The total plastic energy absorbed by the two stationary axisymmetric plastic hinges at a and c in Figure 9.7 during the formation of one wrinkle is

$$D_1 = 2 \times 2\pi R M_0 \pi / 2 \qquad (9.1)$$

where the plastic collapse moment for the cross-section (per unit circumferential length) is

$$M_0 = (2\sigma_0/\sqrt{3})H^2/4. \qquad (9.2)\dagger$$

The radial position of the axisymmetric plastic hinge at b increases from R to $R + l$ during the formation of one complete wrinkle. Thus, the energy absorbed by the axisymmetric central hinge b in Figure 9.7 during an incremental change $d\phi$ is

$$dD_2 = 2\pi(R + l \sin \phi)M_0(2 \, d\phi),$$

which gives the total energy dissipated as

$$D_2 = \int_0^{\pi/2} 4\pi(R + l \sin \phi)M_0 \, d\phi,$$

or

$$D_2 = 4\pi M_0(R\pi/2 + l). \qquad (9.3)$$

It is evident from Figure 9.7 that the axisymmetric portions ab and bc of a wrinkle are stretched circumferentially between ϕ and $\phi + d\phi$ with a mean‡ engineering strain increment

$$d\epsilon_\theta = \frac{2\pi\{(l/2)\sin(\phi + d\phi)\} - 2\pi\{(l/2)\sin \phi\}}{2\pi R},$$

which, expanding the $\sin(\phi + d\phi)$ term using standard trigonometric relations, becomes

$$d\epsilon_\theta = l \cos \phi \, d\phi/2R \qquad (9.4)$$

when $\sin d\phi \to d\phi$ and $\cos d\phi \to 1$. Thus, the energy absorbed in circumferential stretching during an incremental change from ϕ to $\phi + d\phi$ is

$$dD_3 = \sigma_0 \, d\epsilon_\theta 2lH2\pi R,$$

which, using equation (9.4), may be written

$$D_3 = \int_0^{\pi/2} \sigma_0(l \cos \phi \, d\phi)2lH\pi,$$

or

$$D_3 = 2\sigma_0 l^2 H\pi. \qquad (9.5)\ddagger$$

---

† Alexander[9.8] used the von Mises yield condition and assumed that the tube was in a state of plane strain. This is strictly true only for the plastic hinges at a and c in Figure 9.7 because $\epsilon_\theta \geqslant 0$ for the axisymmetric hinge at b.

‡ The assumption of a mean circumferential strain is removed in reference 9.1 and equation (9.5) is then multiplied by $(1 + l/3R)$.

The total energy absorbed during the development of one complete wrinkle in a thin-walled circular tube is

$$D_T = D_1 + D_2 + D_3,$$ (9.6)

which, using equations (9.1), (9.3) and (9.5), becomes

$$D_T = 4\pi M_0(\pi R + l) + 2\sigma_0 l^2 H\pi,$$

or

$$D_T = 2\pi\sigma_0 H^2(\pi R + l)/\sqrt{3} + 2\pi\sigma_0 l^2 H,$$ (9.7)

when eliminating M_0 with equation (9.2).

9.2.2(c) Axial crushing force

Now, in order to conserve energy, the total external work done by a constant axial force during the formation and complete flattening of one wrinkle ($P_m \times 2l$) equals the internal work according to equation (9.7). Thus,

$$P_m 2l = 2\pi\sigma_0 H\{H(\pi R + l)/\sqrt{3} + l^2\},$$

or

$$P_m/\sigma_0 = \pi H\{H(\pi R/l + 1)/\sqrt{3} + l\}.$$ (9.8)

The axial length $l = ab = bc$ between the axisymmetric hinges in Figure 9.7 is unknown but may be obtained by minimising the axial crushing force, or $dP_m/dl = 0$, which gives

$$H(-\pi R/l^2)/\sqrt{3} + 1 = 0,$$

or

$$l = (\pi R H/\sqrt{3})^{1/2}.$$ (9.9)

Substituting equation (9.9) into equation (9.8) gives the axial crushing force

$$P_m/M_0 = 4(3)^{1/4}\pi^{3/2}(R/H)^{1/2} + 2\pi,$$ (9.10a)

or

$$P_m/M_0 = 29.31(R/H)^{1/2} + 6.28,$$ (9.10b)

which is identical to Alexander's theoretical prediction[9.8] when M_0 is defined by equation (9.2).

Equation (9.10b) was obtained using the collapse mechanism in Figure 9.7 which assumes that the convolutions, or wrinkles, form on the outside of a tube. Alexander[9.8] repeated the theoretical analysis and found that the mean crushing force for a tube with convolutions which form internally instead of externally is

$$P_m/M_0 = 4(3)^{1/4}\pi^{3/2}(R/H)^{1/2} - 2\pi.$$ (9.11)

Alexander assumed that the average of equations (9.10a) and (9.11)

$$P_m/M_0 = 4(3)^{1/4}\pi^{3/2}(R/H)^{1/2},$$ (9.12a)†

---

† Alexander[9.8] further modifies the coefficient in equation (9.12a) by assuming that the crushing force is the average of the values predicted by equation (9.12a) and equation (9.8) without the $\pi H^2/\sqrt{3}$ term and with $l = \pi(2RH)^{1/2}/2\{3(1 - v^2)\}^{1/4}$ for the linear elastic case.

or

$$P_m = 2(\pi H)^{3/2} R^{1/2} \sigma_0 / 3^{1/4}, \qquad (9.12b)$$

offers a reasonable approximation to the actual crushing force.

9.2.2(d) Comment

The theoretical procedure in §§ 9.2.2(b) and 9.2.2(c) does not satisfy the upper bound theorem of plasticity which is introduced in § 2.4.3. The collapse mechanism in Figure 9.7 requires finite displacements and finite rotations, whereas the static plastic collapse theorems for a perfectly plastic material were developed using the principle of virtual velocities in Appendix 3 for infinitesimal displacements.

9.3 Dynamic axial crushing of a circular tube

9.3.1 Introduction

The theoretical analysis, which is outlined in § 9.2, was developed for the axisymmetric crushing of a cylindrical tube when subjected to a static axial load. However, it is noted in § 9.1 that this analysis also describes the dynamic progressive buckling of a circular tube which may be considered as a quasi-static problem. Although inertia effects may be neglected, the influence of material strain rate sensitivity must be retained for many materials. Thus, if a circular tube is made from a strain-rate-sensitive material, then it is necessary to modify the plastic flow stress in equation (9.12b) in order to cater for the enhancement of the flow stress with strain rate.

9.3.2 Influence of strain rate sensitivity

Equation (8.3), known as the Cowper–Symonds constitutive equation, gives dynamic flow stresses which agree reasonably well with the dynamic uniaxial tension and compression test results on several materials. Now, if the static flow stress (σ_0) in equation (9.12b) is replaced by the dynamic flow stress according to equation (8.3), then

$$P_m = 2(\pi H)^{3/2} R^{1/2} \sigma_0 \{1 + (\dot{\epsilon}/D)^{1/q}\} / 3^{1/4}, \qquad (9.13)†$$

where D and q are constants for the tube material and are given in Table 8.1.

---

† It should be noted that equation (9.12b) is obtained using equations (9.1), (9.3) and (9.5) for D_1, D_2 and D_3, respectively. D_3 is related to the circumferential membrane stress which would vary with strain rate according to equation (8.3). However, D_1 and D_2 are associated with the bending moments at the plastic hinges which would vary according to equation (8.20). Moreover, it may be shown that $D_1:D_2:D_3::1:1 + (4H/\sqrt{3\pi R})^{1/2}:2$, so that over one-half of the energy is dissipated at the three axisymmetric plastic hinges. Nevertheless, if $q \gg 1$, then $2q/(2q + 1) \cong 1$ (e.g., $q = 5$ for mild steel gives $2q/(2q + 1) = 10/11$) and equations (8.3) and (8.20) would predict a similar increase of the dynamic flow stress, where $H\dot{\kappa}/2$ is the strain rate at the outer surface.

The strain rate ($\dot{\epsilon}$) in equation (9.13) is taken as constant, although it varies spatially and temporally during an impact event. An estimate is now made for $\dot{\epsilon}$ using an approximate procedure which is suggested in reference 9.1.

Equation (9.4) predicts that the mean circumferential strain in a completely flattened buckle of the circular tube in Figure 9.7 is

$$\epsilon_\theta \cong l/2R, \tag{9.14}$$

which may also be obtained by inspection.† The time T to flatten completely a single buckle is now required in order to estimate the mean circumferential strain rate as $\dot{\epsilon}_\theta = \epsilon_\theta/T$. It is assumed that the axial velocity at the struck end of a tube varies linearly with time from the impact velocity V_0 at $t = 0$ until motion ceases when $t > T$ after the development of several wrinkles. This gives a constant deceleration and a constant axial force which is consistent with a mean crushing force P_m. Now the time required to form the first wrinkle, or convolution, is $T = 2l/V_0$ when taking the impact velocity V_0 to remain constant, which would be reasonable when a large number of wrinkles are formed in a tube. Therefore, the average strain rate $\dot{\epsilon}_\theta = \epsilon_\theta/T$ is

$$\dot{\epsilon}_\theta = V_0/4R, \tag{9.15}$$

which is used as an approximation for $\dot{\epsilon}$ in equation (9.13). Equation (9.13) may now be written

$$P_m = 2(\pi H)^{3/2} R^{1/2} \sigma_0 \{1 + (V_0/4RD)^{1/q}\}/3^{1/4}, \tag{9.16}$$

where V_0 is the axial impact velocity at the end of a circular tube. If $D \to \infty$ for a strain-rate-insensitive material, then equation (9.16) reduces to equation (9.12b) as expected.

The lack of accuracy in the simplified estimates for the mean strain rate $\dot{\epsilon} = V_0/4R$ is not as important as appears at first sight because the term $(V_0/4RD)^{1/q}$ is highly non-linear owing to the relatively large values for q in Table 8.1. For example, if the actual mean strain rate in a steel tube with $q = 5$ were twice as large as estimated by equation (9.15), then the actual strain-rate-sensitive term in equation (9.16) would be $(2)^{1/5}(V_0/4RD)^{1/5}$ or $1.15(V_0/4RD)^{1/5}$, which is 15 per cent larger.

9.3.3 Structural effectiveness and solidity ratio

In order to assist in the presentation of experimental results and theoretical predictions for the axial crushing of thin-walled sections, Pugsley[9.2] introduced the two dimensionless ratios known as structural effectiveness and solidity ratio.

The structural effectiveness is defined as

$$\eta = P_m/A\sigma_1, \tag{9.17}$$

where P_m is the mean axial crushing force, A is the cross-sectional area of the thin-

---

† Equation (2.59) with $w = -l/2$ predicts equation (9.14).

walled cross-section and σ_1 is a characteristic stress. If $\sigma_1 = \sigma_0$, where σ_0 is the plastic flow stress, then $A\sigma_0$ is the squash load required to cause uniform plastic flow due to an axial force. The structural effectiveness $\eta = P_m/A\sigma_0$ is, therefore, a ratio between the mean crushing force and the squash load. For the particular case of a thin-walled circular cylindrical tube, $A = 2\pi RH$ and

$$\eta = P_m/2\pi RH\sigma_0. \tag{9.18}$$

The solidity ratio, or relative density, is defined as

$$\phi = A/A_c, \tag{9.19}$$

where A_c is the cross-sectional area which is enclosed by the cross-section. Clearly, $\phi \to 0$ represents a section with very thin walls. For the particular case of a thin-walled circular cylindrical tube, $A = 2\pi RH$ and $A_c = \pi R^2$ so that equation (9.19) becomes

$$\phi = 2\pi RH/\pi R^2,$$

or

$$\phi = 2H/R. \tag{9.20}$$

The dimensionless parameters (9.18) and (9.20) may be used to write equations (9.12b) and (9.16) in the form

$$\eta = (\pi\phi/2\sqrt{3})^{1/2} \tag{9.21}$$

and

$$\eta = (\pi\phi/2\sqrt{3})^{1/2}\{1 + (V_0/4RD)^{1/q}\}, \tag{9.22}$$

respectively. The additional dimensionless parameter $V_0/4RD$ in equation (9.22) caters for the phenomenon of material strain rate sensitivity. If $V_0/4RD = 1$, then the correction factor for material strain rate sensitivity equals 2 and the mean dynamic progressive buckling force is double the corresponding static value regardless of the value of q.

9.3.4 Comparison with experimental results on static crushing

Some experimental results for the static axial crushing of thin-walled cylindrical shells are presented in Figure 9.8 with the aid of the dimensionless parameters η and ϕ which are defined in § 9.3.3. There is considerable scatter in this figure which is due to any initial imperfections in the tubes, different testing arrangements, different material properties, different static loading rates, and varying numbers of fully developed wrinkles, etc. Nevertheless, the theoretical predictions of equation (9.21) tend to under-predict the dimensionless mean crushing force.

It was demonstrated in references 9.1 and 9.12 that the idealised collapse profile in Figure 9.7 might be modified to approximate more closely with the actual deformed shape of a tube such as the one shown in Figure 9.3(b). This is achieved

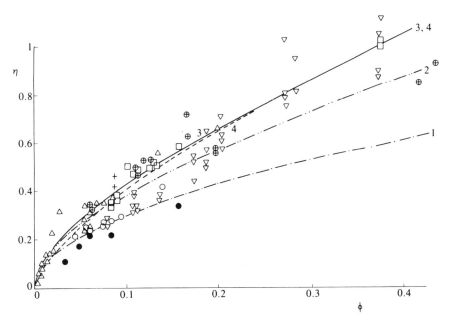

Figure 9.8 Static axial crushing of thin-walled cylindrical shells.
—·—1: equation (9.21). —··—··—2: equation (9.29). ——3: equation
(9.33). ———4: equation (9.35).

Experimental results: +: reference 9.1 (mild steel). ●: Mamalis and Johnson [9.6] (aluminium 6061-T6). ⊕: reference 9.7 (mild steel). ○: Alexander[9.8]
(mild steel). △: Macaulay and Redwood.[9.10] □, ▽: taken from Figure 4.4 of
reference 9.11.

by introducing the effective crushing distance

$$\delta_e = 2l - 2x_m - H, \tag{9.23}$$

which is defined in Figure 9.9. In other words, a circular tube collapses axially
through a distance δ_e during the formation of a complete wrinkle, or buckle, which
gives rise to an external work $P_m\delta_e$ instead of $P_m 2l$ which was used to obtain
equation (9.8).

Abramowicz[9.12] examined the axial crushing behaviour of an inelastic column
of length $2l$ and observed that

$$x_m \cong 0.28(l/2), \tag{9.24}$$

which, substituting into equation (9.23), gives an effective crushing distance

$$\delta_e = 1.72l - H,$$

or

$$\delta_e/2l = 0.86 - H/2l, \tag{9.25}$$

which becomes

$$\delta_e/2l = 0.86 - 0.37(H/R)^{1/2} \tag{9.26}$$

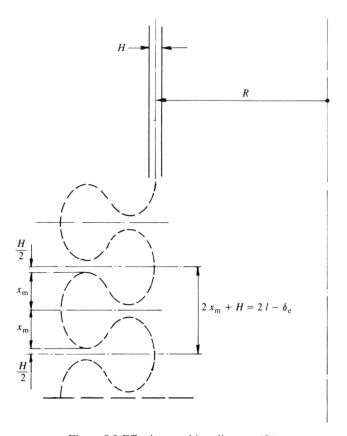

Figure 9.9 Effective crushing distance (δ_e).

when using equation (9.9).† Finally, equations (9.12b) and (9.16) are replaced by

$$P_m = 2(\pi H)^{3/2} R^{1/2} \sigma_0 / [3^{1/4} \{0.86 - 0.37(H/R)^{1/2}\}] \tag{9.27}$$

and

$$P_m = 2(\pi H)^{3/2} R^{1/2} \sigma_0 \{1 + (V_0/4RD)^{1/q}\} / [3^{1/4} \{0.86 - 0.37(H/R)^{1/2}\}], \tag{9.28}$$

respectively.

Equation (9.27) may be written in the dimensionless form

$$\eta = (\pi \phi/2)^{1/2} / [3^{1/4} \{0.86 - 0.37(\phi/2)^{1/2}\}] \tag{9.29}$$

when using equations (9.18) and (9.20) and gives better agreement with the experimental results in Figure 9.8.

If the actual circumferential strain variation in a tube is used instead of the mean value in equation (9.4), then equation (9.5) is replaced by[9.1]

$$D_3 = 2\sigma_0 l^2 H \pi (1 + l/3R). \tag{9.30}$$

† It is assumed that the new shape of a buckle absorbs the same total internal energy.

However, the same theoretical procedure in §§ 9.2.2(b) and 9.2.2(c) leads to a transcendental equation for l. Nevertheless, it is shown in reference 9.1 that

$$l = 1.76(RH/2)^{1/2} \qquad (9.31)$$

gives an acceptable approximation for the axial distance between two adjacent circumferential hinges in Figure 9.7 when $10 \leqslant 2R/H \leqslant 60$, approximately. The mean axial crushing force is now

$$P_{\mathrm{m}}/M_0 = 29.4(R/H)^{1/2} + 11.9 \qquad (9.32)$$

instead of equation (9.10b).

Equation (9.32) may also be modified to cater for the effective crushing length using equation (9.25) with l given by equation (9.31). In this case

$$\eta = 3.36(1 + 0.29\phi^{1/2})/(3.03\phi^{-1/2} - 1) \qquad (9.33)$$

when using equations (9.2), (9.18) and (9.20). Similarly,

$$\eta = 3.36(1 + 0.29\phi^{1/2})\{1 + (V_0/4RD)^{1/q}\}/(3.03\phi^{-1/2} - 1) \qquad (9.34)$$

for the dynamic crushing case.

It is evident from Figure 9.8 that equation (9.33) gives reasonable agreement with the experimental results and would be acceptable for design purposes. An empirical relation[9.11]

$$\eta = 2\phi^{0.7} \qquad (9.35)$$

is also presented in Figure 9.8.

9.3.5 Comparison with experimental results on dynamic crushing

Some experimental results for the dynamic progressive buckling of thin-walled cylindrical tubes subjected to dynamic axial loads are presented in Figure 9.10. The horizontal line is the theoretical prediction for a strain-rate-insensitive material. Clearly, the experimental results lie above this line, which is due largely to the phenomenon of material strain rate sensitivity. In fact, equations (9.18), (9.33) and (9.34) may be cast in the form

$$P_{\mathrm{m}}^{\mathrm{d}}/P_{\mathrm{m}}^{\mathrm{s}} = 1 + (V_0/4RD)^{1/q}, \qquad (9.36)$$

where $P_{\mathrm{m}}^{\mathrm{d}}$ and $P_{\mathrm{m}}^{\mathrm{s}}$ are the dynamic and static progressive buckling forces, respectively. Equation (9.36) gives the increase in axial crushing force due to material strain rate sensitivity.

Now, the parameters D and q in equation (9.36) are listed in Table 8.1 for several strain-rate-sensitive materials. However, these constants were obtained from dynamic tests on materials with strains up to a magnitude of only a few per cent. The average strains in the specimens in Figure 9.10 were of order to 12 per cent[9.1] with an associated plastic flow stress similar to the ultimate tensile stress. The experimental results in Figure 8.1 reveal that the material strain rate characteristics

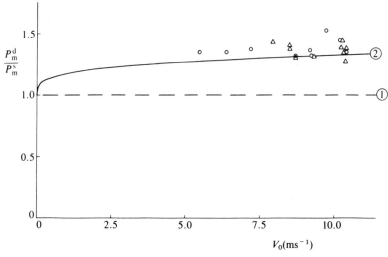

Figure 9.10 Ratio of dynamic axial crushing forces to static axial crushing forces for cylindrical shells. $---$1: equation (9.36) for a strain-rate-insensitive material. $\underline{\quad}$ 2: equation (9.36) with $D = 6844 \text{ s}^{-1}$ and $q = 3.91$. $\bigcirc$: experimental results with an axisymmetric deformation.[9.1] $\triangle$: experimental results with a non-axisymmetric deformation.[9.1]

for mild steel are sensitive to the magnitude of strains larger than about 2–3 per cent. It transpires that $D = 6844 \text{ s}^{-1}$ and $q = 3.91$[9.3] when Campbell and Cooper's[9.13] experimental data for the variation of the ultimate tensile stress with strain rate was replotted using the method in § 8.3.2. Equation (9.36), with these coefficients, gives fair agreement with the experimental results in Figure 9.10 for circular tubes which have an axisymmetric response of the type studied in this chapter.

A few experimental results for the non-axisymmetric, or diamond, mode of crushing are presented in Figure 9.10. These results suggest that the associated dynamic progressive buckling force is slightly smaller than the corresponding axisymmetric value. Approximate theoretical predictions for non-axisymmetric buckling are discussed in reference 9.1.

It should be noted that the experimental results in Figure 9.10 were obtained on mild steel tubes having the same mean radius, wall thickness and material properties. The only variables in these tests were the initial impact velocity, which has an important influence in equation (9.36), and the tube length, which was not long enough to cause any evidence of overall buckling. However, every tube which deformed with a non-axisymmetric response, except one, commenced crushing with an axisymmetric wrinkle, as indicated for two specimens in Figure 9.6.

9.4 Static axial crushing of a square tube

9.4.1 Introduction

The theoretical analyses for the static and dynamic progressive buckling of thin-walled square tubes follows the same general procedure which is outlined in §§ 9.2 and 9.3 for circular tubes. The crushing behaviour and, therefore, the actual details of the analysis are more complex, as anticipated from the photographs of the axially crushed square tubes in Figures 9.4 and 9.11. Nevertheless, Wierzbicki and Abramowicz[9.14-9.16] have identified the two basic collapse elements in Figure 9.12 which they used to examine the static progressive buckling of square tubes with a mean width C and a mean wall thickness H. These basic collapse elements have also been used to study the dynamic progressive buckling of square tubes.[9.3]

It transpires from the geometrical compatibility requirements at the vertical

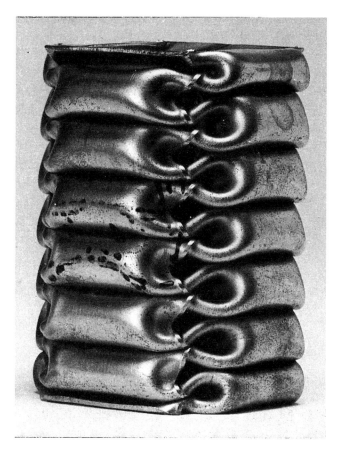

Figure 9.11 Corner view of the axially crushed square tube in Figure 9.4 (symmetric crushing mode).

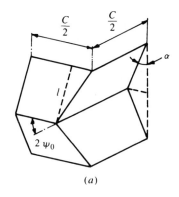

(a)

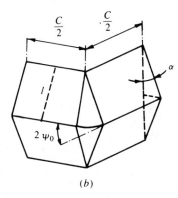

(b)

Figure 9.12 Basic collapse elements. (a) Type I. (b) Type II.

interfaces of the basic elements that there are four different progressive buckling modes.[9.3] The corresponding mean crushing forces were found by equating the external work of the axial crushing force to the internal energy required to form either one complete layer of lobes from four basic collapse elements or two adjacent layers of lobes with eight basic elements. The symmetric crushing mode in Figures 9.4 and 9.11 is idealised with four of the basic collapse elements in Figure 9.12(a) for each layer of lobes, as shown in Figure 9.13 for a paper model. This particular crushing mode is predicted to form in thin square tubes with $C/H > 40.8$, approximately.[9.7]

Each layer of lobes in an extensional mode of crushing is idealised with four of the basic collapse elements in Figure 9.12(b). This type of progressive buckling is shown in Figure 9.14 and is predicted to form in thick square tubes with $C/H < 7.5$, approximately.[9.7]

The asymmetric mixed mode B-type progressive buckling in Figure 9.15 is idealised as two adjacent layers of lobes having seven of the basic elements in

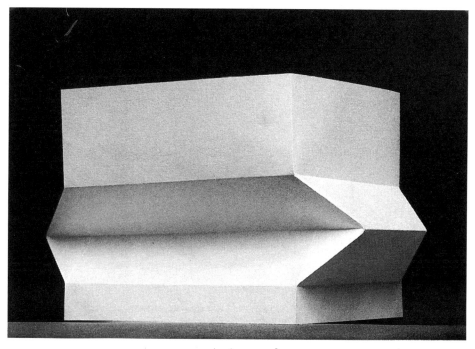

Figure 9.13 Paper model of the symmetric crushing mode for a square tube.[9.3] This is an idealisation of the deformation mode in Figures 9.4 and 9.11 and is constructed using four of the Type I basic collapse elements in Figure 9.12(a).

Figure 9.12(a) and one of the elements in Figure 9.12(b). This type of crushing is predicted[9.7] to form within the range $7.5 \leqslant C/H \leqslant 40.8$. In fact, the difference between the theoretical crushing forces associated with a symmetric mode and an asymmetric mode B is small so that either may develop in an actual square tube specimen which has slight imperfections. Another mode of deformation is possible kinematically, although it has a slightly higher collapse force than any of the three modes discussed above. Nevertheless, this crushing mode did develop in some of the specimens loaded dynamically in reference 9.3.

9.4.2 Static crushing

It is evident from the discussion in § 9.4.1 that the theoretical analysis for symmetric crushing (see Figures 9.4, 9.11 and 9.13) is adequate to predict the behaviour for most thin-walled square tubes and is even acceptable for those with thick walls. As noted previously, the internal energy consumed by four of the basic collapse elements in Figure 9.12(a) in forming one complete layer of lobes is equated to the external work done by the mean axial crushing force. This expression is then

Figure 9.14 Extensional crushing mode for a thin-walled square tube subjected to a dynamic axial force.[9.7]

minimised to predict[9.3,9.7,9.15]

$$P_{\mathrm{m}}/M_0 = 38.12(C/H)^{1/3} \qquad (9.37)$$

with

$$l/H = 0.99(C/H)^{2/3}, \qquad (9.38)$$

where

$$M_0 = \sigma_0 H^2/4. \qquad (9.39)\dagger$$

Figures 9.4 and 9.11 reveal that the lobes do not flatten completely during crushing, as idealised in the theoretical analysis leading to equation (9.37), which uses the basic collapse elements in Figures 9.12(a) and 9.13. This phenomenon was encountered in § 9.3.4 for circular tubes, and it is shown in reference 9.3 that the effective crushing distance for the symmetric crushing of a square tube is

$$\delta_{\mathrm{e}}/2l = 0.73, \qquad (9.40)$$

which allows equation (9.37) to be rewritten

† The von Mises yield condition is used for M_0 in equation (9.2) and in all equations for circular tubes, while the Tresca yield condition is used in equation (9.39) and in all subsequent expressions for square tubes. This difference is historical and artificial but is maintained in this chapter to remain consistent with the literature published on this topic.

Figure 9.15 Asymmetric mixed crushing mode B for a thin-walled square tube subjected to a static axial force.[9.3]

$$P_{\mathrm{m}}/M_0 = 52.22(C/H)^{1/3}. \tag{9.41}$$

Now, the structural effectiveness is defined by equation (9.17) with $A = 4CH$ for a thin-walled square tube, or

$$\eta = P_{\mathrm{m}}/4CH\sigma_0 \tag{9.42}$$

when $\sigma_1 = \sigma_0$. The solidity ratio is

$$\phi = 4H/C \tag{9.43}$$

according to equation (9.19) with $A_{\mathrm{c}} = C^2$. These dimensionless parameters allow equation (9.41) to be written in the form

$$\eta = 1.3\phi^{2/3}. \tag{9.44}$$

The theoretical predictions of equation (9.44) give reasonable agreement with the experimental results in Figure 9.16, which were obtained during several research programmes on square box columns loaded statically. The empirical prediction[9.11]

$$\eta = 1.4\phi^{0.8} \tag{9.45}$$

is also presented in Figure 9.16.

9.5 Dynamic axial crushing of a square tube

The dynamic progressive buckling of a square tube is idealised as a quasi-static response, as discussed for a circular tube in § 9.3.1. Thus, equation (9.41) predicts the dynamic axial crushing force of a square tube made from a strain-rate-insensitive material. If a square tube is made from a strain-rate-sensitive material,

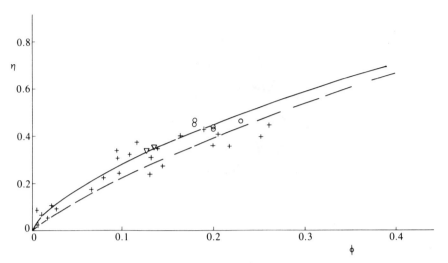

Figure 9.16 Static axial crushing of square box columns. ———: equation (9.44).
— — — —: equation (9.45).
    Experimental results: +: data from three test programmes reported in Figure
3.8 of reference 9.16. ▽: reference 9.3. ○: reference 9.7.

then

$$P_m/M_0 = 52.22(C/H)^{1/3}\{1 + (\dot{\epsilon}/D)^{1/q}\} \qquad (9.46)$$

by analogy with the circular tube in § 9.3.2.

It is difficult to make an accurate estimate of the strain rate $\dot{\epsilon}$ in a square tube because of the complex deformation patterns which leads to the permanent shape shown, for example, in Figures 9.4 and 9.11. However, an estimate

$$\dot{\epsilon} = 0.33V_0/C \qquad (9.47)$$

is obtained in reference 9.3 for the mean strain rate in the toroidal corner regions of the idealised deformation pattern. In this case, equation (9.46) may be written in the form

$$P_m/M_0 = 52.22(C/H)^{1/3}\{1 + (0.33V_0/CD)^{1/q}\}, \qquad (9.48)$$

which, using the dimensionless parameters given by equations (9.42) and (9.43), becomes

$$\eta = 1.3\phi^{2/3}\{1 + (0.33V_0/CD)^{1/q}\}. \qquad (9.49)$$

The theoretical predictions of equations (9.41) and (9.48) give the axial crushing force ratio

$$P_m^d/P_m^s = 1 + (0.33V_0/CD)^{1/q}, \qquad (9.50)$$

where P_m^d and P_m^s are the dynamic and static progressive buckling forces, respectively. Equation (9.50) is compared in Figure 9.17 with some experimental results on steel tubes with square cross-sections. It is evident that the experimental

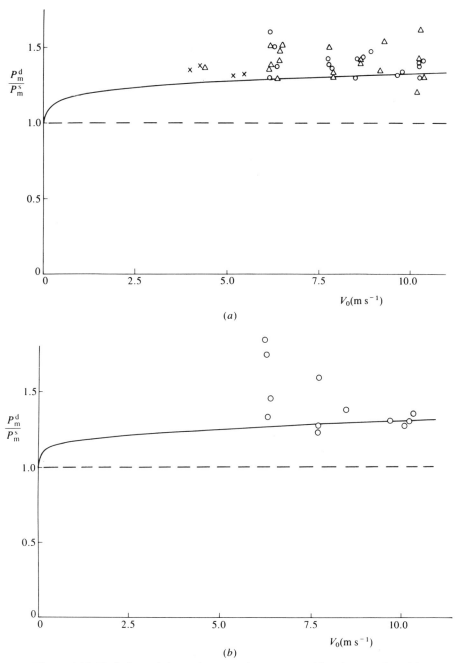

Figure 9.17 Variation of dynamic to static mean crushing force ratio with impact velocity for square tubes. (a) $C = 37.07$ mm, $H = 1.152$ mm. (b) $C = 49.31$ mm, $H = 1.63$ mm. $---$: strain-rate-insensitive material. ———: equation (9.50) for a strain-rate-sensitive material with $D = 6844$ s^{-1} and $q = 3.91$.

Experimental results: $\bigcirc$: symmetric collapse mode.[9.3] $\triangle$: asymmetric collapse mode.[9.3] $\times$: Wierzbicki et al.[9.17]

results exhibit more scatter than the circular tube test results in Figure 9.10. Moreover, the theoretical predictions of equation (9.50) tend towards a lower bound of the experimental values for the dynamic crushing forces. This underestimate may be due partly to the difficulty of calculating a mean strain rate in a square tube and to the paucity of experimental data for the strain-rate-sensitive properties of materials with large strains. Clearly, further studies are required to resolve these difficulties.

9.6 Comparison of the axial crushing characteristics of circular and square tubes

It is interesting to compare the static progressive crushing behaviour of circular and square tubes having the same values of cross-sectional area A_c and wall cross-sectional area† A. Therefore, the solidity ratio ϕ is equal for circular and square tubes, and the corresponding experimental and theoretical values of the structural effectiveness η are given in Figure 9.8 for circular tubes and in Figure 9.16 for square tubes. It is evident, for a given value of ϕ, that a square tube has an associated value of η which is about two-thirds of the corresponding circular tube result. Moreover, the ratio of equations (9.18) and (9.42) gives $P_m^s/P_m^c = \eta^s/\eta^c$, where the supercripts s and c denote square and circular tubes, respectively. Figure 9.18 shows a comparison between the structural effectiveness for square and circular tubes according to the theoretical predictions and empirical relations discussed in earlier sections.

The equality of equations (9.20) and (9.43) demands that $C/R = 2$ when the circular and square tubes have the same wall thickness and solidity ratio and, therefore, the ratio of equations (9.18) and (9.42) in this case gives $\eta^s/\eta^c = (\pi/4)P_m^s/P_m^c$ rather than $\eta^s/\eta^c = P_m^s/P_m^c$ for square and circular tubes having the same values of A and A_c. Thus, if $\eta^s/\eta^c \cong 2/3$, then $P_m^s/P_m^c = (4 \times \frac{2}{3})/\pi = 0.85$. Therefore, the actual static crushing forces are about 15 per cent smaller for square tubes with the same wall thickness and solidity ratio, but the structural effectiveness of a square tube is only about two-thirds that of a circular tube.

The foregoing observations also apply to the dynamic progressive buckling of circular and square tubes which are made from a strain-rate-insensitive material.

9.7 Some comments on energy absorption systems

9.7.1 Introduction

Many practical engineering systems have requirements for absorbing energy during impact events. Energy absorbers have been developed which dissipate

---

† The definitions for A and A_c in §§ 9.3.3 and 9.4.2 show that this condition is achieved when a square tube has a wall thickness of $\sqrt{\pi}H/2$ and a width $C = \sqrt{\pi}R$, where R and H are the mean radius and wall thickness of a circular tube, respectively.

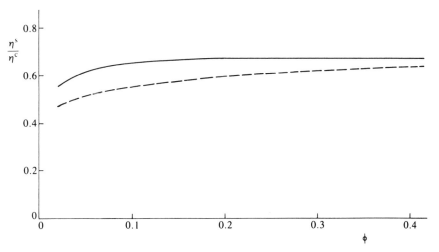

Figure 9.18 Comparison of the structural effectiveness for square (η^s) and circular (η^c) tubes. ———: theoretical; ratio of equations (9.33) and (9.44). – – –: empirical; ratio of equations (9.35) and (9.45).

energy owing to friction, fracture, shear, bending, tension, torsion, crushing, cyclic plastic deformation, metal cutting, extrusion and fluid flow, etc. They are made largely from metal components, although wood, plastics and other materials have also been used. The devices may be reusable, such as a hydraulic damper, rechargeable, with the energy-absorbing component being replaced in a permanent container after an impact event, or expendable, as in the crushing of a vehicle structure during a collision.

There is an extensive literature extant on energy-absorbing devices which is not examined in this section, although an interested reader may find some of the work reviewed in references 9.4 and 9.18 to 9.23. However, a brief introduction to some aspects of the topic is presented in this section because it transpires that the dynamic progressive buckling of a thin-walled tube is a simple and yet efficient energy absorber for many practical engineering problems.

9.7.2 Efficiency

The efficiency of an impact-energy-absorbing device may be specified in several ways to accommodate a wide range of practical applications. One measure of efficiency is the specific energy (S_e) which is defined as the energy absorbed per unit mass, or

$$S_e = D_a/m, \tag{9.51}$$

where D_a is the total energy absorbed and m is the total mass of the energy absorbing device. Another measure is the volumetric efficiency (V_e), or

$$V_e = V_u/V_T, \tag{9.52}$$

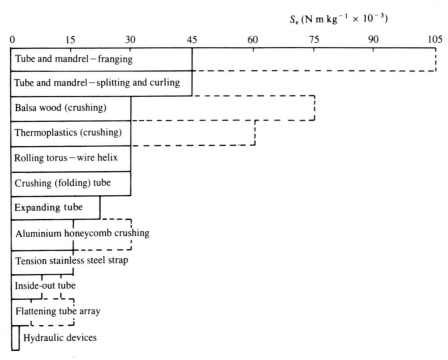

Figure 9.19 Specific energies (S_e) for some energy absorbers.[9.4]

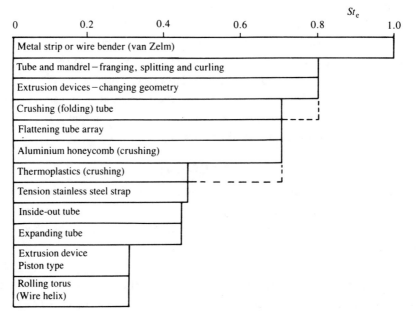

Figure 9.20 Stroke efficiencies (St_e) for some energy absorbers.[9.4]

where V_u is the volume of the energy-absorbing part of a device and V_T is the total volume. For some devices, equation (9.52) simplifies to

$$St_e = S/L, \qquad (9.53)$$

which is known as the stroke efficiency and where S and L are the stroke and total length of an energy absorber, respectively.

An ideal absorber is defined as one which maintains the maximum allowable retarding force throughout the stroke, apart from elastic loading and unloading effects.[9.4, 9.18] However, a designer must often trade off considerations of cost, volume, stroke, weight, deceleration, etc., against efficiency. Above all, an energy absorber must be reliable and in many cases versatile enough to absorb a dynamic load which may strike at a random location.

The specific energies (S_e) and stroke efficiencies (St_e) for several energy absorbers are given in Figures 9.19 and 9.20, respectively. These results were collected by Ezra and Fay,[9.4] together with the summary in Table 9.1.

9.7.3 Circular tube

It is evident from Figures 9.19 and 9.20 and Table 9.1 that an axially crushed cylindrical tube is a suitable energy absorbing device for many practical applications, particularly because tubing is inexpensive and available in a wide range of sizes. Now, it is remarked in § 9.1 that the theoretical methods for dynamic progressive buckling are used to predict a mean value of the actual axial force–axial displacement characteristics which are shown for a particular case in Figure 9.3(a). The energy absorbed (D_a) by an axially crushed cylindrical tube is, therefore,

$$D_a = P_m \Delta, \qquad (9.54)$$

where Δ is the total amount of axial crushing and the mean axial crushing forces (P_m) for static and dynamic loading are predicted by equations (9.33) and (9.34), respectively. The actual energy absorbed in the cylindrical shell with the load–displacement characteristics in Figure 9.3(a) is compared with equation (9.54) in Figure 9.21.

The specific energy for an axially crushed cylindrical tube having a total axial length L is

$$S_e = P_m \Delta / 2\pi RHL\rho \qquad (9.55)$$

according to equations (9.51) and (9.54) for a tube which is made from a material having a density ρ. Equation (9.55) remains valid provided $\Delta \leqslant \Delta_b$, where Δ_b is the maximum amount of crushing when a tube bottoms-out,† as shown in Figure 9.22. Equations (9.25) and (9.31) predict the effective crushing distance

$$\delta_e/2l = 0.86 - 0.568(H/2R)^{1/2}, \qquad (9.56)‡$$

† No undeformed tubing is available for the formation of any further wrinkes when $\Delta \geqslant \Delta_b$.
‡ This relation gives good agreement with both static and dynamic tests on mild steel tubes.[9.7]

Table 9.1 *Energy absorber data*[9.4]

Device	St_e Stroke efficiency (approx.)	Force–stroke characteristics	S_e Specific energy (approx.) N m/kg $\times 10^{-3}$	Comments
Tension stainless steel strap	0.5	Strain dependent	15	Simple, inexpensive, fractures if over extended
Flattening tube array	0.7	Nearly constant for $St_e = 0.7$	4.5–15	Reliable
Inside-out tube	0.5	Constant	6–12	Reliable, can resist rebound loads
Contracting tube	0.5	Constant	—	—
Expanding tube	0.5	Constant	24	Reliable, accepts some off-axis loading, inexpensive
Crushing (folding) tube	0.7–0.8	Periodic	30	Reliable, inexpensive, can accept rebound loads
Tube and mandrel 1. Franging	0.8	Large variations (average nearly constant)	45–105	Large fluctuation in load
2. Splitting and curling	0.8	Nearly constant	45	Reliable, inexpensive if conical mandrel is used

Device		Behaviour		Comments
Metal strip or wire bender	1.0	Constant	—	Commercially available, reliable, long stroke, works in tension
Rolling torus (wire helix)	0.3	Constant	30	Commercially available, worked in tension or compression, each device must be 'tuned' by adding or subtracting wire
Plastic hinge (pin)	0.8	Function of geometry	—	Very simple, reliable, resists rebound
Extrusion devices 1. Piston type	0.3	Velocity sensitive	—	Hydraulic and viscoelastic reusable
2. Changing geometry	0.8	Velocity sensitive	—	Very simple, an example of this is a water bumper
Crushing materials 1. Balsa wood	—	—	30–75	Omni-directional load capacity
2. Aluminium honeycomb	0.7	Constant for $St_e = 0.7$	15–30	Omni-directional load capability, reliable
3. Thermoplastics	0.5–0.7	Exponential	30–60	Omni-directional load capability, reliable

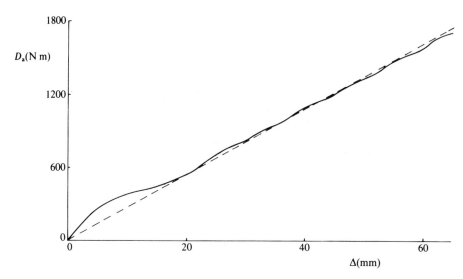

Figure 9.21 Energy absorbed (D_a) versus axial displacement characteristics (Δ) for the axially crushed thin-walled circular tube in Figure 9.3. ———: energy absorbed according to the axial load–axial displacement curve in Figure 9.3(a). ———: equation (9.54) with P_m from Figure 9.3(a).

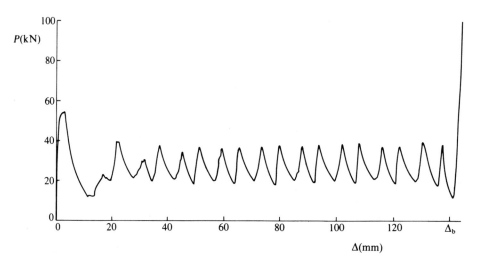

Figure 9.22 Static axial crushing characteristics of a thin-walled mild steel circular tube with a mean radius (R) of 27.98 mm, a mean wall thickness (H) of 1.2 mm and an initial axial length of 178 mm. Δ_b is the axial displacement when the tube bottoms-out.

which was used in the calculations for the mean crushing forces given by equations (9.33) and (9.34). Thus,

$$\Delta_b = L\{0.86 - 0.568(H/2R)^{1/2}\}. \tag{9.57}$$

The maximum specific energy for a circular tube before bottoming-out is, therefore, given by equation (9.55) with $\Delta = \Delta_b$, or

$$S_e = 0.95(1 + 0.29\phi^{1/2})\phi^{1/2}(\sigma_0/\rho) \tag{9.58}$$

when using equations (9.18), (9.20) and (9.33) for the static or dynamic progressive buckling of a tube which is made from a strain-rate-insensitive material.

It is evident that $V_e = St_e$ for an axially crushed cylindrical tube and that the stroke efficiency is

$$St_e = \Delta_b/L, \tag{9.59}$$

which, using equation (9.57), gives

$$St_e \cong 0.75 \tag{9.60}$$

when $15 \leqslant 2R/H \leqslant 60$. A stroke efficiency of $St_e \cong 0.75$ gives good agreement with some experimental results on mild steel tubes which are reported in Figure 24 of reference 9.7 and is in the middle of the range found by Ezra and Fay[9.4] in Table 9.1.

9.7.4 Square tube

Equation (9.54) also gives the energy absorbed by a thin-walled tube with a square cross-section which is subjected to either static or dynamic axial crushing forces, where P_m is defined by equations (9.41) and (9.48), respectively. Thus, the specific energy according to equation (9.51) is

$$S_e = P_m\Delta/4CHL\rho \tag{9.61}$$

provided $\Delta \leqslant \Delta_b$, where

$$\Delta_b = 0.73L \tag{9.62}$$

according to equation (9.40) for the symmetric deformation mode in Figures 9.4 and 9.11. Substituting equation (9.41), together with equation (9.62), into equation (9.61) gives the maximum specific energy

$$S_e = 0.945\phi^{2/3}(\sigma_0/\rho) \tag{9.63}$$

when using equations (9.39) and (9.43).

The stroke efficiency for a square tube is

$$St_e = 0.73 \tag{9.64}$$

according to equations (9.59) and (9.62).

9.8 Structural crashworthiness

9.8.1 Introduction

The term 'structural crashworthiness' is used to describe an investigation into the impact performance of a structure when it collides with another object. A study

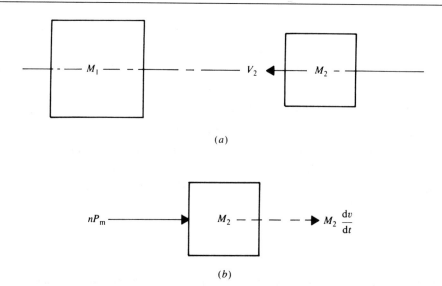

Figure 9.23 (*a*) A mass M_2 travelling with a velocity V_2 towards a stationary mass M_1. (*b*) Horizontal forces acting on mass M_2 during the impact event.

into the structural crashworthiness characteristics of a system is required in order to calculate the forces during a collision which are needed to assess the damage to structures and the survivability of passengers in vehicles, for example. This topic embraces the collision protection of aircraft, buses, cars, trains, ships and offshore platforms, etc.[9.21, 9.24–9.33] and even spacecraft.[9.34] No attempt is made to review the whole field and only that part which is related to dynamic progressive buckling is discussed briefly.

9.8.2 Elementary aspects of inelastic impact

Consider a stationary mass M_1 which is struck by a mass M_2 travelling with an initial velocity V_2, as shown in Figure 9.23(*a*). Conservation of linear momentum demands that

$$M_2 V_2 = (M_1 + M_2)V_3, \tag{9.65}$$

where V_3 is the common velocity of both masses immediately after an inelastic impact.† The loss of kinetic energy is, therefore,

$$K_l = M_2 V_2^2/2 - (M_1 + M_2)V_3^2/2, \tag{9.66}$$

which, when using equation (9.65) for V_3, may be recast into the form

$$K_l = (M_2 V_2^2/2)/(1 + M_2/M_1), \tag{9.67}$$

where $M_2 V_2^2/2$ is the initial kinetic energy of the mass M_2.

---

† The coefficient of restitution is taken as zero ($e = 0$).

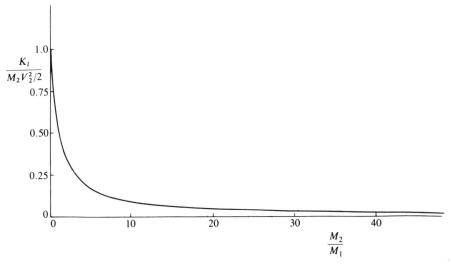

Figure 9.24 Variation of the dimensionless kinetic energy loss given by equation (9.67) with the mass ratio M_2/M_1.

Equation (9.67) gives the energy which must be absorbed by an energy-absorbing system which is interposed between the two masses M_1 and M_2 in Figure 9.23(a). If the striking mass M_2 is much larger than the struck mass M_1 (i.e., $M_2/M_1 \gg 1$), then $K_l \cong 0$, and no kinetic energy is lost during the impact event. In the other extreme case of a striking mass M_2 which is much smaller than the struck mass M_1 (i.e., $M_2/M_1 \ll 1$), then $K_l \cong M_2 V_2^2/2$ and all of the initial kinetic energy of the mass M_2 must be absorbed during the impact. The loss of kinetic energy for an impact between two equal masses is $K_l = M_2 V_2^2/4$, which is one-half of the initial kinetic energy of the striking mass M_2. The variation of the dimensionless kinetic energy loss $K_l/(M_2 V_2^2/2)$ with the mass ratio M_2/M_1 is shown in Figure 9.24.

Frequently, the struck mass M_1 in Figure 9.23(a) is constrained to remain stationary throughout a practical impact event. In other words, $M_1/M_2 \gg 1$ and equation (9.67) gives $K_l = M_2 V_2^2/2$, as expected.

Now, equation (9.54) predicts the energy which is absorbed in axially crushed circular and square tubes and indeed may be used to obtain the energy absorbed in any thin-walled structure which undergoes dynamic progressive buckling with a mean force P_m (e.g., honeycomb structure[9.35]). A nest of n axially loaded thin-walled tubes, for example, may be used to absorb the energy in the impact scenario shown in Figure 9.23(a), and, therefore,

$$D_a = nP_m\Delta = K_l \tag{9.68}$$

provided $\Delta \leqslant \Delta_b$. Thus,

$$nP_m\Delta = M_2 V_2^2/2 \tag{9.69}$$

when one end of the energy-absorbing device remains stationary throughout the impact event.

The mean dynamic crushing force for n tubes is

$$nP_m = -M_2 \, dv/dt \tag{9.70}†$$

according to the free body diagram in Figure 9.23(b), with $v = V_2$ at $t = 0$ and $v = 0$ when the striking mass M_2 has moved a distance Δ at $t = T$, where T is the response duration. Clearly, the deceleration is

$$a = dv/dt = -nP_m/M_2, \tag{9.71}$$

which is constant throughout motion and, therefore, the velocity–time history is

$$v = -nP_m t/M_2 + V_2 \tag{9.72}$$

when satisfying the initial condition. Motion ceases when $v = 0$, which gives the response duration

$$T = M_2 V_2/nP_m, \tag{9.73a}$$

or

$$T = -V_2/a. \tag{9.73b}$$

A further integration of equation (9.72) gives the displacement–time history

$$\delta = -nP_m t^2/2M_2 + V_2 t, \tag{9.74}$$

which, substituting equation (9.73a), predicts that the total crushed distance is

$$\Delta = M_2 V_2^2/2nP_m, \tag{9.75a}$$

or

$$\Delta = -V_2^2/2a, \tag{9.75b}$$

when motion ceases at $t = T$. Equation (9.75a) may be obtained directly from the energy balance equation (9.69). The mean crushing force is often estimated in experimental test programmes and design calculations by dividing the initial kinetic energy $(M_2 V_2^2/2)$ by the total axial crushed distance Δ. This agrees with equations (9.69) or (9.75a).

The impact scenario in Figure 9.23(a) lies in a horizontal plane and is related to situations which arise during bus, car, train and ship collisions, for example. However, another important class of impacts are caused by a mass M_2 which drops vertically onto a mass M_1 with an impact velocity V_2 as shown in Figure 9.25(a). In this case, equation (9.69) for the conservation of energy is replaced by

$$nP_m\Delta = M_2 V_2^2/2 + M_2 g\Delta, \tag{9.76}$$

where $M_2 g\Delta$ is the additional potential energy of the mass M_2 which crushes the energy absorber by an amount Δ during the impact event.

---

† A dynamic compressive crushing force P_m, which is defined as shown in Figure 9.23(b), is taken as positive throughout this chapter.

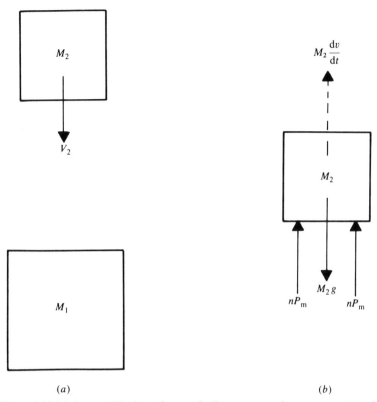

Figure 9.25 (a) A mass M_2 dropping vertically onto a stationary mass M_1 with an impact velocity V_2. (b) Vertical forces acting on a mass M_2 during the impact event.

The vertical equation of motion for the mass M_2 in Figure 9.25(b) during impact is

$$nP_m + M_2\,dv/dt - M_2g = 0, \tag{9.77}$$

which predicts a constant deceleration

$$a = dv/dt = -nP_m/M_2 + g. \tag{9.78}$$

Now, integrating equation (9.78) with respect to time and introducing the initial and terminal conditions gives a response time

$$T = V_2/(nP_m/M_2 - g) \tag{9.79}$$

and a crushed distance

$$\Delta = V_2^2/\{2(nP_m/M_2 - g)\}, \tag{9.80}$$

which may also be obtained directly from the conservation-of-energy equation (9.76).

In many practical impacts, the decelerations $|a| \gg g$ so that equation (9.78) gives

$a \cong -nP_m/M_2$, and equations (9.76) to (9.80) then reduce to equations (9.69) to (9.71), (9.73a) and (9.75a), respectively.

9.8.3 Thin-walled circular tubes

Thin-walled circular tubes have been proposed as energy absorbers behind car bumpers, train buffers and at the bottom of lift shafts to absorb the energy of a runaway lift. They are inexpensive, efficient and versatile and it is, therefore, worthwhile to examine the crushing behaviour of an energy-absorbing system which consists of thin-walled circular tubes.

The constant deceleration during an impact event is predicted by equation (9.71), with the mean dynamic crushing force P_m given by equation (9.34), or

$$P_m = 21.1\sigma_0 RH\{1 + 0.41(H/R)^{1/2}\}\{1 + (V_2/4RD)^{1/q}\}/\{2.14(R/H)^{1/2} - 1\}$$

(9.81)

when using equations (9.18) and (9.20). Thus,

$$a = -21.1n\sigma_0 RH\{1 + 0.41(H/R)^{1/2}\}\{1 + (V_2/4RD)^{1/q}\}/[M_2\{2.14(R/H)^{1/2} - 1\}].$$

(9.82)

Similarly, the response time T and axially crushed distance Δ are predicted by equations (9.73b) and (9.75b), respectively, or

$$T = M_2 V_2\{2.14(R/H)^{1/2} - 1\}/[21.1n\sigma_0 RH\{1 + 0.41(H/R)^{1/2}\}\{1 + (V_2/4RD)^{1/q}\}]$$

(9.83)

and

$$\Delta = M_2 V_2^2\{2.14(R/H)^{1/2} - 1\}/[42.2n\sigma_0 RH\{1 + 0.41(H/R)^{1/2}\}\{1 + (V_2/4RD)^{1/q}\}].$$

(9.84)

Equations (9.81) to (9.84) have been developed for an energy absorbing system which does not bottom-out. Thus, equation (9.84) must satisfy the inequality $\Delta \leqslant \Delta_b$, where Δ_b is shown in Figure 9.22 and defined by equation (9.57).

It is evident that equations (9.82) to (9.84) have been developed for horizontal impact as illustrated in Figure 9.23(a). However, the results are also valid for the vertical impact situation in Figure 9.25(a) provided $|a| \gg g$.

The foregoing equations are now used to examine the dynamic progressive buckling and energy-absorbing characteristics of a thin-walled cylindrical shell. The shell is made from a ductile material with a uniaxial static flow stress (σ_0) of 300 MN/m^2, and has a mean radius $R = 30$ mm and mean wall thickness $H = 1.2$ mm. One end of the shell is fixed while the other end is struck by a mass $M_2 = 100$ kg travelling with an initial velocity† $V_2 = 10$ m/s. Thus, substituting

---

† In the case of a vertical impact, a mass dropped from a height $h = 5.1$ m would give an impact velocity $V_2 = 10$ m/s according to the well-known formula $V_2 = (2gh)^{1/2}$, with $g = 9.81$ m/s^2.

these quantities with $n = 1$ for a single tube into equations (9.82) to (9.84) gives $a = -254$ m/s², $T = 39.3$ ms and $\Delta = 196.7$ mm, respectively, provided the tube is made from a strain-rate-insensitive material (i.e., $D \rightarrow \infty$).

These calculations reveal that the deceleration of the striking mass M_2 is $254/9.81 = 25.9$ times larger than the gravitational acceleration[†] (9.81 m/s²). This suggests that the gravitational terms in equations (9.77) to (9.80) may be neglected for this particular circular tube when impacted vertically. The additional potential energy in equation (9.76) is $M_2 g \Delta = 193$ J, which is about 3.9 per cent of the initial kinetic energy for the mass M_2 (5 kJ).

The response duration is very short (39.3 ms) and the event happens too quickly for the human eye to follow the deformation. Nevertheless, it is long compared with the time required for an elastic stress wave to travel along the length of a tube. Uniaxial tensile or compressive elastic stress waves travel[9.20] at 5150 m/s and 5100 m/s in mild steel and aluminium, respectively, and, therefore, would take about 58 μs to travel along a 300 mm long mild steel or aluminium bar. Thus, the response time T of 39.3 ms is 678 times longer than the time taken for a uniaxial elastic stress wave to traverse a 300 mm long bar.

Now, equation (9.57) predicts that $\Delta_b/L = 0.78$ so that the circular tube must be at least $196.7/0.78 = 252$ mm long in order to prevent the phenomenon of bottoming-out which is illustrated in Figure 9.22.

The foregoing calculations are associated with a strain-rate-insensitive tube. If a tube is made from mild steel with the coefficents $D = 6844$ s⁻¹ and $q = 3.91$ from § 9.3.5, then the term $1 + (V_2/4RD)^{1/q}$, which appears in equations (9.81) to (9.84), equals 1.32. In this circumstance, $a = -335$ m/s², $T = 29.8$ ms and $\Delta = 149$ mm. The influence of material strain rate sensitivity on the values of P_m, a, T and Δ for the above circular tube is shown in Figure 9.26.

9.8.4 Thin-walled square tubes

The general procedure, which is outlined in § 9.8.2 and illustrated for a circular tube in § 9.8.3, may also be used for square tubes having the mean dynamic crushing force given by equation (9.48), or

$$P_m = 13.05\sigma_0 H^2(C/H)^{1/3}\{1 + (0.33V_2/CD)^{1/q}\}, \qquad (9.85)$$

where V_2 is the impact velocity of a striking mass M_2.

---

† It is evident that the gravitational force of the 100 kg striking mass is 981 N, while the mean dynamic crushing force on the tube is 25.9 times larger (i.e., $P_m = 25.4$ kN). Alternatively, the mean dynamic crushing force may be calculated from equation (9.81) or by noting that the initial kinetic energy $= 5$ kJ $= P_m \times \Delta$, or $P_m = 5 \times 10^3/0.1967 = 25.4$ kN. The static uniaxial compressive squash load for this shell is $2\pi RH\sigma_0 = 67.9$ kN. Thus, P_m is only 37 per cent of the squash load. However, the initial peak force is larger than the mean crushing force, as shown in Figure 9.3(a).

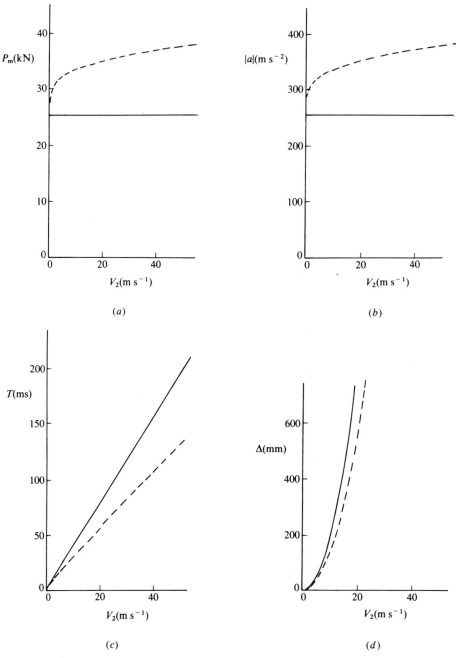

Figure 9.26 (a)–(d) Variation of mean crushing force, deceleration, response time and total crushing distance with impact velocity, respectively, for a thin-walled circular tube having $R = 30$ mm, $H = 1.2$ mm, $\sigma_0 = 300$ MN/m^2 and $M_2 = 100$ kg. ———: equations (9.81) to (9.84) with $D \to \infty$ for a strain-rate-insensitive material. – – –: equations (9.81) to (9.84) with $D = 6844$ s^{-1} and $q = 3.91$.

9.8.5 Impact injury

Theoretical calculations such as those in §§ 9.8.2 to 9.8.4 enable an engineer to assess whether a structural design is capable of withstanding the forces generated during an impact and absorbing the impact energy without excessive damage. However, in the case of passenger transportation systems, it is also necessary to ensure that the passengers can tolerate an impact. Clearly, designers require guidelines and criteria which, however, are difficult to obtain on human beings and are further complicated by the observation that people's response and tolerance to impact varies with size, age and sex, etc. Nevertheless, many studies have been conducted in this field over the years and references 9.21, 9.23, 9.24 and 9.36 to 9.41, together with other articles cited therein, contain some valuable data.

Some typical values[9.23, 9.36] of whole body acceleration (and deceleration), which are associated with various types of impact, are presented in Figure 9.27.

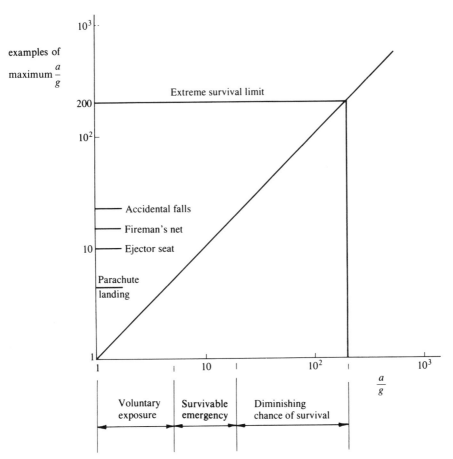

Figure 9.27 Whole-body tolerance of impact.[9.23, 9.36]

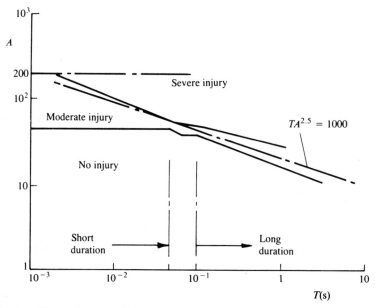

Figure 9.28 Influence of pulse duration on whole-body tolerance of impact.[9.23]

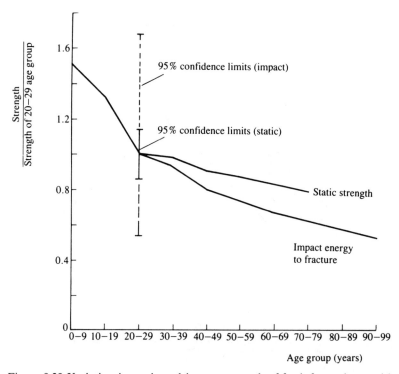

Figure 9.29 Variation in static and impact strength of fresh femur bone with age.[9.23, 9.42]

However, the influence of pulse length plays an important role in the severity of injury, as indicated in Figure 9.28, which is taken from Macaulay.[9.23] A line through the middle of the moderate injury band in Figure 9.28 is approximated by the equation

$$TA^{2.5} = 1000, \tag{9.86}$$

where T is the pulse duration (s) and $A = a/g$ when a is the whole body acceleration (or deceleration).

Equation (9.86) only applies to impacts which involve whole-body accelerations. Other impact situations occur in which specific body components are struck and, therefore, data is required for many parts of the body. For example, Figure 9.29 presents data[9.23, 9.42] on the strength of fresh femur bones. The strength decreases significantly with a person's age and is less for impact loads. Data for other parts of the body under various static and dynamic loads may be found in reference 9.23 and in the articles cited therein.

Head injuries are responsible for a significant loss of life and serious injuries in transportation accidents.[9.39, 9.43] The Wayne State tolerance curve in Figure 9.30 was obtained by dropping embalmed cadaver heads onto hard flat surfaces to determine incipient skull fracture.[9.38] This work has been re-examined by Gadd and others to develop criteria for head injuries which are caused when the front of the head strikes a hard object, or when the head is loaded through the neck by a decelerating body.[9.23, 9.39] Gadd introduced the severity index.[9.36-9.39, 9.43]

$$SI = \int_0^T A_v^{2.5} \, dt, \tag{9.87}$$

where $A_v = a_v/g$, a_v is the average head acceleration (or deceleration) which may vary throughout the loading pulse having a duration T(s) with

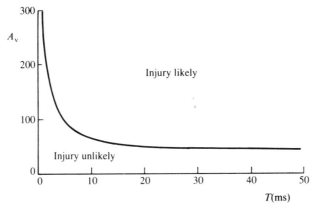

Figure 9.30 Wayne State tolerance curve for head impact.[9.23]

2.5 ms $\leqslant T \leqslant$ 50 ms. Gadd suggested that SI = 1000 marks the threshold conditions between fatal and non-fatal head injuries, although there is a continuing
debate on the choice of a threshold value and about the relevance of this
criterion.[9.44]

The Wayne State tolerance curve in Figure 9.30 was constructed using an
average acceleration for the entire impact event, whereas the acceleration is
allowed to vary with time in equation (9.87). This inconsistency was recognised and
eliminated by replacing the dimensionless acceleration A_v in equation (9.87) by an
average value to produce the Head Injury Criterion[9.21, 9.38]

$$\text{HIC} = (T_2 - T_1) \left\{ \int_{T_1}^{T_2} A_v \, dt / (T_2 - T_1) \right\}^{2.5}, \tag{9.88}$$

where any time interval $T_2 - T_1$ is selected to maximise the right-hand side of
equation (9.88) and a value of HIC = 1000 is considered[9.44] as life-threatening.†

Now, consider the impact situation in § 9.8.2 where a mass M_2, which is
travelling with a velocity V_2, strikes on infinitely large stationary mass (i.e.,
$M_1 \to \infty$ in Figure 9.23(a)). The deceleration during the impact event when an
energy absorber with a constant resistance P_m is interposed between the two
masses M_1 and M_2 is

$$a = -V_2^2/2\Delta \tag{9.89}$$

according to equation (9.75b), where Δ is the total stopping distance. The
associated duration of response from equation (9.73b) is

$$T = 2\Delta/V_2 \tag{9.90}$$

when eliminating a with equation (9.89). Thus, substituting equations (9.89) and
(9.90) into equation (9.86) for whole body accelerations gives

$$V_2^4/\{g(2g\Delta)^{1.5}\} = 1000. \tag{9.91}$$

The calculations for the strain-rate-insensitive circular tube examined in § 9.8.3
predict a crushing displacement $\Delta = 196.7$ mm for an initial impact velocity V_2
= 10 m/s. Thus, the left-hand side of equation (9.91) equals 134. This value is well
below the threshold limit of 1000 which is required for serious injury during whole
body impact.

The acceleration a in equation (9.89) is constant so that equation (9.87) predicts
SI = 134 for the thin-walled cylindrical tube examined in § 9.8.3. It should be noted
that the response time $T = 39.3$ ms lies within the range of validity of equation
(9.87). Similarly, HIC = 134 according to equation (9.88) for the same cylindrical
tube.

Now, substituting equations (9.71) and (9.73b) with $n = 1$ into equation (9.88)

---

† The US regulation FMVSS 208 on occupant crash protection also specifies that the resultant
acceleration at the centre of gravity of the upper thorax shall not exceed 60g except for intervals with a
cumulative duration less than 3 ms.

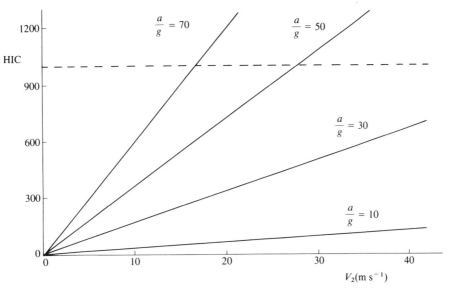

Figure 9.31 Variation of Head Injury Criterion (HIC) with impact velocity V_2 and dimensionless acceleration (a/g) according to equation (9.93) for a constant impact force and acceleration (or deceleration). $---$: threshold between potentially fatal and non-fatal impacts.

gives

$$\text{HIC} = (P_m/M_2 g)^{1.5} V_2/g, \tag{9.92}$$

or

$$\text{HIC} = V_2 a^{1.5}/g^{2.5}, \tag{9.93}$$

which relates the Head Injury Criterion to the impact velocity and acceleration (or deceleration) as shown in Figure 9.31.

9.9 Final remarks

The phenomenon of dynamic progressive buckling has been introduced in this chapter for circular and square thin-walled tubes when subjected to axial impact forces. However, it is anticipated that an overall instability may develop for thin-walled tubes which are long compared with the overall dimensions of the cross-sections. Unfortunately, few authors have explored this aspect of behaviour for impact loads.

Andrews, England and Ghani[9.45] have investigated the influence of tube length on the collapse modes of aluminium alloy circular tubes which are subjected to static axial forces. These experimental results are shown in Figure 9.32 and indicate a transition region above which overall buckling occurs in preference to

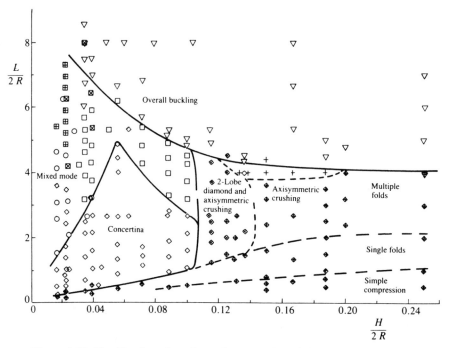

Figure 9.32 Classification chart for collapse modes of thin-walled aluminium alloy circular tubes subjected to static axial loads.[9.45]

Experimental results: ▽: overall buckling. ◇: concertina. □: concertina + 2-lobe diamond. ○: concertina + 3-lobe diamond. ⊞: concertina + 2-lobe and 3-lobe diamond. ⊠: 3-lobe diamond. ⟠: concertina (axisymmetric) crushing. ◈: 2-lobe diamond crushing. +: crushing and tilting of tube axis.

progressive buckling. The overall length-to-diameter ratio decreases along a curve drawn through the middle of this region as the tube wall thickness to mean diameter ratio increases. Thornton, Mahmood and Magee[9.11] have discussed the overall buckling of tubes with thin-walled square or rectangular cross-sections subjected to static axial forces.

It was observed that the overall buckling of thin-walled square tubes[9.3] may develop from imperfections which are produced by dynamic asymmetric progressive buckling modes, as indicated in Figure 9.33.

The crushing force–axial displacement characteristics in Figure 9.3(a) are cyclic, whereas the dynamic progressive buckling analyses in this chapter ignore this phenomenon and use a mean force as shown in Figure 9.3(a). Wierzbicki and Bhat[9.46] have derived a theoretical solution, with moving hinges, which predicts the cyclic crushing force–axial displacement characteristics for a circular tube loaded statically.

The magnitude of the first peak in the crushing force–axial displacement

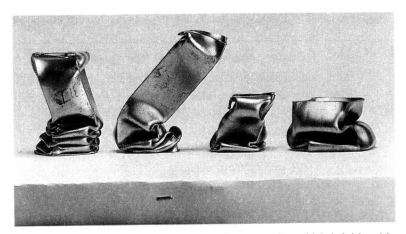

Figure 9.33 Overall buckling of thin-walled square tubes which is initiated by an asymmetric deformation mode associated with dynamic progressive buckling. The tubes are made from mild steel and subjected to axial impact loads.[9.3]

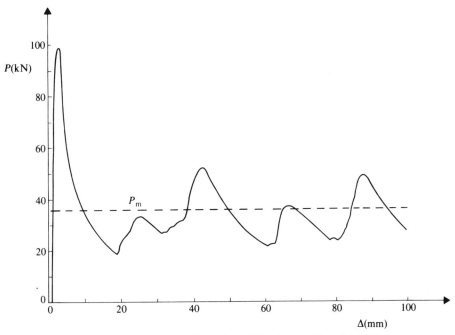

Figure 9.34 Variation of the static axial crushing force with axial displacement for a thin-walled square mild steel tube.[9.3] ($C = 49.3$ mm, $H = 1.625$ mm, $L = 244.1$ mm.)

characteristics is sometimes much larger than the forces associated with subsequent peaks as shown in Figure 9.34 for a statically loaded tube with a thin-walled square cross-section. However, it is possible to manufacture energy-absorbing components with initial imperfections to eliminate this initial peak, which is undesirable because of the large associated decelerations during an impact event. A number of studies have been published on the static plastic buckling of cylindrical shells, which may be used to evaluate this peak, and Andronicou and Walker,[9.47] for example, have developed theoretical and numerical methods to predict the initial peak forces for axially loaded cylindrical shells.

Recently, Abramowicz and Wierzbicki[9.48] have studied the basic collapse elements in Figure 9.12 and made some improvements in the estimates for the energy absorption characteristics which are important when the two adjacent sides are not perpendicular.

Thornton,[9.49] Reid, Reddy and Gray[9.50, 9.51] have explored the effect of foam filling on the overall stability and energy-absorbing characteristics of thin-walled tubes.

The various expressions which are developed in this chapter for progressive buckling may be used for many isotropic ductile materials.[9.7] For example, reasonable agreement has been found between the theoretical predictions and the static tests on PVC circular tubes which were conducted by Johnson, Soden and Al-Hassani.[9.52]

PROBLEMS

9.1 Show that equation (9.2) gives the plastic collapse moment for a unit length of the thin wall of a circular tube with plastic flow controlled by the von Mises yield criterion.

9.2 Show that the mean crushing force for a thin-walled circular tube is given by equation (9.11) when the convolutions in Figure 9.7 form internally instead of externally.

9.3 Use equation (9.4) to show that the mean circumferential strain in a completely flattened buckle of a circular tube is $\epsilon_\theta \simeq l/2R$ when l and R are defined in Figure 9.7.

9.4 Prove that $D_1:D_2:D_3::1:1 + (4H/\sqrt{3}\pi R)^{1/2}:2$. (See footnote to equation (9.13).)

9.5 Prove equation (9.30) for the energy absorbed in circumferential stretching during the development and flattening of a buckle in an axially loaded circular tube.

9.6 Construct the paper model in Figure 9.13 for the symmetric crushing mode in a square tube.

9.7 Show that equation (9.39) gives the plastic collapse moment for a unit length of the thin wall of a square tube with plastic flow controlled by the Tresca yield criterion.

9.8 A thin-walled circular tube has a wall thickness H and a mean radius R. Find the dimensions of a thin-walled square tube with the same values of A and A_c in §§ 9.3.3 and 9.4.2, respectively.

9.9 (a) Derive the axisymmetric axial crushing load of a thin-walled cylindrical tube having a mean radius R and a wall thickness H and made from a rigid, perfectly plastic material with a flow stress σ_0.

(b) An energy-absorbing device is required to arrest a mass M travelling with an impact velocity V_0. If the thin-walled tube examined in (a) is selected, then what is the thickness H required to arrest the mass M with a maximum deceleration αg, where $\alpha > 1$ and g is the gravitational acceleration?

Discuss any assumptions which you introduce.

(c) What is the crushed length of the tube? How does this compare with the length of tubing required for the energy-absorbing system?

9.10 (a) Assume that the circular tube in Problem 9.8 has $R = 30$ mm, $H = 1.2$ mm and is made from a ductile material with a uniaxial static flow stress $\sigma_0 = 300$ MN/m^2. A square tube with the same values of A and A_c is struck by a 100 kg mass travelling with an initial velocity of 10 m/s. Calculate the deceleration, response duration and permanent axial deformation when the tube material is strain-rate-insensitive.

(b) Repeat the calculations for a thin-walled square tube which is made from a strain-rate-sensitive material with $D = 6844$ s^{-1} and $q = 3.91$.

9.11 Estimate the Severity Index (SI) and the Head Injury Criterion (HIC) for the impact event described in Problem 9.10.

9.12 Prove the Head Injury Criterion is

$$\text{HIC} = V_2 a^{1.5} / g^{2.5}$$

for an impact event with an initial impact velocity V_2 and constant deceleration a (g is the gravitational acceleration).

9.13 Consider a fully clamped beam of length $2L$ when struck at the mid-span by a heavy mass M travelling with an initial velocity V_0, as shown in Figure 3.15(a). Use the approximate theoretical method in § 3.8.4 to estimate the deceleration of the mass M and the duration of motion when the beam is made from a rigid, perfectly plastic material and has a solid rectangular cross-section with a plastic bending moment M_0. Calculate the Severity Index (SI) and Head Injury Criterion (HIC) for the mass M.

10

Dynamic plastic buckling

10.1 Introduction

The dynamic plastic progressive buckling of thin-walled tubes subjected to axial impact loads was examined in Chapter 9. The impact loads were applied sufficiently slowly so that neither the axial nor the lateral inertia effects of the tubes played a significant role during the response. The duration of the impact loading was much longer than the transit time of an elastic stress wave which propagates along the length of a tube, as shown in § 9.8.3. A tube was unable, therefore, to support a mean dynamic axial load which was larger than the corresponding static value when disregarding the influence of material strain rate sensitivity examined in Chapter 8. Thus, the deformed profile of a tube is similar in this case for both static buckling and dynamic progressive buckling and a quasi-static theoretical analysis gave satisfactory agreement with the corresponding experimental results, as discussed in Chapter 9.

If a thin-walled tube, or other structural member, is subjected to a sufficiently severe dynamic axial load, then structural inertia effects produce the phenomenon of dynamic plastic buckling. In this circumstance, the deformed shape of the structure may be quite different from the corresponding progressive buckling profile, as illustrated in Figure 10.1 for an axially loaded circular tube. The shell is wrinkled over the entire length when buckled dynamically, unlike the dynamic progressive buckling case with wrinkling confined to one end. This situation should be contrasted with Figure 10.2, which shows the dynamic plastic buckling of a rod subjected to an axial impact load. The wrinkling is confined to the impacted end in this case, whereas a lateral deformation profile with a low mode number would be likely to develop over the entire length for static axial loads.

Generally speaking, the influence of lateral inertia forces on the dynamic plastic buckling of rods, beams, rings, plates and shells favours the development of lateral displacement fields with high mode numbers. This is apparent in Figures 10.1 and

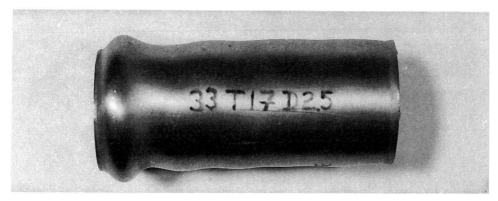

(a)

(b)

Figure 10.1 (a) Dynamic plastic buckling.[10.1] Permanent profile of an aluminium 6061-T6 tube subjected to an axial impulse at the University of Liverpool. (b) Dynamic progressive buckling.[10.2] Permanent profile of a mild steel tube impacted axially at the University of Liverpool.

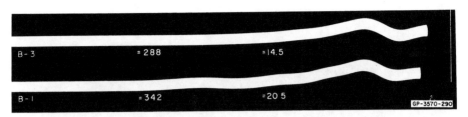

Figure 10.2 Dynamic plastic buckling. Permanent profiles of aluminium 6061-T6 rods impacted axially.[10.3]

10.2, and also in Figure 10.3 for rings with initially circular profiles which were wrinkled by a uniformly distributed external impulsive velocity. By way of contrast, elastic and elastic–plastic perfectly circular rings would buckle into an oval mode when subjected to a uniformly distributed external static pressure.[10.5]

The dynamic loading is applied suddenly to the structures examined in this chapter and is known as pulse loading to distinguish it from oscillatory or repetitive loadings which would give rise to vibration buckling. The dynamic plastic buckling of structures due to pulse, or transient, loads has been encountered in the aerospace, nuclear and petroleum industries,[10.6, 10.7] and is of interest in the chemical industry, many areas of structural safety, crashworthiness and other branches of engineering.

This chapter contains an introduction to the dynamic plastic buckling of structures which is achieved through an examination of some idealised problems. These simple models provide a valuable insight into the phenomenon and are sometimes adequate to assess the dynamic plastic buckling behaviour of an actual engineering structure. However, the phenomenon of dynamic buckling is introduced by first examining the dynamic elastic axial buckling of a rod. It transpires that many of the features of dynamic buckling are common to both dynamic elastic and dynamic plastic buckling.

10.2 Dynamic elastic buckling of a bar

10.2.1 Introduction

Generally speaking, a designer seeks an estimate for the maximum possible static load which can be supported by a structure with a propensity to buckle. However, a dynamic pulse load can exceed the corresponding static buckling load and may even become infinite in the limit of impulsive loading, as found in § 3.5.1 for the stable response of structures. A designer in this circumstance, therefore, requires the response of a structure (e.g., permanent damage) for a specified dynamic load.

The simply supported elastic bar shown in Figure 10.4(a) is examined in this section in order to introduce the phenomenon of dynamic buckling. This bar has

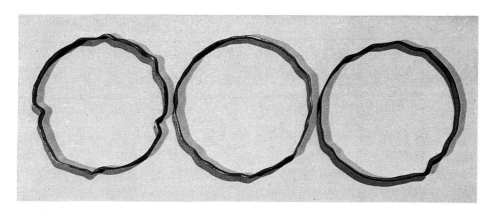

Figure 10.3 Dynamic plastic buckling. Permanent profiles of mild steel rings subjected to an axisymmetric external impulse.[10.4]

an initial imperfect profile $w^i(x)$ and is subjected to a constant axial compressive load P. Static equilibrium is maintained by lateral restraining blocks which are removed suddenly at $t = 0$ when motion commences.

10.2.2 Governing equations

Consider an element of the bar in Figure 10.4(b) after the lateral restraints are removed $(t \geqslant 0)$.

Horizontal equilibrium of this element demands[10.8]

$$\frac{\partial N}{\partial x} = 0 \qquad (10.1)†$$

when the surface of the rod is not loaded in the x sense and the axial inertia forces are neglected, and where N is the membrane, or direct, force per unit length of the beam. Equation (10.1) gives

$$N = -P \qquad (10.2)$$

in order to satisfy the boundary conditions at both ends $(x = 0$ and $x = L)$, where P is the axial compressive load (taken positive).

Vertical equilibrium of the element in Figure 10.4(b) requires[10.8]

$$\frac{\partial Q}{\partial x} + N \frac{\partial^2 (w + w^i)}{\partial x^2} + \frac{\partial N}{\partial x} \frac{\partial (w + w^i)}{\partial x} - \rho A \frac{\partial^2 w}{\partial t^2} = 0,$$

which, using equations (10.1) and (10.2), becomes

$$\frac{\partial Q}{\partial x} - P \frac{\partial^2 (w + w^i)}{\partial x^2} - \rho A \frac{\partial^2 w}{\partial t^2} = 0, \qquad (10.3)$$

---

† See equation (7.13) and equation (A.65) in Appendix 4 without axial inertia forces.

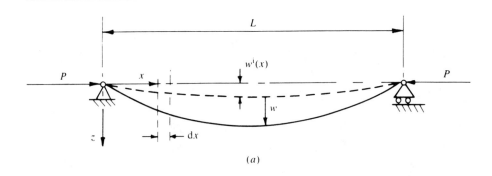

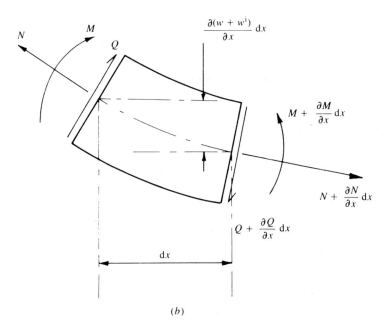

Figure 10.4 (a) Simply supported bar of length L with axial compressive force P.
(b) Element of bar and notation.

when the rod is not loaded laterally and where Q is the transverse shear force per unit length, w^i is the initial lateral displacement, or geometric imperfection, which is a function of x, w is the lateral displacement measured from w^i, and ρ and A are the material density and cross-sectional area of the rod, respectively.

Moment equilibrium of the element in Figure 10.4(b) gives

$$\frac{\partial M}{\partial x} - Q = 0, \tag{10.4}\dagger$$

where M is the bending moment per unit length and the influence of rotatory inertia is neglected.

It is now assumed that the lateral, or bending, deformations dominate the dynamic response. Thus, transverse shear deformations and shortening of the bar axis are neglected and elementary beam theory requires

$$M = -EI\frac{\partial^2 w}{\partial x^2}, \tag{10.5}\ddagger$$

where E is Young's modulus of elasticity and I is the second moment of area for the bar cross-section.

Finally, eliminating M from equation (10.4), using equation (10.5) for a uniform beam and then using the resulting expression to eliminate $\partial Q/\partial x$ from equation (10.3) gives the governing equation

$$EI\frac{\partial^4 w}{\partial x^4} + P\frac{\partial^2}{\partial x^2}(w + w^i) = -\rho A\frac{\partial^2 w}{\partial t^2}. \tag{10.6}$$

If $P = 0$ and $\partial^2 w/\partial t^2 = 0$, then equation (10.6) reduces to the familiar expression for the deflections of elastic beams. Equation (10.6), with $P = 0$, governs the vibrations of elastic beams, while equation (10.6), with $\partial^2 w/\partial t^2 = 0$, is used for the static buckling of elastic columns.

10.2.3 Single mode imperfections

A theoretical solution of equation (10.6) is sought for a simply supported elastic bar subjected to a dynamic compressive force P, as shown in Figure 10.4(a). It is assumed in this section that the initial lateral displacement is described by the expression

$$w^i = W_1^i \sin\left(\frac{\pi x}{L}\right), \tag{10.7}$$

where W_1^i is the maximum value of the initial lateral displacement, or geometric imperfection, and L is the length of the bar. Equation (10.7) suggests that equation

---

† Equation (10.4) is identical to equation (7.14) and equation (A.67) in Appendix 4. The vertical equilibrium equation with $w^i = 0$ is identical to equation (A.66) in Appendix 4 when $p = 0$. If $\partial^2 w/\partial t^2 = 0$ and $w^i = 0$, the vertical equilibrium equation reduces to equation (7.15) with $p = 0$.

‡ The initial imperfect profile of the beam is free of stresses. The curvature equals $-\partial^2 w/\partial x^2$ provided $(\partial w/\partial x)^2 \ll 1$, as noted on p. 527.

(10.6) has a theoretical solution of the form

$$w = W_1(t) \sin\left(\frac{\pi x}{L}\right) \tag{10.8}$$

which satisfies the simple support conditions ($w = 0$ and $M = 0$† at $x = 0$ and $x = L$).

Thus, substituting equations (10.7) and (10.8) into equation (10.6) gives

$$EI\left(\frac{\pi}{L}\right)^4 W_1(t) \sin\left(\frac{\pi x}{L}\right) - P\left(\frac{\pi}{L}\right)^2 (W_1^i + W_1(t)) \sin\left(\frac{\pi x}{L}\right)$$
$$+ \rho A \frac{d^2 W_1(t)}{dt^2} \sin\left(\frac{\pi x}{L}\right) = 0,$$

or

$$\rho A \frac{d^2 W_1(t)}{dt^2} + \left(\frac{\pi}{L}\right)^2 \left\{ EI\left(\frac{\pi}{L}\right)^2 - P \right\} W_1(t) = P\left(\frac{\pi}{L}\right)^2 W_1^i. \tag{10.9}$$

In the particular case of an initially perfect column ($W_1^i = 0$) subjected to a static compressive load ($d^2 W_1(t)/dt^2 = 0$), the coefficient of $W_1(t)$ in equation (10.9) must vanish, which requires that $EI(\pi/L)^2 - P = 0$, or

$$P^e = \frac{\pi^2 EI}{L^2}, \tag{10.10}$$

which is the classical static (Euler) buckling load for a simply supported elastic column.[10.5]

Equation (10.9) can now be written in the form

$$\rho A \frac{d^2 W_1(t)}{dt^2} + \left(\frac{\pi}{L}\right)^2 P^e \left(1 - \frac{P}{P^e}\right) W_1(t) = P\left(\frac{\pi}{L}\right)^2 W_1^i,$$

or

$$\frac{d^2 W_1(t)}{dt^2} + a^2 W_1(t) = b, \tag{10.11}‡$$

where

$$a^2 = \left(\frac{\pi}{L}\right)^2 \left(\frac{P^e}{\rho A}\right)\left(1 - \frac{P}{P^e}\right) \tag{10.12a}$$

and

$$b = \left(\frac{\pi}{L}\right)^2 \left(\frac{P}{\rho A}\right) W_1^i. \tag{10.12b}$$

It is well known that the second-order differential equation (10.11) has a different form of solution depending on the sign of the coefficient a^2. Clearly, $a^2 > 0$ when the compressive axial load (P) is smaller than the Euler buckling load ($P < P^e$)

---

† Equation (10.5) gives $\partial^2 w / \partial x^2 = 0$ when $M = 0$.

‡ The form of this differential equation is similar to equation (5.90).

and $a^2 < 0$ if $P > P^e$. These two cases are now examined in the following two sections.

10.2.3(a) $P < P^e$, oscillatory solution

If $P < P^e$, then the coefficient a^2 defined by equation (10.12a) is positive and equation (10.11) has the solution

$$W_1(t) = \frac{b}{a^2}\{1 - \cos(at)\} \qquad (10.13)$$

when the initial lateral velocity is zero ($dW_1/dt = 0$ at $t = 0$), with zero initial lateral displacement† ($W_1 = 0$ at $t = 0$). Equations (10.8), (10.12) and (10.13) finally give

$$w = \frac{\dfrac{P}{P^e}}{1 - \dfrac{P}{P^e}}\left[1 - \cos\left\{\left(1 - \frac{P}{P^e}\right)^{1/2}\tau\right\}\right]W_1^i \sin\left(\frac{\pi x}{L}\right), \qquad (10.14)$$

where

$$\tau = \left(\frac{\pi^4 EI}{\rho A L^4}\right)^{1/2} t \qquad (10.15)‡$$

is a dimensionless time.

It is evident that the lateral displacement simply oscillates with a dimensionless period $2\pi(1 - P/P^e)^{-1/2}$, as shown in Figure 10.5, and therefore, remains bounded.§ Thus, when $P < P^e$, the lateral displacement varies cyclically with time and the phenomenon known as dynamic elastic buckling does not occur.

10.2.3(b) $P > P^e$, dynamic buckling solution

Equation (10.12a) reveals that $a^2 < 0$ when $P > P^e$. In this case, the solution of the second-order differential equation (10.11) contains hyperbolic functions, so that equations (10.13) and (10.14) are no longer valid.

For convenience, equation (10.11) is rewritten in the form

$$\frac{d^2 W_1(t)}{dt^2} - \bar{a}^2 W_1(t) = b, \qquad (10.16)$$

where

$$\bar{a}^2 = \left(\frac{\pi}{L}\right)^2\left(\frac{P^e}{\rho A}\right)\left(\frac{P}{P^e} - 1\right) \qquad (10.17)$$

---

† The lateral displacements w are measured from the initial imperfect profile w^i, as indicated in Figure 10.4(a).

‡ Equation (10.15) may be written as $\tau = 2\pi t/T$, where T is the fundamental period of elastic vibration for a beam of span L. T is defined by equation (3.135) for a beam which is simply supported across a span $2L$.

§ However, the displacements become very large as P approaches P^e and the period of oscillation is infinite when $P = P^e$.

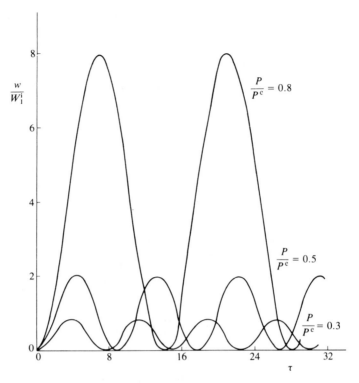

Figure 10.5 Dimensionless lateral displacement history according to equation (10.14) at $x = L/2$ for several values of P/P^e.

and b is defined by equation (10.12b). The solution of the differential equation (10.16), using standard methods, is

$$W_1(t) = \frac{b}{\bar{a}^2} \{\cosh(\bar{a}t) - 1\}, \tag{10.18}$$

which satisfies the initial conditions $W_1(0) = 0$ and $dW_1(0)/dt = 0$. Finally, equations (10.8), (10.12b), (10.17) and (10.18) give

$$w = \frac{\dfrac{P}{P^e}}{\dfrac{P}{P^e} - 1} \left[\cosh\left\{ \left(\frac{P}{P^e} - 1\right)^{1/2} \tau \right\} - 1 \right] W_1^i \sin\left(\frac{\pi x}{L}\right), \tag{10.19}$$

or

$$w = A_1(\tau) W_1^i \sin\left(\frac{\pi x}{L}\right), \tag{10.20a}$$

which, using equation (10.7), can be written

$$w = A_1(\tau) w^i, \tag{10.20b}$$

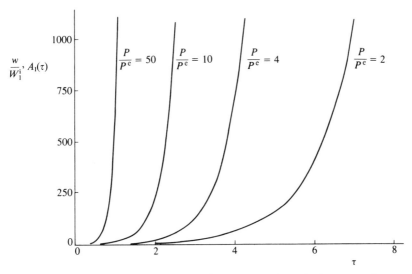

Figure 10.6 Dynamic elastic buckling, for several values of P/P^e, according to equation (10.19) at $x = L/2$.

where the dimensionless time τ is defined by equation (10.15) and

$$A_1(\tau) = \frac{P}{P^e}\left(\frac{P}{P^e} - 1\right)^{-1}\left[\cosh\left\{\left(\frac{P}{P^e} - 1\right)^{1/2}\tau\right\} - 1\right]. \qquad (10.21)$$

It is clear from Figure 10.6 that the lateral displacement w increases with the dimensionless time τ according to equations (10.19) to (10.21) and may become very large.† This phenomenon is known as dynamic buckling and the coefficient $A_1(\tau)$ in equations (10.20) and (10.21), with $A_1(0) = 0$, is an amplification function which gives the temporal growth of the geometric imperfection in the initial lateral displacement profile w^i.

10.2.4 Arbitrary initial imperfections

The theoretical development in § 10.2.3 has examined the particular case when the initial geometric imperfections are distributed in a rod with the sinusoidal form given by equation (10.7). However, initial geometric imperfections may exist in structural members through a variety of different effects so that it is important to examine a more general case to establish the most deleterious distribution.[10.9] Thus, the initial imperfect lateral displacement profile is taken in the form

$$w^i = \sum_{n=1}^{\infty} W_n^i \sin\left(\frac{n\pi x}{L}\right). \qquad (10.22)$$

† The governing equation (10.6) was developed for small displacements and slopes.

where

$$W_n^i = \frac{2}{L} \int_0^L w^i \sin\left(\frac{n\pi x}{L}\right) dx, \tag{10.23}$$

according to the usual procedures of Fourier series.

In view of the form of equation (10.22), and by analogy with equation (10.8), the solution of the governing equation (10.6) is

$$w = \sum_{n=1}^{\infty} W_n(t) \sin\left(\frac{n\pi x}{L}\right), \tag{10.24}$$

which satisfies the simply supported end conditions $w = 0$ and $\partial^2 w/\partial x^2 = M = 0$ at $x = 0$ and $x = L$ for $t \geqslant 0$. Substituting equations (10.22) and (10.24) into equation (10.6) gives

$$\sum_{n=1}^{\infty} \left[\left\{ EI\left(\frac{n\pi}{L}\right)^4 - P\left(\frac{n\pi}{L}\right)^2 \right\} W_n(t) - P\left(\frac{n\pi}{L}\right)^2 W_n^i + \rho A \frac{d^2 W_n(t)}{dt^2} \right]$$

$$\times \sin\left(\frac{n\pi x}{L}\right) = 0,$$

or

$$\rho A \frac{d^2 W_n(t)}{dt^2} + \left(\frac{n\pi}{L}\right)^2 \left\{ EI\left(\frac{n\pi}{L}\right)^2 - P \right\} W_n(t) = P\left(\frac{n\pi}{L}\right)^2 W_n^i, \tag{10.25}$$

for each value of n, where n is any integer lying within the range $1 \leqslant n \leqslant \infty$. Equation (10.25) agrees with equation (10.9) when $n = 1$, as expected.

The coefficient of the $W_n(t)$ term in equation (10.25) is non-negative when

$$P \leqslant \frac{n^2 \pi^2 EI}{L^2}. \tag{10.26}$$

The equality in equation (10.26) gives the eigenvalues, or critical loads, for the static elastic buckling of a simply supported column. These are not of practical interest for static loads except when $n = 1$, for which $P = P^e$, in agreement with equation (10.10). A theoretical analysis for the dynamic elastic case, with inequality (10.26) satisfied for every integer n, would lead to an oscillatory solution, as discussed in section 10.2.3(a) and shown in Figure 10.5 for $n = 1$. Thus, dynamic elastic buckling develops only for those modes n which violate inequality (10.26), that is for

$$n^2 \leqslant P/P^e, \tag{10.27}$$

where P^e is defined by equation (10.10). In this case, equation (10.25) can be written

$$\frac{d^2 W_n(t)}{dt^2} - \left(\frac{1}{\rho A}\right)\left(\frac{n\pi}{L}\right)^2 \left\{ P - EI\left(\frac{n\pi}{L}\right)^2 \right\} W_n(t) = \left(\frac{P}{\rho A}\right)\left(\frac{n\pi}{L}\right)^2 W_n^i$$

and has the solution

$$W_n(\tau) = \frac{\dfrac{P}{n^2 P^e}}{\dfrac{P}{n^2 P^e} - 1} \left[\cosh\left\{ n^2 \left(\frac{P}{n^2 P^e} - 1\right)^{1/2} \tau \right\} - 1 \right] W_n^i \tag{10.28}$$

when satisfying the initial conditions $W_n(0) = dW_n(0)/d\tau = 0$ at $\tau = 0$, where the dimensionless time τ is defined by equation (10.15).

Equation (10.28) can be recast into the form

$$W_n(\tau) = A_n(\tau)W_n^i,\qquad(10.29)$$

where

$$A_n(\tau) = \frac{\dfrac{P}{n^2 P^e}}{\dfrac{P}{n^2 P^e} - 1}\left[\cosh\left\{n^2\left(\frac{P}{n^2 P^e} - 1\right)^{1/2}\tau\right\} - 1\right]\qquad(10.30)$$

is the amplification function of the initial displacement profile. Equation (10.30), with $n = 1$, reduces to $A_1(\tau)$ which is defined by equations (10.19) and (10.20). Finally, if m is the largest integer which satisfies inequality (10.27), then adding the n harmonic contributions of the displacement field yields the total displacement

$$w = \sum_{n=1}^{m} A_n(\tau)W_n^i \sin\left(\frac{n\pi x}{L}\right) + \sum_{n=m+1}^{\infty} W_n(\tau)\sin\left(\frac{n\pi x}{L}\right),\qquad(10.31)$$

where $W_n(\tau)$ are the coefficients in equation (10.24) which are obtained from equation (10.25) for the oscillatory terms with $m + 1 \leqslant n \leqslant \infty$ satisfying inequality (10.26).

It is evident from equation (10.31) that the growth of lateral deflections, or dynamic buckling, of a rod can be determined by examining the characteristics of the amplification function $A_n(\tau)$ which is plotted in Figure 10.7 for several

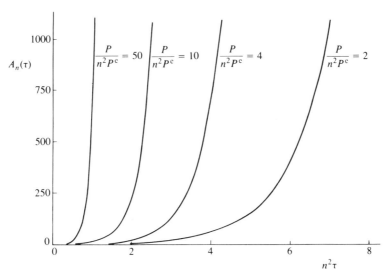

Figure 10.7 Growth of the amplification function $A_n(\tau)$ with dimensionless time $n^2\tau$, according to equation (10.30).

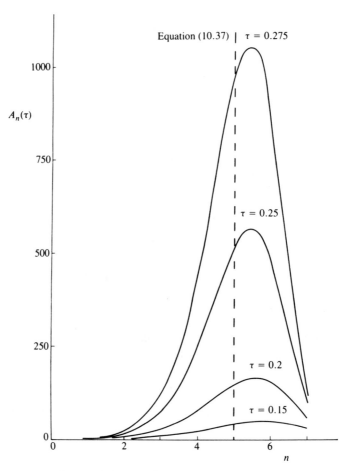

Figure 10.8 Variation of the amplification function $A_n(\tau)$ with mode number n and dimensionless time τ, according to equation (10.30) with $P/P^e = 50$.

combinations of the dimensionless load P/P^e, mode number n and dimensionless time τ. The amplification function in Figure 10.8, for $P/P^e = 50$, grows more rapidly within a narrow band of harmonics, particularly at larger dimensionless times.† This behaviour suggests that initial imperfections, with harmonics lying within this region, would be amplified more than others lying outside. The peak value of this curve occurs at the critical mode number n_c, when $\partial A_n(\tau)/\partial n = 0$, and is of considerable practical interest. Thus, writing equation (10.30) as

$$A_n(\tau) = \left(1 - \frac{n^2 P^e}{P}\right)^{-1} \left[\cosh\left\{\left(\frac{n^2 P}{P^e} - n^4\right)^{1/2} \tau\right\} - 1\right]$$

† Equation (10.27) is not satisfied for integer values $n > 7$ when $P/P^e = 50$.

and differentiating gives

$$\frac{\partial A_n(\tau)}{\partial n} = 2n\frac{P^e}{P}\left(1 - n^2\frac{P^e}{P}\right)^{-2}\left[\cosh\left\{\left(\frac{n^2 P}{P^e} - n^4\right)^{1/2}\tau\right\} - 1\right]$$

$$+ \left(1 - n^2\frac{P^e}{P}\right)^{-1}\left(\frac{nP}{P^e} - 2n^3\right)\left(\frac{n^2 P}{P^e} - n^4\right)^{-1/2}$$

$$\times \tau \sinh\left\{\left(\frac{n^2 P}{P^e} - n^4\right)^{1/2}\tau\right\}.$$

Now, $\partial A_n(\tau)/\partial n = 0$ gives the critical harmonic n_c, or

$$\frac{\cosh(p_c\tau) - 1}{p_c\tau\sinh(p_c\tau)} = 1 - \frac{P}{2n_c^2 P^e}, \tag{10.32}$$

where

$$p_c = \left(\frac{n_c^2 P}{P^e} - n_c^4\right)^{1/2}. \tag{10.33}$$

If only large dimensionless times τ are of interest, then $\sinh(p_c\tau) \cong \cosh(p_c\tau) - 1$, and equation (10.32) can be rewritten in the simpler form

$$\frac{1}{p_c\tau} = 1 - \frac{P}{2n_c^2 P^e}. \tag{10.34}$$

Equation (10.32), or the approximation (equation (10.34)), are transcendental equations for the preferred mode, or critical mode number n_c. Equation (10.34) can be arranged in the form

$$\frac{P}{n_c^2 P^e} = \frac{2(p_c\tau - 1)}{p_c\tau} \tag{10.35}$$

which suggests the crude, but useful, further approximation

$$\frac{P}{n_c^2 P^e} \cong 2 \tag{10.36}$$

when $p_c\tau \gg 1$, or

$$n_c = \left(\frac{P}{2P^e}\right)^{1/2} \tag{10.37}$$

which is shown in Figure 10.8. It is evident from Equation (10.37) that the critical mode number n_c is larger for larger dimensionless dynamic loads P/P^e and a bar, therefore, becomes more highly wrinkled, which is a characteristic of dynamic elastic buckling.

10.3 Dynamic plastic buckling of a bar

10.3.1 Introduction

The phenomenon of dynamic elastic buckling was introduced and explored in § 10.2 for the particular case of an axially loaded bar. The total stress due to the

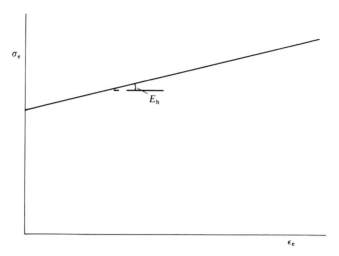

Figure 10.9 Rigid, linear-strain-hardening material.

axial compressive membrane force N and the bending moment M (see Figure 10.4) is less than the yield stress throughout the entire bar $(0 \leqslant x \leqslant L)$ and at all times $(t \geqslant 0)$. However, the load P must satisfy inequality (10.27) for dynamic buckling. Thus material plasticity is likely to develop in many practical dynamic buckling problems, as illustrated in Figures 10.1 to 10.3.

In order to introduce the phenomenon of dynamic plastic buckling, an examination is now made of the bar in Figure 10.4 when made from a rigid, linear-strain-hardening material with a small strain-hardening modulus E_h, as shown in Figure 10.9.

10.3.2 Governing equations

Equations (10.1) to (10.4), for equilibrium of the element in Figure 10.4(b), are valid for both dynamic plastic buckling and dynamic elastic buckling. However, equation (10.5) must be replaced by

$$M = -E_h I \frac{\partial^2 w}{\partial x^2} \qquad (10.38)\dagger$$

for a material with the characteristics shown in Figure 10.9.

Goodier[10.1] and others[10.3, 10.4, 10.9] have used equation (10.38) to simplify the theoretical analyses on the dynamic plastic buckling of structures. The stresses throughout a beam due to the membrane force N and bending moment M must be in the plastic range and lie on the linear-strain-hardening curve in Figure 10.9.

† The yield stress in Figure 10.9 is transmitted instantaneously throughout the entire bar since the wave speed is infinite for a rigid plastic material.

Furthermore, to avoid elastic unloading, all strains must increase as deformation progresses.

Equations (10.3), (10.4) and (10.38) again lead to equation (10.6) but with E replaced by E_h:

$$E_h I \frac{\partial^4 w}{\partial x^4} + P \frac{\partial^2}{\partial x^2}(w + w^i) = -\rho A \frac{\partial^2 w}{\partial t^2}. \tag{10.39}$$

It is now clear that the theoretical development for the dynamic plastic buckling of the bar in Figure 10.4(a), with a constant axial load P, follows closely § 10.2 for the corresponding dynamic elastic case.

10.3.3 Arbitrary initial imperfections

The problem described in § 10.2.2 is examined in this section but for a bar made from the rigid, linear-strain-hardening material shown in Figure 10.9.

Equations (10.22) to (10.24) in § 10.2.4 remain valid for this case, while equation (10.25) is unchanged, except that E is replaced by E_h, according to equation (10.38), to give the governing equation

$$\rho A \frac{d^2 W_n(t)}{dt^2} + \left(\frac{n\pi}{L}\right)^2 \left\{E_h I \left(\frac{n\pi}{L}\right)^2 - P\right\} W_n(t) = P\left(\frac{n\pi}{L}\right)^2 W_n^i. \tag{10.40}$$

The coefficient of $W_n(t)$ is negative, as required for dynamic plastic buckling, provided

$$P \geqslant n^2 \frac{E_h}{E} P^e, \tag{10.41}$$

where P^e is defined by equation (10.10), or

$$P \geqslant n^2 P^h, \tag{10.42}$$

when

$$P^h = \frac{\pi^2 E_h I}{L^2}. \tag{10.43}$$

If equation (10.42) is satisfied, then the solution of equation (10.40) is

$$W_n(\tau) = \frac{\dfrac{P}{n^2 P^h}}{\dfrac{P}{n^2 P^h} - 1}\left[\cosh\left\{n^2\left(\frac{E_h}{E}\right)^{1/2}\left(\frac{P}{n^2 P^h} - 1\right)^{1/2}\right\}\tau - 1\right]W_n^i, \tag{10.44}\dagger$$

when satisfying the initial conditions $W_n(0) = dW_n(0)/d\tau = 0$ at $\tau = 0$ and where the dimensionless time τ is defined by equation (10.15).

---

† The axial compressive force P is assumed to remain constant. However, continuing plastic flow within the bar without elastic unloading (due to bending) requires P to increase with time. This increase is neglected. Thus, the analysis is, therefore, restricted to a material having a slight strain hardening (i.e., small E_h), or $E_h \epsilon_T / 2\bar{\sigma} \ll 1$, where ϵ_T is the total equivalent strain associated with the mean flow stress $\bar{\sigma}$.

Equations (10.24) and (10.44) predict that

$$w = \sum_{n=1}^{m} A_n(\tau) W_n^i \sin\left(\frac{n\pi x}{L}\right) + \sum_{n=m+1}^{\infty} W_n(\tau) \sin\left(\frac{n\pi x}{L}\right), \qquad (10.45)\dagger$$

where the amplification function is

$$A_n(\tau) = \left(1 - \frac{n^2 P^h}{P}\right)^{-1} \left[\cosh\left\{n^2 \left(\frac{E_h}{E}\right)^{1/2} \left(\frac{P}{n^2 P^h} - 1\right)^{1/2}\right\}\tau - 1\right]. \qquad (10.46)$$

It was noted in § 10.2.4 that the amplification function has a peak value for a critical value of n_c given by $\partial A_n(\tau)/\partial n = 0$, which yields

$$\frac{\cosh(p_c^h \tau - 1)}{p_c^h \tau \sinh(p_c^h \tau)} = 1 - \frac{P}{2n_c^2 P^h}, \qquad (10.47)$$

where

$$p_c^h = \left(\frac{E_h}{E}\right)^{1/2} \left(\frac{n_c^2 P}{P^h} - n_c^4\right)^{1/2}. \qquad (10.48)$$

Equations (10.44) to (10.48) with $E_h = E$ reduce to equations (10.28), (10.31), (10.30), (10.32) and (10.33) for the linear elastic case, respectively. Thus if $\sinh(p_c^h \tau) \cong \cosh(p_c^h \tau) - 1$ and $p_c^h \tau \gg 1$, then following § 10.2.4 gives

$$n_c \cong \left(\frac{P}{2P^h}\right)^{1/2}. \qquad (10.49)$$

A comparison of equation (10.49) with equation (10.37) for the linear elastic case shows that the critical mode number is $(E/E_h)^{1/2}$ times larger for dynamic plastic buckling. However, equation (10.40), for static (i.e., $d^2 W_n(t)/dt^2 = 0$) plastic buckling, predicts

$$\left\{E_h I \left(\frac{n\pi}{L}\right)^2 - P\right\} W_n = P W_n^i,$$

or

$$W_n = \frac{W_n^i}{n^2 \dfrac{P^h}{P} - 1} \qquad (10.50)$$

when using equation (10.43). The coefficient of W_n^i on the right-hand side of equation (10.50) is an amplification function which becomes infinite for a critical mode number

$$n_p = \left(\frac{P}{P^h}\right)^{1/2}, \qquad (10.51)$$

which is $\sqrt{2}$ times larger than n_c given by equation (10.49), as pointed out by Goodier.[10.1]

The approximate expression (10.49) for the critical mode number is compared in

† m is the largest integer satisfying inequality (10.42) for a given force P. See also equation (10.31).

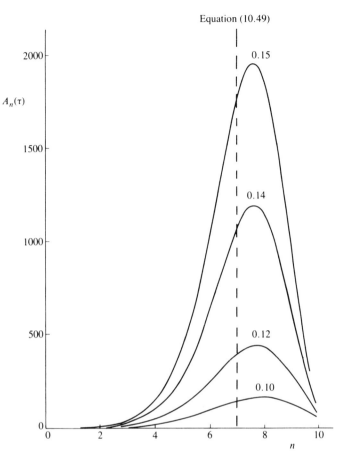

Figure 10.10 Variation of the amplification function $A_n(\tau)$ with mode number n and dimensionless time $(E_h/E)^{1/2}\tau$, according to equation (10.46) with $P/P^h = 100$. Numbers on the curves are values of $(E_h/E)^{1/2}\tau$.

Figure 10.10 with the amplification function $A_n(\tau)$ given by equation (10.46). If this is taken as an acceptable approximation for illustrative purposes, then the amplification function (10.46) simplifies to

$$A_{nc}(\tau) = 2\left[\cosh\left\{\frac{P}{2P^h}\left(\frac{E_h}{E}\right)^{1/2}\tau\right\} - 1\right], \tag{10.52}$$

which is plotted in Figure 10.11.

It is clear that equation (10.52) grows exponentially with dimensionless time (τ) but never reaches a specific value which could be associated with a dynamic buckling load. The definition of dynamic buckling is, therefore, somewhat arbitrary for this class of problem. However, it is evident that the safety of a structure might be imperilled by a response similar to that shown in Figure 10.11. Dynamic buckling, is assumed, therefore, to occur when the maximum deflection reaches a speci-

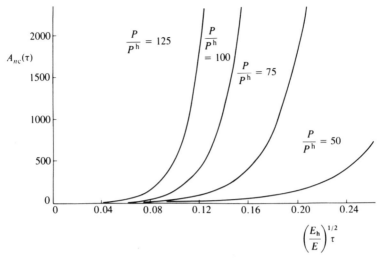

Figure 10.11 Growth of amplification function $A_{nc}(\tau)$ with dimensionless time $(E_h/E)^{1/2}\tau$, according to equation (10.52), for the critical mode number and several values of P/P^h.

fied magnitude. For example, a design might become unserviceable for a maximum displacement which is 100 times larger ($A_{nc}(\tau) = 100$) than the initial imperfection (W_{nc}^i), which is taken to lie in the critical mode only. Equation (10.52), or Figure 10.11, then gives the dimensionless time $(E_h/E)^{1/2}\tau$, when that displacement is reached, and the corresponding value of P gives the dynamic plastic buckling load.

10.4 Dynamic plastic buckling of a circular ring subjected to an external impulse

10.4.1 Introduction

Large external dynamic pressures can produce dynamic plastic instability of circular rings and cylindrical shells, as shown in Figures 10.3 and 10.12, respectively. The wrinkled profiles of buckled circular rings and cylindrical shells exhibit characteristic wavelengths which were found to be reproducible in experimental investigations.[10.4, 10.9] Abrahamson and Goodier[10.11] developed a theoretical procedure which predicted reasonable agreement with the corresponding experimental results. This analytical method assumed that buckling stemmed from the growth of small imperfections in the otherwise uniform initial displacement and velocity fields.

The theoretical analysis for the dynamic plastic buckling of an impulsively loaded circular ring is simplified in this section by assuming that it consists of dominant axisymmetric and perturbed non-axisymmetric contributions, as in-

Figure 10.12 The wrinkling near the base of the cylindrical reactor vessel could be due to dynamic buckling as a result of the pressures released during the Flixborough disaster.[10.10] (Reproduced with the permission of the Controller of Her Majesty's Stationery Office.)

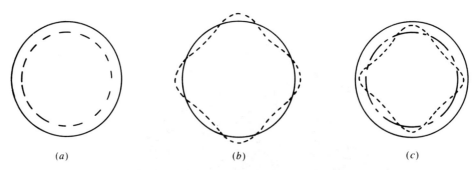

Figure 10.13 Idealised response of a circular ring which is subjected to an external impulsive velocity. (a) dominant radial motion. (b) perturbation behaviour. (c) total radial motion.

dicated in Figure 10.13.† The governing equations are developed in the next section, while the theoretical solutions for the dominant motion and perturbed behaviour are considered separately in the following two sections.

10.4.2 Governing equations

10.4.2(a) Equilibrium equations

A small element of a thin circular ring with a geometrically imperfect initial profile, which is characterised by the radial displacement $w^i(\theta)$, is shown in Figure 10.14. Equilibrium of forces in the radial direction gives

$$\frac{\partial Q}{\partial \theta} + N + N \frac{\partial^2(w + w^i)}{R\,\partial\theta^2} + \frac{\partial N}{\partial \theta} \frac{\partial(w + w^i)}{R\,\partial\theta} - \mu R \frac{\partial^2 w}{\partial t^2} = 0, \qquad (10.53)$$

where μ is the mass per unit circumferential length of a ring with a mean radius R, and w is measured radially inwards from the initial imperfect shape. Similarly, equilibrium in the tangential sense leads to

$$\frac{\partial N}{\partial \theta} - Q - N \frac{\partial(w + w^i)}{R\,\partial\theta} = 0 \qquad (10.54)$$

when disregarding the circumferential inertia force and the non-linear terms containing Q and $\partial Q/\partial\theta$.‡ Finally, moment equilibrium demands

$$\frac{\partial M}{R\,\partial\theta} - Q = 0 \qquad (10.55)$$

when neglecting the influence of rotatory inertia.

---

† This approach differs fundamentally from that in §§ 10.2 and 10.3 for the dynamic axial buckling of a bar. A dominant solution for the bar would give rise to uniform axial deformations which were assumed to be negligible compared with the lateral deflections.

‡ Transverse shear effects are not examined in this chapter so that the non-linear terms containing Q and $\partial Q/\partial\theta$ should not make an important contribution during the response.

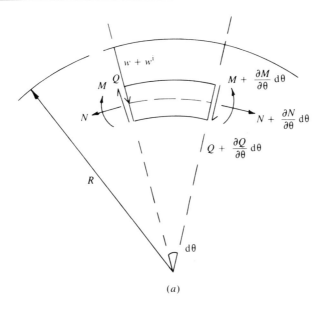

(a)

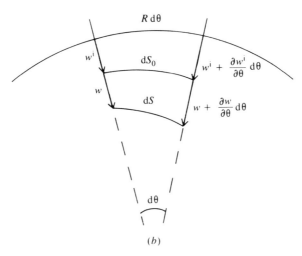

(b)

Figure 10.14 (a) Element of a circular ring and notation. (b) Centre line of element in initial (dS_0) and deformed (dS) configurations.

The transverse shear force Q given by equation (10.55) may be used to rewrite equations (10.53) and (10.54) as

$$\frac{\partial^2 M}{\partial \theta^2} + NR + N\frac{\partial^2 (w + w^i)}{\partial \theta^2} + \frac{\partial N}{\partial \theta}\frac{\partial (w + w^i)}{\partial \theta} - \mu R^2 \frac{\partial^2 w}{\partial t^2} = 0 \qquad (10.56)$$

and

$$\frac{\partial M}{\partial \theta} - R \frac{\partial N}{\partial \theta} + N \frac{\partial(w + w^i)}{\partial \theta} = 0, \tag{10.57}†$$

respectively.

Now, substituting $\partial N/\partial \theta$ from equation (10.57) into equation (10.56) gives the governing equation

$$\frac{\partial^2 M}{\partial \theta^2} + NR + N \frac{\partial^2(w + w^i)}{\partial \theta^2} - \mu R^2 \frac{\partial^2 w}{\partial t^2} = 0 \tag{10.58}$$

when neglecting the non-linear term containing $\partial M/\partial \theta$ in order to remain consistent with the approximations made when deriving equation (10.54). In addition, $(N/R)(\partial(w + w^i)/\partial \theta)^2$ is taken to be small compared with NR.

10.4.2(b) Geometrical relations

The circumferential membrane strain in a thin ring is defined as

$$\epsilon = \frac{dS - dS_0}{dS_0}, \tag{10.59}$$

where dS_0 and dS are shown in Figure 10.14 and are the lengths of an element before and after radial deformation, respectively.‡ Elementary geometry may be used to obtain dS_0 and dS, which substituting into equation (10.59), gives

$$\epsilon = -\frac{w}{R} + \frac{1}{2}\left(\frac{\partial w}{R\,\partial \theta}\right)^2 + \left(\frac{\partial w}{R\,\partial \theta}\right)\left(\frac{\partial w^i}{R\,\partial \theta}\right) \tag{10.60}§$$

when neglecting higher-order terms.

The difference between the curvatures of the initial imperfect profile (dS_0) and the deformed profile (dS) in Figure 10.14 gives the change in curvature

$$\kappa = -\frac{\partial^2 w}{R^2\,\partial \theta^2} \tag{10.61}*$$

when disregarding higher-order terms.

---

† Equations (10.56) and (10.57), for the static case ($\partial^2 w/\partial t^2 = 0$) with $w^i = 0$, agree with the corresponding equilibrium equations for a geometrically perfect circular ring having zero tangential displacements which are derived in references 10.12 and 10.13.

‡ The circumferential displacements are assumed to be small compared with the radial displacements (w) and are, therefore, neglected.

§ Page 109 of reference 10.9 contains a derivation of this expression. Equation (10.60), with $w^i = 0$, agrees with the corresponding expressions in references 10.12 and 10.13 with zero tangential displacements.

* The negative sign in equation (10.61) ensures that $M\dot{\kappa} > 0$ when M is defined in Figure 10.14 (see also § 2.2). Equation (10.61) is identical to the corresponding expressions in references 10.12 and 10.13. However, Abrahamson and Goodier,[10.11] Lindberg and Florence[10.9] and others include a term w/R^2 in equation (10.61). Nevertheless, it may be shown that the equilibrium equations (10.55) and (10.58) and the strain and curvature relations (10.60) and (10.61) are consistent according to the principle of virtual velocities[10.8] (see also Appendix 3) when disregarding a non-linear term which would arise in equation (10.55) and the non-linear term $(\partial N/\partial \theta)(\partial(w + w^i)/\partial \theta)$ in equation (10.58) which is consistent with the simplifications already made in deriving equation (10.58).

It is evident from equations (10.60) and (10.61) that $\epsilon = \kappa = 0$ when $w = 0$ for the initial imperfect profile, which is, therefore, stress-free.

10.4.2(c) Constitutive equations

Abrahamson and Goodier[10.11] simplified their theoretical study on the dynamic plastic buckling of cylindrical shells with the assumption

$$M = E_{\mathrm{h}} I \kappa, \tag{10.62}\dagger$$

where I is the second moment of area for the cross-section. It is assumed that this expression remains valid for a circular ring made from the rigid, linear-strain-hardening material shown in Figure 10.9. In addition, it is assumed that the circumferential membrane force

$$N = (\pm \sigma_0 + E_{\mathrm{h}} \epsilon) H, \tag{10.63}$$

where σ_0 is the uniaxial yield stress and H is the radial thickness of a ring.

Equations (10.62) and (10.63), either individually or superposed, are valid only when the entire cross-section of a circular ring is fully plastic, with no elastic, or rigid, regions and no unloading.

10.4.2(d) Differential equations

Equations (10.60) and (10.63) and equations (10.61) and (10.62) give the membrane force

$$N = \pm \sigma_0 H + E_{\mathrm{h}} H \left\{ -\frac{w}{R} + \frac{1}{2}\left(\frac{\partial w}{R\,\partial\theta}\right)^2 + \left(\frac{\partial w}{R\,\partial\theta}\right)\left(\frac{\partial w^{\mathrm{i}}}{R\,\partial\theta}\right)\right\} \tag{10.64}$$

and bending moment

$$M = -\frac{E_{\mathrm{h}} I}{R^2}\frac{\partial^2 w}{\partial\theta^2}, \tag{10.65}$$

respectively. Thus, equation (10.58) may now be written in the form

$$\frac{\partial^4 w}{\partial\theta^4} - \frac{R^2 H}{I}\left\{-\frac{\sigma_0}{E_{\mathrm{h}}} - \frac{w}{R} + \frac{1}{2}\left(\frac{\partial w}{R\,\partial\theta}\right)^2 + \left(\frac{\partial w}{R\,\partial\theta}\right)\left(\frac{\partial w^{\mathrm{i}}}{R\,\partial\theta}\right)\right\}\frac{\partial^2 w}{\partial\theta^2}$$
$$+ \frac{\mu R^4}{E_{\mathrm{h}} I}\frac{\partial^2 w}{\partial t^2} = \frac{R^2 H}{I}\left\{-\frac{\sigma_0}{E_{\mathrm{h}}} - \frac{w}{R} + \frac{1}{2}\left(\frac{\partial w}{R\,\partial\theta}\right)^2\right.$$
$$\left. + \left(\frac{\partial w}{R\,\partial\theta}\right)\left(\frac{\partial w^{\mathrm{i}}}{R\,\partial\theta}\right)\right\}\left(R + \frac{\partial^2 w^{\mathrm{i}}}{\partial\theta^2}\right). \tag{10.66}$$

It was observed in § 10.4.1 that the dynamic plastic buckling response of an impulsively loaded circular ring is simplified by assuming it consists of two parts: dominant axisymmetric and perturbed non-axisymmetric contributions, as indicated in Figure 10.13. The initial kinetic energy associated with the external impulsive loading is absorbed plastically during the dominant motion, which is taken to remain axisymmetric for an axisymmetric initial impulsive velocity. However, the propensity for a circular ring to buckle, as indicated for the metal

---

† This is similar to equation (10.38) for the dynamic plastic buckling of a bar.

specimens in Figure 10.3, is examined by probing the stability of the dominant axisymmetric response. This is achieved by enquiring whether an arbitrary distribution of small initial imperfections in shape (w^i in Figure 10.14) will grow with time. If the perturbed radial displacements do grow with time, then the circular ring becomes wrinkled and dynamic buckling is said to occur. The growth of the perturbed displacements must remain small during the response in order to prevent local unloading in a ring which would violate the validity of equation (10.62). Thus, the total radial displacement from the initial imperfect profile is

$$w = \bar{w}(t) + w'(\theta, t), \qquad (10.67)$$

where $\bar{w}(t)$ and $w'(\theta, t)$ are the dominant axisymmetric and perturbed non-axisymmetric displacements with $w' \ll \bar{w}$.

Now, substituting equation (10.67) into equation (10.66) gives

$$\frac{\partial^4 w'}{\partial \theta^4} - \frac{R^2 H}{I} \left\{ -\frac{\sigma_0}{E_h} - \frac{(\bar{w} + w')}{R} + \frac{1}{2}\left(\frac{\partial w'}{R\, \partial \theta}\right)^2 + \left(\frac{\partial w'}{R\, \partial \theta}\right)\left(\frac{\partial w^i}{R\, \partial \theta}\right) \right\} \frac{\partial^2 w'}{\partial \theta^2}$$

$$+ \frac{\mu R^4}{E_h I} \frac{\partial^2 (\bar{w} + w')}{\partial t^2} = \frac{R^2 H}{I} \left\{ -\frac{\sigma_0}{E_h} - \frac{(\bar{w} + w')}{R} + \frac{1}{2}\left(\frac{\partial w'}{R\, \partial \theta}\right)^2 \right.$$

$$\left. + \left(\frac{\partial w'}{R\, \partial \theta}\right)\left(\frac{\partial w^i}{R\, \partial \theta}\right) \right\}\left(R + \frac{\partial^2 w^i}{\partial \theta^2} \right). \quad (10.68)$$

It is evident that the dominant (or first-order) terms in equation (10.68) are

$$\frac{d^2 \bar{w}}{dt^2} + \left(\frac{E_h H}{\mu R^2}\right)\bar{w} = -\frac{H \sigma_0}{\mu R}, \qquad (10.69)$$

while the next-higher-order terms (second-order terms) are

$$\frac{\partial^4 w'}{\partial \theta^4} + \frac{R^2 H}{I}\left(\frac{\sigma_0}{E_h} + \frac{\bar{w}}{R}\right)\frac{\partial^2 w'}{\partial \theta^2} + \left(\frac{\mu R^4}{E_h I}\right)\frac{\partial^2 w'}{\partial t^2} =$$

$$\cdot \qquad\qquad -\left(\frac{R^2 H}{I}\right)w' - \frac{R^2 H}{I}\left(\frac{\sigma_0}{E_h} + \frac{\bar{w}}{R}\right)\frac{\partial^2 w^i}{\partial \theta^2}. \quad (10.70)$$

Another higher-order equation could be formed, but with terms which are negligible compared with those in equation (10.70). In any event, the radial displacement w in equation (10.67) has only two components ($\bar{w}$, w'), so that the two equations (10.69) and (10.70) are sufficient. Equations (10.69) and (10.70) are known as the dominant and perturbed equations, respectively, and are examined in the following two sections.

10.4.3 Dominant motion

The axisymmetric dominant response is controlled by equation (10.69), which has the general solution

$$\bar{w} = C_1 \cos\left\{\left(\frac{E_h H}{\mu R^2}\right)^{1/2} t\right\} + C_2 \sin\left\{\left(\frac{E_h H}{\mu R^2}\right)^{1/2} t\right\} - \frac{\sigma_0 R}{E_h}, \qquad (10.71)$$

where C_1 and C_2 are arbitrary constants of integration. However, at $t = 0$, the initial dominant radial displacement $\bar{w} = 0$ and $d\bar{w}/dt = V_0$, where V_0 is the initial uniformly distributed radially inwards impulsive velocity. Thus,

$$\bar{w} = \frac{\sigma_0 R}{E_h} (\cos \tau - 1) + \left(\frac{\mu V_0^2 R^2}{E_h H} \right)^{1/2} \sin \tau, \tag{10.72a}$$

where

$$\tau = \left(\frac{E_h H}{\mu R^2} \right)^{1/2} t. \tag{10.72b}$$

The dominant motion ceases at $t = t_f$, or $\tau = \tau_f$, when $d\bar{w}/dt = 0$, or

$$\tan \tau_f = \lambda^{1/2}, \tag{10.73a}$$

where

$$\lambda = \frac{\mu V_0^2 E_h}{\sigma_0^2 H}. \tag{10.73b}$$

Finally, the maximum permanent dominant radial displacement is

$$\bar{w}_f = \frac{\sigma_0 R}{E_h} \{ (1 + \lambda)^{1/2} - 1 \}. \tag{10.74}$$

10.4.4 Perturbed behaviour

A solution is now sought for the perturbed displacement w' which satisfies equation (10.70). Let

$$w' = \sum_{n=2}^{\infty} W_n'(t) \sin n\theta \tag{10.75a}†‡}$$

and

$$w^i = \sum_{n=2}^{\infty} W_n^i \sin n\theta \tag{10.75b}†‡}$$

which, when substituted into equation (10.70), gives

$$\sum_{n=2}^{\infty} n^4 \, W_n' \sin n\theta - \frac{R^2 H}{I} \left(\frac{\sigma_0}{E_h} + \frac{\bar{w}}{R} \right) \sum_{n=2}^{\infty} n^2 W_n' \sin n\theta + \frac{\mu R^4}{E_h I} \sum_{n=2}^{\infty} \frac{d^2 W_n'}{dt^2} \sin n\theta$$

$$= -\frac{R^2 H}{I} \sum_{n=2}^{\infty} W_n' \sin n\theta + \frac{R^2 H}{I} \left(\frac{\sigma_0}{E_h} + \frac{\bar{w}}{R} \right) \sum_{n=2}^{\infty} n^2 W_n^i \sin n\theta,$$

or

$$\sum_{n=2}^{\infty} \left[n^4 W_n' - \frac{R^2 H}{I} \left(\frac{\sigma_0}{E_h} + \frac{\bar{w}}{R} \right) n^2 (W_n' + W_n^i) + \frac{\mu R^4}{E_h I} \frac{d^2 W_n'}{dt^2} + \frac{R^2 H}{I} W_n' \right] \sin n\theta = 0. \tag{10.76}$$

---

† The $n = 1$ component of radial displacement does not contribute to the distortion of a circular ring since tangential displacements are disregarded.

‡ $\cos n\theta$ components could also be included in equations (10.75). However, equation (10.70) contains only even derivatives, so that the equation governing the $\cos n\theta$ components would be identical to equation (10.77).

Thus, the perturbed displacement component W'_n satisfies

$$\frac{d^2 W'_n}{dt^2} + \frac{E_h I}{\mu R^4} \left\{ \left(\frac{\bar{N} R^2}{E_h I}\right) n^2 + n^4 + \frac{R^2 H}{I} \right\} W'_n = -\left(\frac{\bar{N}}{\mu R^2}\right) n^2 W^i_n, \quad (10.77)\dagger$$

where

$$\bar{N} \cong -\sigma_0 H - E_h H \bar{w}/R \quad (10.78)$$

according to equations (10.64) and (10.67), when the perturbed terms are neglected.

Equation (10.77) can be recast into the simpler form

$$\frac{d^2 W'_n}{dt^2} - R_n^2 W'_n = S_n W^i_n, \quad (10.79)$$

where

$$R_n^2 = -\frac{E_h I}{\mu R^4} \left\{ \left(\frac{\bar{N} R^2}{E_h I}\right) n^2 + n^4 + \frac{R^2 H}{I} \right\} \quad (10.80)$$

and

$$S_n = -\left(\frac{\bar{N}}{\mu R^2}\right) n^2. \quad (10.81)$$

The form of equation (10.79) is similar to equations (10.11), (10.16), (10.25) and (10.40). The perturbed displacement component W'_n, behaves, therefore, in an oscillatory fashion when $R_n^2 < 0$, but would manifest the phenomenon of dynamic plastic buckling when $R_n^2 > 0$. Thus, examining the $R_n^2 > 0$ case,

$$W'_n = A_n \cosh R_n t + B_n \sinh R_n t - \frac{S_n W^i_n}{R_n^2} \quad (10.82)$$

when taking $\bar{N}$ constant.

It is assumed that the initial imperfect profile of a circular ring may be expressed by equation (10.75b), while the external impulsive velocity is $dw/dt = d\bar{w}/dt + dw'/dt$ according to equation (10.67), which, at $t = 0$, becomes

$$\frac{dw(0)}{dt} = V_0 + \sum_{n=2}^{\infty} V^i_n \sin n\theta, \quad (10.83)$$

where V^i_n caters for any deviation from the initial axisymmetric impulsive velocity V_0. Now, $W'_n = 0$ and $dW'_n/dt = V^i_n$ at $t = 0$ when motion commences. Thus,

$$W'_n = \frac{S_n W^i_n}{R_n^2} \cosh R_n t + \frac{V^i_n}{R_n} \sinh R_n t - \frac{S_n W^i_n}{R_n^2}, \quad (10.84)$$

or

$$W'_n = W^i_n E_n(t) + V^i_n F_n(t), \quad (10.85)$$

where

$$E_n(t) = \frac{S_n(\cosh R_n t - 1)}{R_n^2} \quad (10.86a)$$

† See second footnote to equations (10.75).

and

$$F_n(t) = \frac{\sinh R_n t}{R_n} \qquad (10.86b)$$

are known as the displacement and velocity amplification functions, respectively.

It is evident from the evolution of the amplification functions $E_n(\tau)$ and $V_0 F_n(\tau)/R$ in Figure 10.15 that peak values develop for a particular value of n which is known as the critical mode number n_c. Any initial displacement or velocity imperfections lying in this critical mode would, therefore, be amplified more than initial imperfections present in any other mode. The concept of a critical mode number is clearly important, and for velocity imperfections may be obtained from $\partial F_n(t)/\partial n = 0$, which, when using equation (10.86b), gives

$$\frac{\partial F_n(t)}{\partial n} = \frac{1}{R_n^2} \frac{\partial R_n}{\partial n} (R_n t \cosh R_n t - \sinh R_n t). \qquad (10.87)$$

The requirement $\partial F_n(t)/\partial n = 0$ is, therefore, satisfied when $\partial R_n/\partial n = 0$, which, using equation (10.80), predicts a critical mode number†

$$n_c = \left(\frac{-\bar{N}R^2}{2E_h I}\right)^{1/2}. \qquad (10.88)$$

Defining $-\bar{\sigma}$ as an average flow stress throughout the response gives $\bar{N} \cong -\bar{\sigma}H$, and equation (10.88) becomes

$$n_c = \left(\frac{\bar{\sigma}HR^2}{2E_h I}\right)^{1/2}, \qquad (10.89)‡$$

or

$$n_c = \left(\frac{6\bar{\sigma}}{E_h}\right)^{1/2}\left(\frac{R}{H}\right) \qquad (10.90)§$$

when $I = H^3/12$ and where

$$\bar{\sigma} = \sigma_0(1 + \sqrt{1 + \lambda})/2 \qquad (10.91)‡$$

The critical mode number for the displacement amplification function is obtained from $\partial E_n(t)/\partial n = 0$, where $E_n(t)$ is given by equation (10.86a). If $t = t_f$, where t_f is the duration of the dominant motion according to equation (10.73a), then it may be shown that the critical mode number for displacement imperfections equals equation (10.88) (or equations (10.89 and 10.90)) provided $R_n t_f \gg 1$. The

---

† The critical mode number is the integer nearest to the value predicted by equation (10.88).
‡ $\bar{\sigma} \cong \sigma_0 + E_h\bar{w}/R$ according to equation (10.78). Thus, an average value of $\bar{\sigma}$ could be estimated by replacing $\bar{w}$ by $\bar{w}_f/2$, where $\bar{w}_f$ is predicted by equation (10.74).
§ Equation (10.90) is identical to the corresponding theoretical prediction of Abrahamson and Goodier[10.11] for the dynamic plastic buckling of a cylindrical shell with initial velocity imperfections and $(\bar{\sigma}/E_h)(R/H)^2 \gg 1$.

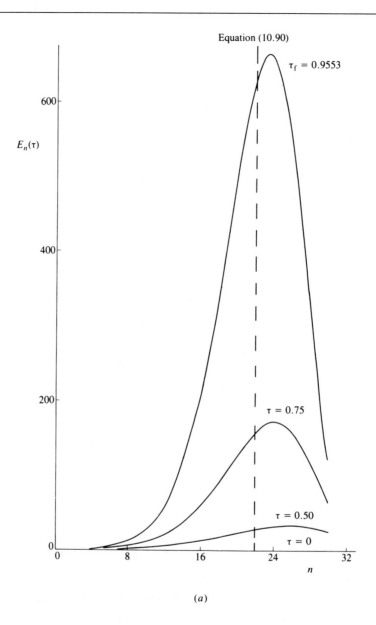

Equation (10.90)

$\tau_f = 0.9553$

$E_n(\tau)$

$\tau = 0.75$

$\tau = 0.50$

$\tau = 0$

n

(a)

nearest integer to the critical mode number predicted by equation (10.90) is indicated in Figures 10.15(a) and 10.15(b).

The relative importance of the displacement and velocity amplification functions are now examined. It is evident from equation (10.85) that the velocity amplification function $F_n(t)$ is not dimensionless. However, equation (10.85) can be written

$$\frac{W'_n}{R} = \frac{W^i_n}{R} E_n(t) + \frac{V^i_n}{V_0} \left\{ \frac{V_0 F_n(t)}{R} \right\}, \tag{10.92}$$

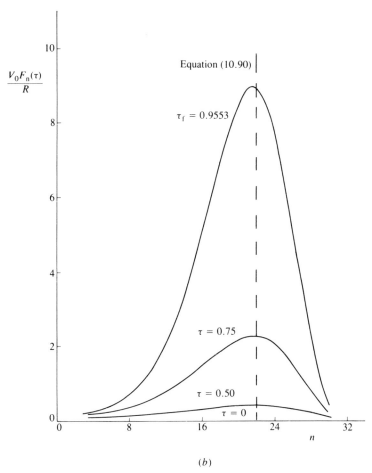

(b)

Figure 10.15 (a) Displacement amplification function $E_n(\tau)$ according to equation (10.86a), with $E_h/\sigma_0 = 7$, $R/H = 20$ and $\lambda = 2$. τ is the dimensionless time defined by equation (10.72b). (b) Dimensionless velocity amplification function $\dfrac{V_0 F_n(\tau)}{R}$ according to equation (10.86b), with $E_h/\sigma_0 = 7$, $R/H = 20$ and $\lambda = 2$. τ is the dimensionless time defined by equation (10.72b).

where $V_0 F_n(t)/R$ is the dimensionless velocity amplification function. Thus, the relative importance of the two amplification functions are given by the ratio $R E_n(t)/V_0 F_n(t)$, which, using equation (10.86), becomes

$$\frac{R E_n(t_f)}{V_0 F_n(t_f)} = \left(\frac{R}{V_0}\right)\left(\frac{S_n}{R_n}\right)\frac{\cosh R_n t_f - 1}{\sinh R_n t_f} \tag{10.93}$$

at the end of the dominant motion $(t = t_f)$. If $R_n t_f \gg 1$, then

$$\frac{R E_n(t_f)}{V_0 F_n(t_f)} \cong \left(\frac{R}{V_0}\right)\left(\frac{S_n}{R_n}\right),$$

or

$$\frac{RE_{nc}(\tau_f)}{V_0 F_{nc}(\tau_f)} = \left(\frac{R}{V_0}\right)\left(\frac{S_{nc}}{R_{nc}}\right) \tag{10.94}$$

when attention is confined to the critical mode number, where R_{nc} and S_{nc} are defined by equations (10.80) and (10.81), respectively. Equation (10.94), together with equation (10.88) for n_c, gives

$$\frac{RE_{nc}(\tau_f)}{V_0 F_{nc}(\tau_f)} = \left(\frac{6}{\sqrt{\lambda}}\right)\left(\frac{\bar{\sigma}}{E_h}\right)\left(\frac{\bar{\sigma}}{\sigma_0}\right)\left(\frac{R}{H}\right)^2\left\{3\left(\frac{\bar{\sigma}}{E_h}\right)^2\left(\frac{R}{H}\right)^2 - 1\right\}^{-1/2}, \tag{10.95}$$

where λ is defined by equation (10.73b) and $I = H^3/12$. This expression further simplifies to

$$\frac{RE_{nc}(\tau_f)}{V_0 F_{nc}(\tau_f)} = \sqrt{\frac{12}{\lambda}}\left(\frac{\bar{\sigma}}{\sigma_0}\right)\left(\frac{R}{H}\right) \tag{10.96}$$

when $(\bar{\sigma}/E_h)^2(R/H)^2 \gg 1$. Finally, using equation (10.91), equation (10.96) may be written

$$\frac{RE_{nc}(\tau_f)}{V_0 F_{nc}(\tau_f)} = \sqrt{\frac{3}{\lambda}}(1 + \sqrt{1 + \lambda})\left(\frac{R}{H}\right). \tag{10.97}$$

Equation (10.97) is plotted in Figure 10.16 and reveals that the influence of the displacement amplification function is more significant than the velocity amplification function. In other words, dynamic plastic buckling for this range of

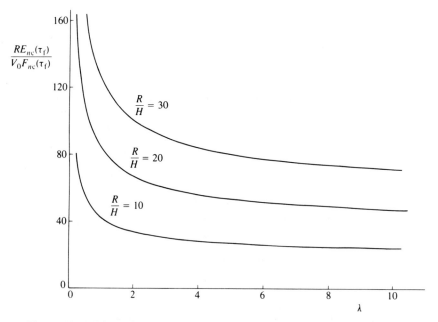

Figure 10.16 Dimensionless ratio of displacement and velocity amplification functions, according to equation (10.97), at cessation of dominant motion ($\tau = \tau_f$), where τ_f and λ are defined by equations (10.73).

parameters is more sensitive to initial imperfections in shape than to imperfections in the initial axisymmetric impulsive velocity. This is fortunate because it is easier to control the initial geometric shape than to ensure uniformity of the external impulsive velocity.

10.4.5 General comments

The specified manufacturing tolerances of a circular ring may be cast into the form of equation (10.75b) and should give the worst initial imperfect shape $w^i(\theta)$. Alternatively, the actual geometric imperfections in a particular ring could be measured and expressed in the form of equation (10.75b) using the standard methods of Fourier series. The deformation of a circular ring at time t is then

$$w(t) = \bar{w}(t) + \sum_{n=2}^{m} E_n(t)W_n^i \sin n\theta \tag{10.98}†$$

when neglecting the influence of any imperfections in the initial velocity which were shown to be less important in Figure 10.16.

In order to simplify subsequent calculations, it is assumed that the initial imperfections w^i according to equation (10.75b) lie only in the mode, with the critical number n_c given by equations (10.88) to (10.90), or

$$w_c(t) = \bar{w}(t) + E_{nc}(t)W_{nc}^i \tag{10.99}$$

when $\sin n_c\theta = 1$. This is the worst possible case.

The temporal growth of the dominant displacement ($\bar{w}$), perturbed displacement (w') and total radial displacement (w_c), according to equations (10.72a), (10.84) and (10.99), respectively, are shown in Figure 10.17 for the critical mode number (n_c) which is predicted by equation (10.89). Clearly, the magnitude of $E_{nc}(t_f)$, when the dominant motion ceases, is important and equals $\{w_c(t_f) - \bar{w}(t_f)\}/W_{nc}^i$.

Now, from equation (10.86a),

$$E_{nc}(t_f) = \frac{S_{nc}}{R_{nc}^2}\{\cosh(R_{nc}t_f) - 1\}, \tag{10.100}$$

which can be written

$$E_{nc}(\tau_f) = \frac{6\left(\dfrac{\bar{\sigma}}{E_h}\right)^2\left(\dfrac{R}{H}\right)^2}{3\left(\dfrac{\bar{\sigma}}{E_h}\right)^2\left(\dfrac{R}{H}\right)^2 - 1}\left[\cosh\left\{\left\{3\left(\dfrac{\bar{\sigma}}{E_h}\right)^2\left(\dfrac{R}{H}\right)^2 - 1\right\}^{1/2}\tau_f\right\} - 1\right] \tag{10.101}$$

when using equation (10.90) for the approximate critical mode number n_c, and equation (10.72b) for the dimensionless time τ. $\bar{\sigma}$ is the mean flow stress which is approximated by equation (10.91). Equation (10.101) simplifies to

$$E_{nc}(\tau_f) = 2\left[\cosh\left\{\sqrt{3}\left(\dfrac{\bar{\sigma}}{E_h}\right)\left(\dfrac{R}{H}\right)\tau_f\right\} - 1\right] \tag{10.102}$$

---

† m is the largest integer giving $R_n^2 > 0$. The contribution of the oscillatory terms having $R_n^2 < 0$ is ignored.

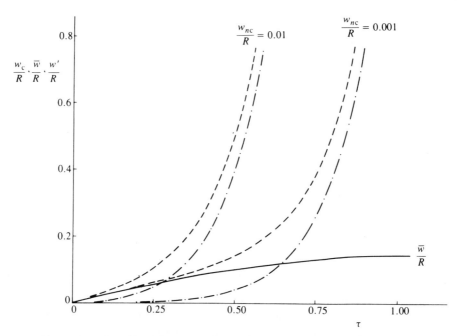

$$\frac{w_c}{R}, \frac{\bar{w}}{R}, \frac{w'}{R}$$

Figure 10.17 Growth of dimensionless radial displacements with dimensionless time τ for a geometrically imperfect circular ring with $E_h/\sigma_0 = 7$, $R/H = 20$ and $\lambda = 3$ ($n_c = 23$, $\tau_f = 1.0472$). ———: dimensionless dominant displacement (equation (10.72a)). —·—·—: dimensionless perturbed displacement for the critical mode number (equation (10.84)). ———: dimensionless total displacement ($n_c = 23$) (equation (10.99)).

when $3(\bar{\sigma}/E_h)^2(R/H)^2 \gg 1$. Substituting for τ_f from equation (10.73a) and rearranging gives the dimensionless impulsive velocity

$$\lambda^{1/2} = \tan\left[\frac{1}{\sqrt{3}}\left(\frac{E_h}{\bar{\sigma}}\right)\left(\frac{H}{R}\right)\left\{\cosh^{-1}\left\{1 + \frac{E_{nc}(\tau_f)}{2}\right\}\right\}\right], \qquad (10.103)\dagger$$

where λ is defined by equation (10.73b).

It is evident from equation (10.103) and the results in Figure 10.18 that, once the material (μ, E_h, σ_0) and geometric (R, H) parameters are specified for a particular circular ring, then the threshold velocity, which is the velocity (V_0) required to produce dynamic plastic buckling, is related directly to the magnitude of the displacement amplification function $E_{nc}(\tau_f)$. Dynamic plastic buckling according to this type of analysis does not occur, therefore, for a particular magnitude of impulsive velocity, since an initial velocity of any magnitude produces some damage. Dynamic plastic buckling is said to occur when the permanent radial deformations are unacceptably large. This leaves a designer free to select a value for

† This is a transcendental equation, since λ appears in $\bar{\sigma}$ defined by equation (10.91).

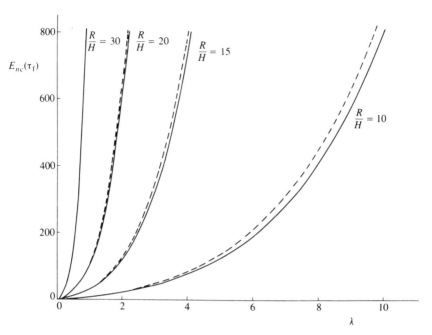

Figure 10.18 Displacement amplification functions, according to equation (10.101) (——) and equations (10.102) or (10.103) (---), for an impulsively loaded circular ring with $E_h/\sigma_0 = 7$.

$E_{nc}(\tau_f)$ in equation (10.103) and define a threshold velocity. However, as noted earlier, equation (10.99) can be used to estimate $E_{nc}(\tau_f) = \{w_c(\tau_f) - \bar{w}(\tau_f)\}/W^i_{nc}$, where w_c is the largest acceptable radial displacement and W^i_{nc} is the largest initial radial deviation from a perfectly circular profile.

If, for convenience, it is assumed that $E_{nc}(\tau_f) = 100$, then equations (10.102) or (10.103) become

$$\cosh^{-1}(51) = \frac{\sqrt{3}}{2}\left(\frac{\sigma_0}{E_h}\right)(1 + \sqrt{1 + \lambda})\left(\frac{R}{H}\right)\sqrt{\lambda} \qquad (10.104)$$

when using equation (10.91) for the mean flow stress $\bar{\sigma}$ and assuming that the tangent modulus E_h is small for a material with little strain hardening in which case $\tan \tau_f \cong \tau_f = \sqrt{\lambda}$ according to equation (10.73a). Equation (10.104) further simplifies to

$$\sqrt{\lambda} = 2.67\left(\frac{E_h}{\sigma_0}\right)\left(\frac{H}{R}\right) \qquad (10.105)$$

when taking $\sqrt{1 + \lambda} \cong 1$, or

$$V_0 = 2.67\left(\frac{H}{R}\right)\left(\frac{E_h}{\rho}\right)^{1/2} \qquad (10.106)$$

when $\mu = \rho H$, where ρ is the density of the material. This is a threshold velocity for the dynamic plastic buckling of a circular ring which satisfies the various restrictions introduced in the derivation of equation (10.106). The associated threshold impulse per unit surface area is $2\pi R H \rho V_0 / 2\pi R$, or

$$I = 2.67 R \sqrt{\rho E_h} \left(\frac{H}{R} \right)^2. \tag{10.107}\dagger$$

10.4.6 Experimental results

Detailed comparisons are made in references 10.4 and 10.9 between various theoretical predictions and experimental results for the dynamic plastic buckling of cylindrical shells and rings subjected to almost axisymmetric lateral impulsive velocities. Generally speaking, the theoretical methods which are developed along lines similar to those in this section capture the essential features of behaviour, and provide reasonable predictions. However, it is necessary to retain the influence of strain rate effects for cylindrical shells made from strain-rate-sensitive materials. Several authors[10.14, 10.15] have examined material strain rate, or viscoplastic, effects and their work is also discussed in reference 10.4.

10.5 Dynamic axial plastic buckling of a long cylindrical shell

10.5.1 Introduction

There are practical situations in which thin-walled members are subjected to axial impacts. Indeed, the impetus for some of the early impact studies on circular tubes was an attempt to reveal the characteristics of structural crashworthiness for railway coaches.[10.16] A more recent example, from the aerospace industry, is the dynamic buckling which might arise from the sudden thrust of a rocket engine ignition.[10.9]

A photograph of the final shape of a long cylindrical shell which has been impacted axially is shown in Figure 10.1(a). The stress field within this shell during deformation was biaxial since both axial and radial deformations are evident. The previous dynamic plastic buckling problems examined in §§ 10.3 and 10.4 were one-dimensional because the membrane force (N) and bending moment (M) produced a stress variation through the structural thickness but did not give rise to stresses on an orthogonal plane. In the present case, however, it is necessary to consider a biaxial stress field which is integrated across the wall thickness of a tube to give the corresponding axial and circumferential membrane forces and bending moments.

The particular problem examined in this section is illustrated in Figure 10.19. In

---

† Lindberg and Florence[10.9] predict a similar threshold impulse, except that the coefficient is $\sqrt{3}$ instead of 2.67. However, $E_{nc}(\tau_f) = 20$ in reference 10.9. If $E_{nc}(\tau_f) = 20$ is used instead of $E_{nc}(\tau_f) = 100$, then the coefficient 2.67 in equations (10.105) to (10.107) is replaced by 1.78.

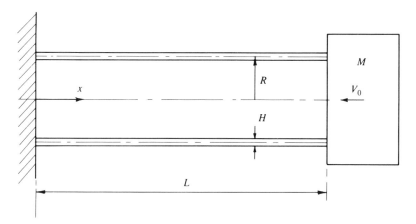

Figure 10.19 Mass M with a constant velocity V_0 impacting axially on a cylindrical shell of length L.

order to simplify the subsequent presentation, the initial axial impact velocity (V_0) of the striking mass (M) is assumed to remain constant throughout the response. This simplification was suggested by the high-speed photographs taken by Florence and Goodier[10.17] during the axial impact of tubes, which indicate selection of the buckling mode early in the deformation process.

The governing equations for the problem in Figure 10.19 are developed in § 10.5.2. In common with the dynamic plastic buckling of rings studied in § 10.4, it is assumed that the dominant solution is axisymmetric and uniform. Dynamic plastic buckling stems from the temporal growth of small axisymmetric initial imperfections in the axial generators of a tube.

10.5.2 Governing equations

10.5.2(a) Geometrical relations

Consider the axisymmetric element of a cylindrical shell shown in Figure 10.20 which has an initial length dS_0 and a deformed length dS. The initial imperfect transverse profile is represented by $w^i(x)$, and w is measured radially inwards from this initial imperfect shape. The longitudinal, or axial, membrane strain is defined as

$$\epsilon_x = \frac{dS - dS_0}{dS_0} \tag{10.108}$$

or

$$\epsilon_x = \frac{\partial u}{\partial x} + \frac{1}{2}\left(\frac{\partial w}{\partial x}\right)^2 + \frac{\partial w^i}{\partial x}\frac{\partial w}{\partial x} \tag{10.109}$$

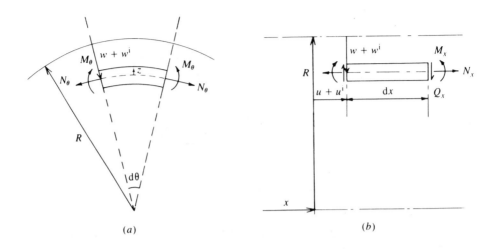

(a)

(b)

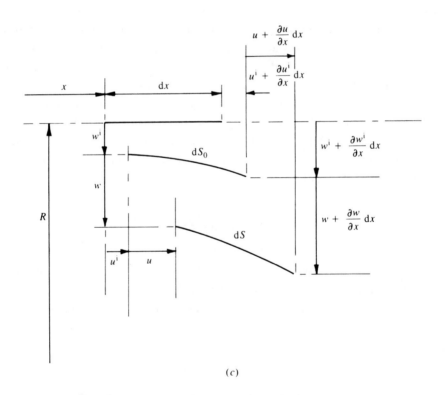

(c)

Figure 10.20 (a) Axial view of element of a circular cylindrical shell and notation. (b) Circumferential view of element of a circular cylindrical shell and notation. (c) Centre line of element in initial (dS_0) and deformed (dS) configurations.

when neglecting higher-order terms and assuming $u^i = 0$. The longitudinal change of curvature is

$$\kappa_x = -\frac{\partial^2 w}{\partial x^2},$$ (10.110)†

which is consistent with the approximations introduced during the derivation of equation (10.109).

If the deformation of a tube is restricted to remain axisymmetric throughout the response, then

$$\epsilon_\theta = \frac{2\pi(R - w^i - w + z) - 2\pi(R - w^i + z)}{2\pi(R - w^i + z)},$$

or

$$\epsilon_\theta \cong -\frac{w}{R}\left(1 + \frac{w^i}{R} - \frac{z}{R}\right)$$ (10.111)‡

and

$$\kappa_\theta = 0,$$ (10.112)

where R is the mean radius of a tube and z is measured through the wall thickness from the mid-surface, as shown in Figure 10.20.

10.5.2(b) Equilibrium equations

The moment, axial and transverse equilibrium equations for a cylindrical shell are

$$Q_x = \frac{\partial M_x}{\partial x}$$ (10.113)

$$\frac{\partial N_x}{\partial x} - \mu \frac{\partial^2 u}{\partial t^2} = 0$$ (10.114)

and

$$\frac{\partial^2 M_x}{\partial x^2} + \frac{\partial}{\partial x}\left\{N_x\left(\frac{\partial w}{\partial x} + \frac{\partial w^i}{\partial x}\right)\right\} + \frac{N_\theta}{R}\left(1 + \frac{w^i}{R}\right) - \mu \frac{\partial^2 w}{\partial t^2} = 0$$ (10.115)

respectively, where $\mu = \rho H$ (ρ is density of material, H is shell thickness), t is time, and the remaining quantities are defined in Figure 10.20.

The geometrical relations (10.109) to (10.112) and equilibrium equations (10.113) to (10.115) are consistent according to the principle of virtual velocities§

---

† The negative sign in equation (10.110) ensures that $M_x\kappa_x > 0$ when M_x is defined in Figure 10.20(b). (See also § 2.2.)

‡ The term z/R is usually neglected in thin-shell theories but is retained here because the shells examined in references 10.9 and 10.17 were fairly thick. (See also equation (A.81) in Appendix 4.)

§ This method[10.8] is discussed in Appendices 3 and 4 and equates the total internal energy dissipation to the external work rate due to the inertia forces and external forces at the boundaries and on the surfaces. The internal energy dissipation has contributions from membrane forces evaluated at the mid-surface (i.e., $z = 0$ in equation (10.111)) and bending moments. Neither transverse shear nor rotatory inertia effects are considered.

and reduce to the corresponding equations in reference 10.8 when $w^i = 0$ for an initially perfect cylindrical shell. The geometrical relations and equilibrium equations in Chapters 2 and 5, for the respective static and dynamic axisymmetric infinitesimal behaviour of cylindrical shells, may also be recovered from equations (10.109) to (10.115).

10.5.2(c) Constitutive equations

It is assumed that the total radial, or transverse, displacement field may be expressed in the form

$$w(x, t) = \bar{w}(t) + w'(x, t), \tag{10.116}$$

where $\bar{w}(t)$ is the axisymmetric dominant radial displacement and $w'(x, t)$ is the perturbed radial displacement. Similarly, the axial, or longitudinal, displacement is

$$u(x, t) = \bar{u}(x, t) + u'(x, t), \tag{10.117}$$

where, again, the $\bar{u}$ and u' terms are the dominant and perturbed components, respectively.

Florence and Goodier[10.17] examined the dominant behaviour of the cylindrical shell in Figure 10.19 and introduced the simplification

$$\frac{\partial \dot{\bar{u}}}{\partial x} = -\frac{V_0}{L}, \tag{10.118}$$

where V_0 is the initial axial impact velocity, which is assumed to remain constant throughout motion, L is the initial length of a tube and $(\dot{\ }) = \partial(\ )/\partial t$. Thus, equations (10.109), (10.116) and (10.117) predict a dominant axial strain $\bar{\epsilon}_x = \partial \bar{u}/\partial x$, which, using equation (10.118), gives

$$\dot{\bar{\epsilon}}_x = -\frac{V_0}{L}. \tag{10.119}$$

Equations (10.111) and (10.116) predict a dominant circumferential strain $\bar{\epsilon}_\theta = -\bar{w}/R$, which Florence and Goodier[10.17] write as

$$\dot{\bar{\epsilon}}_\theta = \frac{V_0}{2L}. \tag{10.120}$$

Finally, an incompressible† material satisfies[10.18] $\dot{\bar{\epsilon}}_x + \dot{\bar{\epsilon}}_\theta + \dot{\bar{\epsilon}}_z = 0$, which, using equations (10.119) and (10.120), predicts a dominant radial, or transverse, strain rate

$$\dot{\bar{\epsilon}}_z = \frac{V_0}{2L}. \tag{10.121}$$

Now, using equations (10.109) to (10.112) and following Florence and Goodier[10.17] and Lindberg and Florence,[10.9] the total strain rates are

$$\dot{\epsilon}_x = -\frac{V_0}{L} + z\frac{\partial^2 \dot{w}'}{\partial x^2} \tag{10.122}$$

† See § 8.3.2.

and

$$\dot{\epsilon}_\theta = \left(1 - \frac{z}{R}\right)\left(\frac{V_0}{2L} - \frac{\dot{w}'}{R}\right) \qquad (10.123)$$

when neglecting products of perturbed quantities, $u' = 0$ and $w^i/R \ll 1$, while

$$\dot{\epsilon}_z = -\dot{\epsilon}_x - \dot{\epsilon}_\theta \qquad (10.124)\dagger$$

for an incompressible material.[10.18]

The Prandtl–Reuss‡ equations for an isotropic plastic material[10.18] predict that

$$\sigma_x = \frac{2\sigma_e}{3\dot{\epsilon}_e}(2\dot{\epsilon}_x + \dot{\epsilon}_\theta) \qquad \text{and} \qquad \sigma_\theta = \frac{2\sigma_e}{3\dot{\epsilon}_e}(2\dot{\epsilon}_\theta + \dot{\epsilon}_x) \qquad (10.125\text{a, b})$$

for a plane stress state with $\sigma_z = 0$, where the equivalent, or effective, stress§ is

$$\sigma_e = \sigma_0 + E_h\epsilon_e \qquad (10.126)$$

for the rigid plastic material with a yield stress σ_0 and a linear-strain-hardening modulus E_h shown in Figure 10.9. The equivalent, or effective, strain rate[10.18]§ is

$$\dot{\epsilon}_e = \frac{2}{\sqrt{3}}(\dot{\epsilon}_x^2 + \dot{\epsilon}_\theta^2 + \dot{\epsilon}_x\dot{\epsilon}_\theta)^{1/2}, \qquad (10.127)$$

where $\dot{\epsilon}_z$ has been eliminated using the material incompressibility condition (equation (10.124)).

Now, substituting equations (10.122) and (10.123) into equation (10.127) and neglecting the z^2-terms and products of the perturbed quantities gives

$$\dot{\epsilon}_e \cong \frac{V_0}{L}\left\{1 + \frac{zL}{V_0}\left(\frac{4\dot{w}'}{3R^2} - 2\frac{\partial^2\dot{w}'}{\partial x^2}\right)\right\}^{1/2}. \qquad (10.128)$$

Thus, the binomial series allows equation (10.128) to be written

$$\dot{\epsilon}_e \cong \frac{V_0}{L}\left\{1 + \frac{zL}{2V_0}\left(\frac{4\dot{w}'}{3R^2} - 2\frac{\partial^2\dot{w}'}{\partial x^2}\right)\right\} \qquad (10.129)$$

and, integrating, therefore, with respect to time

$$\epsilon_e = \frac{V_0}{L}\left\{t + \frac{zL}{2V_0}\left(\frac{4w'}{3R^2} - 2\frac{\partial^2 w'}{\partial x^2}\right)\right\} \qquad (10.130)$$

since the initial imperfect shape is taken as stress-free.

Substituting equation (10.130) into equation (10.126) gives σ_e, which, when substituted into equations (10.125) with equations (10.122), (10.123) and (10.129) and neglecting higher order-terms, yields

$$\sigma_x = -\sigma^0\left(1 + \frac{2L\dot{w}'}{3V_0R}\right) + \frac{z\sigma^0 L}{3V_0}\left(\frac{\partial^2\dot{w}'}{\partial x^2} + \frac{4\dot{w}'}{R^2} - \frac{V_0}{RL}\right) + zE_h\left(\frac{\partial^2 w'}{\partial x^2} - \frac{2w'}{3R^2}\right)$$

$$(10.131)$$

---

† See § 8.3.2.

‡ Equation (8.8).

§ The equivalent stress and strain rate are defined by equations (8.5) and (8.6), respectively.

and

$$\sigma_\theta = -\frac{4L\sigma^0 \dot{w}'}{3V_0 R} + \frac{z2L\sigma^0}{3V_0}\left(\frac{\partial^2 \dot{w}'}{\partial x^2} + \frac{2\dot{w}'}{R^2} - \frac{V_0}{RL}\right), \tag{10.132}$$

where

$$\sigma^0 = \sigma_0 + E_h \frac{V_0 t}{L}. \tag{10.133}$$

The original elemental area over which the axial stress σ_x acts is $2\pi(R - w^i + z)\,dz$, while the axial membrane force N_x is defined with respect to a unit length of the tube circumference. Thus,

$$N_x = \int_{-H/2}^{H/2} \sigma_x \frac{2\pi(R - w^i + z)\,dz}{2\pi(R - w^i)},$$

or, if $w^i/R \ll 1$,

$$N_x \cong \int_{-H/2}^{H/2} \sigma_x (1 + z/R)\,dz, \tag{10.134}$$

which, when substituting equation (10.131) and neglecting z^2-terms in the integrand, gives

$$N_x = \bar{N}_x + N'_x, \tag{10.135}$$

where

$$\bar{N}_x = -\sigma^0 H \tag{10.136a}$$

is the dominant axial membrane force and the perturbed axial membrane force is

$$N'_x = -\frac{\sigma^0 H 2L\dot{w}'}{3V_0 R}. \tag{10.136b}$$

Similarly,

$$M_x \cong -\int_{-H/2}^{H/2} \sigma_x(1 + z/R)z\,dz, \tag{10.137}$$

or

$$M_x = \bar{M}_x + M'_x, \tag{10.138}$$

where

$$\bar{M}_x = \frac{\sigma^0 H^3}{9R} \tag{10.139a}†$$

and

$$M'_x = -\frac{\sigma^0 H^3 L}{36V_0}\left(\frac{\partial^2 \dot{w}'}{\partial x^2} + \frac{2\dot{w}'}{R^2}\right) - \frac{H^3 E_h}{36}\left(3\frac{\partial^2 w'}{\partial x^2} - \frac{2w'}{R^2}\right). \tag{10.139b}†‡$$

---

† Equations (10.139) are slightly different from those presented in references 10.9 and 10.17 because we have retained the z/R term in equation (10.137). It would appear consistent to retain this term when z/R is considered in equation (10.111) for ϵ_θ.

‡ The longitudinal, or axial, bending moment comprises two groups of terms. The first group contains σ^0, and it is evident that $M'_x \neq 0$ even for a perfectly plastic material without any material strain hardening, i.e., $E_h = 0$ and $\sigma^0 \cong \sigma_0$. This group of terms is known as a directional moment.[10.9, 10.17] The last group of terms is related to the material strain hardening modulus E_h and is known as the hardening moment.[10.9]

The circumferential membrane force

$$N_\theta = \int_{-H/2}^{H/2} \sigma_\theta \, dz$$

is

$$N_\theta = \bar{N}_\theta + N'_\theta, \tag{10.140}$$

where

$$\bar{N}_\theta = 0 \tag{10.141a}$$

and

$$N'_\theta = -\frac{4L\sigma^0 H \dot{w}'}{3V_0 R}. \tag{10.141b}$$

Finally,

$$M_\theta = -\int_{-H/2}^{H/2} \sigma_\theta z \, dz$$

can be written

$$M_\theta = \bar{M}_\theta + M'_\theta, \tag{10.142}$$

where

$$\bar{M}_\theta = \frac{\sigma^0 H^3}{18R} \tag{10.143a}$$

and

$$M'_\theta = -\frac{\sigma^0 H^3 L}{18 V_0} \left(\frac{\partial^2 \dot{w}'}{\partial x^2} + \frac{2\dot{w}'}{R^2} \right). \tag{10.143b}$$

10.5.2(d) Differential equations

The axial, or longitudinal, and transverse, or radial, equilibrium equations (10.114) and (10.115) can be rewritten, with the aid of equations (10.116), (10.117), (10.135), (10.138) and (10.140) in the form

$$\frac{\partial}{\partial x} (\bar{N}_x + N'_x) - \mu(\ddot{u} + \ddot{u}') = 0 \tag{10.144a}$$

and

$$\frac{\partial^2}{\partial x^2} (\bar{M}_x + M'_x) + \frac{\partial}{\partial x} \left\{ (\bar{N}_x + N'_x) \left(\frac{\partial w'}{\partial x} + \frac{\partial w^i}{\partial x} \right) \right\}$$
$$+ \left(\frac{\bar{N}_\theta + N'_\theta}{R} \right) \left(1 + \frac{w^i}{R} \right) - \mu(\ddot{w} + \ddot{w}') = 0, \tag{10.144b}$$

respectively. Equation (10.144a) gives the dominant axial equilibrium equation

$$\frac{\partial \bar{N}_x}{\partial x} - \mu \ddot{u} = 0 \tag{10.145a}$$

and the perturbed axial equilibrium equation

$$\frac{\partial N'_x}{\partial x} - \mu \ddot{u}' = 0. \tag{10.145b}\dagger$$

However, $\ddot{u} = 0$ for a constant axial velocity, and equation (10.145a) requires, therefore,

$$\frac{\partial \bar{N}_x}{\partial x} = 0, \tag{10.146}$$

which is also evident from equation (10.136a). Thus, using equations (10.139a) and (10.146), the dominant transverse equilibrium equation according to equation (10.144b) is

$$\frac{\bar{N}_\theta}{R}\left(1 + \frac{w^i}{R}\right) - \mu \ddot{w} = 0, \tag{10.147}\ddagger$$

while the associated perturbed equation is

$$\frac{\partial^2 M'_x}{\partial x^2} + \bar{N}_x\left(\frac{\partial^2 w'}{\partial x^2} + \frac{\partial^2 w^i}{\partial x^2}\right) + \frac{N'_\theta}{R}\left(1 + \frac{w^i}{R}\right) - \mu \ddot{w}' = 0 \tag{10.148}$$

when disregarding higher-order terms.

Now, $\bar{N}_x$, M'_x and N'_θ given by equations (10.136a), (10.139b) and (10.141b), respectively, allow equation (10.148) to be recast in the form

$$\frac{\sigma^0 H^3 L}{36 V_0}\frac{\partial^4 \dot{w}'}{\partial x^4} + \frac{H^3 E_h}{12}\frac{\partial^4 w'}{\partial x^4} + \frac{\sigma^0 H^3 L}{18 V_0 R^2}\frac{\partial^2 \dot{w}'}{\partial x^2} - \frac{H^3 E_h}{18 R^2}\frac{\partial^2 w'}{\partial x^2}$$

$$+ \sigma^0 H\left(\frac{\partial^2 w'}{\partial x^2} + \frac{\partial^2 w^i}{\partial x^2}\right) + \frac{4\sigma^0 H L}{3 V_0 R^2}\left(1 + \frac{w^i}{R}\right)\dot{w}' + \mu\ddot{w}' = 0, \tag{10.149}$$

which, in dimensionless notation, becomes

$$\ddot{u} + S_0\{\beta^2\ddot{u}'''' + 2\alpha^2\beta^2\ddot{u}'' + 48\alpha^2(1 + u_i)\dot{u}\} + \gamma S_0(3u'''' - 2\alpha^2 u'')$$
$$+ 36 S_0(u'' + u_i'') = 0, \tag{10.150}$$

where

$$u_i = w^i/R, \qquad u = w'/R, \qquad \xi = x/L, \qquad \tau = V_0 t/L, \qquad \alpha = L/R, \qquad \beta = H/L,$$
$$\gamma = \beta^2 E_h/\sigma^0, \qquad S_0 = \sigma^0/36\rho V_0^2, \qquad (\dot{}) = \partial()/\partial\tau$$

and

$$()' = \partial()/\partial\xi. \tag{10.151a-j}$$

10.5.3 Dynamic axial plastic buckling

A theoretical solution is now sought to the governing equations developed in the previous sections for the dynamic plastic buckling of the circular tube impacted

---

$\dagger$ The satisfaction of this equation was not examined in references 10.9 and 10.17 and is not considered further here.

$\ddagger$ Equations (10.141a) and (10.147) give $\ddot{w} = 0$.

axially, as shown in Figure 10.19. If it is assumed that the perturbed behaviour of the tube is

$$u(\xi, \tau) = \sum_{n=1}^{\infty} u_n(\tau) \sin(n\pi\xi) \qquad (10.152a)$$

and the initial shape imperfections are

$$u_i(\xi) = \sum_{n=1}^{\infty} a_n \sin(n\pi\xi), \qquad (10.152b)$$

then the dimensionless governing equation (10.150)† yields

$$\sum_{n=1}^{\infty} [\ddot{u}_n + S_0\{48\alpha^2 + \beta^2(n\pi)^2\{(n\pi)^2 - 2\alpha^2\}\}\dot{u}_n$$

$$+ S_0(n\pi)^2(2\alpha^2\gamma + 3\gamma(n\pi)^2 - 36)u_n - 36S_0(n\pi)^2 a_n] \sin(n\pi\xi) = 0.$$

This equation must be satisfied for each value of n, or

$$\ddot{u}_n + Q_n\dot{u}_n + R_n u_n = S_n a_n, \qquad (10.153)$$

where

$$Q_n = S_0[48\alpha^2 + \beta^2(n\pi)^2\{(n\pi)^2 - 2\alpha^2\}]$$
$$R_n = S_0(n\pi)^2\{2\alpha^2\gamma + 3\gamma(n\pi)^2 - 36\}$$

and

$$S_n = 36S_0(n\pi)^2. \qquad (10.154a\text{–c})‡$$

Equation (10.153), for a given value of n, is an ordinary second-order linear differential equation with constant coefficients, when assuming σ^0 is constant, which can be solved using standard methods of analysis. Thus,

$$u_n = E_n(\tau)a_n + F_n(\tau)b_n \qquad (10.155)$$

when satisfying the initial conditions $u_n = 0$ at $\tau = 0$ and

$$\dot{u}(0) = \dot{u}_i = \sum_{n=1}^{\infty} b_n \sin(n\pi\xi) \qquad (10.156)$$

and where

$$E_n(\tau) = \frac{S_n}{R_n}\left(1 + \frac{\lambda n_2\, e^{\lambda n_1\tau} - \lambda n_1\, e^{\lambda n_2\tau}}{\lambda n_1 - \lambda n_2}\right), \qquad (10.157a)§$$

$$F_n(\tau) = \frac{e^{\lambda n_1\tau} - e^{\lambda n_2\tau}}{\lambda n_1 - \lambda n_2}, \qquad (10.157b)$$

† The $48\alpha^2(1 + u_i)\dot{u}$ term in equation (10.150) is taken as $48\alpha^2\dot{u}$.

‡ References 10.9 and 10.17 have a different coefficient for the last term in equation (10.154a). This arises because the z/R term in equation (10.137) is retained in the present analysis, as remarked in the footnote to equations (10.139). Equation (10.154c) is also different from the corresponding equations in references 10.9 and 10.17 owing to different definitions for w. The radial displacement w is measured from the initial imperfect profile, as shown in Figure 10.20, rather than a perfect circular profile in references 10.9 and 10.17.

§ $E_n(0) = 0$ since the radial deflections are measured from the initial imperfect profile, as shown in Figure 10.20(c). However, $E_n(0) = 1$ in reference 10.17 because the radial deflections are measured from a perfect circular reference profile.

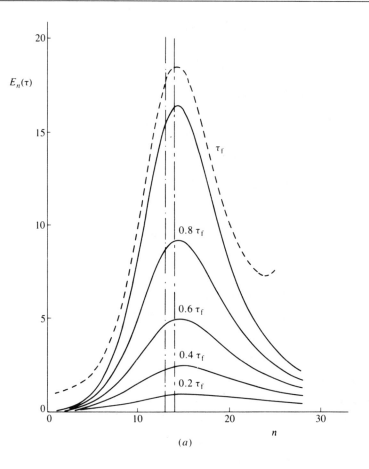

$$\lambda n_1 = \{-Q_n + (Q_n^2 - 4R_n)^{1/2}\}/2 \qquad (10.158a)$$

and

$$\lambda n_2 = -\{Q_n + (Q_n^2 - 4R_n)^{1/2}\}/2. \qquad (10.158b)$$

It is evident from equation (10.158a) that $\lambda n_1 > 0$ when $R_n < 0$ and the displacement and velocity amplification functions $E_n(\tau)$ and $F_n(\tau)$ grow with increase in dimensionless time, as shown in Figure 10.21. This phenomenon is known as dynamic plastic buckling and occurs, therefore, when

$$n < \left(\frac{36 - 2\alpha^2\gamma}{3\gamma\pi^2}\right)^{1/2}, \qquad (10.159)$$

which is the requirement for $R_n < 0$ according to equation (10.154b).

The typical spectral curves in Figure 10.21 exhibit the fastest growth for a specific or critical value of the harmonic n. This critical value (n_c) is now sought when introducing the simplifying assumption

$$4R_n/Q_n^2 \ll 1. \qquad (10.160)\dagger$$

† The largest value of $|4R_n/Q_n^2|$ for the results in Figure 10.21 is 0.063 59 when $n = 11$.

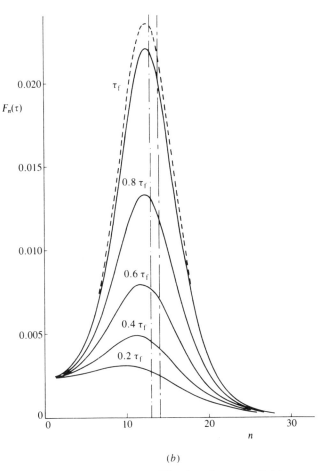

(b)

Figure 10.21 (a) Displacement amplification function $E_n(\tau)$ versus mode number n for a shell with $E_h = 724 \text{ MN/m}^2$, $\sigma^0 = 307 \text{ MN/m}^2$. $\rho = 2685 \text{ kg/m}^3$, $M = 120 \text{ g}$. $L = 101.6 \text{ mm}$, $H = 2.54 \text{ mm}$, $R = 11.43 \text{ mm}$ and $V_0 = 170 \text{ m/s}$. τ is the dimensionless time defined by equation (10.151d) and from equation (10.169) $\tau_f = 0.2975$ when $m = 51 \text{ g}$. ———: equation (10.157a). ———: equation (10.163a). ———: n_c from equation (10.164) (rounded to nearest integer) —·—·—: n_c from equation (10.167a) (rounded to nearest integer). (b) Velocity amplification function $F_n(\tau)$ versus mode number n for a shell having the parameters listed in (a). ———: equation (10.157b). ———: equation (10.163b). ——— and —·—·— are defined in (a).

Expanding equation (10.158a) with the aid of the binomial series gives

$$\lambda n_1 = Q_n(-1 + 1 - 2R_n/Q_n^2 + \cdots)/2,$$

or

$$\lambda n_1 \cong -R_n/Q_n. \qquad (10.161)$$

Similarly, equation (10.158b) reduces to

$$\lambda n_2 \cong -Q_n. \qquad (10.162)$$

In this circumstance, equations (10.157) take the particularly simple forms

$$E_n(\tau) \cong -\frac{S_n}{R_n} e^{-(R_n/Q_n)\tau} \tag{10.163a}$$

and

$$F_n(\tau) \cong \frac{e^{-(R_n/Q_n)\tau}}{Q_n}, \tag{10.163b}$$

which are reasonable estimates of the corresponding exact curves in Figures 10.21(a) and 10.21(b), respectively.

Now, $R_n < 0$ when equation (10.159) is satisfied so that the largest value of R_n/Q_n is assumed to give the fastest growth of $E_n(\tau)$ and $F_n(\tau)$ and, therefore, the critical mode number n_c.[†] Thus, substituting R_n and Q_n from equations (10.154) into the expression $\partial(R_n/Q_n)/\partial n = 0$ leads to

$$\frac{\beta^2(9 - 2\alpha^2\gamma)}{24\alpha^2} (n_c\pi)^4 + 3\gamma(n_c\pi)^2 + \alpha^2\gamma - 18 = 0, \tag{10.164}‡$$

which is a quadratic equation in $(n_c\pi)^2$. However, if

$$\alpha^2\gamma \ll 4.5, \tag{10.165}$$

then equation (10.164) simplifies to

$$n_c = \frac{2\alpha\sqrt{\gamma}}{\pi\beta} \left\{ \left(1 + \frac{3\beta^2}{\alpha^2\gamma^2} \right)^{1/2} - 1 \right\}^{1/2}, \tag{10.166}‡$$

or

$$n_c \cong \frac{2\alpha}{\pi\beta} \left(\frac{\sqrt{3}\beta}{\alpha} - \gamma \right)^{1/2} \tag{10.167a}‡$$

when

$$\frac{3\beta^2}{\alpha^2\gamma^2} \gg 1. \tag{10.167b}$$

The estimates for the critical mode numbers according to equations (10.164) and (10.167a) are compared with the peaks of the exact amplification curves in Figure 10.21.[§]

The displacement amplification function (10.163a) can be rearranged

$$\log_e \left\{ -\left(\frac{R_{nc}}{S_{nc}} \right) E_{nc}(\tau_f) \right\} = -\left(\frac{R_{nc}}{Q_{nc}} \right)\tau_f \tag{10.168}$$

for the critical mode number n_c, where τ_f is the dimensionless time when motion ceases. An estimate for τ_f is made by equating the initial kinetic energy $MV_0^2/2$ to

---

[†] This is not strictly correct because the maxima of the entire expressions for $E_n(\tau)$ and $F_n(\tau)$ should be found.

[‡] The nearest integer value is taken as n_c.

[§] The left-hand sides of inequalities (10.165) and (10.167b) are 0.1165 and 10.91, respectively, for this particular case.

the energy absorbed by plastic deformation in the tube

$$\int_0^{t_f} \sigma_e \dot{\epsilon}_e (2\pi RHL)\, dt.$$

If $\dot{\epsilon}_e \cong V_0/L$, according to equation (10.129), and $\sigma_e \cong \sigma_0$, then

$$\frac{MV_0^2}{2} = 2\pi RHL\sigma_0 \int_0^{t_f} \frac{V_0}{L}\, dt,$$

or

$$\tau_f = \left(\frac{M}{m}\right)\left(\frac{\rho V_0^2}{2\sigma_0}\right), \tag{10.169}$$

where τ is the dimensionless time defined by equation (10.151d) and $m = 2\pi RHL\rho$ is the mass of the tube. Thus, equation (10.168) may be cast in the form

$$V_0^2 = \left(\frac{2\sigma_0}{\rho}\right)\left(\frac{m}{M}\right)\left(-\frac{Q_{nc}}{R_{nc}}\right)\log_e\left\{\left(-\frac{R_{nc}}{S_{nc}}\right)E_{nc}(\tau_f)\right\}. \tag{10.170}$$

The velocity (V_0) which causes a specified growth $E_{nc}(\tau_f)$ of the initial displacement imperfection may be found from equation (10.170). The quantities Q_{nc}/R_{nc} and R_{nc}/S_{nc} are obtained from equations (10.154) for the critical mode number n_c, which is given by the root of equation (10.164), equation (10.166), or equation (10.167a). In the particular case when equation (10.167a) is valid, then

$$\frac{Q_{nc}}{R_{nc}} = \frac{\alpha\beta(12 - \sqrt{3}\alpha\beta + \alpha^2\gamma + 2\delta^2 - 4\sqrt{3}\delta)}{\sqrt{3\alpha^2\gamma - \alpha^2\gamma\delta + 36\delta + 6\delta^3 - 12\sqrt{3}\delta^2 - 18\sqrt{3}}} \tag{10.171a}$$

and

$$\frac{R_{nc}}{S_{nc}} = \frac{\alpha^2\gamma}{18} + \frac{\delta}{\sqrt{3}} - \frac{\delta^2}{3} - 1, \tag{10.171b}$$

where

$$\delta = \alpha\gamma/\beta. \tag{10.171c}$$

The approximate theoretical predictions of equations (10.170) and (10.171) are compared in Figure 10.22 with the exact results according to equation (10.157a) and the approximate ones from equation (10.163a). It is evident that the critical displacement amplification function ($En_c(\tau_f)$) varies relatively slowly for impact velocities up to about 150 m/s.† However, the magnitude of $En_c(\tau_f)$ increases rapidly for relatively small increases of impact velocities larger than about 150 m/s. This suggests the concept of a threshold velocity, which was introduced earlier by equation (10.106), for the dynamic plastic buckling of an impulsively loaded circular ring.

The magnitude of a threshold velocity is arbitrary, since dynamic plastic

---

† With the exception of the impact velocity, the parameters in Figure 10.22 are identical to aluminium 6061-T6 cylindrical shell number 20 examined by Florence and Goodier.[10.17]

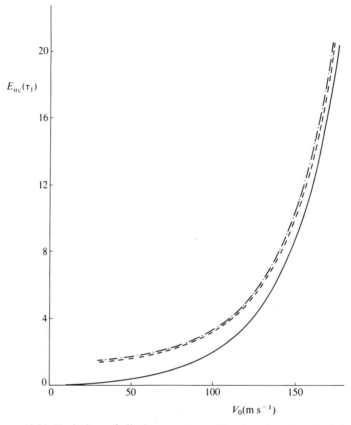

Figure 10.22 Variation of displacement amplification function $E_{nc}(\tau_f)$ with impact velocity V_0. ———: equation (10.157a), $n = 14$. —·—·—: equation (10.163a), $n = 14$. ———: equations (10.170) and (10.171). (The parameters are the same as those listed in the caption to Figure 10.21(a).)

buckling does not develop at a unique value of the impact velocity. However, the results in Figure 10.22 suggests that $En_c(\tau_f) \cong 10$ could be substituted into equation (10.170) to estimate the threshold velocity. Alternatively, $En_c(\tau_f)$ could be estimated from the ratio of the largest permissible displacement, or tolerable damage, to the largest initial geometric imperfection, or manufacturing tolerance.

In view of the foregoing comments, it is evident that equation (10.170) offers a satisfactory approximation of the exact results in Figure 10.22, at least for the range of parameters examined.

The theoretical results presented in Figures 10.21 and 10.22 were calculated for the aluminium 6061-T6 cylindrical shell test specimen number 20 examined by Florence and Goodier,[10.17] except that the axial impact velocity of 170 m/s in

Figure 10.21 is larger than 125.3 m/s in reference 10.17. If $V_0 = 125.3$ m/s, then equations (10.164) and (10.167a) predict critical mode numbers of 14 and 13, respectively, while the peaks of the displacement and velocity amplification functions (equations (10.157a, b) with $\tau = \tau_f$) are 15 and 12, respectively. Florence and Goodier[10.17] predict that the most amplified harmonics for the displacement and velocity perturbations are 14 and 12, respectively, while the critical mode number is 12 for the permanent profile of the corresponding test specimen.

10.5.4 Dynamic axial plastic buckling with non-constant axial velocity

The theoretical analysis in § 10.5.3 was developed for the particular case when the impact velocity V_0, for the problem illustrated in Figure 10.19, remained constant throughout motion. This assumption was introduced to simplify the analysis, although the experimental results of Florence and Goodier[10.17] did show that the critical mode number n_c was selected early in the response before the impact velocity decreased significantly. Vaughan[10.19] removed this assumption and allowed the initial impact velocity V_0 to decrease until motion ceased.

Vaughan[10.19] observed that the axial force which resists motion in a rigid-plastic cylindrical shell is constant when the effects of material strain hardening, material strain rate sensitivity and thickness changes are neglected. This gives rise to a constant deceleration of the striking mass and, therefore, a linear decrease with time of the axial velocity and the other velocity dependent quantities, such as strain rates.

Vaughan[10.19] followed a theoretical procedure which is similar to that developed by Florence and Goodier[10.17] and outlined in § 10.5.3. It transpires that the response duration of the dominant motion might be written in the dimensionless form

$$\tau_f = \left(\frac{\rho V_0^2}{\sigma_e}\right)\left(1 + \frac{M}{m}\right), \tag{10.172}$$

where τ_f is defined by equation (10.151d), σ_e is given by equation (10.126), V_0 is the initial impact velocity, M is the striking mass and $m = 2\pi RHL\rho$. In the particular case when $M/m \gg 1$ (i.e., heavy striking mass), then

$$\tau_f = \left(\frac{\rho V_0^2}{\sigma_e}\right)\left(\frac{M}{m}\right), \tag{10.173}$$

which is twice the theoretical prediction of equation (10.169) if $\sigma_e \cong \sigma_0$.

The theoretical behaviour for the perturbed response is governed by a Bessel equation and Vaughan[10.19] obtained the critical mode number using a numerical scheme. However, a simple expression for the threshold impact velocity (V_0) was obtained using the dominant solution and some experimental observations. Vaughan[10.19] postulated that the maximum circumferential dominant strain is

$(\epsilon_\theta)_{max} = H/4R$, which gives

$$V_0^2 = \frac{H\sigma^0}{\rho R(1 + M/m)},$$ (10.174a)

or

$$V_0^2 \cong \left(\frac{H\sigma^0}{\rho R}\right)\left(\frac{m}{M}\right)$$ (10.174b)

when $M/m \gg 1$.

Vaughan's[10.19] theoretical results do not lead to any improvement in the critical mode numbers predicted for the 3 in (76.2 mm) and 4 in (101.6 mm) long aluminium 6061-T6 tubes tested by Florence and Goodier.[10.17] However, a slight reduction in the mode numbers is found, for the 6 in (152.4 mm) long tubes, which give better agreement with the corresponding experimental results.

Wojewodzki[10.20] has repeated the constant impact velocity analysis of Florence and Goodier[10.17] and examined the influence of material strain rate sensitivity which obeys a linear law (i.e., equation (8.4) with $q = 1$). It turns out that material strain rate sensitivity reduces the critical mode number to give better agreement with the corresponding experimental test results.

The dynamic plastic buckling of a stringer-stiffened cylindrical shell subjected to an axial impact was examined in reference 10.21, using a perturbation method of analysis. It transpired that it is more efficient to place stiffeners with rectangular cross-sections on the outer shell surface than on the inner surface. Further results demonstrate the influence of the second moment of area, eccentricity, cross-sectional area and the number of stiffeners on the dynamic plastic response.

10.6 Critical impulsive radial velocity for collapse of a cylindrical shell without buckling

Equation (10.106) may be used to predict a threshold radial, or uniform transverse, impulsive velocity, which causes a specified growth of the maximum initial radial displacement imperfection in a ring. Impulsive velocities smaller than this specified value cause less growth, while larger velocities lead to unacceptable damage.

However, the high-speed collapse of cylindrical shells occurs in several practical devices which would have an optimum design if dynamic buckling did not develop. The practical applications[10.9] include explosive closure for obtaining gas-tight pipe seals, magnetic field constriction for producing intense transient magnetic fields, rapidly collapsing cylindrical reservoirs which generate high-pressure and high-velocity gas flows in shock tubes, and rapidly collapsing conical shells which produce metallic jets for oil well perforation.

An examination of the available experimental results on rings subjected to radially inwards impulsive velocities (Figures 4 and 11 in reference 10.4) reveals that both the dominant and the perturbed permanent radial displacements are larger for larger initial velocities. However, the ratio of the perturbed to the dominant permanent radial displacements for the two test series on the thickest rings appears to decrease with increase in the initial external velocity. In other words, the dominant motion becomes more important for larger external radial impulsive velocities and, eventually, may dominate the perturbed behaviour at sufficiently large velocities.

Thus, the large dominant displacements are a new feature in this aspect of dynamic buckling and cause a significant increase in the shell thickness. This increase in thickness causes the stability to improve as motion proceeds and, therefore, restrains the departures from circularity. Moreover, the growth of the perturbed deflections requires time, which indicates that insufficient time will elapse for dynamic buckling to develop at sufficiently high radial collapse velocities.

Abrahamson[10.22] developed a theoretical procedure in order to predict the critical velocity above which a cylindrical shell does not collapse. External velocities larger than the critical value lead to acceptable damage, or departures from circularity, while lower velocities give rise to damage which is larger than the allowable amplification of the initial imperfections.

Abrahamson[10.22] simplified the theoretical analysis by assuming that the large initial impulsive radial velocity was constant throughout motion. Thus, the plastic energy absorbed during the deformation of a cylindrical shell is small compared with the initial kinetic energy of the external impulse. The material was assumed to be incompressible, which gave a simple expression for the change in shell thickness. Despite these simplifications, a numerical solution was required. The numerical results presented by Abrahamson[10.22] show that the initial imperfections are amplified less for external velocities which exceed the critical collapse velocity.

Florence and Abrahamson[10.7] have examined the influence of material strain rate sensitivity on the magnitude of the critical collapse velocity. The analysis was simplified with the aid of a linear constitutive equation (equation (8.4) with $q = 1$) and critical collapse velocities are presented for various specified amplifications of the initial radial displacement imperfections. Unfortunately, no experimental results are available to assess the accuracy of the theoretical predictions for cylindrical shells made from strain-rate-sensitive or strain-rate-insensitive materials.[10.9] Moreover, this phenomenon has not been explored for any other structural problems.

10.7 Final comments

This chapter has explored several aspects of the phenomenon of dynamic plastic buckling due to pulse loadings which are defined in § 10.1. To introduce the phenomenon, the behaviour of an elastic bar subjected to an axial impact is examined in § 10.2. It transpires that the transverse deflections may be oscillatory and remain bounded, or, in certain circumstances, may grow from the initial geometric imperfections and become unbounded. This latter behaviour is defined as dynamic buckling, and was also encountered in § 10.3 for the same bar when made from a rigid, linear-strain-hardening material.

The dynamic plastic buckling of a circular ring with radial imperfections in the initial profile and subjected to an external impulsive velocity was examined in § 10.4 with the aid of a perturbation method of analysis. This method assumes that the response consists of the two displacement components illustrated in Figure 10.13. An axisymmetric dominant radial displacement consumes the external dynamic energy, while a non-axisymmetric perturbed radial displacement field grows from the radial imperfections in the initial profile.

A further degree of complexity is introduced in § 10.5, which examines a long cylindrical shell struck axially with a mass, as shown in Figure 10.19. The stress field in this case is biaxial. This arises from the longitudinal, or axial, strain rates due to the axial impact, which also produces circumferential strain rates from the associated change in shell radius.

It was mentioned in § 9.1 that the inertia forces in the wall of a tube do not play an important role during the dynamic progressive buckling of tubes struck by masses having $M \gg m$. For example, the 300 mm long thin-walled circular tube in § 9.8.3 was struck by a 100 kg mass, giving $M/m \cong 549$ when taking $\rho = 2685 \text{ kg/m}^3$ for an aluminium alloy. By way of contrast, a mass ratio of $M/m = 2.35$ is associated with the dynamic axial plastic buckling of the circular tube in Figure 10.21.

It was observed in § 9.8.3 that the response time is 39.3 ms for the dynamic progressive buckling of the circular tube discussed above. This is about 678 times longer than the time required for a uniaxial elastic stress wave to travel along a 300 mm long bar. However, $\tau_f = 0.2975$, or $t_f = 177.8 \text{ μs}$, for the dynamic axial plastic buckling problem in Figure 10.21. This is only about nine times longer than the time required for a uniaxial elastic stress wave to travel along a 101.6 mm long aluminium bar.

Both the above comparisons for the dynamic progressive buckling and dynamic axial plastic buckling of circular tubes illustrate the greater importance of inertia effects in a tube wall during dynamic plastic buckling.

It is evident from Figure 9.8 that $\eta < 1$ for most of the experimental results on the progressive buckling of circular tubes. The structural effectiveness η is defined by equation (9.18) and is the ratio between the mean crushing force and the plastic squash load. Increased strength due to material strain hardening may cause η to exceed unity for the thicker tubes having large values of the solidity ratio (ϕ). The phenomenon of material strain rate sensitivity in Chapter 8 may also cause η to exceed unity in the dynamic progressive buckling case but, again, this would be likely to occur only for thick tubes. However, it is evident from equation (10.136a) that $\eta \geqslant 1$ is always required for the dynamic axial plastic buckling of a long cylindrical shell struck by a mass, as shown in Figure 10.19.

The simple theoretical results presented in this chapter are useful for the preliminary stages of design. Quick estimates may be obtained for the threshold velocity which a structure could tolerate without excessive permanent deflections or damage. It is necessary, however, to estimate the magnitude of the largest initial geometric imperfection having the critical mode number, which, for the particular case of the impulsively loaded circular ring studied in § 10.4, is predicted by equations (10.88) or (10.90). An estimate could be made using the maximum allowable manufacturing tolerances given in the associated engineering drawings and assuming that these lie in the critical mode. The value of the displacement amplification function, $E_{nc}(\tau_f)$, is given by the ratio of the maximum permitted perturbed displacement to the maximum initial geometric imperfection, and the threshold velocity then follows from equation (10.103).

Several other structural problems have been investigated using the theoretical procedure presented in this chapter,[10.23] including the dynamic plastic buckling of rectangular plates.[10.9] The dynamic plastic buckling of complete spherical shells made from a strain-rate-insensitive material has been examined in references 10.24 to 10.27, while Wojewodzki and co-workers[10.28, 10.29] have considered the influence of material strain rate sensitivity. Kao[10.30] has studied the dynamic plastic buckling of spherical caps with the aid of a numerical finite-difference scheme.

Various experimental results, most of which are collated by Lindberg and Florence[10.9] and in reference 10.4 for impulsively loaded rings and cylindrical shells, broadly substantiate the general theoretical procedure for dynamic plastic buckling outlined in this chapter. However, it is important to note that the definition of dynamic buckling is somewhat arbitrary. The threshold velocity for a circular ring according to equation (10.106), for example, is associated with a maximum perturbed radial displacement equal to 100 times the maximum initial radial (geometric) imperfection. Fortunately, the theoretical results in Figure 10.18 indicate that the threshold velocity would be fairly insensitive to the actual value for sufficiently large displacement amplification factors.

The character of dynamic buckling depends, sometimes, on the idealisations introduced when modelling a structure and on the approximations made in obtaining a theoretical solution. The dynamic plastic buckling of the simple elastic-plastic initially imperfect model in reference 10.31 develops for a unique value of the impact load. However, if axial as well as transverse inertia effects are retained in the governing equations for this model, then dynamic plastic buckling, of the kind studied in this chapter, which is known as direct buckling, develops for certain ranges of the dimensionless parameters, while a different kind of dynamic plastic buckling, with growing oscillations (indirect buckling), occurs for the remaining dimensionless parameters. These results show that direct buckling is markedly sensitive to the magnitude of the initial geometric imperfections, while indirect buckling is less imperfection-sensitive. It was observed that different aspects of the numerical scheme in reference 10.31 were important for the two kinds of dynamic plastic buckling: the magnitude of the time step was particularly crucial for direct buckling (since it develops quickly), while the duration of the numerical calculations influenced the magnitude of the indirect dynamic buckling load. Indeed, it is possible that numerical calculations had been terminated prematurely in some cases and wrong conclusions drawn on the dynamic stability of a shell.[10.23] If the numerical programme for such an apparently stable shell were to be run for a longer time, dynamic indirect buckling might be revealed and different conclusions reached. Moreover, the actual dynamic buckling load for this problem depends upon the magnitude of the allowable displacement (cf. $E_{nc}(\tau_f)$) in equation (10.103) which must be specified to obtain a threshold velocity for a circular ring).

The theoretical method discussed in §§ 10.3 to 10.5 is valid for relatively small departures of the perturbed displacement from the dominant displacement field. If the perturbed displacements become too large, then bending (e.g., equation (10.62)) might produce elastic unloading in a part of the shell thickness unless compensated by an increase in the associated compressive membrane force (equation (10.63)).†
In other words, the compressive membrane strain is assumed to dominate the tensile strains which develop as a result of bending associated with the growth of imperfections in the initial profile and the external velocity distribution. Thus, the magnitude of the total strain is increasing throughout the wall thickness, so that a tangent modulus formulation is valid and is largely responsible for the simplicity of the method. In order to examine the influence of strain rate reversal, which is neglected in the simple theory, Lindberg and Kennedy[10.33] used a numerical procedure to obtain the dynamic behaviour of an elastic-plastic cylindrical shell subjected to an external impulsive velocity. It transpires that the simple theory predicts reasonable estimates for the threshold impulses but overestimates the critical mode numbers.

The phenomenon of a critical radial impulsive velocity for a cylindrical shell is introduced in § 10.6. It is possible to achieve relatively undamaged shells for impulsive velocities higher than the critical value because the perturbed displacements are less important than the associated dominant displacements for sufficiently high impulsive velocities. Dynamic stability improves as the shell

---

† Further comments may be found in reference 10.32.

thickness increases, particularly for large radial velocities which do not allow sufficient time to elapse for significant growth of the initial geometric imperfections. Only two theoretical analyses,[10.7, 10.22] and no experimental results, appear to have been published on this phenomenon.

In view of the foregoing comments, there is a pressing need for the development of some basic theorems on dynamic plastic buckling. Simple methods which predict lower bounds to dynamic buckling loads would be of considerable value to a designer whether or not numerical schemes are used.

However, owing to the complexity of the subject, little progress has been achieved on the development of theorems for dynamic plastic buckling. Lee and co-workers have made some valuable contributions in this area, but much more study is required before a useful, reliable and simple method is developed, perhaps as a result of further simplifications originating from an increased understanding of the phenomena.

Lee[10.34] considered bifurcation and uniqueness in an elastic-plastic continuum undergoing dynamic motion. He further developed these ideas and obtained a quasi-bifurcation criterion for an elastic-plastic body[10.35] which is illustrated in reference 10.36 for a four-degrees-of-freedom mass–elastic spring system. The predicted critical time when quasi-bifurcation occurs, and the associated buckling mode shape, compare favourably with exact numerical results. In each of the cases examined, the deviated motion remained small until the critical time, after which it grew rapidly. Similarly, the initial mode of deviated motion undergoes a rapid change to the predicted buckling mode at the critical time. However, the numerical results show that the mode shape changes with time during the post-bifurcation range, which is not considered in the theoretical work. Lee has also used his quasi-bifurcation theory to examine the dynamic plastic buckling of a column[10.37] and a complete spherical shell subjected to an external impulsive velocity[10.38]. However, further investigations are required to clarify the phenomenon of dynamic plastic buckling before Lee's method leads to a suitable design procedure.

Finally, an interesting form of dynamic plastic instability has been observed by Symonds and Yu[10.39] for an elastic, perfectly plastic beam. It was found, for a narrow range of parameters, that an elastic-plastic pin-ended beam, subjected to a uniformly distributed transverse pressure pulse, could snap through the initial position after unloading from the first peak in the transverse displacement–time history. The beam then vibrates about a permanent transverse displacement profile which is in a sense opposite to the direction of the initial transverse pressure loading. This phenomenon occurs because the beam, on unloading from the peak transverse displacement, becomes a shallow arch with a transverse velocity field tending to flatten the arch and inducing compressive forces giving rise to a possible instability.

PROBLEMS

10.1 The theoretical analysis in § 10.2.3 predicts the classical static buckling load (equation (10.10)) for an initially perfect bar which is simply supported at both ends. Obtain the static behaviour when the column is imperfect.

10.2 Obtain the critical mode number (equation (10.37)) for the dynamic elastic buckling of an axially loaded bar with arbitrary initial imperfections and simply supported at both ends. Show that a bar is more highly wrinkled for larger dynamic loads. What is the major limitation of the theoretical method for large dynamic loads?

10.3 Obtain an approximate relationship between the critical mode numbers for the dynamic elastic and dynamic plastic buckling of bars with arbitrary initial imperfections.

10.4 An aluminium alloy circular ring, with a mean radius $R = 44.5$ mm and a mean thickness $H = 1$ mm, is subjected to an external impulsive radial velocity V_0. What is the dimensionless critical impulsive velocity λ to cause dynamic plastic buckling when $E_h/\sigma_0 = 7$? What are the associated dominant radial displacement, critical mode number and dimensionless response duration?

10.5 Prove that equation (10.124) is valid for an incompressible material.

10.6 Show that the Prandtl–Reuss equations reduce to equations (10.125a, b) for a plane stress state in an isotropic plastic material, and that equation (10.127) is the associated effective plastic strain rate.

10.7 A thin-walled circular cylindrical shell, with a mean radius $R = 25$ mm, $H = 2.5$ mm, $L = 100$ mm, $\sigma_0 = 300$ MN/m^2, $\rho = 2700$ kg/m^3 and $E_h/\sigma_0 = 2.5$, is impacted at one end with a mass $M = 100$ g. Use equation (10.170) to determine the impact velocity which produces a tenfold amplification of the initial displacement imperfections lying in the critical mode. What is the critical mode number of the deformed shape and the response duration?

11

Scaling laws

> If the size of a body be diminished, the strength of
> that body is not diminished in the same
> proportion; indeed the smaller the body the greater
> its relative strength. Thus, a small dog could
> probably carry on his back two or three dogs of
> his own size, but I believe that a horse could not
> carry even one of his own size. *Galileo* (1638)

11.1 Introduction

The testing of small-scale models is indispensable for complex structural systems which are difficult to analyse theoretically and numerically or to study experimentally. The dynamic response of underground structures,[11.1] impact of nuclear fuel capsules,[11.2] missile impact of nuclear power installations[11.3] and collision protection of ships[11.4] illustrate several areas which have been studied with the aid of small-scale models.

Dynamic tests are conducted on a small-scale model in order to obtain the response characteristics of a geometrically similar full-scale prototype which is the actual system of interest. This procedure is known as scaling, modelling or similitude and is governed by certain principles. Apart from the obvious purpose of relating the behaviour of a model to that of a prototype, these principles also predict various dimensionless combinations of the governing variables which are valuable for the planning of experimental investigations and the choice of numerical calculations.[11.5]

11.2 Introduction to geometrically similar scaling

Some features of geometrically similar scaling for the dynamic response of structures are introduced in this section using an elementary approach. It is assumed, for simplicity of presentation, that a small-scale model† and a full-scale prototype are made from an elastic material with the same values of Young's modulus of elasticity (E), mass density (ρ) and Poisson's ratio (v). In order to satisfy the requirements of geometrically similar scaling, it is necessary for the external loads to act at homologous points. The lower case variables in Table 11.1

---

† The model is not required to be physically smaller than the prototype, although in practice this is generally the case.

Table 11.1

Small-scale model	Quantity	Full-scale prototype
c	Wave speed	C
l	Length	L
m	Mass	M
p	Pressure	P
t	Time	T
v	Velocity	V
δ	Displacement	$\varDelta$
σ	Stress	$\varSigma$

refer to a small-scale model, while the corresponding upper case variables refer to the full-scale prototype.

The geometric scale factor for a model is

$$\beta = \frac{l}{L},\tag{11.1}$$

where $\beta \leqslant 1$.†

The relationship between the characteristics of a small-scale model and a full-scale prototype is now established for several variables.

11.2.1 Strain

The direct uniaxial engineering strain (ϵ) in a structural member is defined as the change in length divided by the initial length. Thus,

$$\epsilon = \frac{\delta}{l}\tag{11.2}$$

for a small-scale model. However, $l = \beta L$ according to equation (11.1) and $\delta = \beta \varDelta$ from Table 11.1 and equation (11.1). Substituting into equation (11.2) gives

$$\epsilon = \frac{\beta \varDelta}{\beta L} = \frac{\varDelta}{L}.\tag{11.3}$$

Equation (11.3) is identical to the engineering strain in a full-scale prototype.

Equations (11.2) and (11.3) show, therefore, that the engineering strain (ϵ) is the same in a small-scale model and a geometrically similar full-scale prototype. This conclusion is independent of the material properties and is valid for any type of structural material.

11.2.2 Stress

Now, the engineering or nominal stress (σ) in a small-scale model is related to the engineering strain (ϵ) according to Hooke's Law

$$\sigma = E\epsilon\tag{11.4}$$

† More generally, $0 < \beta < \infty$ because a model could be larger than a very small prototype, for instance.

for a linear elastic material. In view of equation (11.3),

$$\Sigma = E\epsilon \qquad (11.5)$$

for a full-scale prototype made from the same linear elastic material. Equations (11.4) and (11.5) require, therefore, $\sigma = \Sigma$. Thus, the stresses in a small-scale model and a geometrically similar full-scale prototype are identical.

11.2.3 Pressure

In order to satisfy local equilibrium (Cauchy's formula) the stress normal to the boundary of a structure must be equal in magnitude and opposite in sign to any pressure acting on that surface. Thus,

$$\sigma = -p \qquad (11.6a)$$

and

$$\Sigma = -P \qquad (11.6b)$$

on the pressurised boundaries of a small-scale model and a full-scale prototype, respectively. However, $\sigma = \Sigma$ according to § 11.2.2 so that equations (11.6) require

$$p = P. \qquad (11.7)$$

A small-scale model and a full-scale prototype are, therefore, subjected to the same magnitudes of pressure in order to satisfy the requirements of geometrically similar scaling. Thus, water could not be used for loading a small-scale model if the full-scale prototype were to be loaded hydrostatically with water. The force on the boundary of a model is related to $pl^2 = p\beta^2 L^2$, which is β^2 smaller than the corresponding force acting on the boundary of a prototype (PL^2).

11.2.4 Wave speed

A tensile or compressive disturbance propagates in a linear elastic rod with a velocity $(E/\rho)^{1/2}$ according to elementary stress wave theory.[11.6]
If $c = (E/\rho)^{1/2}$, then

$$c = C \qquad (11.8)$$

and interior disturbances propagate with the same speed in a small-scale model and a geometrically similar full-scale prototype made from the same linear elastic material.

11.2.5 Time

The time (t_1) for a disturbance to travel a distance l in a small-scale model is l/c. Similarly, $T_1 = L/C$ is the time for a disturbance to propagate a distance L in a full-scale prototype. Thus,

$$\frac{t_1}{T_1} = \frac{l}{c} \cdot \frac{C}{L} = \beta$$

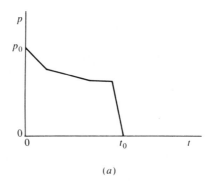

(a)

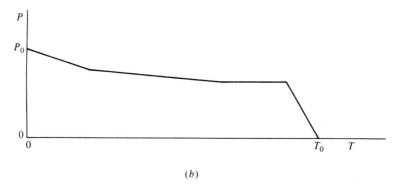

(b)

Figure 11.1 Comparison of pressure-time characteristics for (a) small-scale model and (b) full-scale prototype, with $p = P$ and $t = \beta T$.

or the relation

$$t = \beta T \tag{11.9}$$

must be satisfied for geometrically similar scaling.

Equations (11.7) and (11.9) for geometrically similar scaling require that an external pressure pulse on a small-scale model must have the same magnitude but with temporal characteristics which are β times those of a full-scale prototype, as illustrated in Figure 11.1.

11.2.6 Velocity

A measure of the velocity in a small-scale model is given by the displacement divided by time t, or

$$v = \frac{\delta}{t} \tag{11.10}$$

which, using $\delta = \beta\varDelta$ and $t = \beta T$ from equation (11.9), gives

$$v = \frac{\beta\varDelta}{\beta T} = \frac{\varDelta}{T} = V \tag{11.11}$$

Thus, the structural velocities are independent of the geometric scale factor β.

11.2.7 Drop tests

A drop test is one of the simplest experimental arrangements and has been used extensively for the dynamic testing of structures. If a full-scale prototype is struck by a mass M travelling with a velocity V then the initial kinetic energy of the striking mass M is $MV^2/2$. A geometrically similar small-scale model would be struck by a projectile of mass m travelling with a velocity v giving an initial kinetic energy $mv^2/2$, where $m = \beta^3 M$ and $v = V$ according to equation (11.11). Thus, the striking mass and the initial kinetic energy would be β^3 as large in a small-scale model, while the impact velocity and drop height† would be independent of the scale factor β.

The above remarks apply to any testing method, such as a gas gun, which propels a projectile into the structure or propels the structure into a target.

It is evident from equations (11.9) and (11.11) that the acceleration in a small-scale model is δ/t^2, which is $1/\beta$ times larger than in a full-scale prototype ($\varDelta/T^2$). However, the dynamic force $m\delta/t^2$ is β^2 smaller than the corresponding dynamic force in a full-scale prototype ($M\varDelta/T^2$). This agrees with the observation made in § 11.2.3 on pressure forces.

11.3 Phenomena which do not scale geometrically

The parameters discussed in § 11.2 obey the principles which govern geometrically similar scaling. It is now shown that several phenomena do not scale according to the same principles. A small-scale model, for which at least one of the similarity requirements is not satisfied, is said to be distorted.[11.7]

11.3.1 Gravity

The gravitational force in a full-scale prototype of mass M is Mg, where g is the local acceleration of gravity. If a small-scale model is tested at the same location, then the gravitational force is mg, or $\beta^3 Mg$, which varies with β^3 rather than β^2 as noted in §§ 11.2.3 and 11.2.7.

Thus, it is not possible to scale gravitational forces according to the elementary principles of geometrically similar scaling unless the acceleration of gravity is scaled as $1/\beta$. Fortunately, gravitational forces are not significant when compared with the dynamic forces which are generated in many impact problems. For

---

† Drop height $= v^2/2g$ when assuming $v = (2g \times \text{drop height})^{1/2}$.

example, the average accelerations range from $46g$ to $1507g$ in the plated steel structures with scale factors from 0.25 to 1.00 which were drop tested by Booth, Collier and Miles.[11.8]† However, gravitational forces could be important for the distribution of fragments after the failure of a structural component and for low-velocity impacts.

11.3.2 Material strain rate sensitivity

The Cowper–Symonds constitutive equation for a strain-rate-sensitive material (equation (8.3)) may be written in the form

$$\frac{\Sigma_0'}{\Sigma_0} = 1 + \left(\frac{\dot{\epsilon}}{D}\right)^{1/q},$$ (11.12)

where Σ_0 and Σ_0' are the respective static and dynamic flow stresses and D and q are material constants.

The strain (ϵ) is also invariant for a strain-rate-sensitive material since equations (11.2) and (11.3) are independent of the material characteristics. However, the strain rate in a full-scale prototype is $\dot{\epsilon} = \epsilon/T$, while $\dot{\epsilon} = \epsilon/t$ or $\dot{\epsilon} = \epsilon/\beta T$ for a small-scale model. Thus, the strain rate in a small-scale model is $1/\beta$ times larger than that in a geometrically similar full-scale prototype, and equation (11.12) becomes

$$\frac{\sigma_0'}{\sigma_0} = 1 + \left(\frac{\dot{\epsilon}}{\beta D}\right)^{1/q}.$$ (11.13)

Equations (11.12) and (11.13), with $\Sigma_0 = \sigma_0$, give

$$\frac{\sigma_0'}{\Sigma_0'} = \frac{1 + (\dot{\epsilon}/\beta D)^{1/q}}{1 + (\dot{\epsilon}/D)^{1/q}},$$ (11.14)

which reveals that the dynamic flow stress in a small-scale model (σ_0') is larger than that in a large-scale prototype (Σ_0') and violates, therefore, the stress invariance requirement in § 11.2.2.

It was shown in § 8.3.2 that $D = 40.4 \text{ s}^{-1}$ and $q = 5$ for a typical hot-rolled mild steel and that the dynamic uniaxial tensile flow stress is twice the corresponding static value when $\dot{\epsilon} = D = 40.4 \text{ s}^{-1}$. However, how important is the lack of scaling for material strain rate sensitivity effects, when a small-scale model and a full-scale prototype are made from the same strain rate sensitive material?

If $\dot{\epsilon} = D$, then equation (11.14) becomes

$$\frac{\sigma_0'}{\Sigma_0'} = \frac{1 + (1/\beta)^{1/q}}{2}$$ (11.15)

which, using $\beta = 0.25$, gives $\sigma_0' = 1.16\Sigma_0'$ when $q = 5$. Similarly, equation (11.15) predicts $\sigma_0' = 1.29\Sigma_0'$ when $\beta = 0.1$.

It is conceptually possible to overcome this non-scaling phenomenon by testing

† See also the calculations for the axial impact of circular tubes in § 9.8.3.

small-scale models and full-scale prototypes with different material strain rate properties. However, Nevill[11.9] examined this idea and encountered considerable experimental difficulties, since it is necessary to model properly all material effects, including elasticity, yield stress and strain hardening as well as material strain rate sensitivity. To the author's knowledge, material strain rate effects have not been modelled successfully in small-scale structural model tests.

11.3.3 Fracture

Fast fracture occurs when a crack in a structure suddenly becomes unstable and grows at loads which would otherwise be supported safely within the elastic range.[11.10] According to elementary linear elastic fracture mechanics, the onset of fast fracture in a structural member with a crack of length B and subjected to a stress Σ is governed by the relationship

$$\Sigma\sqrt{\pi B} = \sqrt{EG_c}, \tag{11.16}$$

where E is Young's modulus of elasticity and G_c is the material toughness, or energy absorbed per unit area of crack.

If a small-scale model with a crack length b is subjected to a stress σ, then equation (11.16) predicts the onset of fast fracture when

$$\sigma\sqrt{\pi b} = \sqrt{EG_c}. \tag{11.17}$$

Thus, if a small-scale model and a full-scale prototype are made from the same material, then

$$\frac{\sigma}{\Sigma} = \beta^{-1/2} \tag{11.18}$$

since $\beta = b/B$ for geometrically similar scaling. Equation (11.18) predicts that the stress required for the onset of fast fracture in a small-scale model is larger than in a geometrically similar full-scale prototype which does not, therefore, satisfy the requirements of stress invariance discussed in § 11.2.2. A large structure, or prototype, could crack, therefore, before yielding plastically, whereas the same material could display general yielding before cracking in a small structure or model.

11.4 Dimensional analysis

11.4.1 Introduction

The observations in the previous sections are based on physical *ad hoc* arguments. However, many variables contribute to the response of complex structures subjected to dynamic loads which produce large deformations and inelastic material behaviour. Usually, it is not feasible to investigate the influence of every variable, and a designer requires, therefore, a rational selection method. For example, should the influence of certain variables be tested either experimentally or numerically in preference to others?

It transpires that dimensionless groups of variables govern the structural response, as shown later in this section. Thus, an experimental test programme executed without prior knowledge of these groups could turn out to have dimensionless groups of variables with an inadequate range. In other words, unnecessary repetition of experimental tests, or numerical computations, might be revealed when the data are assembled into dimensionless groups of variables.

Several methods are available for deriving dimensionless groups of variables and these have been studied extensively, particularly in the field of fluid mechanics. The equilibrium equations, constitutive relations and compatibility conditions which govern the dynamic response of a structure may be made dimensionless by the introduction of several characteristic values. Dimensionless groups of variables then occur naturally in the dimensionless governing equations.[11.11] This method, though pedagogically appealing, is not discussed further since its value is somewhat limited in practice. Another procedure, which is known as dimensional analysis, is discussed in this section.

In fact, there are several methods of dimensional analysis,[11.12] but only the Buckingham Π-theorem approach is examined in this section and discussed further in Appendix 5. This theorem proves that, for a given physical problem with K quantities in which there are R dimensions, the quantities can usually be arranged into $K - R$ independent dimensionless parameters.† However, the form of these parameters is not unique.

Nevill[11.9] and Duffey[11.2, 11.14] used the Buckingham Π-theorem to obtain scaling laws which are valid for a wide range of structures subjected to blast, fragment and earth impact as well as thermal loading. This work forms the basis of the next section.

11.4.2 Buckingham Π-theorem

Initially, all the important input and output parameters for the dynamic inelastic response of structures are listed together with the associated physical dimensions. The Buckingham Π-theorem is then used to generate a complete, but non-unique, set of dimensionless groups of variables, or Π-terms. Equality of these Π-terms for a small-scale model and a full-scale prototype produces the similitude requirements, or scaling laws.

11.4.2(a) Input parameters

The input parameters consist of three principal types: geometrical characteristics, material properties and external loads.

---

† A counter-example is shown on page 109 of Hunsaker and Rightmire.[11.13] See Appendix 5 for further details and when R is smaller than the number of dimensions.

The geometry of a structure may be described with the aid of a characteristic linear length L_c (e.g., plate thickness or diameter of a shell). All other lengths in a structure are divided by L_c and expressed as L_k, where k refers to the kth linear dimension. The kth angle in a structure is referred to as ϕ_k.

The material properties play a particularly important role in the response, and many idealisations are possible. However, since the dynamic inelastic response is of primary interest, a time-dependent dimensionless elastic-plastic law[11.2]

$$\frac{\sigma_{ij}}{\sigma_c} = f(\epsilon_{ij}, r\dot{\epsilon}_{ij}, \bar{\alpha}\theta, G_c) \tag{11.19}†$$

will be used, where σ_{ij}, ϵ_{ij} and $\dot{\epsilon}_{ij}$ are the stress, strain and strain rate tensors, respectively, σ_c is a stress constant (e.g. Young's modulus of elasticity or yield stress), r is a strain rate sensitivity characteristic for the material, $\bar{\alpha}$ is a characteristic thermal expansion constant, θ is temperature and G_c is the material toughness.‡

If a structure is made from several different materials, then this is taken into account by expressing the properties of the nth material as a ratio of the corresponding quantities in equation (11.19) for one of the materials (known as the characteristic material, with σ_c, f_c, r_c, $\bar{\alpha}_c$ and G_c), i.e. σ_{cn}, f_n, r_n, $\bar{\alpha}_n$ and G_{cn}. Similarly, ρ_{cn} is the ratio of the density of the nth material divided by the density ρ_c of the characteristic material.

Impact, blast and fragment loadings produce transient surface tractions on the exposed surfaces of a structure which can be written in the form

$$\frac{T}{T_c} = h\left(\frac{x_i}{L_c}, \frac{t}{T_c}\right), \tag{11.20}§$$

where T is the surface traction at a point (x_i) on the surface at time t, T_c is a characteristic stress, t is time (laboratory or real time) and T_c is a characteristic time (e.g. duration of loading).

Similarly, thermal loadings can be expressed in the form

$$\frac{\theta}{\theta_c} = \bar{z}\left(\frac{x_i}{L_c}, \frac{t}{T_c}\right), \tag{11.21}§$$

where θ_c is a characteristic temperature and $\bar{z}$ is a dimensionless function.

---

† σ_{ij}, with both subscripts i and j ranging from 1 to 3, represents the stresses $\sigma_{11}, \sigma_{12}, \sigma_{13}, \sigma_{21}, \sigma_{22}, \sigma_{23},$ σ_{31}, σ_{32} and σ_{33}. In unabridged notation, with x_1, x_2 and x_3 replaced by x, y and z, respectively, σ_{ij} represents the stresses $\sigma_x, \sigma_{xy}, \sigma_{xz}, \sigma_{yx}, \sigma_y, \sigma_{yz}, \sigma_{zx}, \sigma_{zy}$ and σ_z. Similar comments apply to the strain (ϵ_{ij}) and the strain rate $(\dot{\epsilon}_{ij})$ tensors, respectively.

‡ Neither Nevill[11.9] nor Duffey[11.2, 11.14] examined the important phenomenon of fracture considered in § 11.3.3. However, the fracture toughness of a material (G_c) is retained here for completeness.

§ $i = 1, 2, 3$. x_i represents (x_1, x_2, x_3) in unabridged notation.

Table 11.2

Input parameters			
Parameter	Dimension	Parameter	Dimension
L_c	L	$\bar{\alpha}_n$	—
L_k	—	T_c	M/T^2L
ϕ_k	—	h	—
ρ_c	M/L^3	T_c	T
ρ_{cn}	—	θ_c	θ
σ_c	M/T^2L	$\bar{z}$	—
σ_{cn}	—	x_i	L
f_n	—	g	L/T^2
r_c	T	G_c	M/T^2
r_n	—	G_{cn}	—
$\bar{\alpha}_c$	$1/\theta$	t	T

Output parameters	
Parameter	Dimension
σ_{ij}	M/T^2L
ϵ_{ij}	—
δ_i	L
ϕ_i	—

M = mass, L = length,
T = time, θ = temperature.

11.4.2(b) Output parameters

The output parameters of interest might include the spatial (x_i) and temporal (t) variations of the stresses σ_{ij}, strains ϵ_{ij}, deformations δ_i and angular distortions ϕ_i.

11.4.2(c) Dimensionless Π-terms

The 22 input parameters introduced in § 11.4.2(a) are listed in Table 11.2 together with the four output parameters for a wide class of structural impact problems.

The output parameter σ_{ij} is a function of the 22 input parameters, so that $K = 23$ and $R = 4$.† The Buckingham Π-theorem, which is stated in § 11.4.1 and discussed further in Appendix 5, predicts that $K - R = 19$ independent, but not unique,

---

† It may be shown, when using the procedure described in Appendix 5, that the rank of the associated coefficient or dimensional matrix is $R = 4$, which in this case equals the number of dimensions or primary quantities M, L, T and θ.

Table 11.3

$\Pi_1 = L_k$	$\Pi_{12} = (T_c/L_c)(\sigma_c/\rho_c)^{1/2}$
$\Pi_2 = \phi_k$	$\Pi_{13} = T_c/\sigma_c$
$\Pi_3 = \rho_{cn}$	$\Pi_{14} = x_i/L_c$
$\Pi_4 = \sigma_{cn}$	$\Pi_{15} = L_c\rho_c g/\sigma_c$
$\Pi_5 = f_n$	$\Pi_{16} = G_c/\sigma_c L_c$
$\Pi_6 = r_n$	$\Pi_{17} = G_{cn}$
$\Pi_7 = \bar{\alpha}_n$	$\Pi_{18} = (t/L_c)(\sigma_c/\rho_c)^{1/2}$
$\Pi_8 = h$	$\Pi_{19} = \sigma_{ij}/\sigma_c$
$\Pi_9 = \bar{z}$	$\Pi_{20} = \epsilon_{ij}$
$\Pi_{10} = \bar{\alpha}\theta_c$	$\Pi_{21} = \delta_i/L_c$
$\Pi_{11} = (r/L_c)(\sigma_c/\rho_c)^{1/2}$	$\Pi_{22} = \phi_i$

dimensionless parameters exist, as shown in Table 11.3.† Similarly, the three remaining output parameters ϵ_{ij}, δ_i and ϕ_i are each a function of the same 22 input parameters, so that the three additional dimensionless parameters Π_{20} to Π_{22} in Table 11.3 are obtained.

Now, if the input parameters listed in Table 11.2 are scaled geometrically so that the eighteen input Π-terms (Π_1–Π_{18}) are identical for a small-scale model and a full-scale prototype, then the dimensionless response (Π_{19}–Π_{22}) is also identical. It is instructive to examine the physical requirements associated with each of these Π-terms when they are made equal for a small-scale model and a full-scale prototype:

Π_1: linear dimensions in a model are in the same ratio to each other as the linear dimensions in a prototype (i.e. geometrically similar scaling).

Π_2: angles in a model and prototype are the same (i.e. geometrically similar scaling).

Π_3: the ratio of the density of each material in a model to the characteristic density (ρ_c) must be identical to the corresponding ratio for a prototype. This Π-term is satisfied automatically when a model and a prototype are constructed of identical materials at corresponding locations.

Π_4: the ratio of the stress constant for a given material in a model to the characteristic stress constant (σ_c) of the model must be equal to the corresponding ratio in a prototype. This Π-term is satisfied automatically for identical materials at corresponding locations in a model and a prototype.

† The method outlined in Appendix 5 was used to generate the dimensionless Π-terms in Table 11.3. It turns out that Π_{11}, Π_{12}, Π_{15}, and Π_{18} are different from those presented in references 11.2 and 11.14. However, they may be cast into the same form when using Π_{11}/Π_{12}, $(\Pi_{12})^2$, $\Pi_{15}(\Pi_{12})^2$ and Π_{18}/Π_{12} to give r_c/T_c, $\sigma_c T_c^2/\rho_c L_c^2$, gT_c^2/L_c and t/T_c, respectively. The dimensionless terms Π_{16} and Π_{17} were not examined in references 11.2 and 11.14. Moreover, Poisson's ratio has not been considered, but this is a dimensionless number which should be identical for a small-scale model and a full-scale prototype when elastic effects are important.

Π_5: the non-dimensional constitutive relationship is identical for corresponding materials in a model and a prototype. Again this requirement is satisfied automatically for identical corresponding materials in a model and a prototype.

Π_6: the ratio of the strain rate sensitivity constant of the nth material to the characteristic value (r_c) must be the same for corresponding materials in a model and a prototype. This requirement is satisfied automatically for identical materials in a model and in a prototype.

Π_7: the ratios of the thermal expansion constants of the materials in a model are equal to the corresponding ratios in a prototype.

Π_8: the non-dimensional surface traction loading must be similar. Thus, a given surface traction component in a model, when normalised with the characteristic stress or pressure (T_c), must equal the corresponding normalised surface traction component in a prototype at geometrically scaled locations and at times scaled according to the characteristic time (T_c).

Π_9: the non-dimensional temperature distribution ($\bar{z}$) is equal in a model and a prototype.

Π_{10}: the product of the characteristic thermal expansion constant ($\bar{\alpha}$) and characteristic temperature (θ_c) must be equal in a model and a prototype. The characteristic temperature must be scaled in inverse proportion to the ratio of the thermal expansion coefficients of the model and prototype materials. If a model and a prototype are constructed from the same materials, then the characteristic temperature (according to Π_9) and temperature distribution must be the same.

Π_{11}: the ratio of the characteristic strain rate sensitivity constant (r) in a model to a time ($L_c(\rho_c/\sigma_c)^{1/2}$) must equal the corresponding ratio in a prototype. If the same materials are used at corresponding locations in a model and in a prototype, then r is the same for both. However, Π_{11} is different unless a model and a prototype are the same size, which is contrary to the object of scaling studies. Thus, it is not possible to achieve invariance of Π_{11} for geometrically similar scaling unless material strain rate effects can be ignored.

Π_{12}: if a model and a prototype are constructed of identical materials, then σ_c and ρ_c have common values. Thus, invariance of Π_{12} requires that the characteristic times for a model and a prototype are directly proportional to the characteristic length ratio (scale factor).

Π_{13}: the ratio of the traction or pressure loading magnitude to the characteristic stress constant is the same for a model and a prototype. Thus, the magnitude of the surface stress at a given scaled location at any scaled time is identical for a model and a prototype which are constructed from the same material (i.e. have the same characteristic stress constants).

Π_{14}: coordinates of homologous points vary in accordance with the geometric scale factor. Thus, concentrated loads, for example, must act at scaled locations.

Π_{15}: if a model and a prototype are made from identical materials at scaled locations, then Π_{15} reduces to $gL_c = $ constant. This is impossible to satisfy when conducting small-scale and full-scale tests at the same geographic location.

Π_{16}: if a small-scale model and a full-scale prototype are made from the same materials with the same fracture toughness, then Π_{16} is impossible to satisfy unless they are the same size.

Π_{17}: the ratio of fracture toughness of the nth material to the characteristic value (G_c) must be the same for corresponding materials in a model and a prototype. This requirement is satisfied automatically for identical materials in a model and a prototype, provided G_c is independent of the material thickness.

Π_{18}: the response time is directly proportional to the time $(L_c(\rho_c/\sigma_c)^{1/2})$. Alternatively, the response time is directly proportional to the characteristic length when a small-scale model and a full-scale prototype are made from the same material.

Π_{19}: all stresses are proportional to the characteristic stress constant. Stresses are identical at corresponding scaled locations and scaled impact times in models and prototypes made from the same material.

Π_{20}: strains are identical in model and prototype at scaled locations for scaled times.

Π_{21}: deformations at geometrically scaled locations for corresponding scaled times scale as the geometric scale factor.

Π_{22}: angular deformations are the same in a model and a prototype.

11.4.2(d) Comments

Invariance of strain, stress and pressure, which are predicted by Π_{20}, Π_{19}, and Π_{13}, confirms the conclusions reached in §§ 11.2.1 to 11.2.3, respectively, while invariance of Π_{12} leads to the observation noted in § 11.2.5. The invariance of velocities derived in §§ 11.2.4 and 11.2.6 has not been obtained directly in any of the dimensionless Π-terms. However, Π_{12} can be cast into the form $L_c/T_c = (\sigma_c/\rho_c)^{1/2}/\Pi_{12}$ which, for identical materials in a model and a prototype, gives invariance of L_c/T_c, which may be regarded as a characteristic velocity v_c.

In view of the above comments, the dimensionless term $1/\Pi_{12}^2$ can be written in the form $\rho_c v_c^2/\sigma_c$. Johnson[11.6] refers to the quantity $\rho_c v_c^2/\sigma_c$, with σ_c equal to the mean flow stress, as a damage number and uses it to identify various impact regions in metals.

It was noted in §§ 11.3.1 to 11.3.3 that gravitational effects, the influence of material strain rate sensitivity and fast fracture do not obey the principles of geometrically similar scaling. These conclusions were also recovered when studying Π_{15}, Π_{11} and Π_{16}, respectively.

It was shown by Pugsley and Johnson[11.6, 11.15] that the uniform retardation of a tubular structure of length L is $\sigma/\rho L$, where σ is the mean crushing stress and ρ is the notional density.† This expression shows that the retardation is less for longer vehicles, which, when assuming that passenger injury is related to excessive retardation,‡ partly explains why railway coaches are safer than motor cars. The dimensionless retardation is $\sigma/\rho L g$, which is identical to $1/\Pi_{15}$ and is impossible to satisfy for geometrically similar scaling between a small-scale model and a full-scale prototype made from the same material and tested at the same location.

The Buckingham Π-theorem is a rational method for deriving the dimensionless parameters which control the behaviour of a particular problem, but it does not reveal their relative importance. For example, it is shown in Appendix 5 that the three dimensionless parameters Π_1 to Π_3 given by equations (A.95) to (A.97), respectively, describe the permanent deformations of an impulsively loaded rigid-plastic beam. However, the theoretical and experimental results in Figure 7.19 indicate that the two dimensionless parameters $W_f/H = \Pi_3$ and $\lambda = 4\Pi_1^2 \Pi_2^2$ are adequate. The Buckingham Π-theorem procedure is unable, unfortunately, to predict the relative importance of the various dimensionless parameters.

11.5 Crack propagation in elastic structures

11.5.1 Introduction

It was noted in § 11.4.2(d) that the Buckingham Π-theorem may be used to derive the dimensionless parameters which control the behaviour of a particular problem. The laws which govern geometrically similar scaling, or similitude, are based on these dimensionless parameters. However, Duffey[11.2, 11.14] and others did not examine the influence of fracture mechanics which was discussed in § 11.3.3 and is retained in § 11.4.2. In anticipation of the potential importance of this pheno-menon, which does not obey the principles of geometrically similar scaling as shown by Π_{16} in § 11.4.2(c) and in § 11.3.3, the theoretical studies of Atkins, Mai and co-workers[11.16-11.18] are studied in this section.

This theoretical work focuses on the static elastic behaviour because neither dynamic loads nor inelastic materials have been studied in references 11.16 to 11.18. However, Atkins, Mai and co-workers have examined the influence of non-

---

† The resisting force of a vehicle is σA, where A is the cross-sectional area of the resisting structure with a normal crushing stress σ. The tubular structure has a total mass ρAL. Thus, the retardation is equal to $\sigma A/\rho AL$. (Contrast this expression with equation (9.71) for a stationary tube struck by a mass.)
‡ See equations (9.87) and (9.88).

linear elastic materials as well as non-proportional scaling, both of which are discussed in this section.

11.5.2 Linear elastic fracture mechanics

Atkins and Caddell[11.16] examined the two geometrically similar cracked structures shown in Figure 11.2 and used a somewhat different procedure from that in § 11.3.3 to derive the same relationship between the stresses necessary to propagate cracks in a small-scale model and a full-scale prototype:

$$\frac{\sigma}{\Sigma} = \beta^{-1/2}. \tag{11.22}$$

Any area on a model is β^2 times the homologous area of a geometrically similar full-scale prototype. Thus, the fracture loads for a small-scale model (f) and a full-scale prototype (F) are related by the expression

$$f/F = \beta^2 \sigma/\Sigma, \tag{11.23}$$

which, using equation (11.22), predicts

$$f/F = \beta^{3/2}, \tag{11.24}$$

This result should be contrasted with $f/F = \beta^2$, when using Π_1 and Π_{13} in

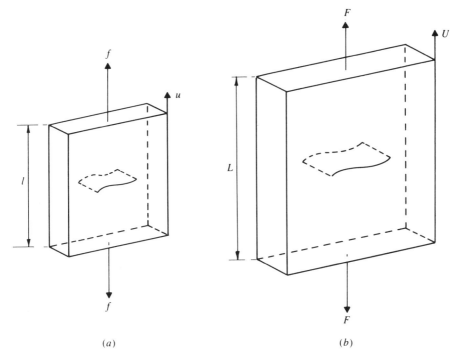

(a)   (b)

Figure 11.2 Geometrically similar small-scale cracked model (a) and full-scale cracked prototype (b), with $l = \beta L$.

§ 11.4.2(c), and the principles of geometrically similar scaling. The fracture load of a full-scale prototype (F) is smaller, therefore, than would be expected from tests conducted on a geometrically similar small-scale model (f).

The overall engineering strain in a full-scale prototype is $\epsilon = U/L$, where U is the overall displacement of a length L shown in Figure 11.2. Young's modulus of elasticity for the full-scale prototype is $E = \Sigma/\epsilon = \Sigma L/U$, while $E = \sigma l/u$ for a small-scale model made from the same material. Thus, equating these two expressions gives $\Sigma L/U = \sigma l/u$, or $u/U = \sigma l/\Sigma L$ which becomes

$$u/U = \beta^{1/2} \tag{11.25}$$

when using equation (11.22). The displacements in a full-scale prototype (U) are, therefore, $\beta^{-1/2}$ times those in a small-scale model (u) rather than β^{-1} times larger as predicted by the Π_{21}-term in Table 11.3 which is discussed in § 11.4.2(c). In other words, the displacements when cracking occurs in a full-scale prototype are smaller than the displacements which would be predicted from small-scale model tests according to geometrically similar scaling laws.

11.5.3 Non-linear elastic fracture mechanics

Mai and Atkins[11.17] have derived scaling laws which govern the non-linear elastic behaviour of a cracked structure having the non-linear load (F)–displacement (U) relation

$$F = k(A)U^n, \tag{11.26}$$

where $k(A)$ is any mathematical function of the crack area (A) and n is the degree of non-linearity. The authors showed that the displacements at fracture in a small-scale model (u) and in a full-scale prototype (U) are related by the equation

$$u/U = \beta^{n/(n+1)} \tag{11.27}$$

when the fracture toughness is invariant, while the fracture stresses satisfy the relation

$$\sigma/\Sigma = \beta^{-n/(n+1)}. \tag{11.28}$$

Equations (11.23) and (11.28) predict the fracture–load ratio

$$f/F = \beta^{(n+2)/(n+1)}. \tag{11.29}$$

It is evident that equations (11.27), (11.28) and (11.29) reduce, respectively, to equations (11.25), (11.22) and (11.24) for a linear elastic material with $n = 1$.

Equation (11.28) indicates that the stresses required to cause cracking in large non-linear structures are even smaller than those in large linear structures. Indeed, equation (11.28), with $n \to \infty$, gives $\sigma/\Sigma = \beta^{-1}$, which is contrasted in Figure 11.3 with $\sigma/\Sigma = \beta^{-1/2}$ from equation (11.22) for a linear elastic material and with $\sigma/\Sigma = 1$ according to elementary scaling laws which disregard the principles of fracture mechanics. (See § 11.2.2 and Π_{19} in § 11.4.2(c).) Thus, brittle fracture can occur

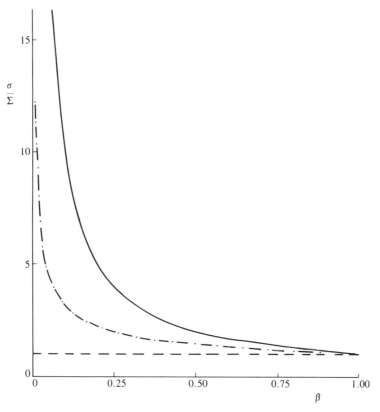

Figure 11.3 Ratio of fracture stresses for a small-scale model (σ) and a full-scale prototype (Σ) – – –: elementary scaling law, – · – · –: equation (11.22) (linear elastic case), ———: equation (11.28), with $n \gg 1$ (non-linear elastic case).

before yielding in large structures of a given material, whereas the same material can display generalised yielding before cracking in small-scale laboratory test-pieces. Moreover, cracking occurs at lower stresses in large non-linear structures than in large linear ones.

11.5.4 Non-proportionally scaled elastic fracture mechanics

Mai and Atkins[11.18] have extended the theoretical study in § 11.5.3 in order to obtain the scaling laws for non-proportionally scaled elastic structures. The scaling factors for height, width, thickness and crack length of non-proportionally scaled structures might all be different in contrast to geometrically similar structures in which all factors are equal.

This situation might arise in the laboratory testing of pipelines for which it is usually impossible to duplicate the actual length even though the full-scale thickness could be examined. The actual thickness of ship plating could be examined in a laboratory, but the height and width dimensions are generally much smaller than the corresponding full-scale values.

It transpires that a generalised approach is not possible for the behaviour of non-proportionally scaled elastic structures. The scaling laws must be formulated for each individual cracked structure. Mai and Atkins examined centre-cracked panels and double-cantilever-beams in reference 11.18, to which an interested reader is referred for further information.

11.6 Ductile–brittle fracture transitions

The existence of ductile–brittle fracture transitions are well known and are due to several factors. They can occur, for example, when the yield stress of a material increases with increase in triaxial stress effects, while the associated fracture stress is relatively unaffected.[11.19] Moreover, the yield stress increases with increase in strain rate as described by equation (11.12), while, again, the fracture stress is far less sensitive to strain rate. The yield stress decreases with increasing temperature and the fracture stress also decreases but, usually, only a small amount for practical values. Thus, a material may respond in a ductile fashion if the stresses associated with a particular loading condition produce a yield stress which is smaller than the corresponding fracture stress. On the other hand, a different set of environmental conditions for the same problem (e.g. increased strain rate, or decreased temperature) might give rise to a brittle fracture if the fracture stress is smaller than the yield stress. Clearly, the environmental conditions associated with a ductile–brittle transition in a material are an important factor in design.

Kendall[11.20] and others have noted that even the hardest and most brittle substances should squash plastically rather than crack if made sufficiently small. Indeed, it is evident from equation (11.28) that the fracture stress increases as body size decreases. It may exceed the corresponding yield stress of a material and allow plastic yielding to intervene before brittle fracture in a sufficiently small body. Thus, ductile–brittle transitions may also be encountered when changing the dimensions of a body. This phenomenon is important for similitude studies and is examined in the remainder of this section.

Kendall[11.20–11.22] has investigated static compression failure of the specimen shown in Figure 11.4 by developing a failure criterion based on an energy balance of fracture and conducting supporting experimental tests using glassy materials. If both arms of the compression split specimen are idealised as cantilever beams and loaded by a static axial end load $F/2$ with an eccentricity $(D - W)/4$, then the elastic strain energy in each is $3F^2(D - W)^2B/4EHD^3$, according to elementary beam theory. All the work in excess of that needed for linear elastic behaviour is grouped within a surface energy term G_cBH, where G_c is the fracture energy

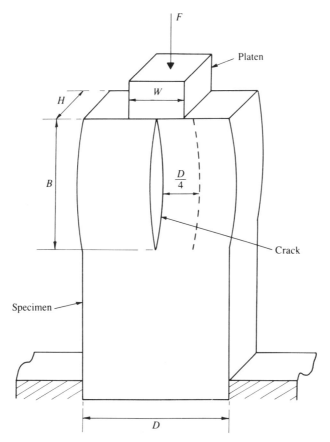

Figure 11.4 Geometry of the compression split specimen examined theoretically by Kendall.[11.22]

required to break a unit area $(A = BH)$ of solid material.[11.23] Thus, the total energy (U_T) can be written

$$U_T = -2 \times 3F^2(D - W)^2B/4EHD^3 + G_cBH, \tag{11.30}$$

from which $dU_T/dB = 0$ gives

$$F_1 = H(2EG_cD/3)^{1/2}/(1 - W/D), \tag{11.31}$$

which is the force necessary to propagate a crack of length B. Thus, the fracture stress (F_1/DH) varies as $D^{-1/2}$ when $W/D \ll 1$, which agrees with equation (11.18).

Plastic yielding occurs immediately under the loading platen in Figure 11.4 when

$$F_2 = HW\Sigma_0, \tag{11.32}$$

where Σ_0 is the yield stress for the material. Kendall[11.22] remarks that a change in failure mode from cracking to yielding occurs when the forces predicted by equations (11.31) and (11.32) are equal.

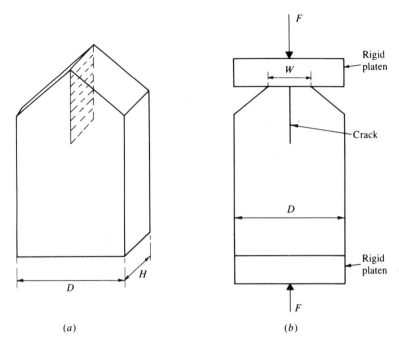

Figure 11.5 (a) Compression test specimen in reference 11.22. (b) Compression test for determining D_c in reference 11.22.

The test specimen in Figure 11.5 yields locally around the contact region when loaded. Equation (11.32) predicts, therefore, that

$$W = F/H\Sigma_0 \tag{11.33}$$

which, substituting into equation (11.31), gives

$$(F/H)^2/\Sigma_0 D - F/H + (2EG_c D/3)^{1/2} = 0, \tag{11.34}$$

or

$$\frac{F}{H} = \frac{1 \pm \left[1 - 4\left(\dfrac{2EG_c D}{3}\right)^{1/2} \Big/ \Sigma_0 D\right]^{1/2}}{(2/\Sigma_0 D)}. \tag{11.35}$$

This expression becomes complex if the specimen size is smaller than the critical specimen size

$$D_c = 32EG_c/3\Sigma_0^2 \tag{11.36}†$$

when cracking becomes impossible and gross yielding occurs. Equations (11.35) and (11.36) indicate, therefore, that even the hardest, most brittle substance should squash plastically rather than crack if made small enough. Kendall presents experimental data, on glassy materials, which support his theoretical predictions. However, the good quantitative agreement in Figure 11.6 occurs partly because it

† This phenomenon had been noted earlier by Gurney and Hunt.[11.23]

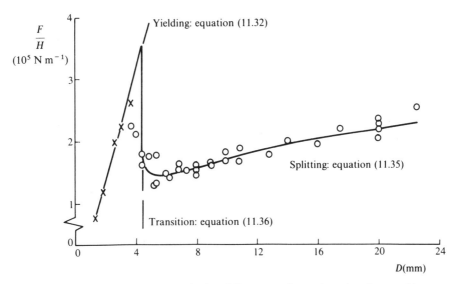

Figure 11.6 Compression results for different specimen sizes showing cracking (○) of large samples and yielding (×) of small samples of polystyrene.[11.22]

appears that equation (11.31) is used to predict the fracture energy G_c from tests on the material.†

Puttick and Yousif[11.25, 11.26] have examined fracture transitions in linear elastic materials and noted that the strain energy release rate in structures having inhomogeneous stress fields is determined by a characteristic length of the strain energy field (L) as well as the crack length. In fact, Puttick[11.25] assumes, for a restricted range of static linear elastic problems, that the characteristic length governing ductile–brittle transitions (L_c) can be written approximately in the form

$$L_c = \alpha(EG_c/\Sigma_0^2),\qquad(11.37)$$

where EG_c/Σ_0^2 is a material parameter and α is a parameter associated with the particular test. It is evident from equation (11.36) that $\alpha = 32/3$ for the specimen in Figure 11.4 when L_c is identified with D_c.

Equation (11.16), for linear elastic fracture mechanics, can be rearranged to give the critical crack length

$$B = EG_c/\pi\Sigma^2.\qquad(11.38)$$

Thus, if B is identified with L_c and Σ with Σ_0, then $\alpha = 1/\pi$ according to equation (11.37). Other examples are given by Atkins and Mai.[11.27]

† Kendall[11.22] assumes that plastic behaviour is confined to the region immediately underneath the platen for an elastic perfectly plastic material, with the rest of the specimen remaining elastic, except possibly at the crack tip. However, it appears that Kendall overlooked the possibility that plastic flow could develop due to bending of the specimen arms in Figure 11.4. It is shown in reference 11.24 that Kendall's analysis is only valid strictly for $0 \leqslant W/D \leqslant \frac{1}{3}$, and that $W/D_c = \frac{1}{2}$.

Brittle fracture occurs when $L \geqslant L_c$, where L is some linear dimension of a structure. Thus, fracture can be initiated by changing L_c. This may be brought about by changing the strain rate, the temperature, or the size of a geometrically similar structure, which is of particular interest in this chapter.

Unfortunately, insufficient experimental evidence is available currently to enable the critical characteristic length (L_c) method to be used in design.[11.25] However, it is evident that this information is required for numerical codes and, when scaling up the results of model tests, to estimate the behaviour of full-scale prototypes.

11.7 Experimental results on the scaling of structures loaded dynamically

11.7.1 Introduction

The dimensional analysis introduced in § 11.4 shows how the parameters of a structural problem must be varied in order to satisfy the requirements for geometrically similar scaling of a small-scale model and a full-scale prototype. However, it is impossible to scale all effects simultaneously and the Buckingham Π-theorem and other dimensional methods are incapable of assessing the potential importance of distortion which is produced when all the variables of a problem are not scaled geometrically. The theoretical methods in § 11.5 were developed using linear elastic fracture mechanics to assess the departure from the predictions of the elementary, geometrically similar scaling laws. Equation (11.28), for example, gives $\sigma/\Sigma = \beta^{-1}$ for a non-linear elastic material with $n \gg 1$. Thus, the stress to initiate fracture in a full-scale prototype is only one-quarter of that required to initiate fracture in a geometrically similar small-scale model with $\beta = 0.25$ as shown in Figure 11.3. This prediction should be contrasted with the geometrically similar scaling requirement $\sigma/\Sigma = 1$, according to § 11.2.2. A scale factor of 10 (i.e., $\beta = 0.1$) (scale factors larger than this are encountered in naval architectural structural research, for example) shows that the stresses to initiate fracture in a full-scale prototype are only one-tenth of the corresponding stresses in a geometrically similar small-scale model. The associated displacements at fracture in a full-scale prototype would be ten times larger than those in a small-scale model with $\beta = 0.1$, according to equation (11.27) with $n \gg 1$. This agrees with the elementary requirements of geometrically similar scaling.

If $n = 1$ for linear elastic behaviour, then, according to equation (11.22), the stresses to initiate fracture in a full-scale prototype are 0.5 and $1/\sqrt{10} = 0.316$ times those in one-quarter and one-tenth small-scale models, respectively. The associated displacements for the full-scale prototypes from equation (11.25) are 2 and 3.16 times larger than those of small-scale models with $\beta = 0.25$ and $\beta = 0.1$, respectively, which are to be contrasted with the corresponding values of 4 and 10 required for elementary geometrically similar scaling.

Unfortunately, no similar general theoretical treatments have been published for the influence of fracture on the static plastic and dynamic plastic behaviour of structures. However, the large departures from elementary geometrically similar scaling observed in the calculations for the static elastic case underline the importance of the ductile–brittle fracture transitions discussed in § 11.6. They show that the stress required to initiate fracture in a sufficiently small-scale model (i.e., small β) will exceed the corresponding yield stress for the material so that the behaviour is ductile, whereas a geometrically similar full-scale prototype could fracture. The testing of small-scale models might not always relate, therefore, to the behaviour of the full-scale prototype because of the possible intervention of a ductile–brittle fracture transition as a specimen is made larger.

Apart from the theoretical work in § 11.5, little information† is available to assess the importance of the various Π-terms listed in Table 11.3. Moreover, the theoretical concepts in § 11.6 are inadequately developed to predict ductile–brittle fracture transitions. Thus, experiments are sometimes required to increase confidence in a particularly crucial design, as exemplified by the full-scale test of a locomotive travelling at 160 km/h and striking a 48 tonne nuclear fuel transportation flask lying across the railway track.[11.28]

11.7.2 Drop tests on welded plated steel structures

Surprisingly few experimental investigations have examined the validity, or otherwise, of the small-scale modelling of structures.† This situation occurs because most structural members were designed, until relatively recently, to remain elastic.

Booth, Collier and Miles[11.8] reported a series of thirteen drop tests on one-quarter-scale to full-scale thin plated mild steel and stainless steel structures. The specimens were geometrically similar, as far as possible, and subjected to the same impact velocity which is required for elementary scaling as noted in § 11.2.6. Therefore, the permanently deformed shapes of the specimens should be geometrically similar according to Π_{21} in Table 11.3, but the photographs in Figure 11.7 show significant departures from similitude. In fact, the authors found that the post-impact deformations might be as much as 2.5 times greater in a full-scale prototype than would have been expected from the extrapolated results obtained on a one-quarter-scale model.

It is evident from the experimental data in reference 11.8 that the specimen thicknesses depart from the nominal values and that the yield stresses are different at the various scales. Thus, the experimental results from reference 11.8 have been

† Reference 11.24 contains a literature review.

$\frac{1}{3}$ scale

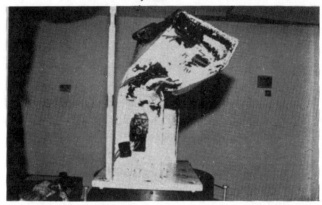

$\frac{2}{3}$ scale

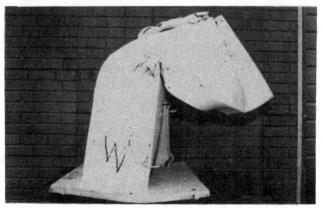

Full scale

Figure 11.7 Photographs of the side view of the permanently deformed type D mild steel plate girder test specimens examined by Booth, Collier and Miles.[11.8]

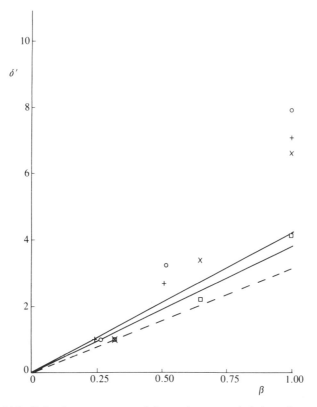

Figure 11.8 δ' is the permanent deformation recorded in reference 11.8, normalised with respect to the permanent deformation at the smallest scale and corrected with respect to the yield stress for the full-scale test specimens and material strain rate sensitivity.[11.24] β is calculated from actual average plate thicknesses. $\bigcirc$, $+$, $\square$ and $\times$ correspond to test numbers 1–3, 4–6, 7–10 and 11–13, respectively. —— and – – – are estimates when scaling geometrically the smallest-scale test results.

re-examined in reference 11.24 and replotted in Figure 11.8 to cater for these various effects and the influence of material strain rate sensitivity. The straight lines in Figure 11.8 are the geometrically similar scaling laws which predict that the maximum final displacements are proportional to β. (See Π_{21} in Table 11.3.) They pass through the origin and the experimental results for the smallest-scale model. The full-scale prototype test results are considerably larger than expected on the basis of elementary geometrically similar scaling from the corresponding smallest-scale model test results.

The experimental results are replotted in Figure 11.9, where the horizontal line represents the predictions of elementary geometrically similar scaling. The experimental results are normalised with respect to the predictions of the

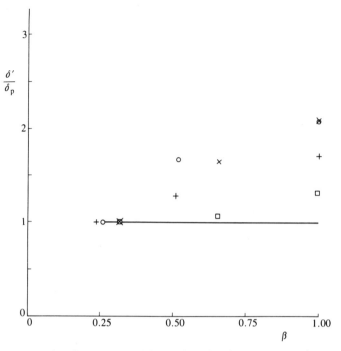

Figure 11.9 Ratio of permanent deformations in Figure 11.8 and permanent deformations predicted from the smallest-scale test results (δ'/δ_p) using geometrically similar scaling based on β. ———: $\delta'/\delta_p = 1$ for geometrically similar scaling.

corresponding smallest-scale model test results which are scaled up using the elementary principles of geometrically similar scaling.

It was observed that weld fractures and tearing were considerably more pronounced in the full-scale eggbox and plate girder specimens than in the smaller ones. Furthermore, the deviation from the elementary scaling laws is greater for the mild steel eggbox-type structures than the stainless steel ones.

11.7.3 Drop tests on steel plates

A series of experiments on the dynamic cutting of mild steel plate was conducted on a drop hammer rig and reported in references 11.24 and 11.29. Heavy solid wedges were attached to the head of a variable mass tup. The mild steel plate specimens were clamped along the bottom edge and both vertical sides in a rigid frame which was secured vertically to the base of the drop hammer rig, as shown schematically in Figure 11.10.

A set of experimental results is taken from references 11.24 and 11.29 and presented in Figure 11.11 for mild steel plates having four different thicknesses (H)

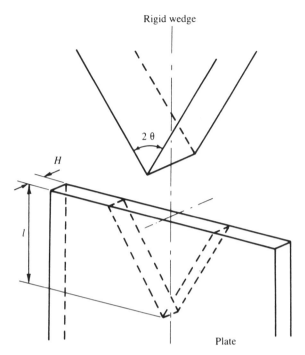

Figure 11.10 Rigid wedge with an included angle of 2θ striking an edge of a vertical plate with a thickness H. l is depth of penetration of wedge.

and struck by a wedge with an included angle (2θ) of 15 degrees. The ordinate is the scaled impact energy (i.e. initial kinetic energy/β^3, as noted in § 11.2.7), while the abscissa is the scaled penetration (l/β according to Π_{21} in Table 11.3, where l is the penetration indicated in Figure 11.10 and β is taken as the ratio of the plate thickness to the largest (full-scale) plate thickness). Therefore, all the straight lines in Figure 11.11 would coincide if the behaviour of the plates obeyed the laws of geometrically similar scaling. However, it is evident that the full-scale prototype test specimens absorb the least scaled kinetic energy for a given scaled penetration, while the smallest-scale models absorb the most.

These results are displayed in an alternative manner in Figure 11.12 for l/β = 120 mm. The lowest line in Figure 11.12 originates at $\beta = 0.248$ ($H = 1.501$ mm) and all the predicted values of energy absorbed (E_p) on this line are scaled, therefore, by multiplying the $\beta = 0.248$ test results by $1/\beta^3$, as observed in § 11.2.7, while the values of energy absorbed (E_a) are the actual experimental results recorded at each scale. Similar remarks apply to the remaining two lines, except that they originate from larger values of β. For example, the predicted values (E_p) for the upper line ($\square$) in Figure 11.12 are obtained from the test results on the 4.958 mm thick plate with $\beta = 0.820$.

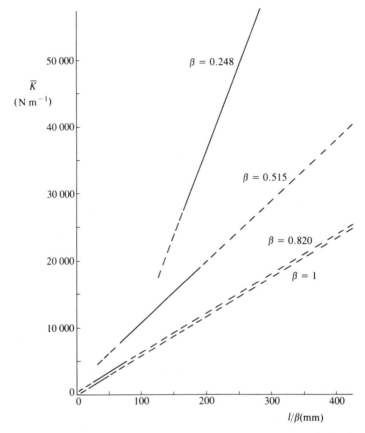

Figure 11.11 Best straight lines drawn through the experimental data for the rigid wedge–plate impact tests in reference 11.24 (see Figure 11.10), with scaled energy ($\bar{K}$) versus scaled penetration (l/β). $\bar{K} = K\beta^{-3}\sigma_u(1)/\sigma_u(\beta)$, where K is the initial kinetic energy. ——: range of experimental data. – – –: extrapolated lines.

It is evident from Figure 11.12 that the principles of elementary scaling predict, from the experimental results, with $\beta = 0.248$ ($H = 1.501$ mm), that $1/0.44 = 2.27$ times more energy would be absorbed than was actually absorbed during the experiments on the thickest plate ($H = 6.045$ mm). In other words, the 6.045 mm plates absorb only 44 per cent of the energy, which would have been expected from small-scale tests on 1.501 mm thick plate.

It is interesting to note that the above observations are in broad agreement with several other dynamic and static studies discussed in reference 11.24. The value $E_p/E_a = 2.27$ (i.e., 0.44^{-1}) for the $\beta = 0.248$ and $\beta = 1$ results in Figure 11.12 was found for a scaled penetration $l/\beta = 120$ mm. If, on the other hand, the energy

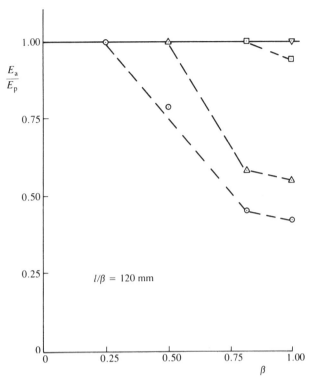

Figure 11.12 Ratio of actual energy absorbed (E_a) and predicted value (E_p) versus β for a scaled penetration of $l/\beta = 120$ mm and $2\theta = 15°$ (see Figure 11.10). E_p for $\bigcirc$, $\triangle$, $\square$ and $\triangledown$ are based on the experimental results for 1.501 mm ($\beta = 0.248$), 3.114 mm ($\beta = 0.515$), 4.958 mm ($\beta = 0.820$) and 6.045 mm ($\beta = 1$) thick plates, respectively. ———: prediction of geometrically similar scaling.

$E_a/\beta^3 = 16\,000$ N m is scaled in these experimental results, then $l/l_p = 2.26$, where $l_p = 120$ mm is given by l/β evaluated from the smallest-scale results with $\beta = 0.248$. These results should be compared with $l/l_p = 2.45$ in Figure 2.5 of reference 11.24 (or $l/l_p = 2.09$ when considering material strain rate effects in Figure 11.9 here) for the dynamic test results, reported by Booth et al.,[11.8] on welded steel plate structures with $\beta \simeq 0.25$ and $\beta = 1$ and having a given scaled initial impact energy. Thus, there is a fair measure of agreement between the significant departures from the principles of geometrically similar scaling observed in the two quite different experimental programmes reported in references 11.8 and 11.24. Furthermore, these conclusions are reinforced by the more recent experimental data reported in reference 11.30 for the same experimental test arrangement in Figure 11.10.

11.8 Final comments

Some of the fundamental concepts of similitude have been introduced in this chapter. It is straightforward to generate the dimensionless Π-terms which govern the behaviour of a dynamic inelastic problem as shown in § 11.4.2 and in Appendix 5.† The magnitude of a dimensionless Π-term must be identical for the geometrically similar behaviours of a small-scale model and a full-scale prototype. However, it was noted in § 11.3 and § 11.4.2(c) that in many practical cases it is impossible to achieve geometrically similar scaling which leads to distortion.‡ Unfortunately, it is difficult to assess the importance of the various dimensionless terms. An attempt is made in § 11.5 to examine fracture effects, although neither material plasticity nor dynamic effects have been examined. In critical applications, recourse is made to full-scale prototype testing. Nevertheless, relatively few such tests have been reported in the literature, though some recent results are discussed in § 11.7. These test results are summarised in Figures 11.8, 11.9, 11.11 and 11.12 and reveal an important departure from the principles of geometrically similar scaling.

Some test results and theoretical predictions are reported in references 11.33 and 11.34 for aluminium alloy and mild steel double-shear specimens subjected to large dynamic loads which produce extensive inelastic behaviour. It was found that the transverse displacements at the initiation of cracking almost obeyed the elementary geometrically similar scaling law. However, the critical displacement for complete transverse severance, or failure, did not, since the fracture displacement as a proportion of the beam thickness was less for the thickest specimens. In addition, the results for the small-scale models, when scaled up using the elementary scaling laws, overpredicted the actual energy absorbed in the experimental tests on the full-scale prototypes. The smallest models overpredicted the energy absorbed in the thickest specimens by a factor of about two for a given scaled deformation.

Notwithstanding the foregoing comments, Duffey, Cheresh and Sutherland[11.35] impacted circular metal plates with small circular punches and observed that the departure from the elementary scaling laws was not significant, although a scale factor of only two was examined for a relatively small number of tests.

It is likely that geometrically similar scaling is satisfied reasonably for structures made from strain-rate-insensitive materials when undergoing a wholly ductile dynamic response with no fracture.[11.24, 11.29] In fact, Thornton[11.36] demonstrated that the dynamic collapse loads of corrugated tubular sections, without any tearing, satisfied the elementary scaling laws. However, it is important to ensure

---

† Emori and Schuring[11.31] have presented a catalogue of the principal named dimensionless numbers used in engineering.

‡ Further comments on distortion, or relaxation, are given in references 11.31 and 11.32.

that a ductile–brittle fracture transition is not encountered when increasing the size of small-scale models, as discussed in § 11.6. Moreover, geometrically similar scaling is not satisfied for strain-rate-sensitive structures with non-linear unstable load–displacement characteristics, as noted by Calladine.[11.37, 11.38]

It is, therefore, necessary to finish this chapter on a cautionary note by concluding that, while similitude does appear useful when applied to structures undergoing plastic deformations, the occurrence of material failure, such as cracking, shearing or tearing, may invalidate the results.

PROBLEMS

11.1 Use the elementary procedure in § 11.2 to obtain the relationships between (a) accelerations, (b) forces and (c) plastic collapse moments of a rectangular cross-section for small-scale models and full-scale prototypes.

11.2 Use the Cowper–Symonds equation (equation (11.12)) to compare the behaviour of a small-scale model and a full-scale prototype made from the same strain-rate-sensitive material. Show that a size effect is present.

11.3 The dimensionless number $\rho V^2/E$ is known as the Cauchy number (Ca) for an elastic system.[11.31] Interpret the physical significance of this dimensionless number when the linear elastic wave speed is $(E/\rho)^{1/2}$.

Obtain the relationships which must be satisfied for similitude of a small-scale model and a full-scale prototype which have the same Ca number but are made from different materials. What are the conditions which must be satisfied for the recovery of equation (11.11)?

Express the Cauchy number in terms of the dimensionless Π-terms in Table 11.3.

11.4 The dimensionless number $\rho v^2/\sigma$ is known as a damage number[11.6] for a structural system or other body with a mean flow stress σ. It is equal to the reciprocal of the Newton number (Ne).[11.31] Show that this dimensionless number is a ratio between the inertial and mean flow forces.

Obtain the similitude requirements for a small-scale model and a full-scale prototype which have the same damage number but are made from different materials.

Express the damage number of terms of the dimensionless Π-terms in Table 11.3.

11.5 Use the Buckingham Π-theorem to obtain a set of dimensionless parameters which govern the maximum permanent transverse displacement of a rigid, perfectly plastic circular plate subjected to an impulsive velocity V_0 distributed uniformly over the entire plate, as shown in Figure 4.14(a). The plate has a radius R, thickness H, density ρ and a yield stress σ_0.

Compare your results with equations (4.93a) and (4.100).

11.6 Repeat Problem 11.5 for a rectangular plate having a breadth $2B$, length $2L$, thickness H, density ρ which is made from a rigid, perfectly plastic material with a yield stress σ_0 and subjected to a uniformly distributed impulsive velocity V_0.

Compare your results with equations (7.94a) and (7.95).

11.7 Use the Buckingham Π-theorem to obtain a set of dimensionless parameters which govern the static axial crushing of the thin-walled circular tube in § 9.2.2, and compare your results with equation (9.21).

APPENDIX 1

Principle of virtual work

The equilibrium equations (1.1) and (1.2), for a beam loaded statically with a pressure p, may be combined to give

$$d^2M/dx^2 + p = 0, \qquad \text{(A.1)†}$$

when the transverse shear force Q is eliminated and taken as a reaction (i.e., no transverse shear deformations). Multiplying equation (A.1) by an arbitrary, or virtual, transverse displacement field, $w(x)$, hereafter labelled as w, gives

$$(d^2M/dx^2 + p)w = 0, \qquad \text{(A.2)‡}$$

which, integrating over the entire length l of a beam, becomes

$$\int_l (d^2M/dx^2 + p)w \, dx = 0, \qquad \text{(A.3)}$$

or

$$-\int_l (d^2M/dx^2)w \, dx = \int_l pw \, dx. \qquad \text{(A.4)}$$

The right-hand side of equation (A.4) is recognised as the external work of the pressure p due to an arbitrary, or virtual, displacement field w. The left-hand side of equation (A.4) requires some further manipulation in order to render a more familiar form. Thus, recasting the term on the left-hand side of equation (A.4) into the form

$$-\int_l (d^2M/dx^2)w \, dx = -\int_l \{d(dM/dx)/dx\}w \, dx,$$

and integrating by parts,§ gives

$$-\int_l (d^2M/dx^2)w \, dx = -[(dM/dx)w] + \int_l (dM/dx)(dw/dx) \, dx. \qquad \text{(A.5)}$$

---

† The quantities M, Q, p and x are defined in Figure 1.1.
‡ The transverse displacement w is defined in Figure 1.1.
§ $\int_l u \, dv = [uv] - \int_l v \, du$, where the brackets [ ] indicate that the quantity uv is evaluated at the boundaries (supports).

The last term on the right-hand side of equation (A.5) is $\int_l d(M)\,dw/dx$, which, when integrating by parts, allows equation (A.5) to be written

$$-\int_l (d^2M/dx^2)w\,dx = -[(dM/dx)w] + [M\,dw/dx] - \int_l M(d^2w/dx^2)\,dx.$$

(A.6)

Thus, equation (A.4) becomes

$$-[Qw] + [M\,dw/dx] + \int_l M\kappa\,dx = \int_l pw\,dx \tag{A.7}$$

when using $dM/dx = Q$ and $\kappa = -d^2w/dx^2$ from equations (1.1) and (1.3), respectively.

Equation (A.7) may be rearranged in the form

$$\int_l M\kappa\,dx = \int_l pw\,dx + [Qw - M\,dw/dx]. \tag{A.8}$$

The left-hand side of this equation gives the total internal energy absorbed by a beam due to a change of curvature which is caused by an arbitrary or virtual transverse displacement field. The first term on the right-hand side of equation (A.8) is the external work due to pressure p, as noted earlier. The remaining quantity $[Qw - M\,dw/dx]$ is the external work done by the transverse shear forces and bending moments at the boundaries of a beam.†

Clearly, equation (A.8) states that the external work equals the corresponding internal work in a beam. However, this equation is more powerful than a simple energy balance. It is recalled that the transverse displacement field w is arbitrary and it is evident that dw/dx and κ in equation (A.8) are derived from w, which forms, therefore, what might be called a kinematically admissible set. This kinematically admissible set is not necessarily the actual kinematic set for a particular problem. Furthermore, the generalised stresses (M, Q) and external pressure (p) are independent of the transverse displacement field w and form a statically admissible, or equilibrium set. In other words, the statically admissible set (M, Q, p) and the kinematically admissible set $(w, dw/dx, \kappa)$ are independent and are only related through the constitutive equations for the actual theoretical solution. This important observation explains why equation (A.8) is called the principle of virtual work rather than being known as a work balance. The independence of the statically admissible and kinematically admissible sets in equation (A.8) is exploited in the proof of many theorems in mechanics, including the limit theorems of plasticity which are introduced in § 1.3.

The principle of virtual work has been proved in many textbooks for a variety of structures and continua.

---

† A bending moment M acting at the support of a beam with the positive sense defined in Figure 1.1 causes a transverse displacement with $dw/dx < 0$ at the support.

Equations (3.1) and (3.2), for a beam loaded dynamically, may be combined to give

$$\partial^2 M/\partial x^2 + p - m\,\partial^2 w/\partial t^2 = 0. \tag{A.9}$$

The theoretical procedure outlined above may be followed and remains identical provided p, in equations (A.2) to (A.4), (A.7) and (A.8), is replaced by $p - m\,\partial^2 w/\partial t^2$, and all the derivatives $d^2(\ )/dx^2$ and $d(\ )/dx$ are replaced by $\partial^2(\ )/\partial x^2$ and $\partial(\ )/\partial x$, respectively. Thus, equation (A.8) becomes

$$\int_l M\kappa\,dx = \int_l (p - m\,\partial^2 w/\partial t^2)w\,dx + [Qw - M\,\partial w/\partial x], \tag{A.10}$$

which is the principle of virtual work for the dynamic loading of beams.

Path-dependence of an inelastic material

The response of an inelastic body depends on the history of the applied loads and not only on the final values. In order to illustrate this phenomenon, consider the rigid-plastic bar in Figure A.1, which is subjected to an axial force N and a bending moment M, which cause plastic flow according to the simplified yield condition in Figure A.2. The behaviour of the bar for two different sequences of loading is now examined and contrasted.

A bar is loaded along the path OA, in Figure A.2, from the unloaded state (0) to the position A on the initial yield condition. If the bar is loaded further along the continuation of path OA, then it stretches axially in order to satisfy the normality requirement of plasticity, which is discussed in § 2.3.4. The yield condition is assumed to expand isotropically, as shown in Figure A.2, until loading ceases when A′ is reached.

The bar remains rigid, when loaded along the path A′B′ in Figure A.2, until the loading path just reaches the subsequent yield condition at B′. If the bar is now unloaded along the path B′O to the origin O, then it again remains rigid and is stretched permanently, therefore, owing to the plastic flow between A and A′ in Figure A.2.

It is evident that, if the loading cycle is reversed (i.e., OBB′A′AO), then a bar made from a virgin material is bent permanently in order to satisfy the normality requirement for plastic flow between B and B′. In the first loading sequence

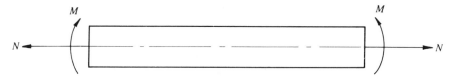

Figure A.1 A long bar which is subjected to an axial force N and a bending moment M.

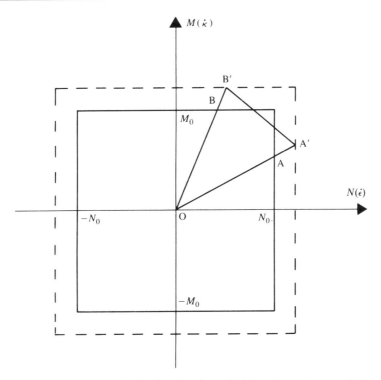

Figure A.2 Simplified yield criterion for a bar which is subjected to an axial load (N) and a bending moment (M). ———: initial yield condition. – – –: subsequent yield condition.

(OAA′B′BO), the bar is stretched and, in the second (OBB′A′AO), it is bent. The order of loading and unloading is, therefore, important. This phenomenon is known as path-dependence and is an important characteristic of a plastic material.

It is evident from the above observation that, unlike linear elastic analysis, for example, the theoretical solution for a plastic structure should proceed incrementally, or step by step, to account properly for any plastic flow which is controlled by the normality requirement of plasticity. This explains why the velocity profile, rather than the displacement profile, is used in the theoretical solutions throughout this book, and why generalised strain rate vectors are associated with the yield criteria.

Principle of virtual velocities

In view of the path-dependence of a plastic material, which is discussed in Appendix 2, another principle of virtual work is derived for an arbitrary or virtual velocity, and is known as the principle of virtual velocities.

The governing equation (A.1) for a beam may be multiplied by an arbitrary transverse velocity field $\dot{w}$, or

$$(d^2M/dx^2 + p)\dot{w} = 0. \tag{A.11}$$

Equations (A.3) to (A.7) may be obtained from equation (A.11) and remain unchanged provided w, dw/dx and κ are replaced by $\dot{w}$, $d\dot{w}/dx$ and $\dot{\kappa}$, respectively. Thus, equation (A.8) becomes

$$\int_l M\dot{\kappa}\,dx = \int_l p\dot{w}\,dx + [Q\dot{w} - M\,d\dot{w}/dx], \tag{A.12}$$

which is known as the principle of virtual velocities. Similarly,

$$\int_l M\dot{\kappa}\,dx = \int_l (p - m\,\partial^2 w/\partial t^2)\dot{w}\,dx + [Q\dot{w} - M\,\partial\dot{w}/\partial x] \tag{A.13}$$

for the dynamic case.

Consistent sets of equilibrium equations and geometrical relations

Introduction

The change of curvature for a beam is $\kappa = -d^2w/dx^2$ according to the elementary theory (see equation (1.3), for example). However, this expression is an approximation of the curvature change $\kappa = -(d^2w/dx^2)\{1 + (dw/dx)^2\}^{-3/2}$, and is valid, therefore, provided $(dw/dx)^2 \ll 1$. It is important that any such approximations in the strain and curvature relations are consistent with those made in the equilibrium equations. This ensures that no terms are neglected in the kinematic relations, which are more important than terms which are retained in the equilibrium equations. Conversely, it is important to guard against neglecting terms in the equilibrium equations which are equally, or more, important than those retained in the kinematic relations.

It is shown in this appendix how the principle of virtual velocities may be employed to obtain a consistent set of equilibrium equations and associated geometrical relations (strains or curvature changes) for a structural problem.

Infinitesimal behaviour of beams

The rate of curvature change for a beam is

$$\dot{\kappa} = -d^2\dot{w}/dx^2 \tag{A.14}$$

from equation (1.3), which, when substituted into the principle of virtual velocities (equation (A.12)), gives

$$-\int_l M(d^2\dot{w}/dx^2)\, dx = \int_l p\dot{w}\, dx + [Q\dot{w} - M\, d\dot{w}/dx].$$

The left-hand side of this equation may be integrated twice by parts to give

$$-[M\, d\dot{w}/dx] + [\dot{w}\, dM/dx] - \int_l (d^2M/dx^2)\dot{w}\, dx$$
$$= \int_l p\dot{w}\, dx + [Q\dot{w} - M\, d\dot{w}/dx],$$

which may be written as

$$\int_l (d^2M/dx^2 + p)\dot{w}\, dx = 0 \tag{A.15}$$

since the terms at the supports cancel when $Q = dM/dx$. Equation (A.15) is satisfied for any arbitrary transverse velocity field $\dot{w}$ provided

$$d^2M/dx^2 + p = 0 \tag{A.16}$$

and

$$Q = dM/dx. \tag{A.17}$$

Thus, equations (A.14), (A.16) and (A.17) satisfy the principle of virtual velocities, and the equilibrium equations (A.16) and (A.17) are consistent, therefore, with the rate of curvature change (equation (A.14)). Equations (A.16) and (A.17) are recognised as the transverse and moment equilibrium equations for static loads, which are given by equations (1.2) and (1.1), respectively.

Similarly, for the dynamic loading of beams

$$\dot{\kappa} = -\partial^2\dot{w}/\partial x^2 \tag{A.18}$$

from equation (3.3), and the principle of virtual velocities according to equation (A.13) predicts that

$$\partial^2M/\partial x^2 + p - m\,\partial^2w/\partial t^2 = 0 \tag{A.19}$$

and

$$Q = \partial M/\partial x, \tag{A.20}$$

which are recognised as the equation of transverse motion and the moment equilibrium equation given by equations (3.2) and (3.1), respectively.

Influence of transverse shear

The elementary theory for the bending of beams in the previous section does not retain the influence of transverse shear displacements. It is now assumed that a beam deforms with both bending curvatures and transverse shear displacements, as shown in Figure 6.2. The angle $\psi(x)$ is the usual rotation of a beam centre line caused by a bending moment M, and $\gamma(x)$ is the shear angle of points along the beam centre line due to a transverse shear force Q. Thus,

$$\dot{\kappa} = d\dot{\psi}/dx, \tag{A.21}$$

according to equation (6.8) for static loads, and the associated transverse shear strain rate is $\dot{\gamma}$ and, therefore,

$$d\dot{w}/dx = \dot{\gamma} - \dot{\psi}. \tag{A.22}$$

The principle of virtual velocities is employed to obtain a consistent set of equilibrium equations. However, an extended version of the principle of virtual velocities is required to cater for the influence of transverse shear effects. Equations (A.2) to (A.5), with w and dw/dx replaced by $\dot{w}$ and $d\dot{w}/dx$, respectively, are relabelled equations (A.2)' to (A.5)', which are valid for the present situation. Substituting equation (A.22) into equation (A.5)' gives

$$- \int_l (\mathrm{d}^2 M/\mathrm{d}x^2)\dot{w}\,\mathrm{d}x = -[\dot{w}\,\mathrm{d}M/\mathrm{d}x] + \int_l (\mathrm{d}M/\mathrm{d}x)(\dot{\gamma} - \dot{\psi})\,\mathrm{d}x,$$

or

$$- \int_l (\mathrm{d}^2 M/\mathrm{d}x^2)\dot{w}\,\mathrm{d}x = -[(\mathrm{d}M/\mathrm{d}x)\dot{w}] + \int_l (\mathrm{d}M/\mathrm{d}x)\dot{\gamma}\,\mathrm{d}x - [M\dot{\psi}]$$

$$+ \int_l M(\mathrm{d}\dot{\psi}/\mathrm{d}x)\,\mathrm{d}x. \qquad (\mathrm{A}.23)$$

Equation (A.4)' becomes, therefore,

$$- [Q\dot{w}] - [M\dot{\psi}] + \int_l Q\dot{\gamma}\,\mathrm{d}x + \int_l M\dot{\kappa}\,\mathrm{d}x = \int_l p\dot{w}\,\mathrm{d}x, \qquad (\mathrm{A}.24)\dagger$$

or

$$\int_l M\dot{\kappa}\,\mathrm{d}x + \int_l Q\dot{\gamma}\,\mathrm{d}x = \int_l p\dot{w}\,\mathrm{d}x + [Q\dot{w} + M\dot{\psi}], \qquad (\mathrm{A}.25)$$

which is the principle of virtual velocities when retaining transverse shear effects.

The geometrical relations (A.21) and (A.22) are now substituted into equation (A.25) to derive a consistent set of equilibrium equations. Thus,

$$\int_l \{M\,\mathrm{d}\dot{\psi}/\mathrm{d}x + Q(\mathrm{d}\dot{w}/\mathrm{d}x + \dot{\psi}) - p\dot{w}\}\,\mathrm{d}x = [Q\dot{w} + M\dot{\psi}], \qquad (\mathrm{A}.26)$$

which, integrating by parts, becomes

$$[M\dot{\psi}] - \int_l (\mathrm{d}M/\mathrm{d}x)\dot{\psi}\,\mathrm{d}x + [Q\dot{w}] - \int_l (\mathrm{d}Q/\mathrm{d}x)\dot{w}\,\mathrm{d}x$$

$$+ \int_l Q\dot{\psi}\,\mathrm{d}x - \int_l p\dot{w}\,\mathrm{d}x = [Q\dot{w} + M\dot{\psi}],$$

or

$$- \int_l (\mathrm{d}M/\mathrm{d}x - Q)\dot{\psi}\,\mathrm{d}x - \int_l (\mathrm{d}Q/\mathrm{d}x + p)\dot{w}\,\mathrm{d}x = 0. \qquad (\mathrm{A}.27)$$

This equation is satisfied for non-zero velocity and rotation rate fields provided

$$\mathrm{d}M/\mathrm{d}x = Q \qquad (\mathrm{A}.28)$$

and

$$\mathrm{d}Q/\mathrm{d}x = -p, \qquad (\mathrm{A}.29)$$

which are the equilibrium equations for the static loading of beams. Equations (6.5a) and (6.6) reduce to equations (A.29) and (A.28), respectively, for the static loading of beams with transverse shear displacements as well as bending curvatures.

† It is observed from equation (A.22) that $\dot{\psi} = -\mathrm{d}\dot{w}/\mathrm{d}x$ when $\dot{\gamma} = 0$.

It is evident by analogy between equations (A.12) and (A.25) and equation (A.13) that the principle of virtual velocities is

$$\int_l M\dot{\kappa}\,dx + \int_l Q\dot{\gamma}\,dx = \int_l (p - m\,\partial^2 w/\partial t^2)\dot{w}\,dx + [Q\dot{w} + M\dot{\psi}] \quad (A.30)$$

for the dynamic case when rotatory inertia effects are neglected.

Influence of transverse shear and rotatory inertia

If rotatory inertia is considered as well as transverse shear effects, then the principle of virtual velocities may be written for a beam as

$$\int_l M\dot{\kappa}\,dx + \int_l Q\dot{\gamma}\,dx = \int_l (p - m\,\partial^2 w/\partial t^2)\dot{w}\,dx$$

$$- \int_l I_r(\partial^2\psi/\partial t^2)\dot{\psi}\,dx + [Q\dot{w} + M\dot{\psi}], \quad (A.31)$$

where $I_r = mk^2$ is the second moment of inertia for the cross-section when rotating about the mid-plane of a beam (per unit length) and k is the radius of gyration. The new term in equation (A.31), which contains I_r, is recognised as the work rate to overcome the rotatory inertia as the beam cross-section rotates about the mid-plane of a beam.

Now, replacing equations (A.21) and (A.22) by

$$\dot{\kappa} = \partial\dot{\psi}/\partial x \quad (A.32)$$

and

$$\partial\dot{w}/\partial x = \dot{\gamma} - \dot{\psi}, \quad (A.33)$$

substituting into equation (A.31), and integrating by parts, gives

$$[M\dot{\psi}] - \int_l (\partial M/\partial x)\dot{\psi}\,dx + [Q\dot{w}] - \int_l (\partial Q/\partial x)\dot{w}\,dx + \int_l Q\dot{\psi}\,dx$$

$$= \int_l (p - m\ddot{w})\dot{w}\,dx - \int_l I_r\ddot{\psi}\dot{\psi}\,dx + [Q\dot{w} + M\dot{\psi}], \quad (A.34)$$

where

$$(\dot{\ }) = \partial(\ )/\partial t. \quad (A.35)$$

Thus,

$$- \int_l (\partial M/\partial x - Q - I_r\ddot{\psi})\dot{\psi}\,dx - \int_l (\partial Q/\partial x + p - m\ddot{w})\dot{w}\,dx = 0, \quad (A.36)$$

which requires

$$\partial M/\partial x = Q + I_r\ddot{\psi} \quad (A.37)$$

and

$$\partial Q/\partial x = -p + m\ddot{w}, \quad (A.38)$$

which are recognised as the equations of motion (6.6) and (6.5a), respectively.

Influence of finite displacements

An important feature of the principle of virtual velocities lies in the derivation of consistent sets of equations for structures which undergo finite-displacements. This is illustrated for the static loading of the beam shown in Figure A.3.

An element of a beam with an initial length dx changes to a new length ds after loading, where

$$ds = \{w'^2 + (1 + u')^2\}^{1/2}\,dx \tag{A.39}$$

and

$$(\;)' = \partial(\;)/\partial x, \tag{A.40}$$

or

$$ds = (1 + u' + w'^2/2)\,dx \tag{A.41}$$

when disregarding higher-order terms. It is evident from Figure A.3 that the direct strain ϵ^η at a distance η from the deformed mid-plane of the beam is given by

$$\epsilon^\eta = \{(R + \eta)(\partial\psi/\partial s)\,ds - dx\}/dx, \tag{A.42}$$

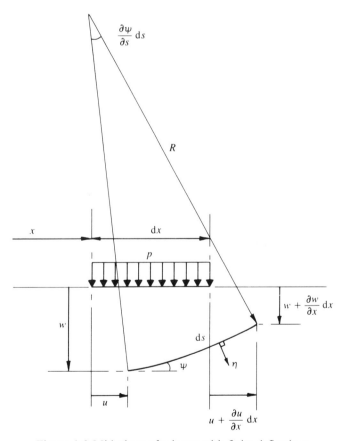

Figure A.3 Mid-plane of a beam with finite-deflections

where the curvature of the mid-plane of the deformed beam is

$$1/R = \partial\psi/\partial s \qquad (A.43)$$

and the plane cross-sections are assumed to remain plane during deformation. Now, from Figure A.3

$$\tan\psi = -w'/(1 + u'), \qquad (A.44)$$

which, differentiating with respect to s, becomes

$$\sec^2\psi \; \partial\psi/\partial s = -\{\partial(w'/(1 + u'))/\partial x\}(dx/ds),$$

or

$$\partial\psi/\partial s = -\{w''(1 + u') - w'u''\} \, (dx/ds)^3 \qquad (A.45)$$

since

$$\cos\psi = (1 + u') \, dx/ds. \qquad (A.46)$$

If our attention is confined to moderate deflections and small strains, then the transverse deflection w may be much larger than the axial displacement u†. Thus, terms which contain, u, u', or u'' multiplied by w, w', or w'' may be neglected when compared with terms containing only w or its various derivatives. Furthermore, it is apparent from equation (A.41) that $ds \cong dx$ for small strains. In view of the foregoing comments, equation (A.45) is simplified, therefore, to

$$\partial\psi/\partial s = -w'' \qquad (A.47)$$

and equation (A.42) may be rewritten as

$$\epsilon'' = \epsilon + \eta\kappa, \qquad (A.48)$$

where

$$\epsilon = u' + w'^2/2 \qquad (A.49)$$

and

$$\kappa = -w''. \qquad (A.50)$$

Equations (A.49) and (A.50) are recognised as the axial strain and change of curvature of the deformed mid-plane of the beam in Figure A.3 and may be written as

$$\dot{\epsilon} = \dot{u}' + w'\dot{w}' \qquad (A.51)$$

and

$$\dot{\kappa} = -\dot{w}'', \qquad (A.52)$$

respectively, where

$$(\dot{\;}) = \partial(\;)/\partial t. \qquad (A.53)$$

In order to complete the kinematic considerations, it is necessary to examine the velocities and rate of rotation at the support of a beam. It is evident from Figure A.4 that

$$\dot{m} = \dot{u}\cos\psi - \dot{w}\sin\psi, \qquad (A.54)$$

$$\dot{n} = \dot{u}\sin\psi + \dot{w}\cos\psi \qquad (A.55)$$

† See reference 7.26.

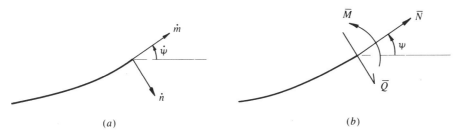

Figure A.4 (a) Velocities and rotation rate on the mid-plane at the boundaries of a beam. (b) Forces and moment on the mid-plane at the boundaries of a beam.

and

$$\dot{\psi} = -\dot{w}', \tag{A.56}$$

where $\dot{m}$ and $\dot{n}$ are the velocity components in the tangential and normal directions, respectively. However, equations (A.41), (A.44) and (A.46) show that $\cos\psi \cong 1$ and $\sin\psi \cong -w'$ when using the simplifications which are introduced earlier. Thus,

$$\dot{m} = \dot{u} + w'\dot{w}, \tag{A.57}$$

$$\dot{n} = \dot{w} \tag{A.58}$$

and

$$\dot{\psi} = -\dot{w}'. \tag{A.59}$$

Now, having completed the geometrical setting for the finite-displacement behaviour of a beam, a consistent set of equilibrium equations is sought. If one pauses here for a moment and attempts to derive the equilibrium equations in the usual manner, one discovers that the various forces produce a number of components in the vertical and horizontal directions and it is not clear which of these contributions may be considered small and, therefore, neglected. On the other hand, if all the components were to be retained, then the equilibrium equations would not only become cumbersome but would contain terms of equal importance to some already disregarded when deriving the associated geometrical relations (A.51) and (A.52). It is now indicated how the principle of virtual velocities may be used to derive the equilibrium equations which are consistent with the geometrical relations (A.51) and (A.52).

It was observed previously that equation (A.8) is a work balance for an arbitrary displacement field. This observation is also true for equation (A.10), which includes dynamic effects, equation (A.25), which considers transverse shear effects, and equation (A.31), which retains both transverse shear and dynamic effects, including rotatory inertia. In general, if a structure is assumed to deform in a manner prescribed by a virtual velocity field, which must be continuous and satisfy the velocity boundary conditions, then the principle of virtual velocities states that

$$\dot{D} = \dot{E}, \tag{A.60}$$

where

$$\dot{D} = \text{total internal energy dissipation}$$

and

$$\dot{E} = \text{total external energy dissipation.}$$

It is important to emphasise that the quantities in equation (A.60) must include all the internal and external energy dissipation in a structure.

Now,

$$\dot{D} = \int (N\dot{\epsilon} + M\dot{\kappa})\,\mathrm{d}s, \tag{A.61}$$

where the integration extends over the deformed mid-plane of the beam and the strain and curvature rates are given by equations (A.51) and (A.52), respectively. It has been noted previously from equation (A.41) that $\mathrm{d}s \cong \mathrm{d}x$ when attention is confined to beams with moderate deflections and small strains. Thus, equation (A.61) becomes

$$\dot{D} = \int \{N(\dot{u}' + w'\dot{w}') - M\dot{w}''\}\,\mathrm{d}x,$$

which may be rearranged in the form

$$\dot{D} = \int N\,\mathrm{d}(\dot{u}) + \int (Nw')\,\mathrm{d}(\dot{w}) - \int M\,\mathrm{d}(\dot{w}')$$

and integrated by parts to give

$$\dot{D} = [N\dot{u}] - \int N'\dot{u}\,\mathrm{d}x + [Nw'\dot{w}] - \int (Nw')'\dot{w}\,\mathrm{d}x - [M\dot{w}'] + \int M'\,\mathrm{d}(\dot{w}),$$

or

$$\dot{D} = [N\dot{u} + (Nw' + M')\dot{w} - M\dot{w}'] - \int N'\dot{u}\,\mathrm{d}x$$

$$- \int \{M'' + (Nw')'\}\dot{w}\,\mathrm{d}x, \tag{A.62}$$

where the terms enclosed within [ ] are to be evaluated at the boundaries.

The external work rate is

$$\dot{E} = \int (p - m\ddot{w})\dot{w}\,\mathrm{d}x - \int m\ddot{u}\dot{u}\,\mathrm{d}x + [\bar{N}\dot{m} + \bar{Q}\dot{n} + \bar{M}\dot{\psi}],$$

or

$$\dot{E} = \int (p - m\ddot{w})\dot{w}\,\mathrm{d}x - \int m\ddot{u}\dot{u}\,\mathrm{d}x + [\bar{N}\dot{u} + (\bar{N}w' + \bar{Q})\dot{w} - \bar{M}\dot{w}'], \tag{A.63}$$

when using equations (A.57) to (A.59).

If the principle of virtual velocities is now invoked and equations (A.62) and (A.63) are substituted into equation (A.60), we obtain

$$\int \{(-N' + m\ddot{u})\dot{u} - (M'' + (Nw')' - m\ddot{w} + p)\dot{w}\} \, \mathrm{d}x$$

$$+ [(N - \bar{N})\dot{u} + (Nw' + M' - \bar{N}w' - \bar{Q})\dot{w} - (M - \bar{M})\dot{w}'] = 0. \quad \text{(A.64)}$$

It is clear that, when N, M and Q act in the same direction as $\bar{N}$, $\bar{M}$ and $\bar{Q}$ shown in Figure A.4, then $N = \bar{N}$, $M = \bar{M}$, and $Q = \bar{Q}$ at the boundaries. Moreover, equation (A.64) can only remain valid for all virtual velocities $\dot{u}$ and $\dot{w}$ when the coefficients of $\dot{u}$ and $\dot{w}$ vanish. Therefore,

$$N' - m\ddot{u} = 0, \quad \text{(A.65)}$$

$$M'' + (Nw')' + p - m\ddot{w} = 0 \quad \text{(A.66)}$$

and

$$Q = M'. \quad \text{(A.67)}$$

Equations (A.65) and (A.66) are recognised as the equilibrium equations in the horizontal ($\dot{u}$) and vertical ($\dot{w}$) directions, respectively, while equation (A.67) ensures moment equilibrium. The strain and curvature relations (A.49) and (A.50) and the equilibrium equations ((A.65) to (A.67)) are, therefore, a consistent set of equations. Equations (A.65) to (A.67) for static loads, which produce finite-displacements, reduce to equations (7.13), (7.15) and (7.14), while, for dynamic infinitesimal displacements, equations (A.37) with $I_r = 0$ and (A.38) are recovered from equations (A.67) and (A.66), respectively.

The general procedure developed in this section may be used to obtain consistent sets of equations for any structure. However, the examples of some common engineering structures which are given in the following sections are described only briefly and developed for infinitesimal displacements.

Axisymmetric behaviour of a circular plate

The curvature changes of the mid-plane of a circular plate are

$$\kappa_r = -w'' \quad \text{(A.68)}$$

and

$$\kappa_\theta = -w'/r \quad \text{(A.69)}$$

for infinitesimal displacements, and

$$\dot{D} = \int (M_r \dot{\kappa}_r + M_\theta \dot{\kappa}_\theta) 2\pi r \, \mathrm{d}r, \quad \text{(A.70)}$$

while

$$\dot{E} = \int (p - \mu \ddot{w}) \dot{w} 2\pi r \, \mathrm{d}r + [2\pi r \bar{Q} \dot{w} - 2\pi r \bar{M} \dot{w}']. \quad \text{(A.71)}$$

Thus, substituting into equation (A.60) and integrating by parts gives the equilibrium equations

$$(rQ_r)' + rp - r\mu\ddot{w} = 0 \tag{A.72}$$

and

$$rQ_r = (rM_r)' - M_\theta. \tag{A.73}$$

Transverse loading of a rectangular plate

The curvature changes of the mid-plane of the rectangular plate in Figure 4.15 are

$$\kappa_x = -w'', \tag{A.74}$$

$$\kappa_y = -w^{**} \tag{A.75}$$

and

$$\kappa_{xy} = -w'^* \tag{A.76}$$

for infinitesimal displacements, where

$$(\ )^* = \partial(\ )/\partial y \tag{A.77}$$

and $(\ )'$ is defined by equation (A.40). The internal energy dissipation is

$$\dot{D} = \iint (M_x\dot{\kappa}_x + M_y\dot{\kappa}_y + 2M_{xy}\dot{\kappa}_{xy})\,dx\,dy,$$

which, when substituted into equation (A.60), together with the external work rate, predicts that

$$Q'_x + Q^*_y + p - \mu\ddot{w} = 0, \tag{A.78}$$

$$Q_x = M'_x + M^*_{xy} \tag{A.79}$$

and

$$Q_y = M^*_y + M'_{xy}, \tag{A.80}$$

where the various quantities are defined in Figure 4.15.

Axisymmetric behaviour of a cylindrical shell

The strain and curvature changes for the cylindrical shell in Figure 2.14 (or Figure 5.1) are

$$\epsilon_\theta = -w/R, \tag{A.81}$$

$$\kappa_x = -w'' \tag{A.82}$$

and

$$\kappa_\theta = 0 \tag{A.83}$$

for static or dynamic transverse loads which produce axisymmetric infinitesimal displacements with no axial displacements. The associated internal energy dissipation is

$$\dot{D} = \int (N_\theta\dot{\epsilon}_\theta + M_\theta\dot{\kappa}_\theta + M_x\dot{\kappa}_x)2\pi R\,dx$$

which, when substituted into equation (A.60), together with the external work rate, predicts that

$$Q'_x + N_\theta/R - p - \mu\ddot{w} = 0 \qquad (A.84)$$

and

$$Q_x = M'_x, \qquad (A.85)$$

where the various quantities are defined in Figure 5.1.

APPENDIX 5

Buckingham Π-theorem

The primary quantities in Table 11.2 are M, L, T, and θ, while all others are secondary (e.g., ρ, δ). It may be shown that any secondary quantity (S_i) can be expressed in terms of the primary quantities as follows:[A.1]

$$S_i = L^a T^b M^c \theta^d, \tag{A.86}$$

where a, b, c and d are exponents which must satisfy dimensional homogeneity. If K variables ζ_1, ζ_2, ..., ζ_K are required to describe the response of a structure, then dimensionless products of these variables could be constructed as

$$\chi = \zeta_1^{\alpha_1} \zeta_2^{\alpha_2} \cdots \zeta_K^{\alpha_K} \tag{A.87}$$

when α_1, α_2, ... α_K are selected to give a dimensionless product.

Now, taking

$$\zeta_i = L^{a_i} T^{b_i} M^{c_i} \theta^{d_i}, \tag{A.88}$$

equation (A.87) becomes

$$\chi = (L^{a_1} T^{b_1} M^{c_1} \theta^{d_1})^{\alpha_1} (L^{a_2} T^{b_2} M^{c_2} \theta^{d_2})^{\alpha_2} \cdots (L^{a_K} T^{b_K} M^{c_K} \theta^{d_K})^{\alpha_K}. \tag{A.89}$$

However, for χ to be dimensionless it is evident that the exponents of the primary quantities on the right-hand side of equation (A.89) must equal zero, or

$$a_1 \alpha_1 + a_2 \alpha_2 + \cdots + a_K \alpha_K = 0,$$
$$b_1 \alpha_1 + b_2 \alpha_2 + \cdots + b_K \alpha_K = 0,$$
$$c_1 \alpha_1 + c_2 \alpha_2 + \cdots + c_K \alpha_K = 0$$

and

$$d_1 \alpha_1 + d_2 \alpha_2 + \cdots + d_K \alpha_K = 0 \tag{A.90}$$

Equations (A.90) contain an equation for each primary quantity (M, L, T and θ in this case), with K unknowns (α_1 to α_K), where K is the number of original variables (ζ_1, ..., ζ_K) for the problem (primary and secondary). It is evident from the theory of simultaneous equations that equations (A.90) have $K - R$ linearly independent equations, where R is the rank of the dimensional matrix

$$\begin{vmatrix} a_1 & a_2 & \cdots & a_K \\ b_1 & b_2 & \cdots & b_K \\ c_1 & c_2 & \cdots & c_K \\ d_1 & d_2 & \cdots & d_K \end{vmatrix}.$$

The rank (R) of a matrix is associated with the highest-order non-zero determinant contained in the matrix. This cannot exceed the number of equations, but may be smaller. Thus, the number of independent dimensionless products which can be formed is equal to the number (K) of the original primary and secondary variables minus the rank (R) of the coefficient, or dimensional, matrix.

The foregoing procedure gives rise to a complete set. All other possible dimensionless combinations are products of powers of the products contained in a complete set. Finally, the Buckingham Π-theorem can be stated:[A.1]

> if an equation containing K variables is dimensionally homogeneous, it can be reduced to a relationship among $K - R$ independent dimensionless products, where R is the rank of the dimensional matrix.

Many books and articles have been published on the Buckingham Π-theorem to which an interested reader is referred for a more rigorous exposition. In particular, Gibbings[A.2] has discussed the various weaknesses in earlier proofs and has presented a proof which avoids many pitfalls.

Illustration of Buckingham Π-theorem

The response of a rigid-plastic beam made from a material with a flow stress σ_0 and density ρ and subjected to an impulsive velocity V_0 distributed uniformly over the entire span was examined in §§ 3.5, 3.7 and 7.6. The beam has a uniform thickness H and a length $2L$. The input parameters for this particular problem are σ_0, ρ, V_0, H and L, while the output parameter of interest is the maximum permanent transverse displacement (W_f). Now, following equation (A.87) gives

$$\sigma_0^{\alpha_1} \rho^{\alpha_2} V_0^{\alpha_3} H^{\alpha_4} L^{\alpha_5} W_f^{\alpha_6} \tag{A.91}$$

which, using equation (A.88), becomes

$$(ML^{-1}T^{-2})^{\alpha_1}(ML^{-3})^{\alpha_2}(LT^{-1})^{\alpha_3}(L)^{\alpha_4}(L)^{\alpha_5}(L)^{\alpha_6} \tag{A.92}$$

and, according to equations (A.90), gives the set of equations

$$\begin{aligned} M: \quad \alpha_1 + \alpha_2 + 0 + 0 + 0 + 0 &= 0 \\ L: \quad -\alpha_1 - 3\alpha_2 + \alpha_3 + \alpha_4 + \alpha_5 + \alpha_6 &= 0 \\ T: \quad -2\alpha_1 + 0 - \alpha_3 + 0 + 0 + 0 &= 0. \end{aligned} \tag{A.93}$$

The dimensional matrix of equations (A.93) is

$$\begin{vmatrix} 1 & 1 & 0 & 0 & 0 & 0 \\ -1 & -3 & 1 & 1 & 1 & 1 \\ -2 & 0 & -1 & 0 & 0 & 0 \end{vmatrix}.$$

The determinant of

$$
\begin{vmatrix}
1 & 1 & 0 \\
-1 & -3 & 1 \\
-2 & 0 & 0
\end{vmatrix} = -2,
$$

so that the rank (R) is three, which is the highest-order non-zero determinant contained in the dimensional matrix. Thus, since $K = 6$ in equations (A.91) and (A.92), there are $6 - 3 = 3$ dimensionless variables required to describe this particular problem.

It is noted that the coefficients of α_1, α_2 and α_4 form the above non-zero determinant so, to obtain the dimensionless variables, it is assumed that

$$\alpha_3 = 1, \alpha_5 = \alpha_6 = 0 \text{ in equations (A.93) to give}$$

$$\alpha_1 + \alpha_2 = 0$$

$$-\alpha_1 - 3\alpha_2 + 1 + \alpha_4 = 0$$

and

$$-2\alpha_1 - 1 = 0. \tag{A.94}$$

Solving equations (A.94) gives $\alpha_1 = -\frac{1}{2}$, $\alpha_2 = \frac{1}{2}$ and $\alpha_4 = 0$. Hence, substituting these values into equation (A.91) yields

$$\Pi_1 = \sigma_0^{-1/2} \rho^{1/2} V_0^1 H^0 L^0 W_f^0,$$

or

$$\Pi_1 = V_0 (\rho/\sigma_0)^{1/2}. \tag{A.95}$$

Similarly, taking $\alpha_3 = 0$, $\alpha_5 = 1$, $\alpha_6 = 0$ gives

$$\alpha_1 + \alpha_2 = 0$$

$$-\alpha_1 - 3\alpha_2 + \alpha_4 + 1 = 0$$

and

$$-2\alpha_1 = 0,$$

from which $\alpha_1 = \alpha_2 = 0$ and $\alpha_4 = -1$ and, therefore,

$$\Pi_2 = L/H. \tag{A.96}$$

The remaining dimensionless term is

$$\Pi_3 = W_f/H \tag{A.97}$$

according to the parameters

$$\alpha_3 = \alpha_5 = 0, \alpha_6 = 1, \quad \text{which predict} \quad \alpha_1 = \alpha_2 = 0 \quad \text{and} \quad \alpha_4 = -1.$$

The dimensionless parameters Π_1 to Π_3 are not unique but they can be used to obtain other dimensionless quantities. For example, $(\Pi_1 \Pi_2)^2 = \rho V_0^2 L^2/\sigma_0 H^2$ which, apart from a numerical factor, was used in Figure 7.19.

Once the rank of the dimensional matrix has been established, then the number of the independent dimensionless Π-terms is known. They may be written down by inspection rather than following the above procedure leading to equations (A.94) to (A.97). However, Gibbings[A.3] has presented an alternative simple rational procedure for generating non-dimensional groups which is illustrated for the above problem.

Now,

$$W_f = F(\sigma_0, \rho, V_0, H, L) \tag{A.98}$$

which might be expressed in the dimensionless form

$$W_f/H = F(\sigma_0/\rho V_0^2, L/H) \tag{A.99}$$

when using the procedure set out in Table A.1.

Table A.1

Variables	W_f	σ_0	ρ	V_0	H	L
Dimensions	L	$ML^{-1}T^{-2}$	ML^{-3}	LT^{-1}	L	L
Variables	W_f	σ_0/V_0^2	ρ		H	L
Dimensions	L	ML^{-3}	ML^{-3}		L	L
Variables	W_f	$\sigma_0/\rho V_0^2$			H	L
Dimensions	L	1			L	L
Variables	W_f/H	$\sigma_0/\rho V_0^2$				L/H
Dimensions	1	1				1

Equation (A.99) may be expressed in the form

$$\Pi_3 = F(1/\Pi_1^2, \Pi_2) \tag{A.100}$$

according to equations (A.95) to (A.97).

The variables governing this particular problem give rise to the independent dimensionless terms Π_1 to Π_3 which can be used to develop similarity principles. Geometrically similar scaling is satisfied when Π_1 to Π_3 are equal for a small-scale model and a full-scale prototype.

References

Chapter 1

1.1 Baker, J. F., Horne, M. R. and Heyman, J., *The Steel Skeleton, vol. 2: Plastic Behaviour and Design*, Cambridge University Press, Cambridge (1956).

1.2 Hodge, P. G., *Plastic Analysis of Structures*, McGraw-Hill, New York (1959).

1.3 Prager, W., *An Introduction to Plasticity*, Addison-Wesley, Boston, Mass. (1959).

1.4 Mendelson, A., *Plasticity: Theory and Application*, Macmillan, New York (1968).

1.5 Baker, J. F. and Heyman, J., *Plastic Design of Frames 1: Fundamentals*, Cambridge University Press, Cambridge (1969).

1.6 Heyman, J., *Plastic Design of Frames 2: Applications*, Cambridge University Press, Cambridge (1971).

1.7 Horne, M. R., *Plastic Theory of Structures*, MIT Press, Cambridge, Mass. (1971).

1.8 Johnson, W. and Mellor, P. B., *Engineering Plasticity*, Van Nostrand Reinhold, London (1973).

1.9 Kachanov, L. M., *Fundamentals of the Theory of Plasticity*, MIR, Moscow (1974).

1.10 Martin, J. B., *Plasticity: Fundamentals and General Results*, MIT Press, Cambridge, Mass. (1975).

1.11 Calladine, C. R., *Plasticity for Engineers*, Ellis Horwood, Chichester, and John Wiley, New York (1985).

1.12 Venkatraman, B. and Patel, S. A., *Structural Mechanics with Introductions to Elasticity and Plasticity*, McGraw-Hill, New York (1970).

1.13 Haythornthwaite, R. M., Beams with full end fixity, *Engineering* **183**, 110–12 (1957).

Chapter 2

2.1 Prager, W., *An Introduction to Plasticity*, Addison-Wesley, Boston, Mass. (1959).

2.2 Hodge, P. G., *Limit Analysis of Rotationally Symmetric Plates and Shells*, Prentice-Hall, Englewood Cliffs, N.J. (1963).

2.3 Drucker, D. C., A more fundamental approach to plastic stress-strain relations, *Proceedings First US Congress of Applied Mechanics*, ASME, 487–91 (1951). (See also *Journal of Applied Mechanics* **26**, 101–6 (1959).)

2.4 Drucker, D. C., Prager, W. and Greenberg, H. J., Extended limit design theorems for continuous media, *Quarterly of Applied Mathematics*, **9**, 381–9 (1952).

2.5 Calladine, C. R., *Plasticity for Engineers*, Ellis Horwood, Chichester, and John Wiley, New York (1985).

2.6 Hopkins, H. G. and Prager, W., The load carrying capacities of circular plates, *Journal of the Mechanics and Physics of Solids*, **2**, 13 (1953).

2.7 Timoshenko, S. and Woinowsky-Krieger, S., *Theory of Plates and Shells*, McGraw-Hill, New York (1959).

2.8 Wood, R. H., *Plastic and Elastic Design of Slabs and Plates*, Thames & Hudson, London (1961).

2.9 Sawczuk, A. and Winnicki, L., Plastic behavior of simply supported reinforced concrete plates at moderately large deflections, *International Journal of Solids and Structures*, **1**, 97–111 (1965).

2.10 Prager, W., The general theory of limit design, *Proceedings of the Eighth International Congress on Theoretical and Applied Mechanics*, **2**, 65–72 (1952).

2.11 Jones, N., *A Lower Bound to the Static Collapse Pressure of a fully Clamped Rectangular Plate*, Department of Ocean Engineering Report 71-20, MIT, Cambridge, Mass., November (1971).

2.12 Fox, E. N., Limit analysis for plates: the exact solution for a clamped square plate of isotropic homogeneous material obeying the square yield criterion and loaded by uniform pressure, *Philosophical Transactions of the Royal Society of London, Series A (Mathematical and Physical Sciences)*, **277**, 121–55 (1974).

2.13 Zaid, M., On the carrying capacity of plates of arbitrary shape and variable fixity under a concentrated load, *Journal of Applied Mechanics*, **25**, 598–602 (1958).

2.14 Kraus, H., *Thin Elastic Shells*, John Wiley, New York (1967).

2.15 Hodge, P. G., *Plastic Analysis of Structures*, McGraw-Hill, New York (1959).

2.16 Calladine, C. R., *Theory of Shell Structures*, Cambridge University Press, Cambridge (1983).

2.17 Onat, E. T. and Haythornthwaite, R. M., 'The load-carrying capacity of circular plates at large deflection', *Journal of Applied Mechanics*, **23**, 49–55 (1956).

2.18 Jones, N., Rigid plastic behaviour of plates, *Bulletin of Mechanical Engineering Education*, **9**, 235–48 (1970).

2.19 Hooke, R. and Rawlings, B., An experimental investigation of the behaviour of clamped, rectangular, mild steel plates subjected to uniform transverse pressure, *Proceedings of the Institution of Civil Engineers*, **42**, 75–103 (1969).

2.20 Jones, N. and Walters, R. M., Large deflections of rectangular plates, *Journal of Ship Research*, **15**, 164–71, 288 (1971).

2.21 Augusti, G. and d'Agostino, S., Experiments on the plastic behavior of short steel cylindrical shells subject to internal pressure, *Proceedings of the First International Conference on Pressure Vessel Technology*, pt 1, 45–57 (1969).

2.22 Perrone, N., An experimental verification of limit analysis of short cylindrical shells, *Journal of Applied Mechanics*, **36**, 362–4 (1969).

2.23 Drucker, D. C., Limit analysis of cylindrical shells under axially-symmetric loading, *Proceedings of the First Midwestern Conference on Solid Mechanics*, 158–63 (1953).

2.24 Demir, H. H. and Drucker, D. C., An experimental study of cylindrical shells under ring loading, *Progress in Applied Mechanics, Prager Anniversary Volume*, Macmillan, New York, 205–20 (1963).

2.25 Eason, G. and Shield, R. T., The influence of free ends on the load-carrying capacities of cylindrical shells, *Journal of the Mechanics and Physics of Solids*, **4**, 17–27 (1955).

2.26 Eason, G., The load carrying capacities of cylindrical shells subjected to a ring of force, *Journal of the Mechanics and Physics of Solids*, **7**, 169–81 (1959).

2.27 Hu, L. W., Design of circular plates based on plastic limit load, *Proceedings of the ASCE*, **86**, EM1, 91–115 (1960).

2.28 Mansfield, E. H., Studies in collapse analysis of rigid-plastic plates with a square yield diagram, *Proceedings of the Royal Society of London, Series A (Mathematical and Physical Sciences)*, **241**, 311–38 (1957).

2.29 Jones, N., Combined distributed loads on rigid-plastic circular plates with large deflections, *International Journal of Solids and Structures*, **5**, 51–64 (1969).

2.30 Olszak, W. and Sawczuk, A., *Inelastic Behaviour in Shells*, Noordhoff, Groningen (1967).

2.31 Duszek, M., Plastic analysis of cylindrical shells subjected to large deflections, *Archiwum Mechaniki Stosowanej* **18**, 599–614 (1966).

2.32 Duszek, M. and Sawczuk, A., 'Load-deflexion relations for rigid-plastic cylindrical shells beyond the incipient collapse load', *International Journal of Mechanical Sciences*, **12**, 839–48 (1970).

2.33 Jones N. and Ich, N. T., The load carrying capacities of symmetrically loaded shallow shells, *International Journal of Solids and Structures*, **8**, 1339–51 (1972).

2.34 Onat, E. T. and Prager, W., Limit analysis of shells of revolution, pts 1 and 2, *Proceedings of the Royal Netherlands Academy of Science*, **57**(B), 534–48 (1954).

2.35 Drucker, D. C. and Shield, R. T., Limit analysis of symmetrically loaded thin shells of revolution, *Journal of Applied Mechanics*, **26**, 61–8 (1959).

2.36 Shield, R. T. and Drucker, D. C., Design of thin-walled torispherical and toriconical pressure-vessel heads, *Journal of Applied Mechanics*, **28**, 292–7 (1961).

2.37 Save, M. and Janas, M., Collapse and bursting pressures of mild-steel vessels, *Archiwum Budowy Maszyn*, **18**, 77–106 (1971).

2.38 Gill, S. S., The limit pressure for a flush cylindrical nozzle in a spherical pressure vessel, *International Journal of Mechanical Sciences*, **6**, 105–15 (1964).

2.39 Cloud, R. L. and Rodabaugh, E. C., Approximate analysis of the plastic limit pressures of nozzles in cylindrical shells, *Transactions of the ASME, Journal of Engineering for Power*, **90**, 171–6 (1968).

2.40 Jones, N., The collapse pressure of a flush cylindrical nozzle intersecting a conical pressure vessel axisymmetrically, *International Journal of Mechanical Sciences*, **11**, 401–15 (1969).

2.41 Gill, S. S. (ed.), *The Stress Analysis of Pressure Vessels and Pressure Vessel Components*, Pergamon, Oxford (1970).

2.42 Hodge, P. G., 'Plastic analysis and pressure-vessel safety', *Applied Mechanics Reviews*, **24**, 741–7 (1971).

2.43 Lance, R. H. and Onat, E. T., A comparison of experiments and theory in the plastic bending of circular plates, *Journal of the Mechanics and Physics of Solids*, **10**, 301–11 (1962).

Chapter 3

3.1 Liu, J. and Jones, N., Experimental investigation of clamped beams struck transversely by a mass, *International Journal of Impact Engineering*, **6**(4), 303–35 (1987).

3.2 Lowe, W. T., Al-Hassani, S. T. S. and Johnson, W., Impact behaviour of small scale model motor coaches, *Proceedings of the Institution of Mechanical Engineers*, **186**, 409–19 (1972).

3.3 Lee, E. H. and Symonds, P. S., Large plastic deformations of beams under transverse impact, *Journal of Applied Mechanics*, **19**, 308–14 (1952).

3.4 Symonds, P. S., Ting, T. C. T. and Robinson, D. N., *Survey of Progress in Plastic Wave Propagation in Solid Bodies*, Brown University Report, Contract DA-19-020-ORD-5453(A) (1967).

3.5 Symonds, P. S., Large plastic deformations of beams under blast type loading, *Proceedings of the Second US National Congress of Applied Mechanics*, 505–15 (1954).

3.6 Selby, S. M., *Standard Mathematical Tables*, CRC Press, Cleveland, Ohio, 22nd edn (1974).

3.7 Parkes, E. W., The permanent deformation of an encastré beam struck transversely at any point in its span, *Proceedings of the Institution of Civil Engineers*, **10**, 277–304 (1958).

3.8 Parkes, E. W., The permanent deformation of a cantilever struck transversely at its tip, *Proceedings of the Royal Society of London, Series A (Mathematical and Physical Sciences)*, **228**, 462–76 (1955).

3.9 Florence, A. L. and Firth, R. D., Rigid-plastic beams under uniformly distributed impulses, *Journal of Applied Mechanics*, **32**, 481–8 (1965).

3.10 Symonds, P. S. and Fleming, W. T., Parkes revisited: on rigid-plastic and elastic–plastic dynamic structural analysis, *International Journal of Impact Engineering*, **2**(1), 1–36 (1984).

3.11 Duwez, P. E., Clark, D. S. and Bohnenblust, H. F., The behaviour of long beams under impact loading, *Journal of Applied Mechanics*, **17**, 27–34 (1950).

3.12 Conroy, M. F., Plastic deformation of semi-infinite beams subject to transverse impact loading at the free end, *Journal of Applied Mechanics*, **23**, 239–43 (1956).

3.13 Symonds, P. S., Survey of methods of analysis for plastic deformation of structures under dynamic loading, Brown University Report, BU/NSRDC/1–67 (1967).

3.14 Symonds, P. S. and Frye, C. W. G., On the Relation Between Rigid-Plastic and Elastic-Plastic Predictions of Response to Pulse Loading, *International Journal of Impact Engineering*, **7**(2), 139–49 (1988).

3.15 Bodner, S. R. and Symonds, P. S., Experimental and theoretical investigation of the plastic deformation of cantilever beams subjected to impulsive loading, *Journal of Applied Mechanics*, **29**, 719–28 (1962).

3.16 Goldsmith, W., *Impact*, Edward Arnold, London (1960).

3.17 Johnson, W., *Impact Strength of Materials*, Edward Arnold, London, and Crane Russak, New York (1972).

3.18 Stronge, W. J. and Shioya, T., Impact and bending of a rigid-plastic fan blade, *Journal of Applied Mechanics*, **51**(3), 501–4 (1984).

3.19 Yu, T. X., Hua, Y. L. and Johnson, W., The plastic hinge position in a circular cantilever when struck normal to its plane by a constant jet at its tip, *International Journal of Impact Engineering*, **3**(3), 143–54 (1985).

3.20 Hua, Y. L., Yu, T. X. and Johnson, W., The plastic hinge position in a bent cantilever struck normal to its plane by a steady jet applied at its tip, *International Journal of Impact Engineering*, **3**(4), 233–41 (1985).

3.21 Yu, T. X., Symonds, P. S. and Johnson, W., A reconsideration and some new results for the circular beam impact problem, *International Journal of Impact Engineering*, **4**(4), 221–8 (1986).

3.22 Reid, S. R. and Gui, X. G., On the elastic–plastic deformation of cantilever beams subjected to tip impact, *International Journal of Impact Engineering*, **6**(2), 109–27 (1987).

3.23 Jones, N and Wierzbicki, T., Dynamic plastic failure of a free–free beam, *International Journal of Impact Engineering*, **6**(3), 225–40 (1987).

3.24 Yu, J. and Jones, N., Numerical simulation of a clamped beam under impact loading, *Computers and Structures*, **32**, 281–93 (1989).

3.25 Wegener, R. B. and Martin, J. B., Predictions of permanent deformation of impulsively loaded simply supported square tube steel beams, *International Journal of Mechanical Sciences*, **27**(1/2), 55–69 (1985).

Chapter 4

4.1 Jones, N., Uran, T. O. and Tekin, S. A., The dynamic plastic behaviour of fully clamped rectangular plates, *International Journal of Solids and Structures*, **6**, 1499–512 (1970).

4.2 Hopkins, H. G. and Prager, W., On the dynamics of plastic circular plates, *Journal of Applied Mathematics and Physics (ZAMP)*, **5**, 317–30 (1954).

4.3 Wang, A. J., The permanent deflection of a plastic plate under blast loading, *Journal of Applied Mechanics*, **22**, 375–6 (1955).

4.4 Florence, A. L., Clamped circular rigid-plastic plates under blast loading, *Journal of Applied Mechanics*, **33**(2), 256–60 (1966).

4.5 Wang, A. J. and Hopkins, H. G., On the plastic deformation of built-in circular plates under impulsive load, *Journal of the Mechanics and Physics of Solids*, **3**, 22–37 (1954).

4.6 Timoshenko, S. and Woinowsky-Krieger, S., *Theory of Plates and Shells*, 2nd edn, McGraw-Hill, New York (1959).

4.7 Cox, A. D. and Morland, L. W., Dynamic plastic deformations of simply supported square plates, *Journal of the Mechanics and Physics of Solids*, **7**, 229–41 (1959).

4.8 Jones, N., Impulsive loading of a simply supported circular rigid-plastic plate, *Journal of Applied Mechanics*, **35**, 59–65 (1968).

4.9 Johnson, W., *Impact Strength of Materials*, Edward Arnold, London, and Crane Russak, New York (1972).

4.10 Shapiro, G. S., On a rigid-plastic annular plate under impulsive load, *Journal of Applied Mathematics and Mechanics (Prik Mat i Mek)* **23**, 234–41 (1959).

4.11 Florence, A. L., Annular plate under a transverse line impulse, *AIAA Journal*, **3**(9), 1726–32 (1965).

4.12 Perrone, N., Impulsively loaded strain-rate-sensitive plates, *Journal of Applied Mechanics*, **34**(2), 380–4 (1967).

4.13 Jones, N., Finite-deflections of a rigid-viscoplastic strain-hardening annular plate loaded impulsively, *Journal of Applied Mechanics*, **35**(2), 349–56 (1968).

4.14 Jones, N., Finite-deflections of a simply supported rigid-plastic annular plate loaded dynamically, *International Journal of Solids and Structures*, **4**, 593–603 (1968).

4.15 Aggarwal, H. R. and Ablow, C. M., Plastic bending of an annular plate by uniform impulse, *International Journal of Non-Linear Mechanics*, **6**, 69–80 (1971).

4.16 Florence, A. L., Clamped circular rigid-plastic plates under central blast loading, *International Journal of Solids and Structures*, **2**, 319–35 (1966).

4.17 Conroy, M. F., Rigid-plastic analysis of a simply supported circular plate due to dynamic circular loading, *Journal of the Franklin Institute*, **288**(2), 121–35 (1969).

4.18 Florence, A. L., Response of circular plates to central pulse loading, *International Journal of Solids and Structures*, **13**, 1091–102 (1977).

4.19 Perzyna, P., Dynamic load carrying capacity of a circular plate, *Archiwum Mechaniki Stosowanej*, **10**(5), 635–47 (1958).

4.20 Symonds, P. S., Large plastic deformations of beams under blast type loading, *Proceedings of the Second US National Congress of Applied Mechanics, ASME*, 505–15 (1954).

4.21 Youngdahl, C. K., Correlation parameters for eliminating the effect of pulse shape on dynamic plastic deformation, *Journal of Applied Mechanics*, **37**, 744–52 (1970).

4.22 Krajcinovic, D., Dynamic analysis of clamped plastic circular plates, *International Journal of Mechanical Sciences*, **14**, 225–34 (1972).

4.23 Krajcinovic, D., Clamped circular rigid-plastic plates subjected to central blast loading, *Computers and Structures*, **2**, 487–96 (1972).

4.24 Stronge, W. J., Efficient pulse shapes to plastically deform beams, *Journal of Applied Mechanics*, **41**(3), 604–8 (1974).

4.25 Hopkins, H. G., On the plastic theory of plates, *Proceedings of the Royal Society of London, Series A (Mathematical and Physical Sciences)*, **241**, 153–79 (1957).

Chapter 5

5.1 Hodge, P. G., *Limit Analysis of Rotationally Symmetric Plates and Shells*, Prentice-Hall, Englewood Cliffs, NJ (1963).

5.2 Hodge, P. G., Impact pressure loading of rigid-plastic cylindrical shells, *Journal of the Mechanics and Physics of Solids*, **3**, 176–88 (1955).

5.3 Baker, W. E., The elastic–plastic response of thin spherical shells to internal blast loading, *Journal of Applied Mechanics*, **27**(E), 139–44 (1960).

5.4 Duffey, T., Significance of strain-hardening and strain-rate effects on the transient response of elastic–plastic spherical shells, *International Journal of Mechanical Sciences*, **12**, 811–25 (1970).

5.5 Flügge, W., *Stresses in Shells*, Springer, New York, 2nd edn (1973).

5.6 Ich, N. T. and Jones, N., The dynamic plastic behaviour of simply supported spherical shells, *International Journal of Solids and Structures*, **9**, 741–60 (1973).

5.7 Jones, N., Consistent equations for the large deflections of structures, *Bulletin of Mechanical Engineering Education*, **10**, 9–20 (1971).

5.8 Onat, E. T. and Prager, W., Limit analysis of shells of revolution, *Proceedings of the Royal Netherlands Academy of Science*, **B57**, 534–41 and 542–8 (1954).

5.9 Jones, N. and Ich, N. T., The load carrying capacities of symmetrically loaded shallow shells, *International Journal of Solids and Structures*, **8**, 1339–51 (1972).

5.10 Hodge, P. G. and Paul, B., Approximate yield conditions in dynamic plasticity, *Proceedings of the Third Midwestern Conference on Solid Mechanics*, University of Michigan, 29–47 (1957).

5.11 Sankaranarayanan, R., On the dynamics of plastic spherical shells, *Journal of Applied Mechanics*, **30**, 87–90 (1963).

5.12 Sankaranarayanan, R., On the impact pressure loading of a plastic spherical cap, *Journal of Applied Mechanics*, **33**, 704–6 (1966).

5.13 Walters, R. M. and Jones, N., An approximate theoretical study of the dynamic plastic behavior of shells, *International Journal of Non-Linear Mechanics*, **7**, 255–73 (1972).

5.14 Jones, N. and Walters, R. M., A comparison of theory and experiments on the dynamic plastic behaviour of shells, *Archives of Mechanics*, **24** (5–6), 701–14 (1972).

5.15 Jones, N., Giannotti, J. G. and Grassit, K. E., An experimental study into the dynamic inelastic behaviour of spherical shells and shell intersections, *Archiwum Budowy Maszyn*, **20**, 33–46 (1973).

5.16 Hodge, P. G., The influence of blast characteristics on the final deformation of circular cylindrical shells, *Journal of Applied Mechanics*, **23**, 617–24 (1956).

5.17 Hodge, P. G., Ultimate dynamic load of a circular cylindrical shell, *Proceedings of the Second Midwestern Conference on Solid Mechanics*, Lafayette, Ind., 150–77 (1956).

5.18 Hodge, P. G., The effect of end conditions on the dynamic loading of plastic shells, *Journal of the Mechanics and Physics of Solids*, **7**, 258–63 (1959).

5.19 Eason, G. and Shield, R. T., Dynamic loading of rigid-plastic cylindrical shells, *Journal of the Mechanics and Physics of Solids*, **4**, 53–71 (1956).

5.20 Kuzin, P. A. and Shapiro, G. S., On dynamic behaviour of plastic structures, *Proceedings of the Eleventh International Congress of Applied Mechanics*, Munich, 1964, ed. H. Gortler, Springer, New York, 629–35 (1966).

5.21 Nemirovsky, Y. V. and Mazalov, V. N., Dynamic behaviour of cylindrical shells strengthened with ring ribs – Pt I: Infinitely Long Shell, *International Journal of Solids and Structures*, **5**, 817–32 (1969).

5.22 Youngdahl, C. K., Correlation parameters for eliminating the effect of pulse shape on dynamic plastic deformation, *Journal of Applied Mechanics*, **37**, 744–52 (1970).

5.23 Youngdahl, C. K., Dynamic plastic deformation of circular cylindrical shells, *Journal of Applied Mechanics*, **39**, 746–50 (1972).

5.24 Youngdahl, C. K. and Krajcinovic, D., Dynamic plastic deformation of an infinite plate, *International Journal of Solids and Structures*, **22**, 859–81 (1986).

5.25 Youngdahl, C. K., Effect of pulse shape and distribution on the plastic deformation of a circular plate, *International Journal of Solids and Structures*, **23**, 1179–89 (1987).

5.26 Zhu, G., Huang, Y.-G., Yu, T. X. and Wang, R., Estimation of the plastic structural response under impact, *International Journal of Impact Engineering*, **4**(4), 271–82 (1986).

5.27 Duffey, T. and Krieg, R., The effects of strain-hardening and strain-rate sensitivity on the transient response of elastic-plastic rings and cylinders, *International Journal of Mechanical Sciences*, **11**, 825–44 (1969).

5.28 Perrone, N., Impulsively loaded strain hardened rate-sensitive rings and tubes, *International Journal of Solids and Structures*, **6**, 1119–32 (1970).

5.29 Jones, N., An approximate rigid-plastic analysis of shell intersections loaded dynamically, *Transactions of the ASME, Journal of Engineering for Industry*, **95**, 321–31 (1973).

5.30 Summers, A. B. and Jones, N., Some experiments on the dynamic plastic behaviour of shell intersections, *Nuclear Engineering and Design*, **26**, 274–81 (1974).

5.31 Jones, N., A literature review of the dynamic plastic response of structures, *The Shock and Vibration Digest*, **7**(8), 89–105 (1975). Recent progress in the dynamic plastic behaviour of structures, *The Shock and Vibration Digest*, pt I, **10**(9), 21–33 (1978); pt II, **10**(10), 13–19 (1978); pt III, **13**(10), 3–16 (1981); pt IV, **17**(2), 35–47 (1985).

Chapter 6

6.1 Horne, M. R., *Plastic Theory of Structures*, MIT Press, Cambridge, Mass. (1971).

6.2 Liu, J. and Jones, N., Experimental investigation of clamped beams struck transversely by a mass, *International Journal of Impact Engineering*, **6**(4), 303–35 (1987).

6.3 Jones, N. and Wierzbicki, T., A study of the higher modal dynamic plastic response of beams, *International Journal of Mechanical Sciences*, **18**, 533–42 (1976).

6.4 Jones, N. and Guedes Soares, C., Higher modal dynamic plastic behaviour of beams loaded impulsively, *International Journal of Mechanical Sciences*, **20**, 135–47 (1978).

6.5 Spencer, A. J. M., Dynamics of ideal fibre-reinforced rigid-plastic beams, *Journal of the Mechanics and Physics of Solids*, **22**, 147–59 (1974).

6.6 Jones, N., Dynamic behaviour of ideal fibre-reinforced rigid-plastic beams, *Journal of Applied Mechanics*, **43**, 319–24 (1976).

6.7 Nonaka, T., Shear and bending response of a rigid-plastic beam to blast-type loading, *Ingenieur-Archiv*, **46**, 35–52 (1977).

6.8 Symonds, P. S., Plastic shear deformations in dynamic load problems, *Engineering Plasticity*, ed. J. Heyman and F. A. Leckie, Cambridge University Press, Cambridge, 647–64 (1968).

6.9 Jones, N. and Song, B. Q., Shear and bending response of a rigid-plastic beam to partly distributed blast-type loading, *Journal of Structural Mechanics*, **14**(3), 275–320 (1986).

6.10 Jones, N. and de Oliveira, J. G., The influence of rotatory inertia and transverse shear on the dynamic plastic behaviour of beams, *Journal of Applied Mechanics*, **46**(2), 303–10 (1979).

6.11 de Oliveira, J. G. and Jones, N., A numerical procedure for the dynamic plastic response of beams with rotatory inertia and transverse shear effects, *Journal of Structural Mechanics*, **7**(2), 193–230 (1979).

6.12 de Oliveira, J. G., Beams under lateral projectile impact, *Proceedings of the ASCE, Journal of the Engineering Mechanics Division*, **108**(EM1), 51–71 (1982).

6.13 Sawczuk, A. and Duszek, M., A note on the interaction of shear and bending in plastic plates, *Archiwum Mechaniki Stosowanej*, **15**(3), 411–26 (1963).

6.14 Jones, N. and de Oliveira, J. G., Dynamic plastic response of circular plates with transverse shear and rotatory inertia, *Journal of Applied Mechanics*, **47**(1), 27–34 (1980).

6.15 Wang, A. J., The permanent deflection of a plastic plate under blast loading, *Journal of Applied Mechanics*, **22**, 375–6 (1955).

6.16 Kumar, A. and Reddy, V. V. K., Dynamic plastic response of circular plates with transverse shear, *Journal of Applied Mechanics*, **53**(4), 952–3 (1986).

6.17 Jones, N. and de Oliveira, J. G., Impulsive loading of a cylindrical shell with transverse shear and rotatory inertia, *International Journal of Solids and Structures*, **19**(3), 263–79 (1983).

6.18 Jones, N., The influence of large deflections on the behaviour of rigid-plastic cylindrical shells loaded impulsively, *Journal of Applied Mechanics*, **37**(2), 416–25 (1970). (See section without geometry changes.)

6.19 Martin, J. B., Impulsive loading theorems for rigid-plastic continua, *Proceedings of the ASCE, Journal of the Engineering Mechanics Division*, **90**(EM5), 27–42 (1964).

6.20 Morales, W. J. and Nevill, G. E., Lower bounds on deformations of dynamically loaded rigid-plastic continua, *AIAA Journal*, **8**(11), 2043–6 (1970).

6.21 Jones, N., Bounds on the dynamic plastic behaviour of structures including transverse shear effects, *International Journal of Impact Engineering*, **3**(4), 273–91 (1985).

6.22 Heyman, J., The full plastic moment of an I-beam in the presence of shear force, *Journal of the Mechanics and Physics of Solids*, **18**, 359–65 (1970).

6.23 de Oliveira, J. G. and Jones, N., Some remarks on the influence of transverse shear on the plastic yielding of structures, *International Journal of Mechanical Sciences*, **20**, 759–65 (1978).

6.24 Hodge, P. G., Interaction curves for shear and bending of plastic beams, *Journal of Applied Mechanics*, **24**, 453–6 (1957).

6.25 Ranshi, A. S., Chitkara, N. R. and Johnson, W., Plastic yielding of I-beams under shear, and shear and axial loading, *International Journal of Mechanical Sciences*, **18**, 375–85 (1976).

6.26 Neal, B. G., Effect of shear force on the fully plastic moment of an I-beam, *Journal of Mechanical Engineering Science*, **3**(3), 258–66 (1961).

6.27 Ilyushin, A. A., *Plasticité* (in French), Eyrolles, Paris (1956).

6.28 Shapiro, G. S., On yield surfaces for ideally plastic shells, *Problems of Continuum Mechanics*, SIAM, Philadelphia, 414–18 (1961).

6.29 Robinson, M., The effect of transverse shear stresses on the yield surface for thin shells, *International Journal of Solids and Structures*, **9**, 819–28 (1973).

6.30 Drucker, D. C., The effect of shear on the plastic bending of beams, *Journal of Applied Mechanics*, **23**, 509–14 (1956).

6.31 Reckling, K. A., Der Ebene Spannungszustand bei der plastischen Balkenbeigung, *Aus theorie und Praxis der Ingenieurwissenschaften*, W. Ernst und Sohn, Berlin, 39–46 (1971).

6.32 Menkes, S. B. and Opat, H. J., Broken beams, *Experimental Mechanics*, **13**, 480–6 (1973).

6.33 Jones, N., Plastic failure of ductile beams loaded dynamically, *Transactions of the ASME, Journal of Engineering for Industry*, **98**(B), 131–6 (1976).

6.34 Jouri, W. S. and Jones, N., The impact behaviour of aluminium alloy and mild steel double-shear specimens, *International Journal of Mechanical Sciences*, **30**(3/4), 153–72 (1988).

6.35 Jones, N., On the dynamic inelastic failure of beams, *Structural Failure*, ed. T. Wierzbicki and N. Jones, John Wiley, New York, 133–59 (1989).

6.36 Jones, N., Some comments on the dynamic plastic behaviour of structures, Keynote Address, *Proceedings of the International Symposium on Intense Dynamic Loading and Its Effects*, ed. Zheng Zhemin and Ding Jing, Science Press, Beijing, China, 49–71 (1986); and Pergamon Press (1988).

6.37 Recht, R. F. and Ipson, T. W., Ballistic perforation dynamics, *Journal of Applied Mechanics*, **30**, 384–90 (1963).

6.38 Liss, J., Goldsmith, W. and Kelly, J. M., A phenomenological penetration model of plates, *International Journal of Impact Engineering*, **1**(4), 321–41 (1983).

Chapter 7

7.1 Haythornthwaite, R. M., Beams with full end fixity, *Engineering*, **183**, 110–12 (1957).

7.2 Onat, E. T. and Haythornthwaite, R. M., The load-carrying capacity of circular plates at large deflection, *Journal of Applied Mechanics*, **23**, 49–55 (1956).

7.3 Sawczuk, A., Large deflections of rigid-plastic plates, *Proceedings of the Eleventh International Congress of Applied Mechanics*, 224–8 (1964).

7.4 Jones, N., A theoretical study of the dynamic plastic behaviour of beams and plates with finite-deflections, *International Journal of Solids and Structures*, **7**, 1007–29 (1971).

7.5 Jones, N. and Walters, R. M., Large deflections of rectangular plates, *Journal of Ship Research*, **15**(2), 164–71 and 288 (1971).

7.6 Taya, M. and Mura, T., Dynamic plastic behaviour of structures under impact loading investigated by the extended Hamilton's principle, *International Journal of Solids and Structures*, **10**, 197–209 (1974).

7.7 Gürkök, A. and Hopkins, H. G., The effect of geometry changes on the load carrying capacity of beams under transverse load, *SIAM Journal of Applied Mathematics*, **25**, 500–21 (1973).

7.8 Jones, N., Plastic behaviour of ship structures, *Transactions of the Society of Naval Architects and Marine Engineers*, **84**, 115–45 (1976).

7.9 Young, A. G., Ship plating loaded beyond the elastic limit, *Transactions of the Institution of Naval Architects*, **101**, 143–62 (1959).

7.10 Hodge, P. G., Limit analysis of rotationally symmetric plates and shells, Prentice-Hall, Englewood Cliffs, N.J. (1963).

7.11 Jones, N., Uran, T. O. and Tekin, S. A., The dynamic plastic behaviour of fully clamped rectangular plates, *International Journal of Solids and Structures*, **6**, 1499–512 (1970).

7.12 Jones, N. and Baeder, R. A., An experimental study of the dynamic plastic behaviour of rectangular plates, *Symposium on Plastic Analysis of Structures*, pub. Ministry of Education, Polytechnic Institute of Jassy, Romania, Civil Engineering Faculty, **1**, 476–97 (1972).

7.13 Jones, N., Griffin, R. N. and Van Duzer, R. E., An experimental study into the dynamic plastic behaviour of wide beams and rectangular plates, *International Journal of Mechanical Sciences*, **13**(8), 721–35 (1971).

7.14 Guedes Soares, C., A mode solution for the finite deflections of a circular plate loaded impulsively, *Rozprawy Inzynierskie, Engineering Transactions, Polish Akademia Nauk*, **29**(1), 99–114 (1981).

7.15 Florence, A. L., Circular plate under a uniformly distributed impulse, *International Journal of Solids and Structures*, **2**, 37–47 (1966).

7.16 Jones, N., Impulsive loading of a simply supported circular rigid plastic plate, *Journal of Applied Mechanics*, **35**(1), 59–65 (1968).

7.17 Jones, N., Consistent equations for the large deflections of structures, *Bulletin of Mechanical Engineering Education*, **10**(1), 9–20 (1971).

7.18 Symonds, P. S. and Wierzbicki, T., Membrane mode solutions for impulsively loaded circular plates, *Journal of Applied Mechanics*, **46**(1), 58–64 (1979).

7.19 Jones, N., Combined distributed loads on rigid-plastic circular plates with large deflections, *International Journal of Solids and Structures*, **5**, 51–64 (1969).

7.20 Jones, N., Finite deflections of a simply supported rigid-plastic annular plate loaded dynamically, *International Journal of Solids and Structures*, **4**, 593–603 (1968).

7.21 Jones, N., Damage estimates for plating of ships and marine vehicles, *International Symposium on Practical Design in Shipbuilding (PRADS)*, Society of Naval Architects of Japan, Tokyo, 121–8 (1977).

7.22 Jones, N., Slamming damage, *Journal of Ship Research*, **17**(2), 80–6 (1973).

7.23 Kaliszky, S., Approximate solutions for impulsively loaded inelastic structures and continua, *International Journal of Non-Linear Mechanics*, **5**, 143–58 (1970).

7.24 Kaliszky, S., Large deformations of rigid-viscoplastic structures under impulsive and pressure loading, *Journal of Structural Mechanics*, **1**(3), 295–317 (1973).

7.25 Kaliszky, S., Dynamic plastic response of structures, *Plasticity Today: Modelling Methods and Applications*, ed. A. Sawczuk and G. Bianchi, Elsevier Applied Science, London, 788–820 (1985).

7.26 Jones, N., Influence of in-plane displacements at the boundaries of rigid-plastic beams and plates, *International Journal of Mechanical Sciences* **15**, 547–61 (1973).

7.27 Duszek, M., Effect of geometry changes on the carrying capacity of cylindrical shells, *Bulletin de l'Academie Polonaise des Sciences, Séries des Sciences Techniques*, **13**(4), 183–91 (1965).

7.28 Jones, N., On the influence of large deflections on the behaviour of rigid-plastic cylindrical shells loaded impulsively, *Journal of Applied Mechanics*, **37**, 416–25 (1970).

7.29 Walters, R. M. and Jones, N., An approximate theoretical study of the dynamic plastic behaviour of shells, *International Journal of Non-Linear Mechanics*, **7**, 255–73 (1972).

7.30 Reid, S. R., Laterally compressed metal tubes as impact energy absorbers, *Structural Crashworthiness*, ed. N. Jones and T. Wierzbicki, Butterworths, London, 1–43 (1983).

7.31 Horne, M. R., *Plastic Theory of Structures*, MIT Press, Cambridge, Mass. (1971).

7.32 Yuhara, T., Fundamental study of wave impact loads on ship bow, *Journal of the Society of Naval Architects of Japan*, **137**, 240–5 (1975).

7.33 Menkes, S. B. and Opat, H. J., Broken beams, *Experimental Mechanics*, **13**, 480–6 (1973).

7.34 Nonaka, T., Some interaction effects in a problem of plastic beam dynamics, Parts 1–3, *Journal of Applied Mechanics*, **34**, 623–43 (1967).

7.35 Symonds, P. S. and Mentel, T. J., Impulsive loading of plastic beams with axial restraints, *Journal of the Mechanics and Physics of Solids*, **6**, 186–202 (1958).

7.36 Jones, N., Plastic failure of ductile beams loaded dynamically, *Transactions of the ASME, Journal of Engineering for Industry*, **98**(B1), 131–6 (1976).

7.37 Jones, N., On the dynamic inelastic failure of beams, *Structural Failure*, ed. T. Wierzbicki and N. Jones, John Wiley, New York Chapter 5 133–59 (1989).

7.38 Duffey, T. A., Dynamic rupture of shells, *Structural Failure*, ed. T. Wierzbicki and N. Jones, John Wiley, New York Chapter 6 161–92 (1989).

7.39 Atkins, A. G., Tearing of thin metal sheets, *Structural Failure*, ed. T. Wierzbicki and N. Jones, John Wiley, New York Chapter 4 107–32 (1989).

7.40 Yu, T. X., Zhang, D. J., Zhang, Y., and Zhou, Q., A study of the quasi-static tearing of thin metal sheets, *International Journal of Mechanical Sciences*, **30**(3/4), 193–202 (1988).

Chapter 8

8.1 Campbell, J. D., *Dynamic Plasticity of Metals*, Springer, Vienna and New York (1972).

8.2 Marsh, K. J. and Campbell, J. D., The effect of strain rate on the post-yield flow of mild steel, *Journal of the Mechanics and Physics of Solids*, **11**, 49–63 (1963).

8.3 Symonds, P. S. and Jones, N., Impulsive loading of fully clamped beams with finite plastic deflections and strain rate sensitivity, *International Journal of Mechanical Sciences*, **14**, 49–69 (1972).

8.4 Bodner, S. R. and Symonds, P. S., Experimental and theoretical investigation of the plastic deformation of cantilever beams subjected to impulsive loading, *Journal of Applied Mechanics*, **29**, 719–28 (1962).

8.5 Perrone, N., Crashworthiness and biomechanics of vehicle impact, *Dynamic Response of Biomechanical Systems*, ed. N. Perrone, ASME, 1–22 (1970).

8.6 Perzyna, P., Fundamental problems in viscoplasticity, *Advances in Applied Mechanics*, Academic Press, vol. 9, 243–377 (1966).

8.7 Campbell, J. D., Dynamic plasticity: macroscopic and microscopic aspects, *Materials Science and Engineering*, **12**, 3–21 (1973).

8.8 Duffy, J., Testing techniques and material behavior at high rates of strain,

Mechanical Properties at High Rates of Strain, ed. J. Harding, Institute of Physics Conference Series No. 47, 1–15 (1979).

8.9 Harding, J., *Testing Techniques at High Rates of Strain*, University of Oxford, Department of Engineering Science Report No. 1308/80 (March 1980).

8.10 Malvern, L. E., Experimental and theoretical approaches to characterisation of material behaviour at high rates of deformation, *Mechanical Properties at High Rates of Strain*, ed. J. Harding, Institute of Physics Conference Series No. 70, 1–20 (1984).

8.11 Goldsmith, W., *Impact*, Edward Arnold, London (1960).

8.12 Nicholas, T., Material behavior at high strain rates, *Impact Dynamics*, ed. J. A. Zukas, *et al.*, John Wiley, New York, ch. 8, 277–332 (1982).

8.13 Maiden, C. J. and Green, S. J., Compressive strain-rate tests on six selected materials at strain rates from 10^{-3} to 10^4 in/in/sec, *Journal of Applied Mechanics*, **33**, 496–504 (1966).

8.14 Hauser, F. E., Techniques for measuring stress–strain relations at high strain rates, *Experimental Mechanics*, **6**, 395–402 (1966).

8.15 Manjoine, M. J., Influence of rate of strain and temperature on yield stresses of mild steel, *Journal of Applied Mechanics*, **11**, 211–18 (1944).

8.16 Campbell, J. D. and Cooper, R. H., Yield and flow of low-carbon steel at medium strain rates, *Proceedings of the Conference on the Physical Basis of Yield and Fracture*, Institute of Physics and Physical Society, London, 77–87 (1966).

8.17 Symonds, P. S., *Survey of Methods of Analysis for Plastic Deformation of Structures under Dynamic Loading*, Brown University, Division of Engineering Report BU/NSRDC/1-67, June (1967).

8.18 Nicholas, T., Tensile testing of materials at high rates of strain, *Experimental Mechanics*, **21**, 177–85 (1981).

8.19 Harding, J. and Huddart, J., The use of the double-notch shear test in determining the mechanical properties of uranium at very high rates of strain, *Mechanical Properties at High Rates of Strain*, ed. J. Harding, Institute of Physics Conference Series No. 47, 49–61 (1979).

8.20 Klepaczko, J., The strain rate behaviour of iron in pure shear, *International Journal of Solids and Structures*, **5**, 533–48 (1969).

8.21 Nicholas, T. and Campbell, J. D., Shear-strain-rate effects in a high-strength aluminium alloy, *Experimental Mechanics*, **12**, 441–7 (1972).

8.22 Tsao, M. C. C. and Campbell, J. D., *Plastic Shear Properties of Metals and Alloys at High Strain Rates*, Oxford University, Department of Engineering Science Report No. 1055/73, March (1973).

8.23 Duffy, J., Some experimental results in dynamic plasticity, *Mechanical Properties at High Rates of Strain*, ed. J. Harding, Institute of Physics Conference Series, No. 21, 72–80 (1974).

8.24 Gerard, G. and Papirno, R., Dynamic biaxial stress–strain characteristics of aluminium and mild steel, *Transactions of the American Society for Metals*, **49**, 132–48 (1957).

8.25 Lindholm, U. S. and Yeakley, L. M., A dynamic biaxial testing machine, *Experimental Mechanics*, **7**, 1–7 (1967).

8.26 Ng, D. H. Y., Delich, M. and Lee, L. H. N., Yielding of 6061-T6 aluminium tubings under dynamic biaxial loadings, *Experimental Mechanics*, **19**, 200–6 (1979).

8.27 Hoge, K. G., Influence of strain rate on mechanical properties of 6061-T6 aluminium under uniaxial and biaxial states of stress, *Experimental Mechanics*, **6**, 204–11 (1966).

8.28 Lewis, J. L. and Goldsmith, W., A biaxial split Hopkinson bar for simultaneous torsion and compression, *Review of Scientific Instruments*, **44**, 811–13 (1973).

8.29 Rawlings, B., The dynamic behaviour of mild steel in pure flexure, *Proceedings of the Royal Society of London, Series A (Mathematical and Physical Sciences)*, **275**, 528–43 (1963).

8.30 Aspden, R. J. and Campbell, J. D., The effect of loading rate on the elasto-plastic flexure of steel beams, *Proceedings of the Royal Society of London, Series A (Mathematical and Physical Sciences)* **290**, 266–85 (1966).

8.31 Davies, R. G. and Magee, C. L., The effect of strain rate upon the bending behaviour of materials, *Transactions of the ASME, Journal of Engineering Materials and Technology*, **99**(H), 47–51 (1977).

8.32 Jones, N., Some remarks on the strain-rate sensitive behaviour of shells, *Problems of Plasticity*, ed. A. Sawczuk, Noordhoff, Groningen, vol. 2, 403–7 (1974).

8.33 Cowper, G. R. and Symonds, P. S., Strain hardening and strain-rate effects in the impact loading of cantilever beams, Brown University Division of Applied Mathematics Report No. 28, September (1957).

8.34 Symonds, P. S. and Chon, C. T., Approximation techniques for impulsive loading of structures of time-dependent plastic behaviour with finite-deflections, *Mechanical Properties of Materials at High Strain Rates*, Institute of Physics Conference Series No. 21, 299–316 (1974).

8.35 Forrestal, M. J. and Sagartz, M. J., Elastic–plastic response of 304 stainless steel beams to impulse loads, *Journal of Applied Mechanics*, **45**, 685–7 (1978).

8.36 Fung, Y. C., *A First Course in Continuum Mechanics*, Prentice-Hall, Englewood Cliffs, N.J. (1969).

8.37 Mendelson, A., *Plasticity: Theory and Application*, Macmillan, New York (1968).

8.38 Malvern, L. E., The propagation of longitudinal waves of plastic deformation in a bar of material exhibiting a strain rate effect, *Journal of Applied Mechanics*, **18**, 203–8 (1951).

8.39 Symonds, P. S., Viscoplastic behavior in response of structures to dynamic loading, *Behavior of Materials Under Dynamic Loading*, ed. N. J. Huffington, ASME, 106–124 (1965).

8.40 Gillis, P. P. and Kelly, J. M., On the determination of stress, strain, strain-rate relations from dynamic beam tests, *Journal of Applied Mechanics*, **36**, 632–34 (1969).

8.41 Perrone, N., A mathematically tractable model of strain-hardening, rate-sensitive plastic flow, *Journal of Applied Mechanics*, **33**, 210–11 (1966).

8.42 Perrone, N., On a simplified method for solving impulsively loaded structures of rate-sensitive materials, *Journal of Applied Mechanics*, **32**, 489–92 (1965).

8.43 Selby, S. M., *Standard Mathematical Tables*, 22nd edn, CRC Press, Cleveland, Ohio (1974).

8.44 Symonds, P. S., Approximation techniques for impulsively loaded structures of rate sensitive plastic behavior, *SIAM Journal of Applied Mathematics*, **25**, 462–73 (1973).

8.45 Perrone, N. and Bhadra, P., A simplified method to account for plastic rate sensitivity with large deformations, *Journal of Applied Mechanics*, **46**, 811–16 (1979).

8.46 Perrone, N. and Bhadra, P., Simplified large deflection mode solutions for impulsively loaded, viscoplastic, circular membranes, *Journal of Applied Mechanics*, **51**, 505–9 (1984).

8.47 Symonds, P. S. and Wierzbicki, T., Membrane mode solutions for impulsively loaded circular plates, *Journal of Applied Mechanics*, **46**, 58–64 (1979).

8.48 Bodner, S. R. and Symonds, P. S., Experiments on viscoplastic response of circular plates to impulsive loading, *Journal of the Mechanics and Physics of Solids*, **27**, 91–113 (1979).

8.49 Jones, N., Uran, T. O. and Tekin, S. A., The dynamic plastic behaviour of fully clamped rectangular plates, *International Journal of Solids and Structures*, **6**, 1499–512 (1970).

8.50 Jones, N., Giannotti, J. G. and Grassit, K. E., An experimental study into the dynamic inelastic behaviour of spherical shells and shell intersections, *Archiwum Budowy Maszyn*, **20** (1), 33–46 (1973).

8.51 Perrone, N., Impulsively loaded strain-rate-sensitive plates, *Journal of Applied Mechanics*, **34**, 380–4 (1967).

8.52 Hashmi, M. S. J., Strain rate sensitivity of a mild steel at room temperature and strain rates of up to 10^5 s^{-1}, *Journal of Strain Analysis*, **15**, 201–7 (1980).

8.53 Jones, N., Structural aspects of ship collisions, *Structural Crashworthiness*, Ed. N. Jones and T. Wierzbicki, Butterworths, London, 308–37 (1983).

8.54 Abramowicz, W. and Jones, N., Dynamic axial crushing of square tubes, *International Journal of Impact Engineering*, **2**, 179–208 (1984).

8.55 Haque, M. M. and Hashmi, M. S. J., Stress–strain properties of structural steel at strain rates of up to 10^5 per second at sub-zero, room and high temperatures, *Mechanics of Materials*, **3**, 245–56 (1984).

8.56 Robotnov, Y. N. and Suvorova, J. V., Dynamic problems for elastic–plastic solids with delayed yielding, *International Journal of Solids and Structures*, **7**, 143–59 (1971).

8.57 Klepaczko, J. R., Loading rate spectra for fracture initiation in metals, *Theoretical and Applied Fracture Mechanics*, **1**, 181–91 (1984).

8.58 Klepaczko, J. R., Fracture initiation under impact, *International Journal of Impact Engineering*, **3**(3), 191–210 (1985).

8.59 Kawata, K., Hashimoto, S. and Kurokawa, K., Analyses of high velocity tension of bars of finite length of BCC and FCC metals with their own constitutive equations, *High Velocity Deformation of Solids*, ed. K. Kawata and J. Shioiri, Springer, New York, 1–15 (1977).

8.60 Fyfe, I. M. and Rajendran, A. M., Dynamic pre-strain and inertia effects on the fracture of metals, *Journal of the Mechanics and Physics of Solids*, **28**, 17–26 (1980).

8.61 Rajendran, A. M. and Fyfe, I. M., Inertia effects on the ductile failure of thin rings, *Journal of Applied Mechanics*, **49**, 31–6 (1982).

8.62 Regazzoni, G., Montheillet, F., Influence of strain rate on the flow stress and ductility of copper and tantalum at room temperature, *Mechanical Properties at High Rates of Strain*, ed. J. Harding, Institute of Physics Conference Series No. 70, 63–70 (1984).

8.63 Soroushian, P. and Choi, K-B., Steel mechanical properties at different strain rates, *Transactions of the ASCE, Journal of Structural Engineering*, **113**(4), 663–72 (1987).

Chapter 9

9.1 Abramowicz, W. and Jones, N., Dynamic axial crushing of circular tubes, *International Journal of Impact Engineering*, **2**, 263–81 (1984).

9.2 Pugsley, Sir A., The crumpling of tubular structures under impact conditions, *Proceedings of the Symposium on The Use of Aluminium in Railway Rolling Stock*, Institute of Locomotive Engineers, The Aluminium Development Association, London, 33–41 (1960).

9.3 Abramowicz, W. and Jones, N., Dynamic axial crushing of square tubes, *International Journal of Impact Engineering*, **2**, 179–208 (1984).

9.4 Ezra, A. A. and Fay, R. J., An assessment of energy absorbing devices for prospective use in aircraft impact situations, *Dynamic Response of Structures*, ed. G. Herrmann and N. Perrone, Pergamon, New York, 225–46 (1972).

9.5 Mallock, A., Note on the instability of tubes subjected to end pressure, and on the folds in a flexible material, *Proceedings of the Royal Society of London, Series A (Mathematical and Physical Sciences)*, **81**, 388–93 (1908).

9.6 Mamalis, A. G. and Johnson, W., The quasi-static crumpling of thin-walled circular cylinders and frusta under axial compression, *International Journal of Mechanical Sciences*, **25**, 713–32 (1983).

9.7 Abramowicz, W. and Jones, N., Dynamic progressive buckling of circular and square tubes, *International Journal of Impact Engineering*, **4**, 243–70 (1986).

9.8 Alexander, J. M., An approximate analysis of the collapse of thin cylindrical shells under axial loading, *Quarterly Journal of Mechanics and Applied Mathematics*, **13**, 10–15 (1960).

9.9 Pugsley, Sir A. and Macaulay, M., The large scale crumpling of thin cylindrical columns, *Quarterly Journal of Mechanics and Applied Mathematics*, **13**, 1–9 (1960).

9.10 Macaulay, M. A. and Redwood, R. G., Small scale model railway coaches under impact, *The Engineer*, 1041–6, 25 Dec. (1964).

9.11 Thornton, P. H., Mahmood, H. F. and Magee, C. L., Energy absorption by structural collapse, *Structural Crashworthiness*, ed. N. Jones and T. Wierzbicki, Butterworths, London, 96–117 (1983).

9.12 Abramowicz, W., The effective crushing distance in axially compressed thin-walled metal columns, *International Journal of Impact Engineering*, **1**, 309–17 (1983).

9.13 Campbell, J. D. and Cooper, R. H., Yield and flow of low-carbon steel at medium strain rates, *Proceedings of the Conference on the Physical Basis of Yield and Fracture*, Institute of Physics and Physical Society, 77–87 (1966).

9.14 Hayduk, R. J. and Wierzbicki, T., Extensional collapse modes of structural members, *Computers and Structures*, **18**(3), 447–58 (1984).

9.15 Wierzbicki, T. and Abramowicz, W., On the crushing mechanics of thin-walled structures, *Journal of Applied Mechanics*, **50**, 727–34 (1983).

9.16 Wierzbicki, T., Crushing behaviour of plate intersections, *Structural Crashworthiness*, ed. N. Jones and T. Wierzbicki, Butterworths, London, 66–95 (1983).

9.17 Wierzbicki, T., Molnar, C. and Matolscy, M., Experimental–theoretical correlation of dynamically crushed components of bus frame structures, *Proceedings of the Seventeenth International FISITA Congress*, Budapest, June 1978.

9.18 Coppa, A. P., New ways to soften shock, *Machine Design*, 130–40, 28 March 1968.

9.19 Shaw, M. C., Designs for safety: the mechanical fuse, *Mechanical Engineering*, ASME, **94**, 23–9, April 1972.

9.20 Johnson, W., *Impact Strength of Materials*, Edward Arnold, London and Crane, Russak, New York (1972).

9.21 Johnson, W. and Mamalis, A. G., *Crashworthiness of Vehicles*, MEP, London (1978).

9.22 Johnson, W. and Reid, S. R., Metallic energy dissipating systems, *Applied Mechanics Reviews*, **31**, 277–88 (1978). Update to this article in **39**, 315–19 (1986).

9.23 Macaulay, M. A., *Introduction to Impact Engineering*, Chapman & Hall, London (1987).

9.24 Singley, G. T., Survey of rotary-wing aircraft crashworthiness, *Dynamic Response of Structures*, ed. G. Herrmann and N. Perrone, Pergamon, Oxford, 179–223 (1972).

9.25 Pugsley, Sir A. and Macaulay, M. A., Cars in collision-safe structures, *New Scientist*, **78**, No. 1105, 596–8, 1 June 1978.

9.26 Thornton, P. H., Energy absorption in composite structures, *Journal of Composite Materials*, **13**, 247–62 (1979).

9.27 Perrone, N., Dynamic plastic energy absorption in vehicle impact, *Rozprawy Inzynierskie, Engineering Transactions*, **29**(1), 83–97 (1981).

9.28 Jones, N. and Wierzbicki, T. (eds) *Structural Crashworthiness*, Butterworths, London (1983).

9.29 Hull, D., Impact response of structural composites, *Metals and Materials*, **1**, 35–8, January (1985).

9.30 Grundy, J. D., Blears, J. and Sneddon, B. C., Assessment of crash-worthy car materials, *Chartered Mechanical Engineer*, **32**(4), 31–5 (1985).

9.31 Scott, G. A., The development of a theoretical technique for rail vehicle structural crashworthiness, *Proceedings of the Institution of Mechanical Engineers*, **201**(D2), 123–8 (1987).

9.32 Scholes, A., Railway passenger vehicle design loads and structural crashworthiness, *Proceedings of the Institution of Mechanical Engineers*, **201**(D3), 201–7 (1987).

9.33 Miles, J. C., Molyneaux, T. C. K. and Dowler, H. J., Analysis of the forces on a nuclear fuel transport flask in an impact by a train, *Proceedings of the Institution of Mechanical Engineers*, **201**(A1), 55–68 (1987).

9.34 Jones, N. and Wierzbicki, T., Dynamic plastic failure of a free–free beam, *International Journal of Impact Engineering*, **6**(3), 225–40 (1987).

9.35 Wierzbicki, T., Crushing analysis of metal honeycombs, *International Journal of Impact Engineering*, **1**(2), 157–74 (1983).

9.36 Snyder, R. G., Human impact tolerance, SAE paper 700398, *International Automobile Safety Conference Compendium*, SAE, 712–82 (1970).

9.37 King, A. I., Human tolerance limitations related to aircraft crashworthiness, *Dynamic Response of Structures*, ed. G. Herrmann and N. Perrone, Pergamon, New York, 247–63 (1972).

9.38 Huston, R. L. and Perrone, N., Dynamic response and protection of the human body and skull in impact situations, *Perspectives in Biomechanics*, ed. H. Reul, D. N. Ghista and G. Rau, Harwood, vol. 1, pt B, 531–71 (1978).

9.39 Perrone, N., Biomechanical problems related to vehicle impact, *Biomechanics: Its Foundations and Objectives*, ed. Y. C. Fung, N. Perrone and M. Anliker, Prentice-Hall, Englewood Cliffs, N.J., 567–83 (1972).

9.40 Perrone, N., Biomechanical and structural aspects of design for vehicle impact, *Human Body Dynamics, Impact, Occupational and Athletic Aspects*, ed. D. N. Ghista, Clarendon, Oxford, 181–200 (1982).

9.41 Johnson, W., Mamalis, A. G. and Reid, S. R., Aspects of car design and human injury, *Human Body Dynamics, Impact, Occupational and Athletic Aspects*, ed. D. N. Ghista, Clarendon, Oxford, 164–80 (1982).

9.42 Currey, J. D., Changes in the impact energy absorption of bone with age, *Journal of Biomechanics*, **12**, 459–69 (1979).

9.43 King, A. I., Crash course, *Mechanical Engineering*, ASME, **108**(6), 58–61 (1986).

9.44 Goldsmith, W., Current controversies in the stipulation of head injury criteria, *Journal of Biomechanics*, **14**, 883–4 (1981).

9.45 Andrews, K. R. F., England, G. L. and Ghani, E., Classification of the axial collapse of cylindrical tubes under quasi-static loading, *International Journal of Mechanical Sciences*, **25**(9–10), 687–96 (1983).

9.46 Wierzbicki, T. and Bhat, S. U., A moving hinge solution for axisymmetric crushing of tubes, *International Journal of Mechanical Sciences*, **28**(3), 135–51 (1986).

9.47 Andronicou, A. and Walker, A. C., A plastic collapse mechanism for cylinders under uniaxial end compression, *Journal of Constructional Steel Research*, **1**(4), 23–34 (1981).

9.48 Abramowicz, W. and Wierzbicki, T., Axial crushing of multi-corner sheet metal columns, *Journal of Applied Mechanics*, **111**, 113–20 (1989).

9.49 Thornton, P. H., *Energy absorption by foam-filled structures*, SAE Paper 800081, SAE Congress, Detroit (1980).

9.50 Reid, S. R, Reddy, T. Y. and Gray, M. D., Static and dynamic axial crushing of foam-filled sheet metal tubes, *International Journal of Mechanical Sciences*, **28**(5), 295–322 (1986).

9.51 Reid, S. R. and Reddy, T. Y., Axial crushing of foam-filled tapered sheet metal tubes, *International Journal of Mechanical Sciences*, **28**(10), 643–56 (1986).

9.52 Johnson, W., Soden, P. D. and Al-Hassani, S. T. S., Inextensional collapse of thin-walled tubes under axial compression, *Journal of Strain Analysis*, **12**, 317–30 (1977).

Chapter 10

10.1 Goodier, J. N., Dynamic plastic buckling, *Proceedings of the International Conference on Dynamic Stability of Structures*, ed. G. Herrmann, Pergamon, New York, 189–211 (1967).

10.2 Abramowicz, W. and Jones, N., Dynamic axial crushing of circular tubes, *International Journal of Impact Engineering*, **2**(3), 263–81 (1984).

10.3 Abrahamson, G. R. and Goodier, J. N., Dynamic flexural buckling of rods within an axial compression wave, *Journal of Applied Mechanics*, **33**(2), 241–47 (1966).

10.4 Jones, N. and Okawa, D. M., Dynamic plastic buckling of rings and cylindrical shells, *Nuclear Engineering and Design*, **37**, 125–47 (1976).

10.5 Timoshenko, S. P. and Gere, J. M., *Theory of Elastic Stability*, 2nd edn, McGraw-Hill, New York (1961).

10.6 Al-Hassani, S. T. S., The plastic buckling of thin-walled tubes subject to magnetomotive forces, *Journal of Mechanical Engineering Science*, **16**, 59–70 (1974).

10.7 Florence, A. L. and Abrahamson, G. R., Critical velocity for collapse of viscoplastic cylindrical shells without buckling, *Transactions of the ASME, Journal of Applied Mechanics*, **44**(1), 89–94 (1977).

10.8 Jones, N., Consistent equations for the large deflections of structures, *Bulletin of Mechanical Engineering Education*, **10**, 9–20 (1971).

10.9 Lindberg, H. E. and Florence, A. L., *Dynamic pulse buckling: theory and experiment*, SRI International Report to Defense Nuclear Agency, Washington, DC, February 1982 and Martinus Nijhoff publishers, Norwell, Mass., 1987.

10.10 *The Flixborough Disaster, Report of the Court of Enquiry*, Department of Employment, HMSO, London (1975).

10.11 Abrahamson, G. R. and Goodier, J. N., Dynamic plastic flow buckling of a cylindrical shell from uniform radial impulse, *Proceedings of the Fourth US National Congress of Applied Mechanics*, ASME, Berkeley, 939–50 (1962).

10.12 Dym, C. L., *Stability Theory and its Applications to Structural Mechanics*, Noordhoff, Leyden (1974).

10.13 Brush, D. O. and Almroth, B. O., *Buckling of Bars, Plates and Shells*, McGraw-Hill, New York (1975).

10.14 Florence, A. L., Dynamic buckling of viscoplastic cylindrical shells, *Inelastic Behaviour of Solids*, ed. M. F. Kanninen, W. F. Adler, A. R. Rosenfield and R. I. Jaffee, McGraw-Hill, New York, 471–99 (1970).

10.15 Wojewodzki, W., Buckling of short viscoplastic cylindrical shells subjected to radial impulse, *International Journal of Non-Linear Mechanics*, **8**, 325–43 (1973).

10.16 Pugsley, A., The crumpling of tubular structures under impact conditions, *Proceedings of the Symposium on the Use of Aluminium in Railway Rolling Stock*, Institution of Locomotive Engineers, The Aluminium Development Association, London, 33–41 (1960).

10.17 Florence, A. L. and Goodier, J. N., Dynamic plastic buckling of cylindrical shells in sustained axial compressive flow, *Transactions of the ASME, Journal of Applied Mechanics*, **35**(1), 80–6 (1968).

10.18 Johnson, W. and Mellor, P. B., *Engineering Plasticity*, Van Nostrand Reinhold, London (1973).

10.19 Vaughan, H., The response of a plastic cylindrical shell to axial impact, *ZAMP*, **20**, 321–8 (1969).

10.20 Wojewodzki, W., Dynamic buckling of a visco-plastic cylindrical shell subjected to axial impact, *Archives for Mechanics*, **23**(1), 73–91 (1971).

10.21 Jones, N. and Papageorgiou, E. A., Dynamic axial plastic buckling of stringer stiffened cylindrical shells, *International Journal of Mechanical Sciences*, **24**, 1–20 (1982), and Dynamic plastic buckling of stiffened cylindrical shells, *Proceedings of the Institute of Acoustics*, Spring Conference, Newcastle upon Tyne, 29–32 (1981).

10.22 Abrahamson, G. R., Critical velocity for collapse of a shell of circular cross section without buckling, *Transactions of the ASME, Journal of Applied Mechanics*, **41**(2), 407–11 (1974).

10.23 Jones, N., Dynamic elastic and inelastic buckling of shells, *Developments in Thin-Walled Structures*, vol. 2, ed. J. Rhodes and A. C. Walker, Elsevier Applied Science, London and New York, 49–91 (1984).

10.24 Jones, N. and Ahn, C. S., Dynamic buckling of complete rigid-plastic spherical shells, *Journal of Applied Mechanics*, **41**, 609–14 (1974).

10.25 Jones, N. and Ahn, C. S., Dynamic elastic and plastic buckling of complete spherical shells, *International Journal of Solids and Structures*, **10**, 1357–74 (1974).

10.26 Song, B. Q. and Jones, N., Dynamic buckling of elastic-plastic complete spherical shells under step loading, *International Journal of Impact Engineering*, **1**(1), 51–71 (1983).

10.27 Ishizaki, T. and Bathe, K-J., On finite element large displacement and elastic-plastic dynamic analysis of shell structures, *Computers and Structures*, **12**, 309–18 (1980).

10.28 Wojewodzki, W. and Lewinski, P., Viscoplastic axisymmetrical buckling of spherical shell subjected to radial pressure impulse, *Engineering Structures*, **3**, 168–74 (1981).

10.29 Bukowski, R. and Wojewodzki, W., Dynamic buckling of viscoplastic spherical shell, *International Journal of Solids and Structures*, **20**(8), 761–76 (1984).

10.30 Kao, R., Nonlinear dynamic buckling of spherical caps with initial imperfections, *Computers and Structures*, **12**, 49–63 (1980).

10.31 Jones, N. and dos Reis, H. L. M., On the dynamic buckling of a simple elastic-plastic model, *International Journal of Solids and Structures*, **16**, 969–89 (1980).

10.32 Wang Ren, Han Mingbao, Huang Zhuping and Yan Qingchun, An experimental study on the dynamic axial plastic buckling of cylindrical shells, *International Journal of Impact Engineering*, **1**(3), 249–56 (1983).

10.33 Lindberg, H. E. and Kennedy, T. C., Dynamic plastic pulse buckling beyond strain-rate reversal, *Journal of Applied Mechanics*, **42**, 411–16 (1975).

10.34 Lee, L. H. N., Bifurcation and uniqueness in dynamics of elastic-plastic continua, *International Journal of Engineering Science*, **13**, 69–76 (1975).

10.35 Lee, L. H. N., Quasi-bifurcation in dynamics of elastic-plastic continua, *Journal of Applied Mechanics*, **44**, 413–18 (1977).

10.36 Lee, L. H. N., On dynamic stability and quasi-bifurcation, *International Journal of Nonlinear Mechanics*, **16**, 79–87 (1981).

10.37 Lee, L. H. N., Dynamic buckling of an inelastic column, *International Journal of Solids and Structures*, **17**, 271–9 (1981).

10.38 Funk, G. E. and Lee, L. H. N., Dynamic buckling of inelastic spherical shells, *Transactions of the ASME, Journal of Pressure Vessel Technology*, **104**, 79–87 (1982).

10.39 Symonds, P. S. and Yu, T. X., Counterintuitive behaviour in a problem of elastic-plastic beam dynamics, *Journal of Applied Mechanics*, **52**(3), 517–22 (1985).

Chapter 11

11.1 Young, D. F. and Murphy, G., Dynamic similitude of underground structures, *Proceedings of the ASCE, Journal of the Engineering Mechanics Division*, **90** (EM3), 111–33 (1964).

11.2 Duffey, T. A., Scaling laws for fuel capsules subjected to blast, impact and thermal loading, SAE paper 719107, *Proceedings of the Intersociety Energy Conversion Engineering Conference*, 775–86 (1971).

11.3 Barr, P., Studies of the effects of missile impacts on structures, *Atom*, No. 318, 6 pp., April 1983.

11.4 Jones, N., Structural aspects of ship collisions, *Structural Crashworthiness*, ed. N. Jones and T. Wierzbicki, Butterworths, London, 308–37 (1983).

11.5 Huffington, N. J. and Wortman, J. D., Parametric influences on the response of structural shells, *Transactions of the ASME, Journal of Engineering for Industry*, **97**(B), 1311–16 (1975).

11.6 Johnson, W., *Impact Strength of Materials*, Edward Arnold, London and Crane Russak, New York (1972).

11.7 Young, D. F., Basic principles and concepts of model analysis, *Experimental Mechanics*, **11**, 325–36 (1971).

11.8 Booth, E., Collier, D. and Miles, J., Impact scalability of plated steel structures, *Structural Crashworthiness*, ed. N. Jones and T. Wierzbicki, Butterworths, London, 136–74 (1983).

11.9 Nevill, G. E., *Similitude Studies of Re-entry Vehicle Response to Impulsive Loading*, Southwest Research Institute Report AF SWC-TDR-63-1, vol. 1, March 1963.

11.10 Ashby, M. F. and Jones, D. R. H., *Engineering Materials: An Introduction to their Properties and Applications*, Pergamon, Oxford (1980).

11.11 Jones, N., Similarity principles in structural mechanics, *International Journal of Mechanical Engineering Education*, **2**, 1–10 (1974).

11.12 Barr, D. I. H., A survey of procedures for dimensional analysis, *International Journal of Mechanical Engineering Education*, **11**, 147–59 (1983).

11.13 Hunsaker, J. C. and Rightmire, B. G., *Engineering applications of fluid mechanics*, McGraw-Hill, New York (1947).

11.14 Duffey, T. A., Cheresh, M. C. and Sutherland, S. H., Experimental verification of scaling laws for punch-impact-loaded structures, *International Journal of Impact Engineering*, **2**(1), 103–17 (1984).

11.15 Johnson, W., Structural damage in airship and rolling stock collisions, *Structural Crashworthiness*, ed. N. Jones and T. Wierzbicki, Butterworths, London, 417–39 (1983).

11.16 Atkins, A. G. and Caddell, R. M., The laws of similitude and crack propagation, *International Journal of Mechanical Sciences*, **16**, 541–8 (1974).

11.17 Mai, Y. W. and Atkins, A. G., Scale effects and crack propagation in non-linear elastic structures, *International Journal of Mechanical Sciences*, **17**, 673–5 (1975).

11.18 Mai, Y. W. and Atkins, A. G., Crack propagation in non-proportionally scaled elastic structures, *International Journal of Mechanical Sciences*, **20**, 437–49 (1978).

11.19 Richards, C. W., *Engineering Materials Science*, Wadsworth, Belmont, Cal. (1961).

11.20 Kendall, K., The impossibility of comminuting small particles by compression, *Nature*, **272**, 710–11, 20 April 1978.

11.21 Kendall, K., Interfacial cracking of a composite, Part 3 Compression, *Journal of Materials Science*, **11**, 1267–9 (1976).

11.22 Kendall, K., Complexities of compression failure, *Proceedings of the Royal Society of London, Series A (Mathematical and Physical Sciences)*, **361**, 245–63 (1978).

11.23 Gurney, C. and Hunt, J., Quasi-static crack propagation, *Proceedings of the Royal Society of London, Series A (Mathematical and Physical Sciences)*, **299**, 508–24 (1967).

11.24 Jones, N., Scaling of inelastic structures loaded dynamically, *Structural Impact and Crashworthiness*, vol. 1, ed. G. A. O. Davies, Elsevier Applied Science, London, 45–74 (1984).

11.25 Puttick, K. E., The correlation of fracture transitions, *Journal of Physics D (Applied Physics)*, **13**, 2249–62 (1980).

11.26 Puttick, K. E. and Yousif, R. H., Indentation fracture transitions in polymethylmethacrylate, *Journal of Physics D (Applied Physics)*, **16**, 621–33 (1983).

11.27 Atkins, A. G. and Mai, Y. W., *Elastic and plastic fracture*, Ellis Horwood, Chichester and John Wiley, New York (1984).

11.28 *The resistance to impact of spent Magnox fuel transport flasks*, MEP, London (1985).

11.29 Jones, N., Jouri, W. and Birch, R., On the scaling of ship collision damage, *International Maritime Association of the East Mediterranean, Third International Congress on Marine Technology*, Athens, vol. 2, 287–94 (1984).

11.30 Jones, N. and Jouri, W. S., A study of plate tearing for ship collision and grounding damage, *Journal of Ship Research*, **31**(4), 253–68 (1987).

11.31 Emori, R. I., and Schuring, D. J., *Scale Models in Engineering*, Pergamon, Oxford (1977).

11.32 Szücs, E., *Similitude and Modelling*, Elsevier, Amsterdam (1980).

11.33 Jouri, W. S. and Jones, N., The impact behaviour of aluminium alloy and mild steel double-shear specimens, *International Journal of Mechanical Sciences*, **30**(3/4), 153–72 (1988).

11.34 Jouri, W. S. and Jones, N., Scaling and impact behaviour of ductile double-shear specimens, University of Liverpool, Impact Research Centre Report ES/45/89 (1989).

11.35 Duffey, T. A., Cheresh, M. C. and Sutherland, S. H., Experimental verification of scaling laws for punch-impact-loaded structures, *International Journal of Impact Engineering* **2**(1), 103–17 (1984).

11.36 Thornton, P. H., Static and dynamic collapse characteristics of scale model corrugated tubular sections, *Transactions of the ASME, Journal of Engineering Materials and Technology*, **97**(H4), 357–62 (1975).

11.37 Calladine, C. R., An investigation of impact scaling theory, *Structural Crashworthiness*, ed. N. Jones and T. Wierzbicki, Butterworths, London, 169–74 (1983).

11.38 Calladine, C. R. and English, R. W., Strain-rate and inertia effects in the collapse of two types of energy-absorbing structure, *International Journal of Mechanical Sciences*, **26**(11/12), 689–701 (1984).

Appendices

A.1 Young, D. F., Basic principles and concepts of model analysis, *Experimental Mechanics*, **11**, 325–36 (1971).

A.2 Gibbings, J. C., A logic of dimensional analysis, *Journal of Physics A (Mathematical and General)*, **15**, 1991–2002 (1982).

A.3 Gibbings, J. C. (ed.), *The systematic experiment*, Cambridge University Press, Cambridge (1986).

Answers to selected problems

Chapter 1

1.2 $\frac{9}{8}$

1.3 $M/M_0 = 1 - (\eta/H)^2/3$, $\kappa/(2\sigma_0/EH) = H/\eta$, where η is total depth of central elastic core. 1.35 per cent.

1.5   Exact collapse load $= 2M_0L/a(2L - a)$.

1.6   Exact collapse load $= 4M_0L/a(2L - a)$.

1.8   Exact collapse pressure $= 4M_0/(2LL_1 - L_1^2)$.

1.9   Exact collapse pressure $= 2M_0/(\eta^2 - a^2)$, where $\eta = a + 2L_1 - L_1^2/L - aL_1/L$.

1.12 $Q_{max}/Q_0 = H/2L$.

Chapter 2

2.3 $3\sigma_0(H/R)^2$.

2.5 $\dot{w} = A \log_e r + B$, $\dot{D} = -M_0A/r^2$ (A, B constants of integration).

2.6 $Q_r = -3M_0r/R^2$, $|Q_r|_{max}/Q_0 \cong 1.5H/R$.

Chapter 3

3.1   272.7 m/s and 150 m/s at $\xi = \xi_0$ and $\xi = 0$, respectively.

3.2 $w = p_0\tau^2\{2\eta(2 + \bar{x})(1 - \bar{x})/3 - 1\}/2m$, $0 \leqslant \bar{x} \leqslant \bar{\xi}_0$,

        $w = p_0\tau^2\{1/(1 - \bar{\xi}_0) + 2\eta(1 + 2\bar{\xi}_0)/3\}(1 - \bar{x})/2m$, $\bar{\xi}_0 \leqslant \bar{x} \leqslant 1$,

        where $\bar{x} = x/L$.

3.3 w_f is given by equation (3.78).

3.4 W_f is given by equation (3.80).

3.5   Equation (3.101).

3.9   (a) $GV_0^2L/2\sigma_0BtH$,

        (b) $GV_0^2/8\sigma_0BtH$, $V_0/4L$,

        (c) GV_0L/M_p.

Chapter 4

4.4 (a) See §§ 4.4 and 4.5 and Figures 4.8 and 4.10.

(b) $(8M_0B/\mu R^2)^{1/2}$.

(c) For $(8M_0B/\mu R^2)^{1/2} \leqslant V_0 \leqslant (12M_0B/\mu R^2)^{1/2}$ the strain gauge would detect rupture at the supports, while for $V_0 \geqslant (12M_0B/\mu R^2)^{1/2}$ rupture would occur before being detected by the strain gauge.

4.5 See footnote to equation (4.143).

4.6 (a) $P_e = p_0$, $t_c = \tau/2$.

(c) Equation (4.48) can be written $w_f = I^2(1 - p_c/P_e)(1 - r/R)/\mu p_c$ where for the triangular pressure pulse $I = p_0 t_0/2$ and $P_e = 3p_0/4$, provided $t_0 < t_f \cong I/p_c$.

Chapter 5

5.6 $0 \leqslant t \leqslant t_1$. See § 5.6.3(a) with t_1 given by equation (5.105)

$t_1 \leqslant t \leqslant \tau$. $w = (2N_0/R - p_0)(t - t_1)^2/2\mu - RN_0(1 - v)/EH -$
$\qquad\qquad - \{b(t - t_1)/a\} \sin at_1$.

$\tau \leqslant t \leqslant T_1$. w is given by equation (5.109) with

$\qquad\qquad A_3 = -p_0\tau/\mu - (b/a) \sin at_1 - (2N_0/R - p_0)t_1/\mu$,

$\qquad\qquad B_3 = p_0(\tau^2 - t_1^2)/2\mu + N_0 t_1^2/\mu R - RN_0(1 - v)/EH$
$\qquad\qquad\qquad + (bt_1/a) \sin at_1$

and $T_1 = t_1 + p_0 R(\tau - t_1 + a^{-1} \sin at_1)/2N_0$.

$t \geqslant T_1$. See § 5.6.4(e) with correct continuity conditions at $t = T_1$.

5.7 (a) $2.47M_0/R^2 \leqslant p_c \leqslant 8M_0/R^2$. As noted in the question, the solution for the two-moment limited interaction yield condition reduces to equation (2.18).

(b) $\sqrt{3}H$, $\sqrt{3}/\{1 + (H/R)^2\}$.

Chapter 6

6.2 $Q/Q_0 = -(\eta + 3)(H/4L)$.

6.6 $w = mL^2 V_0^2 \{2 - (x/L)^2 - x/L\}/6M_0$, $\qquad 0 \leqslant x \leqslant \xi_0$.

$w = mL^2 V_0^2 \{3 + 2v(4v - 3)(1 - x/L)\}/16v^2 M_0$, $\qquad \xi_0 \leqslant x \leqslant L$.

6.8 $Q = -6mM_0/\{M\alpha(2 + \alpha)\} \to -\infty$ at $t = 0$.

6.13 $V_0^s = (4kQ_0H/\mu R)^{1/2}$.

6.14 $V_0^s = \{4kM_0H(v + c^2)/\mu L^2\}^{1/2}$.

Chapter 7

7.2 $\epsilon = du/dx + (dw/dx)^2/2$, $\kappa = -d^2w/dx^2$.

7.8 Equations (7.60a, b) give the finite-deflection behaviour of a simply supported beam.

7.12   (b) $W_f/H = [\{1 + 2\eta(\eta - 1)(1 - \cos a\tau)\}^{1/2} - 1]$ with $\eta = p_0/p_c$ where
$p_c = 12M_0/R^2$.
$W_f/H = \{(1 + \lambda/6)^{1/2} - 1\}$ for impulsive loading.

   (c) $W_f/H = (\lambda/6)^{1/2}$.

Chapter 8

8.3 $0 \leqslant t \leqslant \tau$. $\dot{\epsilon}_\theta = \dot{\epsilon}_\phi = (p_0 - p_c)t/\mu R$, $\dot{\epsilon}_r = -2\dot{\epsilon}_\theta$, $\dot{\epsilon}_e = 2\dot{\epsilon}_\theta$,
$\sigma'_e/\sigma_0 = 1 + \{2(p_0 - p_c)t/\mu RD\}^{1/q}$.
$\tau \leqslant t \leqslant T$. $\dot{\epsilon}_\theta = \dot{\epsilon}_\phi = (p_0\tau - p_ct)/\mu R$, $\dot{\epsilon}_r = -2\dot{\epsilon}_\theta$, $\dot{\epsilon}_e = 2\dot{\epsilon}_\theta$,
$\sigma'_e/\sigma_0 = 1 + \{2(p_0\tau - p_ct)/\mu RD\}^{1/q}$.

8.4 $\dot{\epsilon}_e = (2a/\sqrt{3})(\mu V_0^2 H/6M_0) \sin at \cos at$,
$\sigma'_e/\sigma_0 = 1 + \{(2a/\sqrt{3}D)(\mu V_0^2 H/6M_0) \sin at \cos at\}^{1/q}$.

8.5   See footnote to equation (8.20).

8.6   Annular plate, $0 \leqslant t \leqslant \tau$, $\dot{\kappa}_\theta = (p_0 - p_c)t/\mu r(R - a)$,
$\tau \leqslant t \leqslant T$, $\dot{\kappa}_\theta = (p_0\tau - p_ct)/\mu r(R - a)$.
Circular plate, $0 \leqslant t \leqslant \tau$, $\dot{\kappa}_\theta = 2(p_0 - p_c)t/\mu rR$,
$\tau \leqslant t \leqslant T$, $\dot{\kappa}_\theta = 2(p_0\tau - p_ct)/\mu rR$,
$M'_\theta/M_0 = 1 + \{2q/(q + 1)\}(H\dot{\kappa}_\theta/2D)^{1/q}$.

8.7 $\dot{\epsilon}_x = (V_0 - A\sigma_0 t/M)/L$, $\sigma'_x/\sigma_0 = 1 + \{(V_0/DL)(1 - t/t_f)\}^{1/q}$.

8.10 See footnote on page 381.

8.11 $\dot{\epsilon}_\theta = V_0/R$, $\dot{\epsilon}_x = 0$, $\dot{\epsilon}_r = -V_0/R$, $\dot{\epsilon}_e = 2V_0/\sqrt{3}R$,
$\sigma'_e/\sigma_0 = 1 + (2V_0/\sqrt{3}RD)^{1/q}$.

Chapter 9

9.3   See equation (9.14)

9.8   See footnote on page 408.

9.9   (a) See equation (9.12b).

   (b) $H = (\alpha g M 3^{1/4}/2\sigma_0 R^{1/2})^{2/3}/\pi$.

   (c) $\Delta = V_0^2/2\alpha g$, $L \geqslant \Delta/\{0.86 - 0.568(H/2R)^{1/2}\}$.

9.10 (a) -163.2 m/s^2, 0.061 s, 0.306 m.

   (b) -212.2 m/s^2, 0.047 s, 0.236 m.

9.11 SI = HIC = 102.5.

9.13 From equilibrium of beam, or conservation of angular momentum,
$a = -4M_0/ML$ and $T = MLV_0/4M_0$.
SI = HIC = $(4M_0/ML)^{1.5}V_0/g^{2.5}$.

Chapter 10

10.1 $P/P^e = 1/(1 + W_1^i/W_1)$.

10.3 n_c (plastic)/n_c (elastic) = $(E/E_h)^{1/2}$.

10.4 $\lambda = 0.1764$ (equation (10.105)), $\bar{w}_f = 0.538$ mm, $n_c \cong 42$, $\tau_f = 22.78$.

10.7 $V_0 = 149.8$ m/s, $n_c \cong 10$, $t_f = 63.6$ μs.

Chapter 11

11.1   (a) $1/\beta$,   (b) β^2,   (c) β^3.

11.2   See equations (11.14) and (11.15).

11.3 $\rho_m/\rho_p = (V/v)^2(E_m/E_p)$, $\rho_m/\rho_p = E_m/E_p$, $(\Pi_{21}/\Pi_{18})^2/\Pi_5$.

11.4 $(v/V)^2 = (\rho_p/\rho_m)(\sigma_m/\Sigma_p)$, $1/\Pi_{12}^2$.

11.5 $\Pi_1 = V_0(\rho/\sigma_0)^{1/2}$, $\Pi_2 = R/H$, $\Pi_3 = W_f/H$. $\Pi_3 = (\Pi_1\Pi_2)^2/2$,
         $\Pi_3 = (0.529\Pi_1\Pi_2)^2$.

11.6 $\Pi_1 = V_0(\rho/\sigma_0)^{1/2}$, $\Pi_2 = L/H$, $\Pi_3 = W_f/H$, $\Pi_4 = B/H$.

11.7 $\Pi_1 = R(\sigma_0/P_m)^{1/2}$, $\Pi_2 = H(\sigma_0/P_m)^{1/2}$, $\Pi_1^{-1/2} = 2(\pi\Pi_2)^{3/2}/3^{1/4}$.

Subject index

Author index